Methods in Enzymology

Volume 298
MOLECULAR MOTORS AND THE CYTOSKELETON
Part B

METHODS IN ENZYMOLOGY

EDITORS-IN-CHIEF

John N. Abelson   Melvin I. Simon

DIVISION OF BIOLOGY
CALIFORNIA INSTITUTE OF TECHNOLOGY
PASADENA, CALIFORNIA

FOUNDING EDITORS

Sidney P. Colowick and Nathan O. Kaplan

Methods in Enzymology

Volume 298

Molecular Motors and the Cytoskeleton Part B

EDITED BY

Richard B. Vallee

WORCESTER FOUNDATION FOR BIOMEDICAL RESEARCH
SHREWSBURY, MASSACHUSETTS

ACADEMIC PRESS

San Diego   London   Boston   New York   Sydney   Tokyo   Toronto

Academic Press
a division of Harcourt Brace & Company
525 B Street, Suite 1900, San Diego, California 92101-4495, USA
http://www.academicpress.com

Academic Press Limited
24-28 Oval Road, London NW1 7DX, UK
http://www.hbuk.co.uk/ap/

International Standard Book Number: 0-12-182199-4

PRINTED IN THE UNITED STATES OF AMERICA
98 99 00 01 02 03 MM 9 8 7 6 5 4 3 2 1

Table of Contents

Section I. Analysis of Actomyosin-Related Systems

Section II. Analysis of Microtubule-Related Systems

Section III. Other Cytoskeletal Systems

Section IV. Cell-Free and Genetic Systems

Section V. Force Production Assays

Section VI. General Methods

Contributors to Volume 298

Article numbers are in parentheses following the names of contributors.
Affiliations listed are current.

BRUCE ALBERTS (19), *Department of Biochemistry and Biophysics, National Academy of Sciences, Washington, DC 20418*

VIKI J. ALLAN (28), *School of Biological Sciences, University of Manchester, Manchester M13 9PT, United Kingdom*

KATHRYN AYSCOUGH (2), *Department of Biochemistry, University of Dundee, Dundee DD14HN, Scotland*

RONALD J. BASKIN (35), *Section of Molecular and Cellular Biology, University of California, Davis, California 95616*

MARY C. BECKERLE (7), *Department of Biology, University of Utah, Salt Lake City, Utah 84112-0840*

JAMES B. BINGHAM (15), *Department of Biology, Johns Hopkins University, Baltimore, Maryland 21218*

STEVEN M. BLOCK (38), *Department of Molecular Biology and Princeton Materials Institute, Princeton University, Princeton, New Jersey 08544*

MAUREEN BLOMBERG-WIRSCHELL (20), *Program in Molecular Medicine, UMASS Medical Center, Worcester, Massachusetts 01605*

KERRY BLOOM (26), *Department of Biology, University of North Carolina, Chapel Hill, North Carolina 27599*

GARY G. BORISY (43), *University of Wisconsin, Madison, Wisconsin 53706*

MICHAEL C. BROWN (8), *Department of Anatomy and Cell Biology, State University of New York Health Science Center, Syracuse, New York 13210*

MICHAEL R. BUBB (3), *Department of Medicine, University of Florida College of Medicine, Gainesville, Florida 32610*

JANIS K. BURKHARDT (31), *Department of Pathology and Committee on Immunology, University of Chicago, Chicago, Illinois 60637*

RICHARD E. CHENEY (1), *Department of Physiology, University of North Carolina at Chapel Hill, Chapel Hill, North Carolina 27599-7545*

DUANE A. COMPTON (27), *Department of Biochemistry, Dartmouth Medical School, Hanover, New Hampshire 03755*

MICAH DEMBO (40), *Department of Biomedical Engineering, Boston University, Boston, Massachusetts 02215*

ARSHAD DESAI (12, 23), *Department of Biochemistry and Biophysics, University of California, San Francisco, San Francisco, California 94143*

GEORGIOS S. DIAMANTOPOULOS (17), *Department of Cell Biology, University of Geneva, CH-1211 Geneva 4, Switzerland*

STEPHEN J. DOXSEY (20), *Program in Molecular Medicine, UMASS Medical Center, Worcester, Massachusetts 01605*

HAROLD P. ERICKSON (25), *Department of Cell Biology, Duke University Medical Center, Durham, North Carolina 27710*

JOHN E. ERIKSSON (42), *Turku Centre for Biotechnology, University of Turku and Åbo Akademi University, BioCity, FIN-20521 Turku, Finland*

URSULA EUTENEUER (34), *Adolf-Butenandt-Institute for Cell Biology, Ludwig-Maximilians-University Munich, D-80336 Munich, Germany*

CHRISTINE M. FIELD (23, 41), *Department of Biochemistry and Biophysics, University of California, San Francisco, San Francisco, California 94143*

JEFFREY T. FINER (37), *Department of Biochemistry, Stanford University School of Medicine, Stanford, California 94305*

DAVID C. FUNG (11), *Department of Biochemistry, Stanford University School of Medicine, Stanford, California 94305-5307*

VLADIMIR I. GELFAND (30), *Department of Cell and Structural Biology, University of Illinois at Urbana/Champaign, Urbana, Illinois 61801*

LEAH HAIMO (33), *Department of Biology, University of California, Riverside, Riverside, California 92521*

YUNG JIN HAN (36), *Laboratory of Molecular Cardiology, National Heart, Lung and Blood Institute, National Institutes of Health, Bethesda, Maryland 20892*

ANN-SOFI HÄRMÄLÄ-BRASKÉN (42), *Department of Biochemistry and Pharmacy, Åbo Akademi University, BioCity, FIN-20520 Turku, Finland*

SUSAN BAND HORWITZ (21), *Albert Einstein College of Medicine, Bronx, New York 10461*

KEN JACOBSON (40), *Department of Cell Biology and Anatomy, and Lineburger Comprehensive Cancer Center, University of North Carolina at Chapel Hill, Chapel Hill, North Carolina 27599-7090*

THOMAS JARCHAU (10), *Medizinische Universitätsklinik, Institut für Klinische Biochemie und Pathobiochemie, D-97078 Würzburg, Germany*

MARY ANN JORDAN (22), *Department of Molecular, Cellular, and Developmental Biology, University of California, Santa Barbara, Santa Barbara, California 93106-9610*

ANNA KASHINA (13), *Section of Molecular and Cellular Biology, University of California, Davis, Davis, California 95616*

JOSEPH F. KELLEHER (5), *Department of Cell Biology, Duke University Medical Center, Durham, North Carolina 27710*

STEPHEN J. KING (15), *Department of Biology, Johns Hopkins University, Baltimore, Maryland 21218*

DAVID G. I. KINGSTON (21), *Virginia Polytechnic Institute and State University, Blacksburg, Virginia 24061*

MICHAEL P. KOONCE (34), *Division of Molecular Medicine, Wadsworth Center, Albany, New York 12201-0509*

THOMAS E. KREIS (17, 32), *Département de Biologie Cellulaire, Sciences III, University of Geneva, CH-1211 Geneva 4, Switzerland*

JANARDAN KUMAR (16, 29), *Department of Cell Biology, Duke University Medical Center, Durham, North Carolina 27710*

JUDITH LACOSTE (9), *Department of Microbiology, University of Virginia, Health Sciences Center, Charlottesville, Virginia 22908*

CHUNLIN LU (25), *Department of Cell Biology, Duke University Medical Center, Durham, North Carolina 27710*

ELIZABETH J. LUNA (4), *Department of Cell Biology, University of Massachusetts Medical Center, Shrewsbury, Massachusetts 01545*

JOE LUTKENHAUS (24), *Department of Microbiology, Molecular Genetics and Immunology, University of Kansas Medical Center, Kansas City, Kansas 66160*

AMY MA (9), *Department of Microbiology, University of Virginia, Health Sciences Center, Charlottesville, Virginia 22908*

CAROL MCGOLDRICK (29), *Department of Cell Biology, Duke University Medical Center, Durham, North Carolina 27710*

FRANK MCNALLY (18), *Section of Molecular and Cellular Biology, University of California, Davis, Davis, California 95616*

AMIT D. MEHTA (37), *Department of Biochemistry, Stanford University School of Medicine, Stanford, California 94305*

DAVID MEYER (13), *Section of Molecular and Cellular Biology, University of California, Davis, Davis, California 95616*

ANDREY MIKHAILOV (42), *Turku Centre for Biotechnology, University of Turku and Åbo Akademi University, BioCity, FIN-20521 Turku, Finland*

TIMOTHY J. MITCHISON (6, 12, 19, 23, 41), *Department of Cell Biology, Harvard Medical School, Boston, Massachusetts 02115 and Department of Cellular and Molecular Pharmacology, University of California, San Francisco, San Francisco, California 94143-0450*

AMIT MUKHERJEE (24), *Department of Microbiology, Molecular Genetics and Immunology, University of Kansas Medical Center, Kansas City, Kansas 66160*

R. DYCHE MULLINS (5), *The Salk Institute for Biological Studies, La Jolla, California 92037*

THOMAS MUND (10), *Medizinische Universitätsklinik, Institut für Klinische Biochemie und Pathobiochemie, D-97078 Würzburg, Germany*

KAREN OEGEMA (23, 41), *Department of Cellular and Molecular Pharmacology, University of California, San Francisco, San Francisco, California 94143*

TIMOTHY N. OLIVER (40), *Physiology Department, University of North Carolina at Chapel Hill, Chapel Hill, North Carolina 27599-7545*

GEORGE A. ORR (21), *Albert Einstein College of Medicine, Bronx, New York 10461*

J. THOMAS PARSONS (9), *Department of Microbiology, University of Virginia, Health Sciences Center, Charlottesville, Virginia 22908*

ROBERT J. PELHAM, JR. (39), *Department of Physiology, University of Massachusetts Medical School, Shrewsbury, Massachusetts 01545*

DANIEL W. PIERCE (14), *Department of Cellular and Molecular Pharmacology, Howard Hughes Medical Institute, San Francisco, California 94143*

THOMAS D. POLLARD (5), *The Salk Institute for Biological Studies, La Jolla, California 92037*

SRINIVASA RAO (21), *Albert Einstein College of Medicine, Bronx, New York 10461*

MATTHIAS REINHARD (10), *Medizinische Universitätsklinik, Institut für Klinische Biochemie und Pathobiochemie, D-97078 Würzburg, Germany*

JANET E. RICKARD (17), *Department of Cell Biology, University of Geneva, CH-1211 Geneva 4, Switzerland*

DANIEL R. RINES (13), *Section of Molecular and Cellular Biology, University of California, Davis, Davis, California 95616*

STEPHEN L. ROGERS (30), *Department of Cell and Structural Biology, University of Illinois at Urbana/Champaign, Urbana, Illinois 61801*

CECILIA SAHLGREN (42), *Department of Biology, Åbo Akademi University, BioCity, FIN-20520 Turku, Finland*

E. D. SALMON (26), *Department of Biology, University of North Carolina, Chapel Hill, North Carolina 27599*

JOCHEN SCHEEL (32), *Department of Physical Biology, Max-Planck-Institute for Developmental Biology, Tubingen, Germany*

MANFRED SCHLIWA (34), *Adolf-Butenandt-Institute for Cell Biology, Ludwig-Maximilians-University Munich, D-80336 Munich, Germany*

KAREN L. SCHMEICHEL (7), *Ernest Orlando Berkeley National Laboratory, Berkeley, California 94720*

JONATHAN M. SCHOLEY (13), *Section of Molecular and Cellular Biology, University of California, Davis, Davis, California 95616*

TRINA A. SCHROER (15), *Department of Biology, Johns Hopkins University, Baltimore, Maryland 21218*

JAMES R. SELLERS (36), *Laboratory of Molecular Cardiology, National Heart, Lung and Blood Institute, National Institutes of Health, Bethesda, Maryland 20892*

SID SHAW (26), *Department of Biology, University of North Carolina, Chapel Hill, North Carolina 27599*

MICHAEL P. SHEETZ (16, 29), *Department of Cell Biology, Duke University Medical Center, Durham, North Carolina 27710*

BOB SKIBBENS (26), *Department of Biology, University of North Carolina, Chapel Hill, North Carolina 27599*

ILAN SPECTOR (3), *Department of Physiology and Biophysics, State University of New York, Stony Brook, Stony Brook, New York 11794*

JAMES A. SPUDICH (37), *Department of Biochemistry, Stanford University, School of Medicine, Stanford, California 94305*

BETH E. STRONACH (7), *Department of Genetics, Harvard Medical School, Boston, Massachusetts 02115*

TATYANA M. SVITKINA (43), *University of Wisconsin, Madison, Wisconsin 53706*

CHARLES S. SWINDEL (21), *Bryn Mawr College, Bryn Mawr, Pennsylvania 19010*

JULIE A. THERIOT (11), *Department of Biochemistry, Stanford University School of Medicine, Stanford, California 94305-5307*

IRINA S. TINT (30), *Department of Anatomy and Cell Biology, Temple University School of Medicine, Philadelphia, Pennsylvania 19122*

DIANA M. TOIVOLA (42), *Department of Biology, Åbo Akademi University, BioCity, FIN-20520 Turku, Finland*

CHRISTOPHER E. TURNER (8), *Department of Anatomy and Cell Biology, State University of New York Health Science Center, Syracuse, New York 13210*

RONALD D. VALE (14), *Department of Cellular and Molecular Pharmacology, Howard Hughes Medical Institute, San Francisco, California 94143*

KOEN VISSCHER (38), *Department of Molecular Biology and Princeton Materials Institute, Princeton University, Princeton, New Jersey 08544*

CLAIRE E. WALCZAK (41), *Department of Cellular and Molecular Pharmacology, University of California, San Francisco, San Francisco, California 94143*

ULRICH WALTER (10), *Medizinische Universitätsklinik, Institut für Klinische Biochemie und Pathobiochemie, D-97078 Würzburg, Germany*

YU-LI WANG (39), *Department of Physiology, University of Massachusetts Medical School, Shrewsbury, Massachusetts 01545*

MATTHEW D. WELCH (6), *Department of Cellular and Molecular Pharmacology, University of California, San Francisco, San Francisco, California 94143-0450*

LESLIE WILSON (22), *Department of Molecular, Cellular, and Developmental Biology, University of California, Santa Barbara, Santa Barbara, California 93106-9610*

MEI LIE WONG (19, 23), *Department of Biochemistry and Biophysics, University of California, San Francisco, San Francisco, California 94143*

ELAINE YEH (26), *Department of Biology, University of North Carolina, Chapel Hill, North Carolina 27599*

YIXIAN ZHENG (19, 41), *Department of Embryology, Carnegie Institution of Washington, Baltimore, Maryland 21210*

Preface

This volume is the fourth in the *Methods in Enzymology* series on the same general topic. The first two volumes (85 and 134) were entitled "Structural and Contractile Proteins: The Contractile Apparatus and the Cytoskeleton." The title was changed to its current version with the third volume (196). The structural and functional complexity of cytoplasm was undreamed of a few decades ago and has been one of the great surprises of modern cell biology. These volumes represent a compilation of methodological information on the three major filament systems—F-actin, microtubules, and intermediate filaments—as well as on a number of cytoskeletal structures which lie outside these categories. The topics covered reflect the increasing diversity of isoforms, regulatory factors, and motility proteins that have been identified so far.

The rapid pace of discovery of new proteins and novel activities continues unabated and is reflected in the contents of this volume. Much of this information originated from genetic and molecular analysis, and a clear trend has been to produce and characterize polypeptides identified by these approaches and others by recombinant means. Common methods have emerged for the preparation of diverse proteins, making the art of protein purification and handling seem to be of decreasing value. However, many aspects of the behavior of individual proteins are distinctive, and the need for unique methods for the purification of recombinant proteins is becoming more appreciated, as revealed in this volume.

The need to understand protein–protein interactions has also become increasingly evident. Thus, although many interactions can be reproduced using recombinant proteins, the need to visualize, preserve, and isolate complexes with minimal perturbation has become of great interest. Methods to identify and isolate complexes with minimal manipulation are included in this volume. The case of the kinesin and myosin protein families is particularly revealing. Within the past few years there has been a remarkable proliferation of myosin-related and kinesin-related genes and their products, and analysis of recombinant motor domains has proceeded at a rapid pace. Gradually the newly identified proteins are becoming realities in the test tube, and each has its unique behavior and biochemical complexity. This trend is likely to follow for most, if not all, of the kinesin-, myosin-, and dynein-related genes. Methods to isolate novel myosin- and kinesin-related holoenzyme complexes are described in this volume, and future volumes promise to reflect a dramatic increase in the purification of these proteins from their cell and tissue sources.

Many of the energy-utilizing proteins of the cytoskeleton catalyze novel enzymatic reactions involving force-producing, filament-severing, and other nonconventional activities. New methods to monitor these activities both *in vivo* and *in vitro* have been essential to progress in the field and are included in this and in previous volumes.

A number of colleagues have provided valuable help in surveying the current state of the field and identifying topics of value for this volume. I am particularly grateful to Drs. Elizabeth Luna, Yu-li Wang, Kevin Vaughan, and Jorge Garces for their advice and assistance.

RICHARD B. VALLEE

METHODS IN ENZYMOLOGY

VOLUME XVII. Metabolism of Amino Acids and Amines (Parts A and B)
Edited by HERBERT TABOR AND CELIA WHITE TABOR

VOLUME XVIII. Vitamins and Coenzymes (Parts A, B, and C)
Edited by DONALD B. MCCORMICK AND LEMUEL D. WRIGHT

VOLUME XIX. Proteolytic Enzymes
Edited by GERTRUDE E. PERLMANN AND LASZLO LORAND

VOLUME XX. Nucleic Acids and Protein Synthesis (Part C)
Edited by KIVIE MOLDAVE AND LAWRENCE GROSSMAN

VOLUME XXI. Nucleic Acids (Part D)
Edited by LAWRENCE GROSSMAN AND KIVIE MOLDAVE

VOLUME XXII. Enzyme Purification and Related Techniques
Edited by WILLIAM B. JAKOBY

VOLUME XXIII. Photosynthesis (Part A)
Edited by ANTHONY SAN PIETRO

VOLUME XXIV. Photosynthesis and Nitrogen Fixation (Part B)
Edited by ANTHONY SAN PIETRO

VOLUME XXV. Enzyme Structure (Part B)
Edited by C. H. W. HIRS AND SERGE N. TIMASHEFF

VOLUME XXVI. Enzyme Structure (Part C)
Edited by C. H. W. HIRS AND SERGE N. TIMASHEFF

VOLUME XXVII. Enzyme Structure (Part D)
Edited by C. H. W. HIRS AND SERGE N. TIMASHEFF

VOLUME XXVIII. Complex Carbohydrates (Part B)
Edited by VICTOR GINSBURG

VOLUME XXIX. Nucleic Acids and Protein Synthesis (Part E)
Edited by LAWRENCE GROSSMAN AND KIVIE MOLDAVE

VOLUME XXX. Nucleic Acids and Protein Synthesis (Part F)
Edited by KIVIE MOLDAVE AND LAWRENCE GROSSMAN

VOLUME XXXI. Biomembranes (Part A)
Edited by SIDNEY FLEISCHER AND LESTER PACKER

VOLUME XXXII. Biomembranes (Part B)
Edited by SIDNEY FLEISCHER AND LESTER PACKER

VOLUME XXXIII. Cumulative Subject Index Volumes I–XXX
Edited by MARTHA G. DENNIS AND EDWARD A. DENNIS

VOLUME XXXIV. Affinity Techniques (Enzyme Purification: Part B)
Edited by WILLIAM B. JAKOBY AND MEIR WILCHEK

VOLUME XXXV. Lipids (Part B)
Edited by JOHN M. LOWENSTEIN

VOLUME 266. Computer Methods for Macromolecular Sequence Analysis
Edited by RUSSELL F. DOOLITTLE

VOLUME 267. Combinatorial Chemistry
Edited by JOHN N. ABELSON

VOLUME 268. Nitric Oxide (Part A: Sources and Detection of NO; NO Synthase)
Edited by LESTER PACKER

VOLUME 269. Nitric Oxide (Part B: Physiological and Pathological Processes)
Edited by LESTER PACKER

VOLUME 270. High Resolution Separation and Analysis of Biological Macromole-cules (Part A: Fundamentals)
Edited by BARRY L. KARGER AND WILLIAM S. HANCOCK

VOLUME 271. High Resolution Separation and Analysis of Biological Macromole-cules (Part B: Applications)
Edited by BARRY L. KARGER AND WILLIAM S. HANCOCK

VOLUME 272. Cytochrome P450 (Part B)
Edited by ERIC F. JOHNSON AND MICHAEL R. WATERMAN

VOLUME 273. RNA Polymerase and Associated Factors (Part A)
Edited by SANKAR ADHYA

VOLUME 274. RNA Polymerase and Associated Factors (Part B)
Edited by SANKAR ADHYA

VOLUME 275. Viral Polymerases and Related Proteins
Edited by LAWRENCE C. KUO, DAVID B. OLSEN, AND STEVEN S. CARROLL

VOLUME 276. Macromolecular Crystallography (Part A)
Edited by CHARLES W. CARTER, JR., AND ROBERT M. SWEET

VOLUME 277. Macromolecular Crystallography (Part B)
Edited by CHARLES W. CARTER, JR., AND ROBERT M. SWEET

VOLUME 278. Fluorescence Spectroscopy
Edited by LUDWIG BRAND AND MICHAEL L. JOHNSON

VOLUME 279. Vitamins and Coenzymes (Part I)
Edited by DONALD B. MCCORMICK, JOHN W. SUTTIE, AND CONRAD WAGNER

VOLUME 280. Vitamins and Coenzymes (Part J)
Edited by DONALD B. MCCORMICK, JOHN W. SUTTIE, AND CONRAD WAGNER

VOLUME 281. Vitamins and Coenzymes (Part K)
Edited by DONALD B. MCCORMICK, JOHN W. SUTTIE, AND CONRAD WAGNER

VOLUME 282. Vitamins and Coenzymes (Part L)
Edited by DONALD B. MCCORMICK, JOHN W. SUTTIE, AND CONRAD WAGNER

VOLUME 283. Cell Cycle Control
Edited by WILLIAM G. DUNPHY

Section I

Analysis of Actomyosin-Related Systems

[1] Purification and Assay of Myosin V

By RICHARD E. CHENEY

Introduction

The class V myosins[1–3] constitute one of the most widely distributed classes of the myosin superfamily and are found in organisms ranging from yeast to man. Although their precise functions are unclear, the class V myosins have been widely hypothesized to act as motors for actin-based organelle transport. Myosin V was initially characterized as p190, a calmodulin-binding protein present in precipitates of actomyosin from brain.[4–6] It was independently discovered in mouse as the product of the *dilute* coat color gene.[7] In addition to the mouse and chicken proteins, which exhibit 91% identity, the sequence of human myosin V is also available (MHC12 or "myoxin"; accession no. Y07759). At least two different genes encode class V myosins in mammals, with the second gene (*myr6*) sharing only 61% amino acid identity to mouse *dilute*/myosin V.[8]

Dilute mutations cause a "washed out" coat color, which appears to be due to a defect in the transport of melanosomes.[9,10] Null mutations that affect expression in all tissues also lead to seizures and death,[7] indicating that *dilute*/myosin V is essential for survival. Although myosin V is expressed at low levels in many cell types, it is particularly abundant in neurons, neurosecretory cells, and melanocytes.[5–7] In brain it constitutes ~0.2% of total protein[11] and is thus approximately as abundant as kinesin. Several

[1] R. E. Larson, *Braz. J. Med. Biol. Res.* **29,** 309 (1996).
[2] M. A. Titus, *Curr. Biol.* **7,** R301 (1997).
[3] M. S. Mooseker and R. E. Cheney, *Ann. Rev. Cell Dev. Biol.* **11,** 633 (1995).
[4] R. E. Larson, D. E. Pitta, and J. A. Ferro, *Braz. J. Biol. Med. Res.* **21,** 213 (1988).
[5] F. S. Espindola, E. M. Espreafico, M. V. Coelho, A. R. Martins, F. R. C. Costa, M. S. Mooseker, and R. E. Larson, *J. Cell Biol.* **118,** 359 (1992).
[6] E. M. Espreafico, R. E. Cheney, M. Matteoli, A. A. C. Nascimento, P. V. De Camilli, R. E. Larson, and M. Mooseker, *J. Cell Biol.* **119,** 1541 (1992).
[7] J. A. Mercer, P. K. Seperack, M. C. Strobel, N. G. Copeland, and N. A. Jenkins, *Nature* **349,** 709 (1991).
[8] L-P. Zhao, J. S. Koslovsky, J. Reinhard, M. Bahler, A. E. Witt, D. W. Provance, and J. A. Mercer, *Proc. Natl. Acad. Sci. U.S.A.* **93,** 10826 (1996).
[9] D. W. Provance, M. Wei, V. Ipe, and J. A. Mercer, *Proc. Natl. Acad. Sci. U.S.A.* **93,** 14554 (1996).
[10] Q. Wei, X. Wu, and J. A. Hammer, *J. Muscle Res. Cell Motil.* **18,** 517 (1997).
[11] R. E. Cheney, M. K. O'Shea, J. E. Heuser, M. V. Coelho, J. S. Wolenski, E. M. Espreafico, P. Forscher, R. E. Larson, and M. S. Mooseker, *Cell* **75,** 13 (1993).

TABLE I

SUBUNIT COMPOSITION OF MYOSIN V ISOLATED FROM CHICK BRAIN[a]

Subunit	Deduced MW	Number of amino acids	Accession number	Apparent stoichiometry
Heavy chain, chicken	212,509	1830	Z11718	2
Calmodulin, chicken	16,837	149	M36167	~10
17 kDa Essential light chain, chicken	16,987	151	J02786	~1.4
23 kDa Essential light chain, chicken	20,619	185	J02823	~0.6
"8" k Dynein light chain, *human*	10,365	89	Q15701	~2

[a] In some SDS–PAGE systems the 23-kDa essential light chain (ELC) is obscured by the much more prominent calmodulin band. It is not clear if individual myosin V molecules contain two 17 kDa ELC, two 23 kDa ELC, or one of each type. The "8 k" dynein light chain has not yet been cloned from chicken.

tissue-specific splice forms have been identified in mammals, one of which leads to the insertion of an additional ~28 amino acids in the tail.[12]

The myosin V heavy chain[6] has an 88-kDa head domain, which shares 41% identity with the motor domain of skeletal muscle myosin. This is followed by a neck domain consisting of six IQ motifs of ~24 amino acids each. Since each IQ motif constitutes a putative binding site for calmodulin or a related protein of the calmodulin superfamily, the neck domain of myosin V is expected to bind six calmodulin or calmodulin-like light chains. Following the neck is a region of predicted coiled coil structure responsible for dimerization and the 30-nm stalk visualized by electron microscopy. The heavy chain ends in a ~400-amino-acid globular tail domain of unknown function. Although myosin V heavy chains form dimers, they have not been observed to form myosin filaments.[11]

Native chick brain myosin V is isolated with multiple calmodulin (16,837 Da) light chains as well as additional light chains of 23, 17, and 10 kDa (Table I).[11,13] Peptide sequences from the 23- and 17-kDa myosin V light chains are identical to sequences from the 23- and 17-kDa essential light chains (ELCs) of nonmuscle and smooth muscle myosin II.[13] In addition to the light chains bound to its neck, myosin V contains a 10-kDa light chain bound to its tail domain. Surprisingly, peptide sequences from this light chain are identical[13] to sequences from the "8 k" light chain[14] of the

[12] P. K. Seperack, J. A. Mercer, M. C. Strobel, N. G. Copeland, and N. A. Jenkins, *EMBO J.* **14,** 2326 (1995).

[13] F. S. Espindola, R. E. Cheney, S. M. King, D. M. Suter, and M. S. Mooseker, *Mol. Biol. Cell* **7s,** 372a (1996).

[14] S. E. Benashski, A. Harrison, R. S. Patel-King, and S. S. King, *J. Biol. Chem.* **272,** 20929 (1997).

microtubule motor dynein. Current data suggest that each myosin V heavy chain is associated with approximately one ELC (17 or 23 kDa), up to five calmodulin light chains, and an "8 k" dynein light chain. Thus a dimer of myosin V heavy chains may have as many as ~14 light chains of which ~12 are associated with the neck and ~2 with the tail. This stoichiometry would lead to a native molecular weight of approximately 650,000.

Like other well-characterized myosins, purified myosin V binds to F-actin. It is unusual, however, in exhibiting a significant affinity for F-actin even in the presence of adenosine triphosphate (ATP) (apparent K_d of ~1 μM in 75 mM KCl).[15] Although purified myosin V has a relatively low actin-activated Mg^{2+}-ATPase in the presence of ethylene-bis(oxyethylene-nitrilo)tetraacetic acid (EGTA), in the presence of Ca^{2+} this ATPase is greatly stimulated to ~25 or more ATP/s per head,[15] a value comparable to that of skeletal muscle myosin. Half-maximal activation occurs at ~1 μM Ca^{2+},[15] similar to the K_d of calmodulin for Ca^{2+}. In the sliding filament motility assay myosin V moves at ~300 nm/s in the presence of EGTA but only at about half this velocity in the presence of Ca^{2+}.[11] The basis for these differential effects of Ca^{2+} on solution ATPase and *in vitro* motility is unclear but could involve uncoupling ATP hydrolysis from force production or activation of the motor by adsorption to a surface.

Purification of Myosin V from Chick Brain

General Considerations

The key to the purification is a precipitation of myosin V, actin, and membranous organelles that is induced by adding NaCl to a vesicle rich supernatant containing myosin V, ethylenediaminetetraacetic acid (EDTA), and ATP. Although the exact basis of this precipitation is unclear, it probably involves the binding of myosin V and associated vesicles to actin filaments. After extraction of the precipitate with Triton X-100 to remove membranous components, myosin V and actin are the major proteins remaining (Figs. 1 and 2). The actomyosin pellet is washed to remove Triton X-100 and then the myosin V is solubilized in NaCl and Mg^{2+}-ATP. Contaminants such as brain spectrin and tubulin are removed by gel filtration (Fig. 3) and then ion-exchange chromatography (Fig. 4) is used as a final purification step to remove residual contaminants such as actin and an actin-binding protein known as drebrin.[16]

[15] A. A. C. Nascimento, R. E. Cheney, S. B. F. Tauhata, R. E. Larson, and M. S. Moosker, *J. Biol. Chem.* **271**, 17561 (1996).
[16] T. Shirao, *J. Biochem.* **117**, 231 (1995).

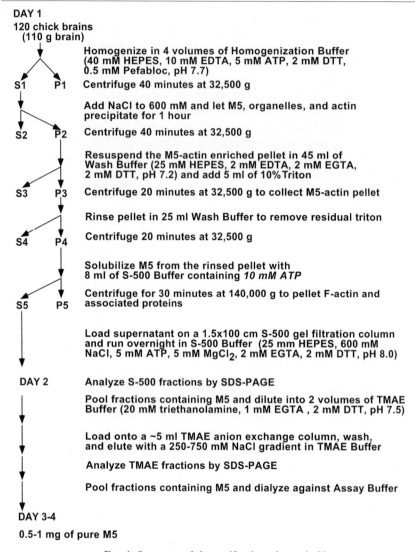

DAY 1

120 chick brains
  (110 g brain)

Homogenize in 4 volumes of Homogenization Buffer
(40 mM HEPES, 10 mM EDTA, 5 mM ATP, 2 mM DTT,
0.5 mM Pefabloc, pH 7.7)

S1   P1   Centrifuge 40 minutes at 32,500 g

Add NaCl to 600 mM and let M5, organelles, and actin
precipitate for 1 hour

S2   P2   Centrifuge 40 minutes at 32,500 g

Resuspend the M5-actin enriched pellet in 45 ml of
Wash Buffer (25 mM HEPES, 2 mM EDTA, 2 mM EGTA,
2 mM DTT, pH 7.2) and add 5 ml of 10% Triton

S3   P3   Centrifuge 20 minutes at 32,500 g to collect M5-actin pellet

Rinse pellet in 25 ml Wash Buffer to remove residual triton

S4   P4   Centrifuge 20 minutes at 32,500 g

Solubilize M5 from the rinsed pellet with
8 ml of S-500 Buffer containing *10 mM ATP*

S5   P5   Centrifuge for 30 minutes at 140,000 g to pellet F-actin and
associated proteins

Load supernatant on a 1.5x100 cm S-500 gel filtration column
and run overnight in S-500 Buffer (25 mm HEPES, 600 mM
NaCl, 5 mM ATP, 5 mM MgCl$_2$, 2 mM EGTA, 2 mM DTT, pH 8.0)

DAY 2   Analyze S-500 fractions by SDS-PAGE

Pool fractions containing M5 and dilute into 2 volumes of TMAE
Buffer (20 mM triethanolamine, 1 mM EGTA , 2 mM DTT, pH 7.5)

Load onto a ~5 ml TMAE anion exchange column, wash,
and elute with a 250-750 mM NaCl gradient in TMAE Buffer

Analyze TMAE fractions by SDS-PAGE

Pool fractions containing M5 and dialyze against Assay Buffer

DAY 3-4

0.5-1 mg of pure M5

Fig. 1. Summary of the purification of myosin V.

Because Ca^{2+} can induce the dissociation of some of myosin V's calmodulin light chains and this can inactivate myosin V,[11,15] it is important to avoid exposure to Ca^{2+} at all stages of purification and storage. Chelators such as EGTA or EDTA are thus included in all buffers. These chelators will also inhibit Ca^{2+}- or Mg^{2+}-dependent proteases. Reducing agents such

A. Coomassie Blue

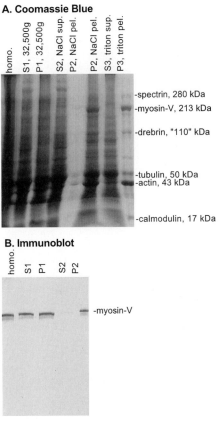

B. Immunoblot

FIG. 2. Fractions from the initial steps of the purification of myosin V. (A) Fractions from the purification of myosin V were analyzed on a 4–15% SDS–PAGE gradient gel and stained with Coomassie Blue. The first five lanes were loaded equally with aliquots corresponding to 0.001% of each fraction's volume (6 μl for the 600 ml homogenate). The last three lanes were loaded 25-fold more heavily and clearly show that myosin V and actin are the major proteins remaining after Triton X-100 extraction of the NaCl precipitate. (B) An immunoblot of similarly loaded fractions stained with anti-myosin V. Note that a large amount of myosin V is discarded in P1. Note also that virtually all of the remaining myosin V precipitates in P2 whereas little of the other protein does so.

as dithiothreitol (DTT) are also included to prevent oxidative damage. ATP is included in several buffers to prevent the rigor binding of myosin V to actin filaments. It is good practice to save a small aliquot from each step for later analysis by sodium dodecyl sulfate–polyacrylamide gel electrophoresis (SDS–PAGE). As with most protein chemistry, it is important to

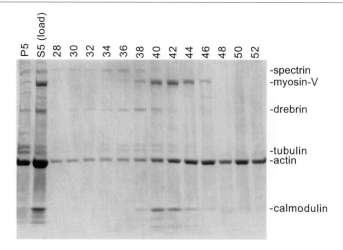

Fig. 3. Sephacryl S-500 gel filtration of myosin V. Samples (15 μl) of the indicated fractions (2.8 ml each) from a 1.5- × 100-cm Sephacryl S-500 column were analyzed by SDS–PAGE on a 4–15% gradient minigel and stained with Coomassie Blue. The polypeptide composition of 15-μl samples of S5 and P5 are also shown. Note the copurification of the calmodulin light chains with the myosin V heavy chain. Fractions 39–45 were pooled.

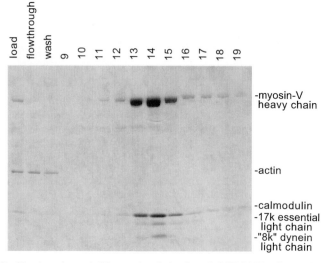

Fig. 4. Purification of myosin V on a trimethylaminoethyl (TMAE) anion-exchange column for purification of myosin V. Aliquots (15 μl) of the indicated fractions (1 ml each) were analyzed by SDS–PAGE on a 4–15% gradient minigel and stained with Coomassie Blue. Aliquots (15 μl) of the column load (~60 ml), flow-through (~60 ml), and wash (20 ml) are also shown. Note that residual actin is removed at this step and that the calmodulin, 17 kDa ELC, and "8 k" dynein light chains are all clearly visible.

keep myosin V cold. Keep all bottles and tubes on ice, and run the columns and centrifuges at 4° with precooled buffers and rotors.

Starting Tissue

We routinely use chick brain as a tissue source. Chicks or fertilized eggs are available at modest cost from poultry suppliers and chick brains can be harvested quickly and easily within seconds of sacrifice. Although chick brains quick-frozen in liquid nitrogen can be used, the key NaCl precipitation step is much more reliable with fresh tissue and thus most of our experiments have been done with myosin V purified from fresh tissue. Myosin V enriched precipitates have also been obtained from rabbit and rat brains.[4,5] We have been repeatedly unable to obtain good precipitates from frozen calf brain, possibly due to interference from the large amounts of myelin present in whole calf brains.

Materials

120 chicks (0–5 days posthatch; ~110 g brain).
1.5- × 100-cm, ~170-ml Sephacryl S-500 (Pharmacia, Piscataway, NJ) gel filtration column equipped with a flow adapter. Pour the column ahead of time in S-500 buffer *without ATP*.
0.7- × 5-cm, ~5-ml TMAE (trimethylaminoethyl) anion-exchange column equipped with a flow adapters. Either TMAE Fractogel 650-S (EM Separations, Gibbstown, NJ) or Q-Sepharose (Pharmacia, Piscataway, NJ) can be used.

Buffers

All buffers should be made with deionized water. Adjust the pH of all buffers at room temperature and then cool to 0–4° prior to use. Because ATP, DTT, and Pefabloc are moderately labile, buffers containing them are made shortly before use. Because DTT can cause pH measurements to drift, it should be added after pH adjustment.

Homogenization buffer (500 ml): 40 mM HEPES, 10 mM EDTA-K, 5 mM sodium ATP (#A-3377, Sigma, St. Louis, MO), 2 mM DTT, 0.5 mM Pefabloc-SC (Boehringer Mannheim, Indianapolis, IN), pH to 7.7 with 1 N NaOH.

Protease inhibitor: We include the fast acting and relatively stable serine protease inhibitor Pefabloc-SC (also known as AEBSF or 4-[2-aminoethyl]benzenesulfonyl fluoride) in homogenization buffer. If Pefabloc-SC is unavailable or too expensive, PMSF (phenylmethylsulfonyl fluoride) at a final concentration of ~0.5 mM can be used instead. A 200 mM stock of PMSF is made in absolute ethanol and a 125-μl aliquot is added just prior to homogenizing each batch of 10 brains.

4.0 M NaCl (100 ml).

Wash buffer (1000 ml): 25 mM HEPES, 2 mM EDTA-K, 2 mM EGTA-K, 2 mM DTT, pH 7.2 with 1 M KOH.

S-500 buffer (1000 ml): 25 mM HEPES, 600 mM NaCl, 5 mM MgCl$_2$, 2 mM EGTA-Na, 5 mM ATP-Na$_2$, 2 mM DTT, pH 8.0, with 1 M NaOH.

Myosin V solubilization buffer (10 ml): Identical to the S-500 buffer except that the ATP concentration is increased to 10 mM

TMAE buffer (500 ml): 20 mM triethanolamine, 1 mM EGTA-Na, 2 mM DTT, pH 7.5, with 5 M HCl.

TMAE buffer + 250 mM NaCl (200 ml).

TMAE buffer + 750 mM NaCl (100 ml).

Assay buffer (1000 ml): 10 mM imidazole, 75 mM KCl, 2.5 mM MgCl$_2$, 2 mM DTT, 0.1 mM EGTA-K, pH 7.2, with HCl or KOH.

Method

Preparations. Begin by equilibrating the gel filtration column with S-500 buffer *containing ATP* such that ~1 column volume (~170 ml) will have passed through when the column is loaded. Precool two centrifuges and two rotors such as the Sorvall SS-34 or Beckman JA-20. Larger capacity rotors such as the Sorvall GSA and Beckman JA-14 do not work well. Start relatively early because the first day requires a minimum of 7–10 hr of work.

Dissection and Homogenization (~1 hr). Sacrifice each chick just prior to dissection by rapid cervical dissociation followed by decapitation. Immediately split the skull in half along the midline using a sharp single-edge razor blade. Scoop out the brain (including the cerebellum) using curved forceps and collect in a small container on ice until 10 brains have been obtained. Transfer the 10 brains to a 55-ml Teflon-on-glass Potter–Elvehjem homogenizer (Wheaton, Millville, NJ) and add 40 ml of ice cold homogenization buffer. Homogenize on ice using 10 slow up-and-down strokes of the pestle attached to an electric drill operated at maximum speed (approximately 500–3000 rpm). Wear goggles and gloves and avoid electrical hazards. Also avoid generating a foam because this can denature proteins. Pool the homogenates from each batch of 10 brains and store on ice until all the brains have been homogenized. It is best if one worker homogenizes while one or two others assist by dissecting. The effectiveness of homogenization can be easily checked by placing 1–10 μl of the homogenate on a microscope slide and examining it with a 60–100 × oil immersion lens.

Although this protocol was optimized and is most reliable using the homogenization procedure just described, alternative methods of homogenization have been used. We have used a 40-ml Dounce homogenizer (~25 strokes of the tight fitting pestle) as well as a Sorvall Omni-Mixer (2–4 min at maximum speed). If using a homogenizer with rotating metal blades, fill the homogenization bucket completely to prevent foaming and homogenize on ice in 1-min bursts with 1 min of cooling between bursts. These homogenization methods, like the use of frozen brains, tend to result in P4 fractions containing greater amounts of fodrin and two proteins with apparent molecular weights of 70 and ~170 kDa.

Centrifugation and NaCl Precipitation (~3 hr). Centrifuge the homogenate for 40 min at 32,500g and 4°. We use two Sorvall SS-34 or Beckman JA-20 rotors operated at 16,500 rpm with 50-ml polycarbonate tubes (Nalgene, Rochester, NY), which have a 35-ml capacity in these rotors. The pellets (P1), which are discarded, should be firm enough that the supernatants (S1) can be collected by pouring them into a graduated cylinder. Note that even with centrifugation at the relatively modest force of 32,500g, approximately one-half of the myosin V in the homogenate is discarded with the particulate and membranous materials in P1 (see Figs. 1 and 2).

Measure the volume of the supernatant and add NaCl to 600 mM by adding 0.177 ml of 4 M NaCl per milliliter of supernatant. Mix gently by inversion and incubate on ice without stirring for 1 hr. Centrifuge for 40 min at 32,500g and 4° to collect the myosin V-rich precipitate. The pellets are usually gelatinous with a whitish tinge and should have a volume of ~0.5 ml per tube. In a typical precipitation essentially all of the myosin V present in the initial supernatant is recovered in this NaCl precipitate (P2, see Fig. 2B). Although virtually no myosin II precipitates under these conditions, myosin II will precipitate if the supernatant is incubated overnight. This is probable due to ATP depletion followed by rigor binding to F-actin. If myosin II is present it migrates slightly slower than myosin V by SDS–PAGE. An efficient myosin V precipitation requires a pH of 7.2–7.8 and appears to be inhibited by Mg^{2+}.

Triton X-100 Extraction and Centrifugation (~1 hr). Pour off the NaCl supernatants (S2) and measure out 45 ml of myosin V wash buffer to resuspend the pellets. Using a 6-inch glass Pasteur pipette, vigorously resuspend each pellet in ~2–3 ml of this wash buffer by repeated pipetting. It is important that *all* of the pellet is collected—the gelatinous pellet sometimes adheres tightly to the side of the tube. Pool the resuspended pellets in a 40-ml glass Dounce homogenizer and rinse each centrifuge tube with an additional 1–2 ml of the wash buffer. Pool the rinses with the resuspended pellets and homogenize with several strokes of the tight pestle. Add 5 ml of 10% Triton X-100 (SurfactAmps, Pierce Chemicals, Rockford, IL) to yield a final concentration of ~1%. Warm this solution by immersing it in a

37° water bath for 2–3 min while gently homogenizing to extract membrane components. Centrifuge in two centrifuge tubes for 20 min at 32,500g and 4° to collect the Triton X-100 extracted pellet (P3). Myosin V and actin should each constitute at least 10–30% of the total protein in the extracted pellet (Fig. 2A, lane P3).

Removing Residual Detergent from the Triton Extracted Pellet (~0.5 hr). Pour off the Triton X-100 supernatant (S3), which contains little or no myosin V. The total volume of the pellet (P3) should be reduced 5- to 10-fold relative to P2 due to extraction of membranes. Using a Pasteur pipette and 2–3 ml of wash buffer, carefully detach the extracted pellets from the walls of the centrifuge tubes. Pour the detached pellets into a 40-ml Dounce homogenizer and add wash buffer to a final volume of ~25 ml. Resuspend the pellets using the tight fitting pestle and then centrifuge at 32,500g for 20 min at 4°. Discard the supernatant (S4) by pouring it off of the pellet (P4). Myosin V remains in the pellet during these steps most likely because it is bound to F-actin.

Solubilizing the Washed Pellet in S-500 Buffer and Ultracentrifugation (~1.5 hr). Measure out 8 ml of S-500 buffer containing 10 mM ATP. Use a Pasteur pipette and ~3 ml of this buffer to detach all of the pellet (P4), and then pour the pellet into a 15-ml Dounce homogenizer. Add the remaining 5 ml of S-500 buffer and resuspend with several strokes of the tight pestle. Let the suspension incubate for 20 min on ice to solubilize the myosin V. Transfer the solution to a ~10-ml ultracentrifuge tube and centrifuge for 30 min at 140,000g and 4° (40,000 rpm in a Beckman type 65 rotor). After centrifugation collect the myosin V-rich supernatant (S5) with a Pasteur pipette and save the pellet (P5) on ice in case the myosin V was not efficiently solubilized. It is extremely helpful to save ~100-μl aliquots of S5 and P5 (resuspend P5 to its original volume in S-500 buffer) for later analysis by SDS–PAGE.

Although the P5 pellet is often large (1–2-ml) and gelatinous it should contain relatively little myosin V. The P5 pellet does contain large amounts of F-actin and actin-binding proteins such as brain spectrin and drebrin. The solubilization step can be prolonged to 1–2 hr without obvious effect, but it should not be continued so long that ATP is depleted. Because ATP is required for efficient solubilization, this step should select for active myosin V molecules capable of ATP-dependent release from F-actin. Note that myosin V is not efficiently solubilized if Mg^{2+} is absent or if it is chelated by EDTA.

S-500 Gel Filtration Column (~1 hr to Load; Run Overnight). Load the supernatant (S5) onto the S-500 gel filtration column freshly equilibrated with S-500 buffer containing ATP. If the column is equipped with a flow adapter it can be loaded by siphoning the myosin V onto the column—but do not allow any air bubbles into the column and never let the column run dry. We elevate the buffer reservoir to generate a flow rate of ~15 ml/hr and collect approximately 70 fractions of ~2.8 ml each. A peristaltic pump can also be used to increase the flow rate to run the column in ~3 hr.

SDS–PAGE for Analysis of S-500 Fractions (~4 hr). The gel filtration fractions are analyzed by SDS–PAGE due to the high absorbance of the ATP in the S-500 buffer. We usually analyze 15-μl aliquots of even numbered fractions 28–52 on 4–15% gradient minigels and stain with Coomassie Blue. Analyzing 15-μl samples of S5 (the column load) and P5 (resuspended to the same volume as S5) provides a good test of the efficiency of the NaCl precipitation and the Mg^{2+}-ATP solubilization. Myosin V usually elutes in a peak between fractions 38 and 45. Brain spectrin, drebrin, and any residual myosin II all elute slightly ahead of the myosin V peak. The myosin V-rich fractions should be pooled, keeping in mind that for the purest preparations of myosin V (for motility or viscometry), fractions contaminated with drebrin should be avoided. After use the column should be washed with two column volumes of S-500 buffer (without ATP) plus 0.02% sodium azide as a preservative for long-term storage.

Dilution into TMAE Buffer and TMAE Ion-Exchange Column (~4 hr). As a final purification step, the myosin V is subjected to ion-exchange chromatography. This step removes actin, which should not bind to the column at ~250 mM NaCl, and also reduces or removes contamination with drebrin. This column also efficiently concentrates the myosin V 5- to 10-fold. Just prior to running the column, add 2 volumes of (zero salt) TMAE buffer to the pooled S-500 fractions to lower the NaCl concentration to 200 mM. Add ATP to a final concentration of 1 mM to prevent rigor binding to actin and adjust the pH of the diluted myosin V to 7.5 using a pH 6.7 stock of 10% triethanolamine.

Load the diluted myosin V onto a ~5-ml TMAE column (1 × 5 cm) equilibrated in TMAE buffer + 250 mM NaCl at a flow rate of 1–2 ml/min using gravity or a peristaltic pump. Wash the column with 20-ml of TMAE buffer containing 250 mM NaCl and 1 mM ATP to remove actin. Save the flow-through and wash solutions in case of error. Elute the column with a 60-ml gradient of 250–750 mM NaCl in TMAE buffer. Use 30 ml of TMAE buffer containing 250 mM NaCl and 30 ml of TMAE buffer containing 750 mM NaCl. We use a ~50-ml gradient former (Sigma) and a stir bar the size of a rice grain. Larger stir bars can cause significant backmixing and may yield nonlinear gradients. We place the gradient pourer on a stir plate and use a peristaltic pump to yield a flow rate of roughly 0.5 ml/min. Collect 1-ml fractions. The myosin V usually elutes at 350–420 mM NaCl.

Some myosin V fractions contain the actin-binding protein drebrin.[17] Drebrin has an apparent molecular weight of 110,000 by SDS–PAGE and often runs as a closely spaced doublet.[16] Drebrin should elute slightly ahead of myosin V on the TMAE column. If contamination with drebrin is a problem (or if the myosin V does not elute as expected), check the slope and linearity of the gradient using a conductivity meter or a vapor pressure osmometer. Using a shallower gradient (~250–500 mM) will give better separation but will also result in greater dilution of the myosin V.

SDS–PAGE for Analysis of TMAE Fractions. SDS–PAGE and Coomassie Blue staining are used to determine which fractions to pool from the TMAE column. This also provides a quality check for contamination with drebrin or myosin V breakdown products.

To identify the subset of TMAE fractions to be analyzed by SDS–PAGE, we add 10-μl aliquots of each fraction to 0.6-ml microfuge tubes containing 40 μl of improved Bradford reagent.[18] The aliquots corresponding to the myosin V peak should turn light blue immediately after mixing. For SDS–PAGE of the peak fractions we load 15 μl per lane. When pooling the TMAE fractions, it is often helped to pool the two or three most concentrated fractions separately from the more dilute fractions to avoid the losses in time and protein entailed by an additional concentration step. If highly concentrated myosin V is required, the pooled TMAE fractions can be concentrated to 500–1000 μg/ml by dialysis for a few hours against a dry absorbant such as Ficoll 500,000 prior to dialysis against assay buffer.

Dialysis of Pooled Myosin V against Assay Buffer. Dialyze the pooled myosin V against prechilled assay buffer containing 1 mM EGTA and 2 mM DTT for at least three changes of ~12 hr each. Using assay buffer containing 0.1 mM EGTA in the final change will facilitate any subsequent experiments that require adding Ca^{2+}.

Determining the Yield and Concentration. Centrifuge the dialyzed myosin V for 10 min at 10,000g to remove any insoluble (light scattering) material. Save some of the final dialysis solution for use as an absorbance blank. The extinction coefficient of myosin V at 280 nm is 1.04 for a 1.0 mg/ml solution.[11] The yield of a typical preparation is 500–1000 μg at a concentration of ~0.2 mg/ml. A centrifugal concentrator such as a Centricon-30 can be used to concentrate the myosin V, although this usually entails significant handling losses. Dilute myosin V can also be concentrated on a small TMAE column following careful step elution with 600 mM NaCl

[17] E. J. Luna, K. N. Pestonjamasp, R. E. Cheney, C. P. Strassel, T. H. Lu, C. P. Chia, A. L. Hitt, M. Fechheimer, and M. S. Mooseker, *Curr. Topics Membr.* **38**, 4 (1997).
[18] S. M. Read and D. H. Northcote, *Anal. Biochem.* **116**, 53 (1981).

(use a ~1-ml column and collect ~0.2-ml fractions). If the experiments require a buffer with low absorbance in the far UV, the imidazole in the assay buffer should be replaced with 10 mM MOPS.

Time Considerations

Day 1, which requires 7–10 hr, must end with the loading of the gel filtration column. On day 2 the fractions from the gel filtration column should be analyzed by SDS–PAGE and it is best to run the TMAE column. The fractions from the TMAE column can be analyzed by SDS–PAGE either at the end of day 2 or at the beginning of day 3. The pooled fractions are dialyzed for a least 1.5 days to obtain accurate A_{280} measurements.

Storage

Storage on Ice. Purified myosin V can be stored on ice in assay buffer containing 0.1 or 1 mM K-EGTA for days to weeks. Although some preparations have retained ATPase and motility activity for a month or more when stored on ice, we prefer to use myosin V within ~1 week of purification. This is especially true for motility experiments, because motility is more labile than ATPase activity. Because oxidation inactivates myosin V, fresh DTT should be added every few days during storage on ice. Note that accidental incubation overnight at room temperature can reduce ATPase activity by 50% or more. After prolonged storage, breakdown products of 140, 80, and 65 kDa often accumulate. These bands appear to correspond to the proteolytic fragments generated after digestion with calpain. Although myosin V can be inactivated by oxidation, proteolysis, or loss of calmodulin, in our hands it is less labile than skeletal muscle II or brush border myosin I.

Freezing. Freezing myosin V in the absence of cryoprotectants completely destroys ATPase and motility activity. In the presence of cryoprotectants such as 35% (w/v) sucrose or 50% (v/v) glycerol, frozen myosin V retains ATPase activity for months to years. To freeze myosin V, we mix equal volumes of myosin V and 70% (w/v) sucrose (both in assay buffer containing 0.1 mM K-EGTA and 10 mM DTT) in microfuge tubes and store at either $-20°$ or $-70°$.

Miscellaneous Considerations

Since brain is approximately 10% protein by weight and myosin V constitutes ~0.2% of total protein, the average concentration of myosin V in brain should be ~200 μg/ml. Thus there should be ~20 mg of myosin V in 100 g of brain, while the purification yields only ~1 mg. A major loss clearly occurs in the first centrifugation and routine handling losses at each step probably account for significant additional losses. The limiting factors for scaling up are rotor capacity and the capacity of the gel filtration column. For acceptable resolution on the gel filtration column no more than 1–5% of the column volume should be loaded (5% of 170 ml = 8.5 ml). On some SDS–PAGE gels, heavily loaded samples of the myosin V heavy chain may appear to run as a doublet. This gel artifact is particularly common on prepoured gels that have one plastic plate and one glass plate and can be easily recognized by cutting a lane in half and examining the cut surface of the band.

Assay of Myosin V

Steady-State ATPase

Myosin V prepared by this method should exhibit an actin-activated Mg^{2+}-ATPase of 25–40 ATP/s per myosin head.[15] Our standard assay conditions currently involve a 5-min incubation at 37° in assay buffer containing 1 mM calcium, 2 mM ATP, 870 μg/ml F-actin (20 μM), 100 μg/ml calmodulin (6 μM), and 5 μg/ml myosin V (7.8 nM). Although many highly sensitive ATPase assays are available, we use a simple colorimetric assay.[19] We prepare 430-μl samples for each condition and add myosin V just prior to initiating the reaction with ATP. A 200-μl aliquot is immediately removed and mixed with a tube containing 20 μl of 50% (w/v) TCA (trichloroacetic acid) to stop the reaction and yield a zero time point sample. The remainder of each 430-μl sample is incubated in a 37° water bath. At precisely 5 min a 200-μl aliquot is removed and added to a tube containing 20 μl of 50% TCA. The samples are incubated on ice until all have been collected and are then centrifuged for 2 min in a microfuge to remove the precipitated protein. Then 180 μl of each supernatant is collected and vortexed briefly in a 0.5-ml microfuge tube containing 50 μl of 2% w/v ammonium molybdate in 1 M H_2SO_4. Color development is initiated by adding 20 μl of freshly prepared 40% (w/v) ferrous sulfate in 0.5 M H_2SO_4 to each tube. We initiate color development at 1-min intervals, and then measure each sample's absorbance at 700 nm after ~15 min of color development. Samples for a phosphate standard curve (~10 samples ranging from 0 to 0.5 mM potassium phosphate) should be prepared similarly in assay buffer and developed for the same period of time. Controls such as actin alone and myosin V alone should exhibit little or no ATPase.

---

[19] H. H. Taussky and E. Shorr, *J. Biol. Chem.* **202**, 675 (1953).

Motility Assay

Using the standard sliding filament motility assay with rhodamine-phalloidin labeled actin,[20] myosin V should translocate filaments at ~300 nm/s in the presence of EGTA.[11] We generally construct flow cells using nitrocellulose-coated coverslips and spacers of double-stick tape. Purified myosin V is absorbed for 2–3 min at a concentration of 5–10 μg/ml chamber and then the chamber is blocked with 1–2 mg/ml bovine serum albumin in assay buffer. To minimize photobleaching, a glucose oxidase/catalase should be used to scavenge free radicals.[20] We use a SIT camera and collect four to eight frames for averaging at 2- to 3-sec intervals. An electronic shutter should be used to minimize photobleaching. Myosin V is somewhat unusual in that it rapidly shreds actin filaments into small fragments. This tendency is especially pronounced at higher concentrations of myosin V (over 10 μg/ml). To prevent the filaments from being shredded before they can be observed and recorded, it is often best to absorb the filaments in the absence of ATP and then later trigger movement by adding ATP. It is also helpful to use a high enough concentration of actin filaments that fresh filaments constantly fall onto the surface to replace those that have been shredded. We have not been able to reduce shredding by preincubating myosin V with actin filaments in the presence of ATP and then centrifuging at ~150,000g to remove inactive motors. In the presence of Ca^{2+}, myosin V translocates filaments more slowly. This slower movement will eventually cease altogether unless calmodulin is present.[11] Care should be taken to avoid accidental exposure to Ca^{2+} when motility chambers are blocked or washed.

Cosedimentation Assays

For cosedimentation assays with F-actin[15] we typically prepare ~100-μl aliquots of ~100 μg/ml myosin V and ~300 μg/ml of F-actin in ~200-μl centrifuge tubes. Samples are incubated for 5 min and then centrifuged for 10–20 min at ~150,000g. Aliquots of the pellets (resuspended to the initial volume) and supernatants are then analyzed by SDS–PAGE and Coomassie Blue staining. Cosedimentation assays with liposomes[21] are performed similarly except that ~0.5 mM phospholipid should be used. Note that all samples must have identical buffer compositions and that the pellets can be difficult to resuspend. If a significant amount of myosin V is pelleted in the absence of F-actin, the myosin V should be centrifuged before use to remove insoluble material.

[20] J. R. Sellers, G. Cuda, F. Wang, and E. Homsher, *Methods Cell Biol.* **39**, 23 (1993).
[21] S. M. Hayden, J. S. Wolenski, and M. S. Mooseker, *J. Cell Biol.* **111**, 443 (1990).

Calpain Cleavage

Myosin V is a good substrate for the calcium-dependent protease calpain and appears to have two major calpain cleavage sites.[15] One site is located adjacent to the PEST region and cleavage at this site yields a 130-kDa head fragment and an 80-kDa tail fragment that begins at residue 1142.[15] The major products after complete digestion are the 80-kDa tail fragment and a 65-kDa head fragment.[15] To digest myosin V in assay buffer, add $CaCl_2$ to obtain 2 mM free Ca^{2+} and then add ~1 μg of calpain per 80 μg of myosin V. Calpain is available from Sigma and is prepared by mixing 1 unit (~30 μg) in 199 μl of deionized water immediately before use. Digest the myosin V at 25° for 10 min to 2 hr depending on the degree of cleavage desired.[15] Terminate the digestion by adding EGTA to 10 mM. The 80-kDa tail fragment can be purified from the other fragments using a 1.0-ml Q-Sepharose ion-exchange column and a 75–1000 mM gradient of NaCl in pH 7.4 assay buffer.[15]

Acknowledgments

The author gratefully acknowledges the many colleagues whose work has contributed to the development of this protocol. Acknowledgment is particularly due to Roy Larson, Mark Mooseker, Milton Coelho, Joe Wolenski, Maura O'Shea, Enilza Espreafico, Alexandra Nascimento, Foued Espindola, Lisa Evans, and the students of the 1994–1997 physiology courses at the Marine Biological Laboratory. Special thanks are also due to Olga Rodriguez and Michele Wing who also provided samples used in the figures. Work in the author's laboratory is supported by National Institutes of Health grant DC0322901.

[2] Use of Latrunculin-A, an Actin Monomer-Binding Drug

By KATHRYN AYSCOUGH

Introduction

Latrunculin-A (LAT-A) is a marine toxin, first isolated from the Red Sea sponge, *Latrunculia magnifica*.[1] Its effects on the actin cytoskeleton in cultured mammalian cells were first noted by Spector and colleagues.[2] They demonstrated that in both neuroblastoma and fibroblast mouse cells, submicromolar levels of the drug caused changes in cell morphology that

[1] Y. Kashman, A. Groweiss, and U. Shmueli, *Tetrahedron Lett.* **21,** 3629 (1980).
[2] I. Spector, N. R. Shochett, Y. Kashman, and A. Groweiss, *Science* **219,** 493 (1983).

correlated with major alterations in the organization of the actin, but not the microtubular cytoskeleton. Subsequent experiments in other cell types indicate that LAT-A is a potent, rapid, and reversible inhibitor of the actin cytoskeleton, which can provide an additional tool for investigations into the functioning and regulation of this complex cell structure.

Comparison with other Actin-Disrupting Drugs

Although LAT-A causes a disruption of the actin cytoskeleton, its mode of action is strikingly different from the most commonly used actin-disrupting drugs, the cytochalasins. *In vitro* the cytochalasin toxins bind to the barbed end of actin filaments, which inhibits both association and dissociation of subunits at those ends.[3,4] Although the detailed mechanisms by which cytochalasins inhibit cell movements and affect actin organization *in vivo* remain unclear, it is assumed that they compete with endogenous barbed end capping agents and interfere with the dynamic equilibrium between filamentous actin (F-actin) and a pool of monomeric actin (G-actin).[5] In contrast, experiments in which rabbit muscle actin was incubated with increasing concentrations of LAT-A have demonstrated disruption of filament formation consistent with the formation of a 1 : 1 complex between LAT-A and actin monomer.[6] Thus, latrunculin-A appears to bind only to the monomeric form of actin, thereby rendering it incompetent for filament formation.

Specificity of the LAT-A–Actin Interaction

Of central importance to studies involving use of toxins is a consideration of their specificity. This point was addressed in studies we conducted using *Saccharomyces cerevisiae* in which we analyzed the LAT-A sensitivity of a collection of actin mutants that had been generated in a congenic background.[7] Three out of the 23 mutants were resistant to the effects of the drug. The demonstration that mutation of actin was sufficient to make a strain resistant to LAT-A provided powerful evidence for a highly specific interaction between LAT-A and actin. Furthermore, the amino acid substitutions in the three resistant alleles mapped to a distinct region on the actin molecule, implicating this region as part of the binding site for LAT-A. This region was adjacent to the nucleotide-binding cleft. *In vitro* assays

[3] S. S. Brown and J. A. Spudich, *J. Cell Biol.* **88,** 487 (1981).
[4] D. Godette and C. Frieden, *J. Biol. Chem.* **261,** 15974 (1986).
[5] J. Cooper, *J. Cell Biol.* **105,** 1473 (1987).
[6] M. Coué, S. Brenner, I. Spector, and E. Korn, *FEBS Lett.* **213,** 316 (1987).
[7] K. R. Ayscough, J. Stryker, N. Pokala, M. Sanders, P. Crews, and D. G. Drubin, *J. Cell Biol.* **137,** 399 (1997).

tested the prediction that LAT-A might impair nucleotide exchange and demonstrated that this was the case. In addition, actin purified from a LAT-A-resistant mutant was resistant to the action of LAT-A *in vitro*, adding further support to the proposal that the mutations identified the LAT-A binding site on actin.

While binding sites on actin have been identified for other drugs, only in the case of LAT-A has it been demonstrated that mutations in actin alone can abolish this interaction *in vivo*. For other actin interacting drugs it remains a possibility that they can affect not only actin but also other cell proteins to cause their *in vivo* phenotypes. Of particular relevance is the identification of a family of actin-related proteins (for a recent review of actin-related proteins, see Frankel and Mooseker.[8] These have about 50% sequence identity to conventional actin and, it has been proposed, a similar tertiary structure. It is conceivable that certain drugs will interact not only with actin but also with one or more of these actin-related proteins. Such interactions might cause a wide array of phenotypes, making interpretation of results problematic.

Using LAT-A for Studies in Yeast

LAT-A is used for studies on the actin cytoskeleton in *S. cerevisiae* for several reasons. Foremost is the observation that cytochalasins do not affect actin in growing yeast cells. The reason for this is not clear since isolated yeast actin is affected by the drug.[9] The most likely explanation, however, is that the dense yeast cell wall somehow prevents drug uptake into the cell. To an extent, the need for actin-disrupting drugs was circumvented by the relative ease of generating conditional lethal mutants in the single yeast actin gene, *ACT1*.[10,11] When strains containing certain mutations in their actin gene were shifted to the nonpermissive temperature, the cells could no longer direct their growth to the required regions on the cell surface; instead, cells grew isotropically. However, while the filamentous actin in these mutant strains was delocalized it was not disassembled and to date no viable mutant has been generated that can lead to complete loss of F-actin.

We have demonstrated that the drug LAT-A can cause the rapid and reversible disassembly of filamentous actin structures in cells *in vivo*.[7] Following addition of the drug to cells in suspension, all F-actin structures are

[8] S. Frankel and M. Mooseker, *Curr. Opin. Cell Biol.* **8**, 30 (1996).
[9] C. Greer and R. Schekman, *Mol. Cell Biol.* **2**, 1270 (1982).
[10] P. Novick and D. Botstein, *Cell* **40**, 405 (1985).
[11] K. F. Wertman, D. G. Drubin, and D. Botstein, *Genetics* **132**, 337 (1992).

disrupted in just a few minutes. Washing the cells with new medium and releasing them in the absence of the drug leads to a full recovery of an intact cytoskeleton in about 45 min. Incubation with LAT-A for several hours does not appear to be detrimental with 90% of cells still viable after 6 hr at 30°.

Addition of LAT-A to Yeast Cells

Latrunculin-A is a potent toxin so care should be taken when handling the drug. It is currently available from Molecular Probes (Eugene, OR). A stock should be made of LAT-A in dimethyl sulfoxide (DMSO) at an appropriate concentration and stored at −80°. LAT-A has a molecular weight of 421 and we have found that for work with *S. cerevisiae* a convenient stock concentration is 20 mM (8.42 mg/ml). Working stocks can be made from this and stored at 4° or −20° for a few weeks.

We most frequently use LAT-A at 200 μM to disrupt the actin cytoskeleton in haploid cells and at 100 μM to generate the same effect in diploid cells (differences in sensitivity between haploid and diploids reported by Ayscough and colleagues[7]). The LAT-A is added directly to log-phase cells $(2 \times 10^6 - 2 \times 10^7$ cells/ml) in suspension. An equivalent volume of DMSO is added to a control population of cells. These levels of LAT-A will inhibit the growth of cells grown either in rich media (YPD; 1% yeast extract, 2% peptone, 2% glucose) or in defined minimal media (described by Kaiser and colleagues[12]). For cells in suspension, visualization of the actin cytoskeleton using rhodamine-phalloidin or by immunofluorescence indicates that actin cables are completely disrupted within 2 min following LAT-A addition, and cortical actin patches are no longer present after 10 min.

To observe the processes involved in the reassembly of the actin cytoskeleton following LAT-A disruption, the drug can simply be washed out. We would normally treat cells for at least 10 min at the levels indicated earlier to ensure complete F-actin disruption. Cells can then be harvested by centrifugation (3000 rpm, 5 min), washed twice with YPD, and then resuspended in fresh YPD medium at about 1×10^7 cells/ml. Reassembly of the actin cytoskeleton is slower than disruption, requiring about 45 min before the wild-type polarized distribution of cortical actin patches can be visualized.

Testing and Comparing LAT-A Sensitivities Using Halo Assays

We routinely use halo assays to assess the sensitivity of a strain to LAT-A. This might be appropriate before commencing studies with the drug

---

[12] C. Kaiser, S. Michaelis, and A. Mitchell, *in* "Methods in Yeast Genetics: A Laboratory Course Manual," p. 208. Cold Spring Harbor Laboratory Press, New York, 1994.

since we have noticed variation among commonly used laboratory strain backgrounds of *S. cerevisiae*. In addition, if the protein of interest has a possible interaction with the actin cytoskeleton, such an assay might provide further insights into its role. For example, both Cap2p and Sac6p interact with the actin cytoskeleton,[13,14] and deletion of the genes encoding these proteins lead to increased sensitivity to LAT-A.[7] This evidence has provided further support for the notion that deletion of these genes causes destabilization of the actin cytoskeleton, indicating a role for both of the proteins in maintaining and stabilizing normal actin structure in cells.

Cultures of yeast cells are grown overnight in YPD to early stationary phase. Sterile agar (1%) is melted and cooled to <55°. Then 10 μl of the stationary phase culture is diluted into 2 ml of 2 × YPD. Two milliliters of the molten agar is then added, gently mixed, and poured onto the surface of a YPD plate. The plate should be gently moved to ensure the agar covers the surface. This can then be left for a few minutes for the agar to solidify. Dilutions of LAT-A should be made in DMSO. We have found halos forming from 100 μM to 2 mM LAT-A. An appropriate dilution of DMSO itself should also be made to use as a control. Note that the concentration of LAT-A required for halo formation is about two- to fivefold higher than the level necessary for complete disruption of the actin cytoskeleton of cells in suspension. This is presumably due to the diffusion of the drug in the agar such that even cells relatively close to the disk are exposed to significantly lower levels of LAT-A than were actually placed onto the disk. Using up to three concentration disks per plate, pipette 10 μl of each dilution of the drug onto the center of a sterile concentration disk (available from Difco Laboratories, Detroit, Michigan). Then with sterile forceps, gently place the disks on the top agar in a triangular pattern. Invert the plate and leave at 25°. Halos should develop after 24–48 hr. The size of the halos can then be measured and recorded. The calculation of the relative apparent sensitivity[15] allows comparison of LAT-A sensitivity for a number of strains. This requires that the halo sizes for at least three concentrations of the drug be measured and recorded. A graph is then plotted for the diameter of these halos against the log of the drug concentration. For a yeast strain of interest, the concentration of LAT-A required to give a halo of a specific size can be calculated and compared to the concentration of LAT-A required to give a halo of the same size for a wild-type strain.

[13] A. E. M. Adams, D. Botstein, and D. G. Drubin, *Science* **243,** 231 (1989).
[14] J. F. Amatruda, J. F. Cannon, K. Tatchell, C. Hug, and J. A. Cooper, *Nature* **344,** 352 (1990).
[15] J. Reneke, K. Blumer, W. Courchesne, and J. Thorner, *Cell* **55,** 221 (1988).

Visualization of the Disruption of F-Actin Structures by LAT-A using Rhodamine-Phalloidin

If the gene of interest has a marked effect on the sensitivity of cells to LAT-A then it will be of interest to observe its effects on the actin cytoskeleton, possibly in the presence and absence of the drug. Actin can be observed either by staining using fluorescently labeled phalloidin or by immunofluorescence using antibodies against actin. The use of rhodamine-phalloidin is relatively straightforward and the procedure is given below. However, if double labeling is required between actin and a particular protein of interest most often an immunofluorescence approach is taken such as that described by Pringle and colleagues[16] or by Ayscough and Drubin.[17]

Cultures of yeast cells are grown to exponential phase (2×10^6–2×10^7 cells/ml) and a 1-ml aliquot is taken. LAT-A (100–200 μM) is added to the cells for the required time before fixing for 1 hr by addition of 0.134 ml of 37% formaldehyde. The cells are then pelleted (3000 rpm, 5 min) and washed twice with PBS + 1 mg/ml bovine serum albumin (BSA; Boehringer Mannheim, Indianapolis, Indiana) + 0.1% Triton X-100. The pellet is finally resuspended in 50 μl of the wash buffer and to this is added 20 μl Rd-phalloidin. (Rd-phalloidin is purchased from Molecular Probes and before use resuspended in methanol according to instructions; the 20 μl added can be taken straight from this MeOH stock.) The cells are then incubated for 30 min in the dark before washing three times with phosphate-buffered saline (PBS) containing 1 mg/ml BSA (but no Triton). Finally, the cells are allowed to settle for 10 min on a polylysine-coated slide (Sigma, St. Louis, MO)[16,17] before washing for a final time. Mounting solution is then added and the cells covered with a coverslip before observing using the rhodamine channel on a suitable microscope.

Mounting solution can be bought commercially, for example, ProLong™ antifade solution is available from Molecular Probes, or can be made in the lab using an antifade reagent *p*-phenylenediamine (Sigma). Because this chemical is toxic, we prefer to weigh it out infrequently and make large batches that can be stored at −80° for long periods of time. Add 100 mg *p*-phenylenediamine to 10 ml PBS. If the pH is below 9.0, bring it to pH 9.0 by adding NaOH while stirring. Add 90 ml glycerol and stir until homogeneous. If desired, add 2.25 μl of DAPI (diamidino-2-phenylindole from Sigma; 1 mg/ml in water), which will allow nuclei to be visualized.

---

[16] J. Pringle, A. Adams, D. Drubin, and B. Haarer, *Methods Enzymol.* **194,** 565 (1991).
[17] K. R. Ayscough and D. G. Drubin, *in* "Cell Biology: A Laboratory Handbook" (J. Celis, ed.), p. 477. Academic Press, San Diego, 1997.

TABLE I
LEVELS OF LAT-A OBSERVED TO EFFECT VARIOUS TISSUE CULTURE CELL LINES

Cell type	Concentration of LAT-A used	Phenotype	Reference
3T3 Fibroblasts	0.83 μM (350 ng/ml)	Rounding up and arborization though lower levels (50–150 ng/ml) only give partial effects	2
N1E-115 Mouse neuroblastoma cells	83 nM (35 ng/ml)	Immunofluorescence shows loss of actin filaments and bundles in the cell body and in the growth cone	2
NIL8 Hamster fibroblasts	0.475 μM (0.2 μg/ml)	Rounding up of cells; inhibition of cytokinesis in synchronized cells	18
SpC2 Mouse mammary epithelial cells	0.5 μM (0.21 μg/ml)	Rounding up of cells; loss of normal actin staining visualized by Rd-phalloidin; reversible after 12 hr	19

Store mounting medium at $-80°$ in the dark. The solution should be discarded when it loses its clear color and turns brown.

Studies in Other Yeast

In the course of our studies we have demonstrated that the fission yeast, *Schizosaccharomyces pombe*, is also sensitive to the effects of LAT-A. Using halo assays it appears to be slightly more sensitive to the drug than cells of *S. cerevisiae*.[7] Following addition of the drug at 100 μM we used rhodamine-phalloidin staining to visualize the actin cytoskeleton and observed disruption of all F-actin structures that are normally present.

Using LAT-A for Studies in Mammalian Cells

A number of studies in mammalian cells have demonstrated the efficacy of LAT-A for disrupting the actin cytoskeleton. Interestingly, one study compared the effect of LAT-A to cytochalasin D and showed striking differences in the effects of the two drugs when added to cells in culture.[18] First, the concentration of LAT-A required to achieve complete rounding up of cells was 10–20 times lower than the level of cytochalasin D required for the same effect. Second, with regards to actin organization different patterns of changes were observed with the two drugs indicating that they act in distinct ways on the actin cytoskeleton. Third, after prolonged expo-

[18] I. Spector, N. Shochet, D. Blasberger, and Y. Kashman, *Cell Motil. Cytoskel.* **13,** 127 (1989).
[19] M. J. Close, personal communication (1996).

sure to LAT-A the culture cells maintained the round shape and aberrant actin structures that were visible after 1 hr of treatment. However, the cytochalasin-D-induced changes progressed with time in culture making it difficult to determine a maximal stable effect. These comparisons indicate that LAT-A constitutes a powerful alternative actin-disrupting agent that might provide further insights into the functioning of the actin cytoskeleton both *in vivo* and *in vitro*.

Addition of LAT-A to Mammalian Culture Cells

As mentioned in the previous section, "Addition of LAT-A to Yeast Cells," a stock should be made of LAT-A (Molecular Probes) in DMSO at an appropriate concentration and stored at −80°. LAT-A has a molecular weight of 421 and for most work with mammalian cells a convenient stock concentration is 1 mM (0.42 mg/ml). Working stocks can be made from this and stored at 4° or −20° for a few weeks.

LAT-A can be added directly to the media of subconfluent tissue culture cells and effects have normally been observed 1 hr following drug addition. Furthermore, removal of LAT-A by pelleting the cells and resuspending the cells in fresh medium indicates its effects to be completely reversible. However, some differences have been noted in the concentration of LAT-A added in order to observe maximal effects which appears to vary according to the cell type so we would recommend that the LAT-A concentration be optimized in each situation. Table I summarizes cell types that have been exposed to LAT-A and the concentration of the drug required to cause these changes.

Concluding Remarks

LAT-A is a drug that causes the rapid and reversible disruption of the action cytoskeleton in a wide range of eukaryotic cells. The specificity of the LAT-A interaction with actin and its mode of action as an actin monomer-binding molecule indicating that LAT-A represents a powerful actin-disrupting agent that is likely to be a useful alternative to the cytochalasins in studies of actin organization and function in living cells.

Acknowledgment

I would like to thank Doug Stirling, Tim Rayner, and James Close for a critical reading of these methods. Kathryn Ayscough is an International Prize Travelling Fellow of the Wellcome Trust (038110/Z/93/Z).

[3] Use of the F-Actin-Binding Drugs, Misakinolide A and Swinholide A

By MICHAEL R. BUBB and ILAN SPECTOR

Introduction

The recent availability of new agents that affect the polymerization of actin greatly enhances the precision with which the dynamics and functions of the actin cytoskeleton can be dissected. In this chapter we summarize the effects of two marine natural products that bind directly to actin, misakinolide A and swinholide A, *in vitro* and on the actin cytoskeleton of living cells. We describe how these agents could best be used to achieve these effects, and suggest additional uses for these agents. These drugs have recently been shown to interact with filamentous actin (F-actin), and have been used to compare the known *in vitro* properties of the agents with their *in vivo* effects on cell morphology and actin distribution.[1,2]

Actin-Binding Activities of Misakinolide A and Swinholide A

Both misakinolide A (also named bistheonellide A) and swinholide A are unusual dimeric macrolides isolated from the marine sponges, *Theonella* sp. and *Theonella swinhoei*, respectively.[3,4] The two fold axis of symmetry seen in the two-dimensional chemical structures as a result of their dimeric structure is well preserved in the three-dimensional structures generated by energy minimization techniques. The structures of the two compounds are very similar. Misakinolide A is a 40-membered dilactone, and swinholide A a 44-membered dilactone, only differing from misakinolide A by an addition of two double-bonded carbons in each repeated lactone moiety. The 14-carbon branching side chains are identical for both compounds. The total synthesis of swinholide A has been reported.[5]

In spite of their similar structures, the two compounds have surprisingly different effects on actin filament dynamics *in vitro*. Both compounds bind

[1] D. R. Terry, I. Spector and M. R. Bubb, *J. Biol. Chem.* **272,** 7841 (1997).
[2] M. R. Bubb, I. Spector, A. D. Bershadsky and E. D. Korn, *J. Biol. Chem.* **270,** 3463 (1995).
[3] Y. Kato, N. Fusetani, S. Matsunaga, K. Hashimoto, R. Sakai, T. Higa, and Y. Kashman, *Tetrahedron Lett.* **28,** 6225 (1987).
[4] S. Carmely and Y. Kashman, *Tetrahedron Lett.* **26,** 511 (1985).
[5] I. Paterson, K. Yeung, R. A. Ward, J. D. Smith, J. G. Cumming, and S. Lamboley, *Tetrahedron* **51,** 9467 (1995).

with high affinity to two actin subunits in a configuration that prevents the subunits from participating in either actin-filament elongation or nucleation reactions, but differ in that swinholide A severs actin filaments while misakinolide A caps the barbed end of filaments.

Swinholide A binds cooperatively to two actin subunits with an approximate equilibrium association constant of $K_a = 9 \times 10^{13}$ M^{-2} in a low salt buffer with 2 mM MgCl$_2$.[2] Addition of high salt had little effect on affinity. Swinholide A rapidly severed F-actin in a highly cooperative manner. For efficient severing of actin *in vitro*, the concentration of swinholide A must approach that of the actin subunit concentration. Analysis of depolymerization rates in the presence of swinholide A suggested that the barbed ends of broken actin filaments were not capped by swinholide A, but the effect of actin–swinholide A complexes on actin filament elongation is unknown.

Misakinolide A binds to two actin monomers with K_d of approximately 50 nM at each binding site.[1] Binding at each of the two actin-binding sites was independent as determined by analytical ultracentrifugation in a non-polymerizing buffer. The differences between the monomer-binding properties of misakinolide A and swinholide A may simply reflect differences in buffer conditions, as the buffer used to assess swinholide A–actin binding favored actin polymerization, and therefore favored actin–actin interactions that would be expected to enhance a tendency toward cooperative binding.

In contrast to swinholide A, misakinolide A had no detectable severing activity, but prevented elongation of actin filaments consistent with the addition of misakinolide–actin complex to the barbed end of filaments with a K_d of 50 nM.[1] A compound such as misakinolide A that binds to the barbed end with higher affinity than actin itself, i.e., with K_d of less than the critical concentration of actin, would be expected to inhibit actin depolymerization,[6] yet no such effect was observed. Perhaps disassociation of misakinolide from the barbed end is accompanied by the loss of more than one actin subunit, resulting in depolymerization rates similar to that seen in the absence of drug.

Effects on Actin Organization in Living Cells

Swinholide A and misakinolide A are extremely potent agents and within 1 hr can cause substantial destruction of the actin cytoskeleton in a variety of cell types at concentrations ranging between 10 and 50 nM depending on cell type. However, there are important differences in their effects on living cells, which appear to reflect their different actin-binding activities. For example, treatment of NRK epithelial cells with 10 nM swin-

---

[6] A. Weber, M. Pring, S. L. Lin, and J. Bryan, *Biochemistry* **30**, 9334 (1991).

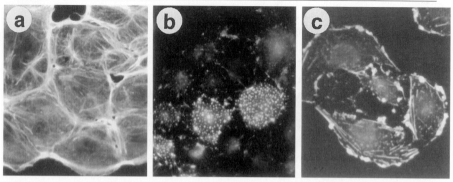

Fig. 1. Distribution of F-actin in NRK epithelial cells in the (a) absence and (b) in the presence of 10 nM swinholide A or (c) misakinolide A after a 1-hr incubation.

holide A for 1 hr results in cell contraction and a complete loss of actin filaments and bundles as assayed by fluorescence microscopy using rhoda-mine-phalloidin staining (Fig. 1b). Furthermore, while some treated cells still show punctate staining patterns, others show only very faint staining, indicating massive depletion of F-actin. Twenty-four hours after drug addition all cells round up, become binucleate, and F-actin staining is hardly visible. Instead, a marked increase in unpolymerized G-actin is observed as monitored by fluorescein-DNase I staining. Treatment of NRK cells with 10 nM misakinolide A also causes marked changes in cell shape and actin organization within 1 hr (Fig. 1c). Light microscopy of cells fixed after 24 hr in the presence of misakinolide A shows that the cells round up, become binucleate, and actin bundles disintegrate with the condensation of actin filaments into amorphous perinuclear aggregates.

Applications

The ability of swinholide A to sever actin filaments *in vitro* and to deplete F-actin *in vivo* provides a powerful and inexpensive means for the quantitative and specific removal of actin filaments from cytoskeletons of detergent-permeabilized cells. Swinholide A can therefore be used to ascertain which cytoskeletal proteins and other cellular components (e.g., membrane receptors, adhesion junction proteins, and components of signal transduction pathways) are connected directly and exclusively to actin filaments. Also, this technique can be utilized to better visualize those cytoskeletal proteins that are not associated exclusively with actin filaments (e.g., myosin, intermediate filaments, and microtubules).[7] Additionally, *in vitro*

[7] A. B. Verkhovsky, T. M. Svitkina, and G. G. Borisy, *J. Cell Biol.* **131,** 989 (1995).

depletion of F-actin may be useful to clean up a preparation (e.g., to strip F-actin and all F-actin-associated proteins from a plasma membrane preparation). By determining if loss of F-actin correlates with the loss of a protein of interest, swinholide A may be used to determine whether a membrane-fractionated or other compartmentalized protein is mostly or exclusively associated with actin filaments.

Despite the extensive use of cytochalasins in attempts to alter actin structure, function, and dynamics, the effects of these fungal metabolites on actin are complex and difficult to interpret. To various degrees, they can cap actin filaments, sever actin filaments, sequester actin monomers, induce the formation of actin dimers, promote nucleation, inhibit polymerization at the pointed ends, and stimulate the ATPase activity of G-actin.[8] These multiple activities have unexpected and various consequences on living cells, including stimulation of actin polymerization, depolymerization of F-actin, redistribution of actin without change in the polymerization state, or a lack of effect.[9–11] Newly characterized agents such as the marine macrolide latrunculins, with actin-binding activity limited to that of monomer sequestration, have already been exploited on the basis of their specificity to characterize actin function in living cells and to affect actin structures resistant to cytochalasins.[11–13] Because of the specific and consistent activities of swinholide A and misakinolide A, their use has provided information about the roles of actin polymer and actin polymerization in cellular processes, and also should provide information regarding the function of specific actin-binding proteins in cells.[14] For example, does a drug that specifically inhibits barbed-end elongation of filaments block actin subunit turnover in stress fibers or, as has been suggested, does annealing of short filaments contribute to stress fiber formation in a process that would be relatively unaffected by this drug?[15] Can a drug that mimics only the severing effects of gelsolin compensate for gelsolin deficiency in cells (or animals) in which gelsolin expression is defective?

Conclusions regarding different effects observed for drugs differing in actin-disrupting activities must be cautiously postulated. If the half-lifetime

[8] J. A. Cooper, *J. Cell Biol.* **105**, 1473 (1987).

[9] J. A. Wilder and R. F. Ashman, *Cell. Immunol.* **137**, 514 (1991).

[10] K. M. K. Rao, J. Padmanabhan, and H. J. Cohen, *Cell Motil. Cytoskel.* **21**, 58 (1992).

[11] A. D. Bershadsky, U. Gluck, O. N. Denisenko, T. V. Sklyarova, I. Spector, and A. Ben-Ze'ev, *J. Cell Sci.* **108**, 1183 (1995).

[12] K. R. Ayscough, J. Stryker, N. Pokala, M. Sanders, P. Crews, and D. G. Drubin, *J. Cell Biol.* **137**, 399 (1997).

[13] C. Lamaze, L. M. Fujimoto, H. L. Yin, and S. L. Schmid, *J. Biol. Chem.* **272**, 20332 (1997).

[14] A. Lyubimova, A. D. Bershadsky, and A. Ben-Ze'ev, *J. Cell. Biochem.* **65**, 469 (1997).

[15] Y-L. Wang, *J. Cell Biol.* **105**, 2811 (1987).

for an actin filament in a stress fiber is only several minutes[16,17] (and less for more dynamic populations of F-actin such as that in lamellipodia) then a scale of observation measured in hours may not be satisfactory to detect differences in cellular effects that reflect differences in actin-binding activities. Observed differences could reflect other, unknown functions of the drug unrelated to actin, although as of yet, no such functions have been identified. Alternatively, entry into cells for these drugs may be rate limiting, so that the observed differences do indeed reflect differences in the mechanism by which they depolymerize actin.

Procedures

Evaluation of Drug–Actin Interactions

The solubility of both misakinolide A and swinholide A exceeds 10 mM in DMSO but is 1–10 μM in aqueous solvent. Actin solubilizes the drug so that if the solution is a uniform suspension prior to pipetting and if adequate mixing is provided, higher concentrations of drug can be tested in the presence of adequate actin concentrations.

Assays for the actin monomer- and filament-binding properties of swinholide A and misakinolide A employ both fluorescence and analytical ultracentrifugation techniques. Monomer-binding activity can be measured at steady-state employing pyrene-labeled rabbit skeletal muscle actin.[1,2,18] Fluorescence is measured as a function of actin concentration, and because polymerization augments steady-state fluorescence of pyrene-labeled actin by 30-fold, the critical concentration is the abscissa value for which the measured fluorescence begins to deviate from that expected for monomeric actin. Artifacts due to problems with actin can be eliminated in part by confirming that the same steady-state fluorescence value is reached whether starting within monomeric or polymerized actin. The apparent change in the critical concentration after adding drug, Δ, is related to the K_d by the equation, $\Delta = (d\bar{a})/(\bar{a} + K_d)$, where d is the concentration of drug and $\bar{a}$ is the critical concentration (appropriate changes are required to reflect a number of actin-binding sites per drug molecule different than unity).

Actin dimer formation induced by either drug is best measured by sedimentation equilibrium in an analytical ultracentrifuge.[1] An actin concentration of 15 μM is optimal for use with absorption optics in a double-sector cell containing 100–120 μl of sample. Equilibrium is reached within

[16] T. E. Kreis, B. Geiger, and J. Schlessinger, *Cell* **29**, 835 (1982).
[17] P. A. Amato and L. Taylor, *J. Cell Biol.* **102**, 1074 (1986).
[18] T. Kouyama and K. Mihashi, *Eur. J. Biochem.* **114**, 33 (1981).

24 hr at 14,500 rpm, with absorption data collected at 290 nm in 0.003-cm increments. At this wavelength, neither drug has significant absorption so that the shape of the concentration gradient reflects only the oligomeric distribution of actin. An equilibrium dissociation constant can then be calculated assuming that binding of drug to actin causes formation of an actin dimer. The severing activity of swinholide A is best measurd with a stopped-flow fluorimeter, using three syringes.[2] Syringe 1 contains stock filamentous actin at a concentration of 10 μM, syringe 2 contains a buffer of choice for dilution, and syringe 3 contains swinholide A at a stock concentration of 2.0 μM. The injection rate will have to be optimized for the individual device to minimize filament fragmentation and was 0.5 sec in the Bio-Logic SFM-3 stopped-flow module. To optimize signal intensity, a 380-nM cutoff filter should be used in the emission path, and angled at approximately 5° from perpendicular in order to slightly red-shift the cutoff limit.

In vivo Applications

Swinholide A and misakinolide A readily enter living mammalian cells when added to culture media. Lower eukaryotes or other cells have not been tested. Final concentration of 10–50 nM for incubation of cells are effective *in vivo*. Changes in actin filament structure are detectable within minutes of drug application. No special incubation conditions or media are necessary, and labeling of F-actin and G-actin with fluorescent derivatives of phalloidin and DNase I, respectively, are performed as in the absence of drug as described elsewhere.[19,20]

Swinholide A is cytotoxic except after very brief application. The effects of misakinolide A, on the other hand, are fully reversible and the drug can be washed out simply by exchange of media. Particularly in light of the cytotoxicity of swinholide A, and the membrane permeability of both agents, these drugs should be handled and disposed of with caution.

In vitro Applications

Plasma membrane,[21] Triton-insoluble cytoskeleton,[22] or membrane skeleton[23] preparations are performed as in the absence of drug. Swinholide

[19] L. S. Barak, R. R. Yocum, E. A. Nothnagel, and W. W. Webb, *Proc. Natl. Acad. Sci. U.S.A.* **77,** 980 (1980).
[20] R. P. Haugland, W. You, V. B. Paragas, K. S. Wells, and D. A. Dubose, *J. Histochem. Cytochem.* **42,** 345 (1994).
[21] B. J. Clarke, T. C. Hohman, and B. Bowers, *J. Protozool.* **35,** 408 (1988).
[22] J. H. Hartwig and M. DeSisto, *J. Cell Biol.* **112,** 407 (1991).
[23] J. R. Apgar, *J. Immunol.* **145,** 3814 (1990).

A can be added to any of these preparations to effect the removal of F-actin and F-actin-associated proteins, or misakinolide A could be used to cap existing barbed ends at any stage of the preparation to prevent additional actin polymerization during the preparation process. Both agents are unaffected by pyrene labeling of F-actin, so they should be equally effective for treating cells or cell preparations that have been previously investigated using this actin derivative. Video microscopy suggests that swinholide A severs phalloidin-labeled F-actin, but it is not known whether the efficiency of severing is affected. The suggested *in vitro* applications for misakinolide A and swinholide A have not been systematically evaluated, but using the results of the affinity measurements described earlier as a guideline, the concentration of swinholide A should be stoichiometric to or greater than the actin concentration of the treated preparation in order to eliminate F-actin entirely, whereas existing filaments can be severed but not eliminated with 100 nM swinholide A. The concentration of misakinolide A necessary in order to cap most barbed ends must exceed 100 nM.

Acknowledgments

M.R.B. is a Pfizer Scholar and I.S. is supported by the NOAA award NA46RG0090 (New York Sea Grant Project R/XBP-5).

[4] F-Actin Blot Overlays

By Elizabeth J. Luna

Introduction

Proteins that bind F-actin are commonly identified by their ability to cosediment with either endogenous actin filaments or with prepolymerized rabbit muscle actin (for example, see Refs. 1 and 2). Alternatively, F-actin binding proteins in cell lysates or detergent extracts of membrane fractions can be identified by their affinity for actin filaments bound to solid supports.[3,4] Both of these approaches are extremely useful since they facilitate the purification of F-actin binding proteins in a native conformation. How-

[1] M. Fechheimer and R. Furukawa, *Methods Enzymol.* **196,** 84 (1991).

[2] J. H. Hartwig and T. P. Stossel, *Methods Enzymol.* **85B,** 480 (1982).

[3] K. G. Miller, C. M. Field, B. M. Alberts, and D. R. Kellogg, *Methods Enzymol.* **196,** 303 (1991).

[4] L. J. Wuestehube, D. W. Speicher, A. Shariff, and E. J. Luna, *Methods Enzymol.* **196,** 47 (1991).

ever, neither approach can differentiate proteins that bind directly to actin filaments from those that are bound only indirectly through interactions with other components of a multisubunit complex. Also, both cosedimentation assays and F-actin affinity columns necessarily involve high local concentrations of F-actin, which is a relatively acidic protein. Thus, some proteins may bind through nonspecific electrostatic interactions, a possibility that should always be excluded through experiments with other electronegatively charged biopolymers, such as heparin, polyaspartic acid, and/or Taxol-stabilized microtubules.

As a complementary technique, my colleagues and I have devised an approach in which nitrocellulose blots containing sodium dodecyl sulfate–polyacrylamide gel electrophoresis (SDS–PAGE) purified proteins are probed with a solution containing dilute, radiolabeled actin filaments.[5–9] The chief advantage of this approach is that there is no ambiguity about whether a protein recognized in this assay binds directly or indirectly to F-actin. Also, F-actin binding activities can be demonstrated for hard-to-solubilize proteins. In addition, all proteins identified to date on F-actin blot overlays bind specifically to F-actin, as compared to other electronegative biopolymers.[7] A potential drawback to this technique is that many actin-binding proteins—including myosin, spectrin, and actin itself—are not recognized in the simplest version of this assay due to denaturation of their actin-binding sites by SDS. In fact, this circumstance potentiates the study of proteins that retain, or can regain, actin-binding activity after SDS denaturation, because most cells exhibit a relatively simple profile of F-actin binding proteins in this assay (Fig. 1). Once the identities of the proteins that are recognized in the cell type of interest have been established, all of these proteins can be followed simultaneously in a single assay with a sensitivity that rivals or surpasses that of immunoblot analyses.[5,8]

Interestingly, the proteins identified so far by the F-actin blot overlay approach have all been directly or indirectly associated with membranes. These "membrane skeleton" proteins include the 17-kDa transmembrane protein ponticulin,[5] a 34-kDa filopodial protein from *Dictyostelium discoideum*,[6] the ~120-kDa neuronal protein drebrin,[8] and the ~80-kDa ezrin-

[5] C. P. Chia, A. L. Hitt, and E. J. Luna, *Cell Motil. Cytoskel.* **18**, 164 (1991).
[6] M. Fechheimer, H. M. Ingalls, R. Furukawa, and E. J. Luna, *J. Cell Sci.* **107**, 2393 (1994).
[7] K. Pestonjamasp, M. R. Amieva, C. P. Strassel, W. M. Nauseef, H. Furthmayr, and E. J. Luna, *Mol. Biol. Cell* **6**, 247 (1995).
[8] E. J. Luna, K. N. Pestonjamasp, R. E. Cheney, C. P. Strassel, T. H. Lu, C. P. Chia, A. L. Hitt, M. Fechheimer, H. Furthmayr, and M. S. Mooseker, *in* "Cytoskeletal Regulation of Membrane Functions" (S. C. Froehner and V. Bennett, eds.), p. 3. Rockefeller University Press, New York, 1997.
[9] K. N. Pestonjamasp, R. K. Pope, J. D. Wulfkuhle, and E. J. Luna, *J. Cell Biol.* **139**, 1255 (1997).

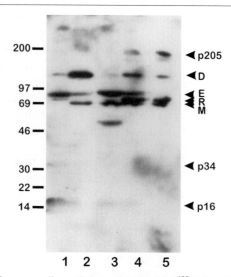

Fig. 1. Proteins in mammalian cell lines visualized by ^{125}I-labeled F-actin blot overlay. Autoradiogram of whole-cell extracts from MCF-7 breast carcinoma (lane 1), SHSY5Y neuroblastoma (lane 2), NIH 3T3 fibroblasts (lane 3), NRK fibroblasts (lane 4), and HeLa S3 cervical carcinoma (lane 5) cells. Migration positions of molecular mass standards, in kilodaltons, are denoted on the left. Arrows on the right denote the positions of a 16-kDa protein (p16), a 34-kDa protein (p34), moesin (M), radixin (R), ezrin (E), drebrin (D), and supervillin (p205). Whole-cell extracts (10^6 cells/lane) were denatured for 10 min at 70° in Laemmli sample buffer containing 3.2 mM DTT, electrophoresed into a 6–16% gradient polyacrylamide SDS gel,[17] and electrotransferred to nitrocellulose at 6 V/cm for 20 hr at 4°.[18] F-actin binding proteins were heat-fixed onto the nitrocellulose at 50° for 1 hr in PBS before visualization with 50 μg/ml ^{125}I-labeled F-actin, as described in the text. [Reproduced with permission from Luna *et al.*, *in* "Cytoskeletal Regulation of Membrane Functions" (S. C. Froehner and V. Bennett, eds.), p. 3. Rockefeller University Press, New York, 1997.]

radixin-moesin (ERM) family of actin-membrane linking proteins.[7] Most recently, a 205-kDa protein called supervillin, which binds F-actin on blot overlays (p205 in Fig. 1), has been shown to colocalize at the light level with cadherin-based sites of cell–cell contact.[9] Supervillin contains sequences that are highly homologous to actin-binding sites in villin and gelsolin,[9] raising the possibility that proteins in this superfamily may be generally reactive in this assay. This prospect is supported by the observation of a major F-actin binding band in human erythrocytes (data not shown) at the position of dematin/band 4.9, another member of the villin superfamily.[10]

[10] A. P. Rana, P. Ruff, G. J. Maalouf, D. W. Speicher, and A. H. Chishti, *Proc. Natl. Acad. Sci. U.S.A.* **90**, 6651 (1993).

Procedure

Materials

Highly purified actin, 3 mg: We use rabbit skeletal muscle actin[11] that has been further purified by gel filtration chromatography,[12] but any highly purified actin preparation could be used.[13,14]

[125]I-Labeled monoiodinated Bolton–Hunter reagent, 250–500 μCi: e.g., product NEX-120 (DuPont–NEN, Wilmington, DE). Significant savings can be realized by making this reagent immediately before use[15] if the expertise and facilities for handling elemental ^{125}I are available.

Buffer A: 2 mM Tris-HCl, 0.2 mM CaCl$_2$, 0.2 mM Na$_2$ATP, 0.5 mM dithiothreitol (DTT), pH ~8.3 at 0°.

Dounce tissue grinder, 1 ml: For example, product 357538 (Wheaton, Millville, NJ).

Actin labeling buffer: 2 mM sodium phosphate, 0.1 mM MgCl$_2$, 0.1 mM Na$_2$ATP, pH 8.5.

5× Actin polymerization buffer: 250 mM KCl, 10 mM MgCl$_2$, 0.25 mM CaCl$_2$, 0.02% NaN$_3$, 100 mM PIPES-NaOH, pH 7.0.

Thick-walled 3.5-ml polycarbonate or polyallomer ultracentrifuge tubes: For example, part 349622 (Beckman Instruments, Palo Alto, CA) for use in an SW50.1 rotor (Beckman Instruments) or equivalent.

(*Optional*) Brevin (plasma gelsolin), ~5 μg: A very simple isolation procedure has been described[16]; brevin stored at −80° as aliquots of 2.5 mg/ml protein in 150 mM NaCl, 0.1 mM Na$_2$ EGTA, 5 mM Tris-HCl, pH 7.5 (at 4°) retains actin-nucleating activity for ≥10 yr.

Phalloidin (Boehringer Mannheim, Germany): 1 mM in 2% dimethyl sulfoxide (DMSO); dissolve 1 mg in bottle first with 25 μl DMSO, then with 1.24 ml water; store at −20°.

Tris-buffered saline plus Tween-20 (TBST), 1 liter: 10 mM Tris-HCl, 90 mM NaCl, 0.5% Tween-20, pH 7.5.

5% (w/v) milk/TBST, 30–60 ml: Carnation nonfat powdered milk stirred overnight in TBST at ~20° and stored at 4°.

[11] J. A. Spudich and S. Watt, *J. Biol. Chem.* **246,** 4866 (1971).
[12] S. MacLean-Fletcher and T. D. Pollard, *Biochem. Biophys. Res. Commun.* **96,** 18 (1980).
[13] J. D. Pardee and J. A. Spudich, *Methods Enzymol.* **85B,** 164 (1982).
[14] S. R. Schaier, *Methods Enzymol.* **215,** 58 (1992).
[15] M. A. Schwartz and E. J. Luna, *J. Cell Biol.* **102,** 2067 (1986).
[16] H. Kurokawa, W. Fujii, K. Ohmi, T. Sakurai, and Y. Nonomura, *Biochem. Biophys. Res. Commun.* **168,** 451 (1990).

Laemmli sample buffer, 100 μl per sample of 100 μg protein or 10^6 whole cells: 62.5 mM Tris-HCl, 2% SDS, 10% glycerol, 0.005% bromphenol blue, pH 6.8, with and without 3.2 mM DTT.

Polyacrylamide slab gel apparatus and Laemmli gel solutions: As described by Laemmli.[17]

Towbin transfer buffer and a tank electrophoretic transfer apparatus: 192 mM glycine, 25 mM Tris-HCl, pH 8.3, 20% methanol[18]; ~3 liters needed for use with the Transblot Cell from Bio-Rad Laboratories (Richmond, CA).

Container with a tightly-fitting lid that will snugly hold a nitrocellulose blot.

(*Optional*) Phosphate-buffered saline (PBS), ~500 ml: 150 mM NaCl, 10 mM sodium phosphate, pH 7.5.

(*Optional*) Heated water bath with adjustable temperature settings.

^{125}I Bolton–Hunter Labeling of G-Actin

Gel-purified actin (3 mg in 500 μl), stored as F-actin at 0–4° or after lyophilization of G-actin with 2 mg sucrose per milligram actin, is dialyzed for 24–36 hr at 0° against 500 ml of buffer A. Lyophilized actin is first hydrated by homogenization in buffer A using a 1-ml Dounce tissue grinder (Wheaton). This amount of actin provides a compromise between specific activity of the radiolabel, which will be higher if starting with less actin (1 mg), and efficient recovery, which is better for larger amounts of actin (6 mg). Although the volume can be scaled up or down, the actin concentration should be maintained at 4–6 mg/ml.

Dialyzed actin is clarified by centrifugation at 40,000 rpm (192,000g) for 40 min in thick-walled polycarbonate or polyallomer tubes in an SW 50.1 rotor (Beckman Instruments) at 3°. The G-actin concentration in the supernatant is determined from the absorbance at 290 nm ($E^{1\%} = 0.62$ cm^{-1}). The volume of actin to be labeled is then dialyzed for 4–16 hr at 0° against actin labeling buffer to lower the concentration of DTT.

^{125}I-Labeled Bolton–Hunter reagent[19] is dried at 0° for 30–60 min under a gentle stream of nitrogen, as described by the supplier. G-actin in actin labeling buffer is pipetted into the bottom of the reagent container, slowly pipetted up and down for ~5 min to promote mixing, and then allowed to react for 2–4 hr at 0°. (*Note*: A separate reaction tube may be required for volumes >500 μl, depending on the capacity of the reagent vial.)

---

[17] U. K. Laemmli, *Nature (Lond.)* **227**, 680 (1970).
[18] H. Towbin, T. Stahelin, and J. Gordon, *Proc. Natl. Acad. Sci. U.S.A.* **76**, 4350 (1979).
[19] A. E. Bolton and W. M. Hunter, *Biochem. J.* **133**, 529 (1973).

Radiolabeled G-actin is separated from unincorporated ^{125}I-labeled Bolton–Hunter reagent by gel filtration at 20° through a 5-ml Sephadex G-25 column equilibrated with buffer A. (The volume of the Sephadex G-25 column is 10× the volume of the labeled G-actin.) Fractions (~500 μl, i.e., ~0.1× column volume) are collected dropwise (10–20 drops/fraction) by hand and assayed for the presence of radioactivity with a handheld Geiger counter, e.g., a Model SD10 Rad-Monitor™ (Research Products International Corp., Mount Prospect, IL). The ^{125}I-labeled G-actin will elute at 30–50% of the total column volume, assuming that the amount of tubing is kept to a minimum; unincorporated Bolton–Hunter reagent will elute with the included volume at ~100% column volume. The fractions containing high concentrations of radiolabeled G-actin are combined in a thick-walled 3.5-ml polycarbonate or polyallomer ultracentrifuge tube (Beckman Instruments). The pooled volume is estimated.

Radiolabeled actin is polymerized in the centrifuge tube by gentle but thorough mixing with ca. 0.25 volume 5× actin polymerization buffer, e.g., 375 μl for ~1.5 ml pooled gel-filtered G-actin (three fractions), and by incubating at room temperature (20–25°) for 30–60 min. ^{125}I-Labeled F-actin is collected by centrifugation at 40,000 rpm (192,000g) for 2.5 hr at 3° in an SW50.1 rotor. The supernatant, which should contain <10% of the radioactivity, is discarded. The pellet is resuspended with 500 μl buffer A at 0°. (If only 1 mg of actin is labeled, this volume is decreased to 250 μl.) While repeated gentle pipetting is usually sufficient, a 1-ml Dounce tissue grinder (Wheaton) is sometimes required to suspend larger amounts of actin. A second, equal volume of ice-cold Buffer A is used to wash the centrifuge tube, pipette, and tissue grinder (if used).

To complete the depolymerization of the radiolabeled actin, the resuspended pellet and wash solution (1-ml total volume) are dialyzed for 16–40 hr against 40 ml of buffer A in a 50-ml screw-top test tube at 0°. One buffer change is recommended, but little radioactivity should dialyze out.

Denatured actin is removed by clarifying the dialyzed actin at 40,000 rpm (192,000g) for 30 min at 3° in an SW50.1 rotor. If more than 10–15% of the radioactivity pellets at this step, then either the previous manual manipulations were too harsh or the actin was not dialyzed long enough.

The amount of ^{125}I-labeled G-actin recovered at this point can be determined by measuring the absorbance of the solution at 290 nm, as described earlier. However, actin recovery is maximized and personnel exposure to radioactivity is lessened by performing a Lowry assay,[20] or other protein determination assay, on 5- to 10-μl portions of the dialyzed G-actin and

[20] O. H. Lowry, N. Rosebrough, A. L. Farr, and R. J. Randall, *J. Biol. Chem.* **193,** 265 (1951).

buffer A. The amount of radiolabeled actin is determined by comparison with a standard curve generated with bovine serum albumin.

To determine the specific activity, 1- or 2-μl portions of the ^{125}I-labeled G-actin are pipetted into tubes containing 1 ml 0.1% SDS, counted in a gamma counter, and the amount of radioactivity determined from the known efficiency of the gamma counter. For preparations with lower specific activities, the 5-μl Lowry analysis samples are sometimes within the range of the gamma counter and can be used for both assays if counted before color development. By comparing the molar specific activity of the radiolabeled actin with that of the ^{125}I-labeled Bolton–Hunter reagent, the percent of labeled actin monomers is determined. Usually, about 0.002–0.005% of the actin monomers are radiolabeled with the protocol recommended earlier.

The ^{125}I-labeled G-actin can be stored in dialysis against buffer A for 2 weeks if buffer A is changed every second day. However, it is usually more convenient to polymerize the actin before long-term storage (see later discussion).

Actin Polymerization

Actin filaments are much more stable than actin monomers, particularly if the F-actin is stored at ≥ 1 mg/ml in the presence of equimolar phalloidin (reviewed in Ref. 21). Unless ^{125}I-labeled G-actin is required for some other purpose, such as coassembly in a sedimentation binding assay,[6,15] the radiolabeled actin should be diluted to a concentration of 1.25 mg/ml with buffer A and polymerized as described later.

The actin nucleating protein, brevin (recently reviewed in Refs. 22 and 23), is optionally added for 10 min at 0° to the ^{125}I-labeled G-actin in buffer A in order to control the average lengths of the actin filaments and the physical behavior of the suspension. In the absence of brevin or other filament-capping protein, actin polymerizes as long filaments in highly viscous solutions that are difficult to pipette accurately. Such F-actin suspensions usually must be vortexed briefly before each pipetting, a procedure that breaks the actin filaments at random.

Brevin generates actin filaments of known average length and specific activity by potentiating the formation of actin nuclei, which elongate much faster than the rate at which actin nuclei form spontaneously. Each brevin molecule nucleates the formation of a single filament. Thus, the number

[21] J. A. Cooper, *J. Cell Biol.* **105**, 1473 (1987).

[22] D. A. Schafer and J. A. Cooper, *Ann. Rev. Cell Dev. Biol.* **11**, 497 (1995).

[23] A. Wegner, K. Aktories, A. Ditsch, I. Just, B. Schoepper, N. Selve, and M. Wille, *Adv. Exp. Med. Biol.* **358**, 97 (1994).

of actin filaments is essentially identical to the number of brevin molecules, and the average filament length is the ratio of the actin monomer concentration to the concentration of brevin. These actin filaments contain brevin bound to their fast-polymerizing (barbed) ends and are inhibited with respect to binding to other barbed end-interacting proteins. The amount of brevin added is determined by the specific activity of the ^{125}I-labeled G-actin and is usually one brevin molecule (molecular mass of ~86,000) per 200 to 1000 actin monomers (molecular mass of ~43,000), so that the resulting filaments each contain at least one ^{125}I-labeled monomer.

To initiate polymerization, 0.25 volume of 5× actin polymerization buffer is added to the brevin/actin mixture. After ~10 min at 0°, the suspension is warmed to room temperature, and sufficient 1 mM phalloidin is added to generate a final concentration of 40 μM. Polymerization is complete after an additional 20–30 min at room temperature. The resulting 1 mg/ml (~23 μM) F-actin suspension can be stored for months at 0° or immediately diluted into 5% (w/v) milk/TBST for use in blot overlays.

What to Expect

The recovery of actin after radiolabeling, polymerization, and depolymerization should be 30–50% of initial amounts. Most of the loss is caused by denaturation since the critical concentration of ^{125}I-labeled Bolton–Hunter-labeled actin is 5–8 μg/ml,[15] which is about the same as unlabeled actin. The average specific activity of the radiolabeled actin is ~17 μCi/mg, with a range from ~7 to ~34 μCi/mg.[5] On a molar basis, 1 out of every 200–400 actin monomers is generally labeled in reactions with 3 mg of initial G-actin and 250 μCi of fresh ^{125}I-labeled Bolton–Hunter reagent. Thus, sufficient radiolabeled actin should be recovered from 3 mg of initial G-actin to prepare 10–15 ml of 100 μg/ml ^{125}I-labeled F-actin in 5% (w/v) milk, TBST, with which nitrocellulose blots from full-size gels of 10 × 10 cm can be probed for several weeks (see later discussion).

F-Actin Blot Overlays

Many actin-binding proteins recognized by ^{125}I-labeled F-actin on blot overlays (Fig. 1) can be solubilized in SDS and electrophoresed into polyacrylamide gels, essentially as described by Laemmli.[17] Others, such as *Dictyostelium discoideum* ponticulin, are denatured by disulfide reducing agents and/or by boiling.[5] If nothing is known *a priori* about the properties of a protein or tissue of interest, it is useful to try a number of SDS-solubilization conditions, including 20 min at 37°, 15 min at 50°, 10 min at 70°, and 2 min at 100°, in Laemmli sample buffer with and without DTT. Electrophoresis conditions also can be varied, but it is prudent to avoid high currents or resistance that cause excess warming of the gel. Similarly, electrotransfer onto nitrocellulose (0.45-μm porosity; Schleicher & Schuell, Keene, NH) in Towbin transfer buffer[18] is usually optimal in a tank apparatus at 4°, as opposed to a "semidry" apparatus.

Because polypeptides with masses of ≤50,000 Da are frequently eluted, either partially or totally, from nitrocellulose during the blocking of unoccupied sites,[5,24,25] it is also a good idea to determine whether fixation of electrotransferred proteins improves their ability to be recognized by F-actin. For this optional procedure, nitrocellulose blot strips are washed five times with PBS at ~20° (2–5 min/wash) and fixed under a variety of conditions before blocking. One useful fixation procedure is 0.1% glutaraldehyde, PBS for 45 min at 20° (Ref. 24). Another good strategy is to heat blot strips in PBS for 1 hr at temperatures ranging from 30 to 90°.[5,26] An initial experiment should include an unfixed blot strip and otherwise identical blot strips that have been fixed using several different protocols.

Nitrocellulose blots containing proteins that have been SDS solubilized, electrophoresed, electrotransferred, and (optionally) fixed under optimal conditions, determined as described earlier, are incubated with 5% (w/v) milk/TBST for 1–2 hr at ~20° or for 16–24 hr at 4°. In addition to blocking unoccupied sites on the nitrocellulose, this step may contribute to localized renaturation of protein structure.[24] Myosin binds F-actin on blot overlays if renatured on the nitrocellulose with a series of urea solutions,[27] a technique that also may work for other proteins with SDS-sensitive actin-binding sites.

For the incubation with ^{125}I-labeled F-actin, the blocked nitrocellulose blot is placed into a box with a tightly fitting lid and interior dimensions that are only slightly larger than the blot itself. If necessary, Parafilm (American National Can, Neenah, WI) can be used to limit the area accessible to the ^{125}I-labeled F-actin solution. Sealable plastic bags, which are often used for incubations with antibodies, tend to trap air bubbles and to result in unaesthetic background staining, perhaps due to actin denaturation. The nitrocellulose is then covered completely with ^{125}I-labeled F-actin that has been diluted to 50–100 μg/ml with 40 μM phalloidin in 5% (w/v) milk/TBST. In general, at least 1 ml of ^{125}I-labeled F-actin solution is required for every 10 cm^2 of nitrocellulose. The blot is incubated with

[24] L. J. Van Eldik and S. R. Solchok, *Biochem. Biophys. Res. Commun.* **124,** 752 (1984).
[25] L. J. Wuestehube, C. P. Chia, and E. J. Luna, *Cell Motil. Cytoskel.* **13,** (1989).
[26] V. M. Fowler, *J. Biol. Chem.* **262,** 12792 (1987).
[27] P. Eldin, M. Le Cunff, K. W. Diederich, T. Jaenicke, B. Cornillon, D. Mornet, H.-P. Vosberg, and J. J. Léger, *J. Muscle Res. Cell Motil.* **11,** 378 (1990).

the ^{125}I-labeled F-actin solution in the sealed box for 2 hr at ~20° without disturbance, i.e., without swirling or rocking.

At the end of the incubation, the ^{125}I-labeled F-actin solution is carefully retrieved and stored at 0–4°. Although sensitivity declines with use and storage time, acceptable results are usually obtained for 6–12 weeks with a 75–100 μg/ml solution of ^{125}I-labeled F-actin. The nitrocellulose blot is washed five times, for 2 min each, with TBST. (Longer washes depolymerize bound F-actin and reduce the signal.) The blot is dried on filter paper and exposed to film. With a freshly prepared ^{125}I-labeled F-actin solution, major actin-binding proteins are visible after a 2- to 24-hr exposure at −80° with an intensifying screen.

Blot Overlays with G-Actin

Blot overlays with ^{125}I-labeled G-actin are performed similarly, except that solutions are prepared within 5–10 min of each experiment. ^{125}I-labeled G-actin is removed from dialysis against buffer A and is quickly diluted to a final concentration of 10 μg/ml with 5% (w/v) milk, TBST, containing 0.2 mM ATP to stabilize the actin. Because 10 μg/ml is close to the critical concentration for assembly, the G-actin should not polymerize appreciably during the 1- to 2-hr incubation with the nitrocellulose blot. To demonstrate that the ^{125}I-labeled G-actin retains its native conformation after this incubation, the overlay solution is removed from contact with the blot and is incubated for 20–24 hr at 20° with 1/4 volume 5× actin polymerization buffer and with 1/25 volume 1 mM phalloidin. This treatment with phalloidin lowers the critical concentration for actin assembly to unmeasurable levels.[21] If 80–90% of the ^{125}I-labeled actin sediments after 2.5 hr at 40,000 rpm (192,000g) in an SW50.1 rotor, then the ^{125}I-labeled G-actin is assumed to have been in the native conformation during the blot overlay. Also, the presence of denatured actin is usually signaled by the presence of a uniform dark gray background on autoradiographic exposures of the nitrocellulose blot.

Alternative Protocols

In a recently published paper, Mackay *et al.*[28] describe a variant of the F-actin blot overlay technique in which [α-^{32}P]ATP is bound to G-actin before polymerization with salt and phalloidin. The apparent specific activity of the resulting F-actin–[α-^{32}P]ADP complex appears to be higher than

[28] D. J. G. Mackay, F. Esch, H. Furthmayr, and A. Hall, *J. Cell Biol.* **138**, 927 (1997).

that usually obtained with ^{125}I-labeled Bolton–Hunter reagent, but the short half-life of ^{32}P will decrease the useful lifetime of the labeling solution. Another experimental strategy is to employ unlabeled, phalloidin-stabilized actin filaments in the overlay solution and to covalently cross-link the actin before visualization with anti-actin antibodies.[29] This approach eliminates the need for radioisotopic labeling but requires several additional experimental steps.

Acknowledgments

Work in this laboratory received financial support from National Institutes of Health grants GM33048 and CA54885. This research also benefited from grants to the Worcester Foundation for Biomedical Research from the J. Aron Charitable Foundation and the Stork Foundation.

[29] C. B. Shuster, A. Y. Lin, R. Nayak, and I. M. Herman, *Cell Motil. Cytoskel.* **35**, 175 (1996).

[5] Purification and Assay of the Arp2/3 Complex from *Acanthamoeba castellanii*

By JOSEPH F. KELLEHER, R. DYCHE MULLINS, and THOMAS D. POLLARD

Introduction

The Arp2/3 complex is an ubiquitous actin filament nucleating and cross-linking factor composed of seven subunits in a stable complex. The complex contains two actin-related proteins (Arp2 and Arp3) and five novel subunits.[1–3] We purify two fractions of Arp2/3 complex from the same extract of *Acanthamoeba*: (A) by poly-L-proline affinity chromatography[1,4] and (B) by conventional column chromatography using specific polyclonal antisera against Arp2 and Arp3[5] to assay for the presence of complex. The

[1] L. M. Machesky, S. J. Atkinson, C. Ampe, J. Vandekerckhove, and T. D. Pollard, *J. Cell Biol.* **127**(1), 107 (1994).
[2] R. D. Mullins, J. F. Kelleher, J. Xu, and T. D. Pollard, *Mol. Biol. Cell* **9**, 841 (1998).
[3] M. D. Welch, A. H. DePace, S. Verma, A. Iwamatsu, and T. J. Mitchison, *J. Cell Biol.* **138**(2), 375 (1997).
[4] R. D. Mullins, W. F. Stafford, and T. D. Pollard, *J. Cell Biol.* **136**(2), 331 (1997).
[5] J. F. Kelleher, S. J. Atkinson, and T. D. Pollard, *J. Cell Biol.* **131**(2), 385 (1995).

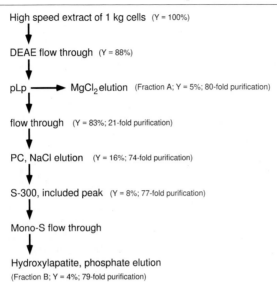

High speed extract of 1 kg cells (Y = 100%)

DEAE flow through (Y = 88%)

pLp ⟶ MgCl$_2$ elution (Fraction A; Y = 5%; 80-fold purification)

flow through (Y = 83%; 21-fold purification)

PC, NaCl elution (Y = 16%; 74-fold purification)

S-300, included peak (Y = 8%; 77-fold purification)

Mono-S flow through

Hydroxylapatite, phosphate elution
(Fraction B; Y = 4%; 79-fold purification)

FIG. 1. Schematic representation of Arp2/3 complex preparation from *Acanthamoeba castelanii*. Yield (Y) and fold purification at each step as determined by quantitative kinetic enzyme-linked immunosorbent assay (ELISA) is indicated in parentheses. During affinity chromatography on poly-L-proline 5% of the total Arp2 and Arp3 present in the DE-52 flowthrough binds to the column (fraction A) and 95% flows through (fraction B). Fraction A (5–10 mg) is purified in one step by selective elution from the polyproline column.

purification scheme outlined in Fig. 1 yields 10–20 mg of Arp2/3 complex from a kilogram of packed *Acanthamoeba*.

Description of Procedures

Assay of Arp2/3 Complex

Western Blot Quantification. Separate polypeptides by 12.5% sodium dodecyl sulfate–polyacrylamide gel electrophoresis (SDS–PAGE),[6] transfer to nitrocellulose and react[7] with rabbit polyclonal antibodies JH-46 for Arp2 or JH-47 for Arp3[5] using high stringency wash conditions (2 M urea, 100 mM glycine 1% Triton X-100). Detect bound antibodies with ^{125}I-labeled protein A or chemiluminescence (Amersham, Arlington Heights, IL) and autoradiography. Digitize autoradiograms using a flatbed scanner

[6] U. K. Laemmli, *Nature* **227,** 680 (1970).
[7] H. Towbin, T. Stahelin, and J. Gordon, *Proc. Natl. Acad. Sci. U.S.A.* **76,** 4350 (1979).

TABLE I
QUANTIFICATION OF ARP2/3 FRACTION B PURIFICATION BY
QUANTITATIVE KINETIC ELISA FOR ARP2[a]

Sample	Recovery (%)	Purification (-fold)
High-speed extract	100	1.0
Dialyzed extract	88.2	0.9
DE-52 flow-through	87.6	18
PLP flow-through	82.8	21
PC flow-through	0.6	0.4
PC pool	15.7	74
S-300 pool	7.9	77
Final pool	4.1	79

[a] Percent recovery is calculated as moles Arp2/3 complex in a sample divided by moles Arp2/3 complex in the high-speed extract. Fold purification is calculated as percent Arp2/3 complex in a sample divided by the fraction of Arp2/3 complex in the high-speed extract.

and either Collage (Fotodyne, New Berlin, WI) or NIH Image 1.60 software. Quantify Arp2 and Arp3 in experimental samples by comparison with pure Arp2/3 complex standards.

Quantitative ELISA. Quantitative determination of Arp2/3 complex concentration (Table I) is performed using a kinetic ELISA protocol.[8] Prepare serial dilutions of experimental samples and a standard curve of pure Arp2/3 complex and incubate in 96-well vinyl microtiter plates (Costar, Cambridge, MA) in 300 mM NaCl, 10 mM imidazole, pH 7.0, at 37° for 1 hr. Wash wells twice with DK buffer (150 mM NaCl, 0.1% Tween 20, and 10 mM Tris-HCl, pH 7.5) and block with 1% bovine serum albumin (BSA) in DK (DKB) at 37° for 30 min. Wash wells four times with DK, incubate with Arp2 antiserum diluted 1 : 500 in DKB at 37° for 1 hr, wash four more times with DK and incubate with horseradish peroxidase conjugated goat anti-rabbit secondary antiserum (Hyclone) diluted 1 : 2000 in DKB at 37° for 1 hr. Wash wells four times with DK and develop with 1 mg/ml *o*-phenylenediamine in phosphate buffer. Determine the absorbance at 450 nm for each well at 14 intervals in a SpectraMax 250 plate reader (Molecular Devices, Sunnyvale, CA) and calculate the rate of appearance of yellow product for each sample using SoftmaxPro software. This assay can be performed on one sample at a time in a standard spectrophotometer equipped for time-based acquisition either digitally or with a chart recorder. This procedure can be adapted for endpoint ELISA by stopping reac-

[8] V. C. Tsang, B. C. Wilson, and J. M. Peralta, *Methods Enzymol.* **92,** 391 (1983).

tions with 0.5 N H_2SO_4 and reading at 450 nm, but this is not linear or quantitative and is useful only for determining peak fractions from columns.

Buffers

For pyrophosphate lysis prep only.

    Ameba wash buffer: 100 mM NaCl, 10 mM imidazole HCl, pH 7.5. Prepare 8 liters.

    Extraction buffer: 75 mM KCl, 12 mM sodium pyrophosphate, 1 mM dithiothreitol (DTT), 1 mM phenylmethylsulfonyl fluoride (PMSF), 0.1 mM benzamidine, 1× inhibitor cocktail (see below), 30 mM imidazole HCl, pH 7.5. Prepare 2 liters (*Note*: For this and any subsequent buffers, add DTT and protease inhibitors immediately prior to use.)

    DE dialysis buffer: 7.5 mM sodium pyrophosphate, 25 mM Tris-HCl, pH 8.0. Prepare 4 liters of 7.5× stock; dilute prior to use.

    DE column buffer: 10 mM KCl, 0.5 mM DTT, 25 mM Tris-HCl, pH 8.0. Prepare 2 liters of 10× stock; dilute prior to use.

For sucrose lysis prep only.

    Ameba wash buffer: 100 mM NaCl, 10 mM Tris-HCl, pH 8.0. Prepare 8 liters.

    Extraction buffer: 11.6% sucrose, 1 mM ethylene-bis(oxyethylene-nitrilo)tetraacetic acid (EGTA), 1 mM adenosine triphosphate (ATP), 1 mM DTT, 1 mM PMSF, 0.1 mM benzamidine, 1× inhibitor cocktail (see below), 10 mM Tris-HCl, pH 8.0. Prepare 2 liters.

    DE column buffer: 0.2 mM CaCl2, 0.5 mM DTT, 0.5 mM ATP, 0.5 mM PMSF, 0.1 mM benzamidine, 10 mM Tris-HCl, pH 8.0. Prepare 3.5 liters of 4× stock; dilute prior to use.

Common to both preps:

    100× Inhibitor cocktail: 0.5 mg/ml leupeptin, 2 mg/ml soy trypsin inhibitor, 0.5 mg/ml pepstatin A, 0.5 mg/ml aprotinin. Fifty milliliters is sufficient for two preparations. Store unused portion at $-20°$.

    Bead buffer: 100 mM NaCl, 0.2% NaN_3, 0.5 mM DTT, 10 mM Tris-HCl, 100 mM glycine, pH 7.0. Prepare 200 ml 10× stock; dilute prior to use.

    PC buffer: 0.2 mM $CaCl_2$, 0.2 mM ATP, 0.5 mM DTT, 25 mM 2-[N-morpholino]ethanesulfonic acid (MES) HCl, pH 6.0. Prepare 200 ml 10× stock. Dilute prior to use, adding solid NaCl to necessary final concentration.

    S-300 buffer: 300 mM NaCl, 0.2 mM $CaCl_2$, 0.2 mM ATP, 0.5 mM DTT, 1 mM NaN_3, 25 mM imidazole HCl, pH 7.0. Prepare 1 liter.

    Mono-S buffer A: 0.2 mM $CaCl_2$, 0.2 mM ATP, 0.5 mM DTT, 25 mM imidazole HCl, pH 7.0. Prepare 1 liter.

Mono-S buffer B: 1 M NaCl, 0.2 mM CaCl$_2$, 0.2 mM ATP, 0.5 mM DTT, 25 mM imidazole HCl, pH 7.0. Prepare 500 ml.

HA elution buffer: 100 mM NaCl, 100 mM NaP$_2$HPO$_4$/NaH$_2$PO$_4$, 0.2 mM CaCl$_2$, 2 mM EGTA, 2 mM MgCl$_2$, 0.2 mM ATP, 0.5 mM DTT, 25 mM imidazole HCl, pH 7.0. Prepare 50 ml from Mono-S buffers.

Complex storage buffer: 150 mM NaCl, 0.2 mM MgCl$_2$, 0.2 mM ATP, 1.0 mM DTT, 10 mM imidazole HCl, pH 7.5. Prepare 300 ml of 10× stock; dilute prior to use.

Preparation of DEAE-Cellulose

Suspend approximately 1 kg of dry new DE-52 or 2.5 liters settled volume of swollen, previously used DE-52 (Whatman, Maidstone, UK) in water to a final volume of 3 liters. Add either 84 g solid KOH or 60 g solid NaOH (0.5 M final concentration). Stir this slurry 5 min then rinse in a large Büchner funnel over 15-cm-diameter Whatman 541 filter paper with 4 liters of water. Resuspend the resin in water to 3 liters, add solid KOH or NaOH to 0.5 M, and stir 30 min. Rinse the resin as just described with 8 liters of water, resuspend in water to 3 liters, add 130 ml concentrated HCl (0.5 M final concentration), and stir 30 min. Wash the resin as described with 8 liters of water and resuspend in water to approximately 3.5 liters. For pyrophosphate extraction, adjust the pH of the resin to 8.0 by addition of 0.2 M unbuffered sodium pyrophosphate. Determine the pH of the slurry either by using pH indicator paper or by filtering 1 ml of the slurry with Whatman paper and reading the filtrate with a standard pH electrode. For sucrose extraction, adjust the pH of the resin to 8.0 with 1.0 M unbuffered Tris base, add 60 ml 0.1 M ATP (pH 7) per liter of resin, and stir for 30 min. For either prep, wash the resin in a funnel with 3 liters of ice-cold DE buffer immediately prior to use, resuspend in DE buffer to ~3 liters and chill to 4°. Pour 1 liter of DE resin into an 8- × 60-cm column and allow to settle. Run the buffer head to the packed bed prior to loading. The DEAE cellulose can be reused many times.

Preparation of Poly-L-Proline-Agarose

For poly-L-proline chromatography,[9,10] dissolve 1 g of 40,000 molecular weight poly-L-proline (Sigma, St. Louis, MO) in conjugation buffer (100 mM sodium carbonate, pH 9–10). Activate Sepharose CL-4B (Pharmacia,

---

[9] D. A. Kaiser, P. J. Goldschmidt-Clermont, B. A. Levine, and T. D. Pollard, *Cell Motil.* **14,** 251 (1989).

[10] L. Tuderman, E.-R. Kuutti, and K. Kivirkko, *Eur. J. Biochem.* **52,** 9 (1975).

Piscataway, NJ) with cyanogen bromide using any standard protocol.[11] Wash the activated beads with 2 M sodium carbonate (pH 9–10) and couple to poly-L-proline by gently shaking for 20 hr at 4°. Rinse beads extensively with bead buffer. Pour a 2.5- × 10-cm column, wash with 8 M urea, and equilibrate with bead buffer prior to use. Poly-L-proline-agarose can be reused many times.

Preparation of Phosphocellulose

Suspend 8–10 g of P-11 phosphocellulose (Whatman) in 1 liter of water, allow to swell, and cycle as done by Lynch.[12] After ~15 min, decant fines and resuspend the resin to 1-liter final volume. Decant fines in this manner three to four more times. Wash the resin on Whatman 541 filter paper in a Büchner funnel sequentially with 400-ml volumes each of 0.5 M NaOH, 0.5 M HCl, 0.5 M NaOH, and 0.5 M acetic acid, washing the resin with 2 liters of water after each solution. Resuspend the resin in 200 ml 75 mM MES HCl, pH 6.0, and adjust the pH to 6.0 by dropwise addition of concentrated HCl. Determine the pH of the slurry with pH indicator paper. Immediately before use, decant the buffer to a clean beaker and dry the resin briefly on a Büchner funnel. Set aside two-thirds of the resin, resuspend the remaining one-third in buffer, and pour into a 5- × 15-cm column. Allow the buffer head to run to the top of the packed resin and clamp the column until loading.

Preparation of Mono-S Column

Cycle a 1-ml column of Mono-S HR 5/5 (Pharmacia) according to manufacturer's cleaning instructions with 500 μl 2M NaCl, 500 μl 2 M NaOH, 500 μl 1 M HCl, 500 μl 75% acetic acid, and 500 μl 2 M NaCl, washing with 5 ml of Mono-S buffer A between solutions. Prepare the column for loading by washing with 10 ml Mono-S buffer B followed by 10 ml Mono-S buffer A containing 10% buffer B to give 100 mM NaCl final.

Preparation of Hydroxylapatite Column

Resuspend hydrated, settled hydroxyapatite Bio-Gel HT (Bio-Rad, Hercules, CA) gently by slowly rolling the bottle. Pour approximately 3 ml of the resulting slurry into a 0.75- × 10-cm column and allow to settle without flow. After settling, allow the storage buffer to drain to the top of the resin bed, wash the resin in place with 20 ml of 100 mM NaCl in Mono-

[11] E. Harlow and D. Lane, "Antibodies: A Laboratory Manual." Cold Spring Harbor Laboratory Press, New York, 1988.
[12] T. J. Lynch, H. Brzeska, I. C. Baines, and E. D. Korn, *Methods Enzymol.* **196,** 12 (1991).

S buffer A (90% buffer A + 10% buffer B), and clamp the column prior to loading.

Purification

Grow two 15-liter carboys of *Acanthamoeba* to late-log phase[13,14] and harvest by centrifuging in a 6 × 1-liter rotor at 3000g for 7 min. These cultures routinely yield 700–1200 g packed *Acanthamoeba*. Weigh cells wet, suspend in three volumes ameba wash buffer and centrifuge at 3000g for 7 min at 4°. Wash cells once more and resuspend in 2 ml extraction buffer per gram of cells.

Lyse cells in either pyrophosphate extraction buffer[12] or sucrose extraction buffer (modified from Ref. 15) at 4° in a Parr bomb at 375 psi for 5 min. Centrifuge extracts at 20,000g for 30 min to remove cell debris and unlysed cells. Centrifuge the supernatant from the first spin at 100,000g for 2 hr. Carefully remove the lipid layer by aspiration and decant the cleared cytosolic extract, leaving behind the loose pellet. Pool the high-speed extract.

For preparations using pyrophosphate extraction, dialyze the high-speed extract for 10–12 hr against 30 liters DE dialysis buffer. Centrifuge the dialyzed extract at 17,700 rpm for 30 min at 4° to remove precipitated protein. Mix the extract in batch for 30 min at 4° with 2 liters of the resuspended DE-52 resin. Carefully layer the extract plus resin onto the column, being careful not to disturb the clean resin already in the column (see chromatography resin instructions provided earlier), and allow the extract to flow until the meniscus almost touches the top of the packed bed. (*Note*: Attempting to adsorb this amount of protein by flowing it through a prepoured column generally results in mechanical failures of the column bed and is not recommended.) Top the bed off with 1 in. of DE buffer, fit the column with a flow adaptor, and wash with 3 liters of DE buffer. A variety of cytoskeletal proteins including actin,[16] myosin I,[12,14] myosin II,[17] α-actinin,[18] and capping protein[19] bind the DE-52 column and can be eluted in a 4-liter gradient from 50 to 500 mM KCl in DE buffer.

[13] E. D. Korn, *in* "Methods in Cell Biology," Vol. 25, pp. 313–331. Academic Press, San Diego, 1982.

[14] T. D. Pollard and E. D. Korn, *J. Biol. Chem.* **248,** 4682 (1973).

[15] P. C. H. Tseng, M. S. Runge, J. A. Cooper, R. C. Williams, Jr., and T. D. Pollard, *J. Cell Biol.* **98,** 214 (1984).

[16] T. D. Pollard, *J. Cell Biol.* **99,** 769 (1984).

[17] T. D. Pollard, W. F. Stafford, and M. E. Porter, *J. Biol. Chem.* **253,** 4798 (1978).

[18] T. D. Pollard, P. C.-H. Tseng, D. L. Rimm, D. P. Bichell, R. C. Williams, and J. Sinard, *Cell Motil.* **6,** 649 (1986).

[19] J. A. Cooper, J. D. Blum, and T. D. Pollard, *J. Cell Biol.* **99,** 217 (1984).

For preparations using sucrose extraction, mix the high-speed extract with 2 liters of DE resin for 30 min at 4°. Layer the extract plus resin carefully onto the column, as earlier, and allow to flow. Again, batch adsorption is the only reliable method to load the column. Subsequent DE-52 column steps are performed as described earlier.

Collect the flow-through from the DEAE-cellulose column load and wash, approximately 4–5 liters total volume, and fractionate by poly-L-proline Sepharose affinity chromatography.[9] Load the column at 2–3 ml/min and wash with 300 ml bead buffer. The first fraction (A) of Arp2/3 complex binds to the poly-L-proline column and is eluted with 0.4 M MgCl$_2$ in bead buffer (Fig. 2). Profilin is subsequently eluted from the poly-L-proline with 8 M urea. Dialyze the purified Arp2/3 complex into complex storage buffer and concentrate to a final volume of 2 ml by dialysis against solid sucrose. Arp2/3 complex fraction A is stored in 50% sucrose on ice with little loss of activity for 30 days. Sucrose concentration is determined using an optical refractometer (American Optical, Keene NH).

Purify the second fraction (B) of Arp2/3 complex from the poly-L-proline column flow-through as follows. Adjust the pH of the flow-through to 6.0 by the addition of solid MES at 4° to a final concentration of 25 mM. Determine the pH using pH indicator paper. Mix two-thirds of the PC resin with this pool for 30 min at 4° and then allow the resin to settle for 15 min. Decant the gravity-cleared protein solution to a clean beaker, leaving a small volume of liquid behind with the settled resin. Layer the

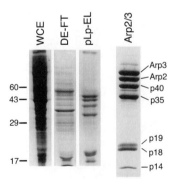

FIG. 2. Purification of Arp2/3 complex by poly-L-proline affinity chromatography. SDS–PAGE of samples from steps of Arp2/3 complex purification from *Acanthamoeba*. WCE, whole-cell extract from *Acanthamoeba*; DE-FT, flow-through fraction from DEAE column; pLp-EL, purified Arp2/3 complex eluted from poly-L-proline column; Arp2/3, purified Arp2/3 complex overloaded on gel to demonstrate purity. All seven subunits are labeled. Conditions: 14.5% SDS–PAGE of samples stained with Coomassie Brilliant Blue. Numbers on the left are positions of molecular weight markers for the first three lanes.

remaining slurry onto the clean resin in the column. Allow the buffer head to run to the top of the resin and fit the column with a flow adapter. Load the decanted protein solution at 6–7 ml/min. Once loaded, wash the PC column with 100 mM NaCl in PC buffer, then with 300 mM NaCl in PC buffer, and finally elute with 500 mM NaCl in PC buffer.

Pool the peak fractions from the 500 mM NaCl step elution, concentrate to a volume of 17 ml by dialysis against solid sucrose, and chromatograph on a 2.5- × 88-cm column of Sephacryl S-300 equilibrated in S-300 buffer. Dilute the Arp-containing pool from this column threefold with Mono-S buffer A to a final NaCl concentration of 100 mM and load onto a 1 ml Mono-S fast protein liquid chromatography (FPLC) column (Pharmacia Biotech, Piscataway, NJ). Most of the Arp2/3 complex does not bind to Mono-S under these conditions but the major remaining contaminants do. Apply the flow-through pool from Mono-S to a 2-ml column of hydroxyapatite Bio-Gel HT (BioRad) and elute the Arp2/3 complex in a concentrated peak with 100 mM sodium phosphate and 100 mM NaCl in Mono-S buffer A (Fig. 3). Dialyze Arp2/3 complex fraction B into complex storage buffer,

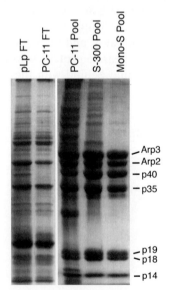

FIG. 3. Purification of Arp2/3 complex from the fraction that does not bind to the poly-L-proline column. SDS–PAGE of fractions from Arp2/3 complex purification from *Acanthamoeba*. pLp FT, material that flows through poly-L-proline Sepharose; PC-11 FT, material that flows through phosphocellulose column; PC-11 pool, fraction of protein that binds to PC-11 and elutes with 500 mM NaCl; S-300 Pool, pooled peak fractions from gel filtration; Mono-S Pool, Arp2/3 pool from Mono-S FLPC. Conditions: 14.5% SDS–PAGE of samples stained with Coomassie Brilliant Blue.

concentrate to 10–20 μM by dialysis against solid sucrose, and store on ice in at least 70% sucrose. Arp2/3 complex fraction B stored in this manner is active for a week or more.

Arp2/3 complex B is present at 200 nM in the poly-L-proline flow-through is purified by the further chromatographic steps. At pH 6.0, 99% of the Arp2/3 complex binds to phosphocellulose, whereas more than 50% of the contaminating polypeptides do not (Table I). Step elution with 500 mM NaCl provides a 3.6-fold purification and an 18-fold concentration, despite a relatively large loss of total Arp2/3 complex due to low recovery from phosphocellulose. Gel filtration on Sephacryl-300 separates Arp2/3 complex from most contaminants: the main peak is largely Arp2/3 complex, and contaminating polypeptides run as shoulders on this main peak. FPLC on a Mono-S column achieves 97% purity of complex B (Table I). The Mono-S flow through is concentrated by binding to hydroxyapatite and step elution with 100 mM NaPO$_4$.

Recovery of Arp2/3 complex fraction A and fraction B depends on the method of preparation. Pyrophosphate extraction and overnight dialysis into low salt buffer without ATP yields 1–5 mg fraction A and 6–12 mg fraction B. Extraction with sucrose and ATP followed immediately by chromatography on DEAE-cellulose in a buffer with ATP yields 20 mg fraction A and 0–6 mg fraction B. Both the ATP and the sucrose in the sucrose extraction preparation seem to stabilize Arp2/3 complex, with 30–45% more Arp2/3 complex present in the DEAE flow-through in a sucrose preparation than in a pyrophosphate preparation. This leads to a much higher recovery of fraction A. Arp2/3 complex fraction B is largely lost at the phosphocellulose chromatography step in sucrose extraction preparations, binding irreversibly, possibly due to aggregation at high concentration on the column.

[6] Purification and Assay of the Platelet Arp2/3 Complex

By Matthew D. Welch and Timothy J. Mitchison

Introduction

The Arp2/3 complex was first purified by profilin affinity chromatography of cell extracts from *Acanthamoeba castellanii*[1] and has subsequently been purified from human platelets[2] and from the budding yeast *Saccharomyces cerevisiae*.[3] It consists of six (*S. cerevisiae*) or seven (*A. castellanii* and human) polypeptide subunits, all present in approximately equal stoichiometry (see Fig. 2 in a later section). The two largest subunits (50 and 43 kDa) are actin-related proteins in the Arp3 and Arp2 families.[4,5] These define the name Arp2/3. The five remaining subunits of the human complex have molecular masses of 40, 34, 23, 22, and 19 kDa by sodium dodecyl sulfate–polyacrylamide gel electrophoresis (SDS–PAGE) and are similar in size and amino acid sequence to the corresponding subunits in the *A. castellanii* and *S. cerevisiae* complexes.[1,3,6] Other than Arp2 and Arp3, only the 40 kDa polypeptide harbors a recognizable sequence motif. It contains WD (β-transducin) repeats[7] and is a member of the Sop2 (suppressor of profilin) family of proteins.[6,8,9]

The Arp2/3 complex appears to be an essential component of the actin cytoskeleton. Subunits of the complex are necessary for cell viability and actin cytoskeletal function in *S. cerevisiae* and *Schizosaccharomyces pombe*.[3,8–12] Furthermore, subunits of the yeast,[3,8,9,12] *A. castel-*

[1] L. M. Machesky, S. J. Atkinson, C. Ampe, J. Vandekerckhove, and T. D. Pollard, *J. Cell Biol.* **127,** 107 (1994).

[2] M. D. Welch, A. Iwamatsu, and T. J. Mitchison, *Nature* **385,** 265 (1997).

[3] D. Winter, A. V. Podtelejnikov, M. Mann, and R. Li, *Curr. Biol.* **7,** 519 (1997).

[4] S. Frankel and M. Mooseker, *Curr. Opin. Cell Biol.* **8,** 30 (1996).

[5] R. D. Mullins, J. F. Kelleher, and T. D. Pollard, *Trends Cell Biol.* **6,** 208 (1996).

[6] M. D. Welch, A. H. DePace, S. Verma, A. Iwamatsu, and T. J. Mitchison, *J. Cell Biol.* **138,** 375 (1997).

[7] E. J. Neer, C. J. Schmidt, R. Nambudripad, and T. F. Smith, *Nature* **371,** 297 (1994).

[8] D. McCollum, A. Feoktistova, M. Morphew, M. Balasubramanian, and K. L. Gould, *EMBO J.* **15,** 6438 (1996).

[9] M. K. Balasubramanian, A. Feoktistova, D. McCollum, and K. L. Gould, *EMBO J.* **15,** 6426 (1996).

[10] J. P. Lees-Miller, G. Henry, and D. M. Helfman, *Proc. Natl. Acad. Sci. U.S.A.* **89,** 80 (1992).

[11] E. Schwob and R. P. Martin, *Nature* **355,** 179 (1992).

[12] V. Moreau, A. Madania, R. P. Martin, and B. Winsor, *J. Cell Biol.* **134,** 117 (1996).

lanii,[1,13,14] and human[6] complexes are localized to dynamic populations of actin filaments in cells.

The Arp2/3 complex has been purified using profilin affinity chromatography,[1,14] antibody affinity chromatography,[3] and conventional chromatography.[1,2] This article describes the purification of the Arp2/3 complex from human platelets using conventional methods and the assay for Arp2/3 complex activity. Both the purification procedure and the activity assay were devised to identify factors that function in the actin polymerization-based motility of the pathogenic bacterium *Listeria monocytogenes*.[2] The purified Arp2/3 complex induces actin polymerization at the *L. monocytogenes* cell surface, where it is thought to play a role in bacterial motility.[2] In addition, the pure complex from *A. castellanii* binds filaments and bundles actin filaments *in vitro*,[14] although these activities have not yet been tested for the human and *S. cerevisiae* complexes.

Purification of the Arp2/3 Complex from Platelets

The Arp2/3 complex is purified from a soluble extract derived from the Triton insoluble cytoskeleton of human platelets, a starting material that is highly enriched in actin cytoskeletal proteins. The entire procedure takes 2–3 days and involves three principal steps. The first two steps, preparation of the Triton-insoluble cytoskeleton and preparation of the cytoskeleton extract, generally take 2 days, although they can be accomplished in 1 day. The third step, purification of the Arp2/3 complex from the cytoskeleton extract by column chromatography, takes 1 day. A flowchart of the entire purification procedure is shown in Fig. 1.

Preparation of the Platelet Triton Insoluble Cytoskeleton

Solutions

Wash buffer: 20 mM PIPES, pH 6.8, 40 mM KCl, 5 mM ethylene-bis(oxyethylenenitrilo)tetraacetic acid (EGTA), 1 mM ethylenediaminetetraacetic acid (EDTA).

Lysis buffer: Wash buffer plus 10 μg/ml leupeptin, pepstatin, and chymostatin (together called LPC protease inhibitors), 1 mM benzamidine, 1 mM phenylmethylsulfonyl fluoride (PMSF), 1% Triton X-100, and 0.05 mM adenosine triphosphate (ATP).

Resuspension buffer: Wash buffer plus LPC protease inhibitors, 100 mM sucrose, 0.05 mM ATP, and 1 mM dithiothreitol (DTT).

[13] J. F. Kelleher, S. J. Atkinson, and T. D. Pollard, *J. Cell Biol.* **131**, 385 (1995).
[14] R. D. Mullins, W. F. Stafford, and T. P. Pollard, *J. Cell Biol.* **136**, 331 (1997).

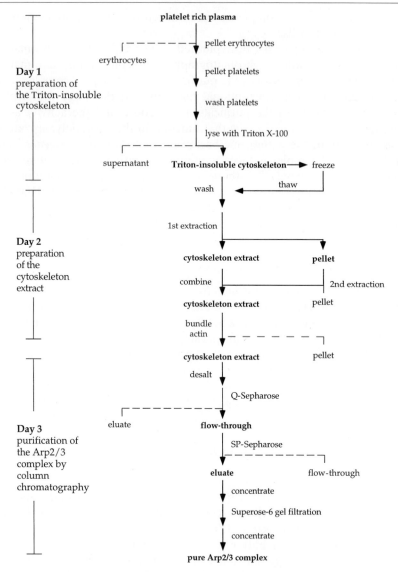

FIG. 1. Steps in the purification of the Arp2/3 complex from platelets.

Procedure. Obtain from a blood bank freshly outdated human platelet enriched plasma that has expired for transfusion purposes within the previous 24 hr. About 500 ml of platelet-rich plasma will yield enough Triton-insoluble cytoskeleton for two Arp2/3 complex purification runs. Store the platelets at room temperature prior to preparing the Triton-insoluble cytoskeleton to prevent them from activating. Furthermore, use only plastic vessels for holding platelets because contact with glass can cause them to activate prematurely. Tips on working with platelets can be found in a previous volume of this series.[15] Rid the platelet-rich plasma of residual erythrocytes and other larger cells by centrifuging at 160g for 15 min at room temperature. Isolate the platelets by centrifuging the platelet-rich supernatant at 2000g for 15 min at room temperature. Resuspend the platelet pellet in 20 volumes of wash buffer per volume of packed platelets and then pellet again at 2000g for 15 min. Repeat this wash step one or two more times. After the final spin, resuspend the platelets in five volumes of wash buffer and chill them on ice for 10 min. To lyse the cells, add an equal volume of ice-cold lysis buffer per volume of resuspended platelets and incubate on ice for 5 min. Centrifuge the lysate at 2000g for 2 min at 4° to pellet the Triton-insoluble cytoskeleton and resuspend the pellet in 5 volumes of resuspension buffer. The resuspended cytoskeleton can then be pelleted and salt extracted as described later or flash frozen in liquid nitrogen and stored at −80° for several months.

Preparation of the Cytoskeleton Extract

Solutions

Low salt wash buffer: 20 mM PIPES, pH 6.8, 10 mM KCl, 5 mM EGTA, 1 mM EDTA, 1 mM DTT, LPC protease inhibitors.

Extraction buffer: 20 mM PIPES, pH 6.8, 0.6 M KCl, 5 mM EGTA, 1 mM EDTA, 1 mM DTT, 0.2 mM ATP, LPC protease inhibitors.

Special Equipment. Teflon tissue homogenizer.

Procedure. The cytoskeleton extract should be prepared 1 day prior to initiating the purification of the Arp2/3 complex. All steps should be carried out at 4°. The procedure described here is based on a previously described protocol for the purification of actin and actin-binding proteins from platelets.[16] Pellet approximately 10 ml of the resuspended fresh Triton-insoluble cytoskeleton or thawed frozen cytoskeleton by centrifuging at 2000g for 2 min. Gently resuspend the pellet in 10 volumes of low salt buffer and then repellet the cytoskeleton by centrifuging at 2000g for 2 min. Resuspend the pellet in 5 volumes of extraction buffer and homogenize for 2 min using a Teflon tissue homogenizer. Incubate the homogenate on ice for 30 min, then spin at 25,000g for 15 min. Remove the supernatant from the first centrifugation and save it—this is the first fraction of the cytoskeleton

[15] J. E. B. Fox, C. C. Reynolds, and J. K. Boyles, *Methods Enzymol.* **215,** 42 (1992).
[16] S. R. Schaier, *Methods Enzymol.* **215,** 58 (1992).

extract. Resuspend the pellet in 5 volumes of extraction buffer. Homogenize for 1 min, then incubate the homogenate on ice for 2 hr. Repeat this step two to three times. After a final 1-min homogenization, centrifuge the homogenate at 25,000g for 15 min. Remove the supernatant from this second spin and combine it with the supernatant from the first spin. This is the cytoskeleton extract. At this point, much of the actin can be removed by adding ATP to 5 mM final and EGTA to 10 mM, and incubating the solution at 4° for 16 hr. Under these conditions, platelet actin and other nonmuscle actins will self-associate into bundles.[17–19] These are removed from the solution by centrifugation at 25,000g for 15 min. The purpose of removing much of the actin by the bundling and centrifugation is to reduce the risk of clogging chromatography columns with actin filaments. In addition, the actin pellet isolated in this step can be used to purify platelet actin.[16]

Purification of the Arp2/3 Complex by Column Chromatography

Solutions

Q-buffer A: 20 mM Tris, pH 8.0, 2 mM $MgCl_2$, 5 mM EGTA, 1 mM EDTA, 0.5 mM DTT, 0.2 mM ATP, 2.5% v/v glycerol.

Q-buffer B: Q-buffer A with 1 M KCl.

S-buffer A: 20 mM 2-[N-Morpholino]ethanesulfonic acid (MES), pH 6.1, 2 mM $MgCl_2$, 5 mM EGTA, 1 mM EDTA, 0.5 mM DTT, 0.2 mM ATP, 5–10% v/v glycerol.

S-buffer B: S-buffer A with 1 M KCl.

Gel filtration buffer: 20 mM MOPS, pH 7.0, 100 mM KCl, 2 mM $MgCl_2$, 5 mM EGTA, 1 mM EDTA, 0.5 mM DTT, 0.2 mM ATP, 5–10% v/v glycerol.

Special Equipment. FPLC (Amersham Pharmacia Biotech, Piscataway, NJ) or equivalent chromatography apparatus.

Procedure. It is important to note at the beginning of this section that the activity of the Arp2/3 complex is quite labile. It is therefore essential that the purification procedure be completed in the shortest possible time. The procedure, starting with the actin-depleted cytoskeleton extract, can be carried out in 1 long day. To achieve this, prepare all buffers and equilibrate the gel filtration column into gel filtration buffer on the day before beginning the purification. It is also important to note that this procedure has been optimized for starting with approximately 10 ml of high salt extract prepared as described earlier. However, it can be scaled up or down appropriately.

[17] J. Bryan and R. E. Kane, *J. Mol. Biol.* **125,** 207 (1978).
[18] R. E. Kane, *J. Cell Biol.* **66,** 305 (1975).
[19] S. Rosenberg, A. Stracher, and R. C. Lucas, *J. Cell Biol.* **91,** 201 (1981).

Desalt the high salt extract by passing it over a gel filtration column preequilibrated with Q-buffer A supplemented with 100 mM KCl. Prepacked 10-ml columns specifically designed for desalting applications are available from Bio-Rad Laboratories (10DG columns, Richmond, CA) or Amersham Pharmacia Biotech (PD10 columns). Add LPC protease inhibitors to the desalted extract. Figure 2, lane 1, shows the protein composition of the starting material.

All subsequent steps can be carried out using an FPLC (Pharmacia) or equivalent chromatography system. Equilibrate a 5-ml Hi-trap Q-Sepharose HP column (Pharmacia) or equivalent anion-exchange column with Q-buffer plus 100 mM KCl. Equilibration should be carried out until the column is saturated with ATP, which will bind to the Q-Sepharose resin. To determine when the column has been equilibrated, monitor the elution of ATP from the column by measuring the A_{254} (or A_{280}) of the eluate. The A_{280} will rise slowly and then plateau after 5–10 volumes of buffer A have passed over the column, indicating that it has become equilibrated. Pass the desalted extract over the Q-column at a flow rate of 0.5 ml/min in Q-buffer A with 100 mM KCl, collecting 4–5 ml fractions. The Arp2/3 complex will pass through the column and is isolated in the flow-through fractions. Figure 2, lane 2, shows the protein composition of the Q-flow-through fraction. Pool the flow-through fractions and adjust their pH to 6.1 by adding MES, pH 6.1, to a final concentration of 40 mM. Add glycerol to 10% v/v and add LPC protease inhibitors. Adjust the KCl concentration to 50 mM by diluting the sample 1:1 into S-buffer A.

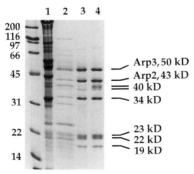

FIG. 2. The protein composition of fractions from the purification of the Arp2/3 complex visualized by 13% SDS–PAGE and Coomassie Blue staining. Lane 1, cytoskeleton extract (30 μg); lane 2, Q-Sepharose flow-through (6 μg); lane 3, SP-Sepharose peak (6 μg); and lane 4, Superose-6 peak (6 μg). [Reprinted with permission from M. D. Welch et al., Nature 385, 265 (1997). Copyright 1997 Macmillan Magazines Limited.]

Equilibrate a 1-ml Hi-trap SP-Sepharose HP column or equivalent cation-exchange column with S-buffer plus 50 mM KCl. Apply the diluted Q-flow-through to the column at a flow rate of 0.5 ml/min. Wash the column with 10 volumes of S-buffer with 50 mM KCl. Elute the Arp2/3 complex with a linear increasing gradient of KCl from 50 to 500 mM over 20 column volumes. The complex elutes at 175–200 mM KCl and represents the major peak of protein eluting from the SP-Sepharose column by A_{280}, as shown in Fig. 3. Pool the peak fractions, usually 2–3 ml total volume, and concentrate to 0.5 ml. Concentration can be achieved quickly by centrifugation through an ultrafiltration membrane device such as Microcon-30 (Amicon, Inc., Beverly, MA). The protein composition of the concentrated peak fractions from the SP-Sepharose column is shown in Fig. 2, lane 3.

Load the concentrated fractions onto a Superose-6 HR 10/30 gel filtration column (Pharmacia) or equivalent column equilibrated with gel filtration buffer. To save time, the column should be equilibrated 1 day prior to its use. Run the column at 0.3–0.5 ml/min, collecting 0.5-ml fractions. The molecular mass of the Arp2/3 complex is approximately 220 kDa and it elutes with a Stokes radius of approximately 5.5 nm. It usually represents the only detectable peak eluting from the gel filtration column by A_{280}.

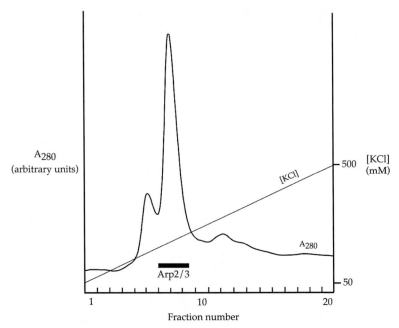

FIG. 3. The elution profile of proteins from the SP-Sepharose column (A_{280}) and the increasing linear gradient of KCl concentration are illustrated. The Arp2/3 complex elutes at 175–200 mM KCl and represents the major protein peak.

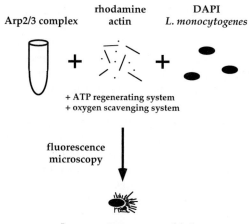

Arp2/3 complex   rhodamine   DAPI
                   actin   L. monocytogenes

+ ATP regenerating system
+ oxygen scavenging system

fluorescence
microscopy

L. monocytogenes assembled
actin structures

FIG. 4. A diagram of the assay used to test the capacity of the Arp2/3 complex to induce actin polymerization at the cell surface of the pathogenic bacterium *L. monocytogenes*.

Pool the peak fractions (usually 2–3 ml) and concentrate using a Microcon-30 concentrator. Figure 2, lane 4, shows the subunit composition of the purified Arp2/3 complex. Aliquot and freeze the pure complex in liquid N_2, and store in liquid N_2. Total yield from 10 ml of cytoskeleton extract is approximately 500 μg of protein.

Assaying Arp2/3 Complex Activity

The Arp2/3 complex promotes actin polymerization at the cell surface of the pathogenic bacterium *Listeria monocytogenes*. This activity provides the basis for the visual assay diagrammed in Fig. 4. Fluorescently labeled *L. monocytogenes* and fluorescently labeled actin are mixed together in the presence of the pure Arp2/3 complex. To regenerate ATP and to prevent photodamage to the fluorescently labeled actin filaments, special enzyme–substrate mixtures are also added. Fluorescence microscopy is used to visualize actin assembly at the surface of the bacteria.

Preparation of Killed DAPI-Labeled Listeria monocytogenes

Bacterial Strains. Listeria monocytogenes strain SLCC-5764 has been used extensively to study bacterial motility cell extracts.[2,20,21] This strain is

[20] J. A. Theriot, J. Rosenblatt, D. A. Portnoy, P. J. Goldschmidt-Clermont, and T. J. Mitchison, *Cell* **76,** 505 (1994).
[21] G. A. Smith, J. A. Theriot, and D. A. Portnoy, *J. Cell Biol.* **135,** 647 (1996).

also used for assays with pure Arp2/3 complex.[2] It expresses high levels of virulence gene products including the ActA protein,[22] which is required for actin assembly by *L. monocytogenes*[23,24] and is sufficient to direct actin assembly in the absence of other *L. monocytogenes* gene products.[25] A control strain DPL-1955, a derivative of SLCC-5764 that carries an in-frame deletion of the *actA* gene,[26,27] is also useful in the Arp2/3 activity assay. A more detailed treatment of *L. monocytogenes* strains and their use in motility assays can be found elsewhere in this volume.[28]

It is important to note that *L. monocytogenes* is a pathogenic organism that can cause serious illness mainly in pregnant women, newborns, and immunocompromised individuals. People handling *L. monocytogenes* should strictly adhere to Biosafety Level 2 (BSL2) guidelines.

Solutions

Brain–heart infusion (BHI) media (DIFCO Laboratories, Detroit, MI).

4′,6-diamidino-2-phenylindole (DAPI): 10 mg/ml stock in water.

Iodoacetic acid: 1 *M*, made fresh.

Wash buffer: 10 m*M* HEPES, pH 7.7, 50 m*M* KCl, 5 m*M* EGTA, 1 m*M* EDTA, 100 m*M* sucrose.

Procedure. Inoculate a 2-ml culture of *L. monocytogenes* in BHI and grow overnight at 37°. In the morning, dilute the culture 1:10 in BHI and grow for 4 hr at 37°. To fluorescently label the bacteria, add DAPI to 10 μg/ml and grow for an additional 1.5 hr. Kill the cells in culture by adding iodoacetic acid to 10 m*M* and incubating for 20 min. After this step, the bacteria no longer pose a health hazard. Pellet 0.75-ml cells in a microfuge for 5 sec at 16,000*g*, remove the supernatant, and then wash the pellet gently with wash buffer without resuspending. Resuspend the pellet in 200 μl wash buffer and add glycerol to 20% v/v. Aliquot into small portions and freeze cells by placing them at −80° (do not quick freeze). Store cells at −80°.

[22] M. Leimeister-Wachter and T. Chakraborty, *Infect. Immun.* **57,** 2350 (1989).

[23] E. Domann, J. Wehland, M. Rhode, S. Pistor, M. Hartl, W. Goebel, M. Leimester-Wachter, M. Wuenscher, and T. Chakraborty, *EMBO J.* **11,** 1981 (1992).

[24] C. Kocks, E. Gouin, M. Tabouret, P. Berche, H. Ohayon, and P. Cossart, *Cell* **68,** 521 (1992).

[25] G. A. Smith, D. A. Portnoy, and J. A. Theriot, *Mol. Microbiol.* **17,** 945 (1995).

[26] R. A. Brundage, G. A. Smith, A. Camilli, J. A. Theriot, and D. A. Portnoy, *Proc. Natl. Acad. Sci. U.S.A.* **90,** 11890 (1993).

[27] G. A. Smith and D. A. Portnoy, (1992) unpublished data.

[28] J. A. Theriot and D. C. Fung, *Methods Enzymol.* **298,** [11], 1998 (this volume).

Assaying for Actin Assembly at the Listeria Surface

Solutions and Reagents

DAPI-labeled *L. monocytogenes* prepared as described earlier.

N-Hydroxysuccinimidyl 5-carboxytetramethyl rhodamine-labeled rabbit muscle actin (TMR-actin).[29]

20× ATP regenerating mix: 150 mM creatine phosphate, 20 mM ATP, 2 mM EGTA, pH 7.7, 20 mM MgCl$_2$ (store at $-20°$).[30]

Creatine kinase, 5 mg/ml in water (store at $-80°$).

Triton X-100, 1%.

Glucose oxidase, 10 mg/ml in 30% glycerol (store at $-20°$).

Catalase, 10 mg/ml in 30% glycerol (store at $-20°$).

Glucose, 1 M.

Special Equipment. Microscope equipped for epi-fluorescence.

Procedure. Begin with 10 μl of pure Arp2/3 complex or fractions from earlier in the purification scheme. Add 0.5 μl 20× ATP regenerating mix, 0.1 μl creatine kinase, and TMR-actin to 1 μM final, Triton X-100 to 0.1% final, and 0.5 μl DAPI-labeled *L. monocytogenes*. To prevent photodamage to the TMR-actin filaments, add catalase to 0.1 mg/ml final, glucose oxidase to 0.1 mg/ml final, and glucose to 10 mM final. These can be combined beforehand into a concentrated mix and then added together. Squash the assay mixture between a glass microscope slide and a glass coverslip, seal with nail polish, and incubate at room temperature for 20 min. View by fluorescence microscopy.

Acknowledgment

We thank Jody Rosenblatt for comments on the manuscript.

[29] D. R. Kellogg, T. J. Mitchison, and B. M. Alberts, *Development* **103,** 675 (1988).
[30] A. W. Murray, *Methods Cell Biol.* **36,** 581 (1991).

[7] Purification and Assay of Zyxin

By Karen L. Schmeichel, Beth E. Stronach, and Mary C. Beckerle

Introduction

Zyxin was first characterized as an 82-kDa protein that localizes to sites of cell adhesion in a variety of cell types.[1] Zyxin is a low abundance phosphoprotein whose sequence features include an extensive N-terminal proline-rich region, a leucine-rich nuclear export signal, and a C-terminal cluster of three zinc-binding [LIM] domains.[2–4] Collectively, these regions support a variety of specific protein–protein interactions that may allow zyxin to participate in the organization or assembly of the actin cytoskeleton or in nuclear-cytoplasmic communication.[5] In this discussion, we catalog a number of molecular features that are characteristic of the zyxin molecule and provide protocols for zyxin's purification and functional analysis.

Diagnostic Features of Zyxin

Subcellular and Tissue Distribution

The zyxin protein was originally identified as an antigen that was recognized by a nonimmune rabbit serum.[1] Immunocytochemical staining with this antibody, as well as other antibodies raised against purified zyxin protein, reveals the presence of zyxin at sites of cell–cell and cell–substratum contact.[1,2,6,7] Within fibroblasts, zyxin is most prominent at cell–substratum attachment sites, or focal adhesions (Figs. 1A and 1B); zyxin is also detected in association with the terminal portions of actin stress fibers near where they associate with focal adhesions at the plasma membrane.[1,2,6] Consistent with zyxin's ability to associate with the actin cross-linking protein α-actinin, zyxin often exhibits a periodic stress fiber staining pattern that is reminiscent of α-actinin distribution (Fig. 1C).[6] In epithelial cells, zyxin is localized not

[1] M. C. Beckerle, *J. Cell Biol.* **103**, 1679 (1986).
[2] A. W. Crawford and M. C. Beckerle, *J. Biol. Chem.* **266**, 5847 (1992).
[3] I. Sadler, A. W. Crawford, J. W. Michelsen, and M. C. Beckerle, *J. Cell Biol.* **119**, 1573 (1992).
[4] D. A. Nix and M. C. Beckerle, *J. Cell Biol.* **138**, 1139 (1997).
[5] M. C. Beckerle, *BioEssays* **19**, 949 (1997).
[6] A. W. Crawford, J. W. Michelsen, and M. C. Beckerle, *J. Cell Biol.* **116**, 1381 (1992).
[7] T. Macalma, J. Otte, M. E. Hensler, S. M. Bockholt, H. A. Louis, M. Kalff-Suske, K.-H. Grzeschik, D. von der Ahe, and M. C. Beckerle, *J. Biol. Chem.* **271**, 31470 (1996).

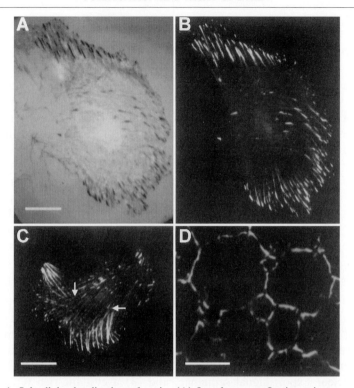

Fig. 1. Subcellular localization of zyxin. (A) Interference reflection microscopy of an adherent chicken embryo fibroblast reveals focal adhesions as dark punctate patches that are concentrated near the cell border. (B) Immunofluorescent staining of the cell in panel A with an anti-zyxin antibody demonstrates that zyxin is localized to focal adhesions. (C) Zyxin is also localized in fibroblasts along actin filaments in a pattern that is reminiscent of that observed for α-actin (see arrows). (D) In epithelial cells, zyxin is also found at sites of cell–cell contact. Bar, 20 μm. [Reprinted with permission from A. W. Crawford, J. W. Michelson, and M. C. Beckerle, *J. Cell Biol.* **271**, 31470 (1996). © 1992 Rockefeller University Press.

only to the basal focal adhesions, but also to lateral cell–cell contacts (Fig. 1D).[2,6] At steady state, zyxin staining is not observable in cell nuclei by indirect immunofluorescence techniques.

A collection of antibodies has been developed for use in immunological studies of zyxin. Four rabbit polyclonal antisera, referred to as B21–B24, have been raised against purified chicken zyxin.[8] Another chicken-specific monoclonal anti-zyxin antibody is available from Transduction Laboratories (Lexington, KY). A polyclonal anti-peptide antibody generated against human zyxin, called B38, also recognizes zyxin from mouse and

[8] A. W. Crawford and M. C. Beckerle, (1991).

rat.[7] Polyclonal antisera have been raised against purified porcine zyxin as well.[9]

Both Northern and Western blot analyses of tissue samples from a variety of species show that zyxin is ubiquitously expressed; however, the relative amounts of zyxin transcripts and protein vary between different organs.[1,7,10,11] For example, avian zyxin, while found at higher levels in smooth muscle tissues like gizzard and intestine, is expressed only at moderate levels in cardiac and skeletal muscle and at low levels in brain and liver. Human blood cells display significant levels of zyxin[12]; human platelets are a proven source for purification of the protein (see procedures described later). In human fibroblasts, viral transformation results in a reduction in zyxin transcript levels.[13] In all organisms studied, the temporal expression profile of zyxin appears to be relatively uniform throughout development.[10,14] In chick and mouse, zyxin expression commences very early in embryogenesis and persists throughout development into adulthood.

Sequence Features of Zyxin

Comparison of zyxin sequences from avian, human, murine, porcine, and *Drosophila* sources illustrates the conserved nature of the zyxin protein (Fig. 2).[3,7,9,13,15] Sequence databases also contain related clones from rabbit and nematode worms.[16] Zyxin displays three notable protein features: (1) an extensive proline-rich N terminus, (2) a short ~10-amino-acid leucine-rich nuclear export signal sequence, and (3) a cysteine-rich C terminus exhibiting three tandem copies of the LIM motif (Fig. 2A).

Both avian and human zyxin display an overall proline content of 18%, which is significantly higher than that observed in a typical eukaryotic protein.[3,7] The majority of proline residues are concentrated in the N-terminal two-thirds of the protein, where they are frequently found in stretches of up to seven contiguous prolines. Proline-rich segments, similar to those found in zyxin, have been described as binding sites for SH3

[9] M. Reinhard, K. Jouvenal, D. Tripier, and U. Walter, *Proc. Natl. Acad. Sci. U.S.A.* **92,** 7956 (1995).
[10] A. W. Crawford, J. D. Pino, and M. C. Beckerle, *J. Cell Biol.* **124,** 117 (1994).
[11] H. A. Louis, J. D. Pino, K. L. Schmeichel, P. Pomiès, and M. C. Beckerle, *J. Biol. Chem.* **272,** 27484 (1997).
[12] O. Hobert, J. W. Schilling, M. C. Beckerle, A. Ullrich, and B. Jallal, *Oncogene* **12,** 1577 (1996).
[13] J. Zumbrunn and B. Treub, *Eur. J. Biochem.* **241,** 657 (1996).
[14] B. A. Benson and M. C. Beckerle, (1997).
[15] B. E. Stronach, T. Macalma, and M. C. Beckerle, *Mol. Biol. Cell* **4S,** 59a (1993).
[16] L. F. Wang, S. Y. Miao, S. D. Zong, Y. Bai, and S. S. Koide, *Biochem. Mol. Biol. Int.* **34,** 1131 (1994).

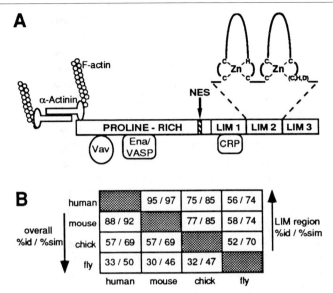

FIG. 2. Zyxin molecules exhibit a similar overall molecular architecture and high levels of sequence conservation. (A) Zyxin displays three distinct domains. The N-terminal proline-rich region supports interactions with α-actinin, vav, and members of the Ena/VASP family. A nuclear export signal (NES) is embedded in the proline-rich region. The first of zyxin's three zinc-binding LIM domains interacts with CRP. (Molecules not drawn to scale.) (B) Pairwise comparisons of amino acid identity and similarity between zyxin molecules. Bottom left boxes display comparisons over the entire length of the protein. Top right boxes display comparisons of the LIM region only. Percentage identity and similarity were derived using the GAP program in the GCG software package version 7.0.[39] Zyxin sequences are available in the GenBank™/EMBL Data Bank under the following accession numbers: human—X94991 and X95735, mouse—Y07711 and X99063, chicken—X69190.

domain-containing proteins[17] and are also found in the bacterial Act A protein, a molecule implicated in the regulation of actin polymerization during *Listeria monocytogenes* infection.[18] Between different species, the proline-rich region of zyxin is typically more divergent than the rest of the molecule.

A conserved leucine-rich sequence resides within the large proline-rich region of zyxin from all species examined to date. In avian zyxin, this sequence has been shown to act as a nuclear export signal (NES).[4] Based on the sequence conservation within this region, as well as the similarity

[17] R. Ren, B. J. Mayer, P. Cicchetti, and D. Baltimore, *Science* **259**, 1157 (1993).
[18] R. Golsteyn, M. C. Beckerle, T. Koay, D. Louvard, and E. Friederich, *J. Cell Sci.* **110**, 1893 (1997).

to NES motifs described in other proteins,[19–21] it appears likely that the presence of a functional NES is a conserved feature of zyxin. The exact position of the NES motif within the proline-rich regions of zyxin molecules is variable.[7,15]

The C-terminal third of zyxin is often referred to as the LIM region because it consists of three repeats of an ~60-amino-acid motif called a LIM domain, which conforms to the consensus $CX_2CX_{16-21}HX_2CX_2CX_2CX_{16-23}CX_2(C,H,D)$ (Fig. 2A).[3,22,23] The conserved cysteine, histidine, and aspartate residues of the LIM motif coordinate two atoms of zinc, giving rise to a structure displaying a double zinc-fingered fold.[24–26] Within zyxin molecules from different species, the LIM region is the most highly conserved part of the protein (Fig. 2B); among the three copies of the LIM domain, the third, or most C-terminal LIM repeat, shows the greatest sequence conservation.[7] Comparing avian and human zyxin, for instance, the third LIM repeat is 92% identical at the amino acid level while the first and second are only 62 and 71% identical, respectively.

The molecular architecture of zyxin serves as a hallmark for identifying related proteins. In humans, two proteins, called LPP (Lipoma Preferred Partner) and Trip6 (Thyroid receptor interacting protein 6), display sequences related to zyxin suggesting zyxin may be one member of a more extensive protein family.[7,27–30] Like zyxin, LPP and Trip6 exhibit three C-terminal LIM domains and long N-terminal regions with several stretches of proline residues. However, LPP and Trip6 are notably divergent from zyxin with respect to overall proline content and the sequence of the nuclear export signal. Zyxin can be further distinguished by its genomic location

[19] L. Gerace, *Cell* **82**, 341 (1995).

[20] U. Fischer, W. M. Michael, and R. Lührmann, *Trends Cell Biol.* **6**, 290 (1996).

[21] M. S. Moore, *Curr. Biol.* **6**, 137 (1996).

[22] I. B. Dawid, R. Toyama, and M. Taira, *C. R. Acad. Sci. Paris, Sci./Life Sci.* **318**, 295 (1995).

[23] G. N. Gill, *Structure* **3**, 1285 (1995).

[24] J. W. Michelsen, K. L. Schmeichel, M. C. Beckerle, and D. W. Winge, *Proc. Natl. Acad. Sci. U.S.A.* **90**, 4404 (1993).

[25] G. C. Pérez-Alvarado, C. Miles, J. W. Michelsen, H. A. Louis, D. R. Winge, M. C. Beckerle, and M. F. Summers, *Nat. Struct. Biol.* **1**, 388 (1994).

[26] G. C. Pérez-Alvarado, J. L. Kosa, H. A. Louis, M. C. Beckerle, D. R. Winge, and M. F. Summers, *J. Mol. Biol.* **257**, 153 (1996).

[27] E. F. P. M. Schoenmakers, S. Wanschura, R. Mols, J. Bullerdiek, H. Van den Berghe, and W. J. M. Van de Ven, *Nat. Genetics* **10**, 436 (1995).

[28] H. R. Ashar, M. S. Fejzo, A. Tkachenko, S. Zhou, J. A. Fletcher, S. Weremowicz, and C. C. Morton, *Cell (Cambridge)* **82**, 57 (1995).

[29] M. M. R. Petit, R. Mols, E. F. P. M. Schoenmakers, N. Mandahl, and W. J. M. Van de Ven, *Genomics* **36**, 118 (1996).

[30] J. W. Lee, H.-S. Choi, J. Gyuris, R. Brent, and D. D. Moore, *Mol. Endocrinol.* **9**, 243 (1995).

on human chromosome 7q32–q36.[7] Functional studies will be necessary to clarify the degree of relatedness of zyxin family members.

Biochemical Properties

Extensive biochemical analyses of the zyxin protein have revealed many properties of the molecule that distinguish it from other proteins localized to focal adhesions.[2,3] When synthesized by *in vitro* transcription/translation, the full-length avian zyxin protein migrates at 82 kDa via sodium dodecyl sulfate–polyacrylamide gel electrophoresis (SDS–PAGE). The avian zyxin cDNA sequence predicts a 542-amino-acid protein with a molecular weight of only ~58 kDa; the high proline content of zyxin contributes to its anomalously slow migration during electrophoresis. Zyxin often migrates as a doublet on denaturing gels. For example, avian zyxin migrates at 82 and 84 kDa; the 84-kDa species is likely to represent a posttranslationally modified form of the protein.

Biophysical studies have been performed on zyxin protein purified from avian smooth muscle to characterize its hydrodynamic properties.[2] The Stokes radius of avian zyxin is 5.6 nm. Zyxin's sedimentation coefficient is 3.0 S. Collectively, these data indicate that, in solution, zyxin behaves as a monomer.

Additional distinctive features of the purified avian protein include a heterogeneous isoelectric point between 6.4 and 7.2 with an average at 6.9.[2] The isoelectric point heterogeneity may be indicative of posttranslational modifications of the protein, as zyxin immunoprecipitated from ^{32}P-labeled chicken embryo fibroblasts is multiply phosphorylated.[2] Phosphoamino acid analysis of protein purified from endogenous sources indicates that stable phosphorylation of zyxin occurs primarily on serine and threonine residues.[8] As predicted for a protein with LIM domains, zyxin is a zinc-binding metalloprotein.[3]

Biochemical Purification Protocols

Extraction and Purification of Avian Zyxin

Protocols have been developed for the isolation of zyxin from avian smooth muscle using chicken gizzard as a starting material.[2] The extraction and purification procedures described next follow the flowchart depicted in Fig. 3A. For best results, all steps in the purification should be executed in immediate succession. All procedures are performed at 4° unless otherwise stated. Sources of specialty reagents are listed; other materials are available from a variety of vendors.

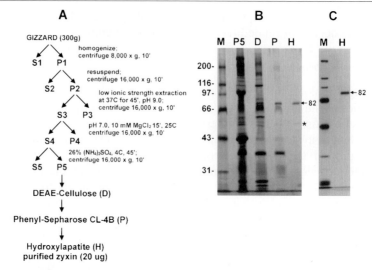

FIG. 3. Purification of avian zyxin. (A) Flowchart outlining the extraction and chromatography steps utilized in the purification of zyxin from chicken gizzard smooth muscle. (B, C) Silver-stained SDS–PAGE gels containing representative samples from steps during zyxin purification; M, molecular weight markers; P5, 26% ammonium sulfate precipitate from the gizzard extract; D, pool from DEAE-cellulose column elution; P, pool from phenyl-Sepharose chromatography; H, final purified zyxin fraction from hydroxylapatite column. Note that in some cases an ~55-kDa protein copurifies with the zyxin protein (see asterisk in panel B, lane H). [Reprinted with permission from A. W. Crawford and M. C. Beckerle, *J. Biol. Chem.* **266,** 5847 (1992). © 1991 The American Society for Biochemistry and Molecular Biology, Inc.]

Reagents

BUFFERS

300 g Wet weight chicken gizzard muscle: fresh or frozen; see discussion in general considerations below

Lysis solution: 0.7 mM phenylmethylsulfonyl fluoride (PMSF) in 4° deionized H$_2$0 (*Caution*: Protease inhibitors, such as PMSF, are highly toxic and should be handled with gloves.)

Buffer A: 2 mM Tris-HCl, pH 8.8, 1.0 mM ethyleneglycol-bis[β-aminoethyl ether]-N,N,N',N'-tetraacetic acid (EGTA), 0.7 mM PMSF at 37°

Buffer B-10: 20 mM Tris-acetate, pH 7.6, 10 mM NaCl, 0.1% 2-mercaptoethanol, 0.1 mM ethylenediaminetetraacetic acid (EDTA)

10× Protease cocktail: 1.0 mM PMSF, 1.0 mM benzamidine HCl, 10 ng/ml pepstatin A, 10 ng/ml 1,10-phenanthroline in 100% ethanol

Phenyl-Sepharose buffer: 12.5% ammonium sulfate in buffer B-10
HAP buffer: 1.0 mM potassium phosphate (mono- and dibasic), pH
   7.2, 10 mM NaCl, 0.1 mM EDTA, 0.1% 2-mercaptoethanol

COLUMNS

DEAE-cellulose column: 8- × 2.5-cm column of DEAE-cellulose
   (DE52 matrix, Whatman, Clifton, NJ), preequilibrated in buffer B-10
Phenyl-Sepharose column: 10- × 0.7-cm column of phenyl-Sepharose
   CL-4B (Pharmacia Biotech, Piscataway, NJ), pre-equilibrated in
   phenyl-Sepharose buffer
Hydroxylapatite (HAP) column: 0.5 ml hydroxylapatite HPLC column
   (Bio-rad, Richmond, CA), preequilibrated in HAP buffer

Tissue Extraction. Gizzard muscles, dissected free of associated connective tissue, are chopped into small pieces and stored on ice. The chopped
tissue (300 g) is placed into a Waring blender, along with 8.5 volumes of
cold lysis solution; homogenization of the tissue is achieved via three 10-
sec high-speed bursts in the blender. The homogenized material is centrifuged for 10 min at 8000g. To reduce further the ionic strength of the
preparation, the supernatant (S1) is discarded and the pellet (P1) is resuspended in another 8.5 volumes of lysis solution with a brief pulse in the
blender. (It is important to avoid excessive foaming during homogenization
because this can cause protein denaturation and disruption of the myofi-
brils.) The second homogenate is centrifuged for 10 min at 16,000g and the
pellet (P2) is retained. P2 is resuspended in 8.5 volumes of buffer A that
has been prewarmed to 37°. The mixture is adjusted to pH 9.0 using 2 M
NaOH and is placed in a 37° water bath. Proteins are extracted at 37° by
stirring the mixture gently and continuously with a plastic pipette for 45
min. The extract is centrifuged for 10 min at 16,000g; the supernatant (S3)
is saved and the pellet (P3) is discarded. The pH of S3 is restored to 7.0 using
0.5 M acetic acid and 1.0 M MgCl$_2$ is then added to a final concentration of
10 mM; the solution is stirred at room temperature for 15 min. (This step
induces the polymerization of actin, which will appear as a white precipitate.) The extract is centrifuged for 10 min at 16,000g to collect the polymerized actin; the pellet (P4) is discarded and the supernatant (S4) is saved.

S4 is enriched in a variety of cytoskeletal proteins, including vinculin,
talin, α-actinin, and zyxin. Initial fractionation of these proteins can be
achieved by differential ammonium sulfate precipitation. To precipitate
zyxin, the final volume of S4 is measured and 15 g of ammonium sulfate/
100 ml of supernatant is added slowly (i.e., to give a 26% saturated solution).
For optimal precipitation, the solution is stirred at 4° for 45 min. The sample
is then centrifuged for 10 min at 16,000g; the supernatant (S5) is discarded
and the pellet (P5) is retained. (All detectable zyxin is contained within
the 26% ammonium sulfate precipitated P5 fraction.) To prepare for column

chromatography, P5 is resuspended in 3 ml of buffer B-10 and dialyzed overnight at 4° against 2 liters of buffer B-10. A protein profile illustrating the complexity of a typical 26% ammonium sulfate fraction is shown in the silver-stained gel in Fig. 3B, lane P5.

Column Chromatography. Because zyxin has a tendency to adhere non-specifically to glass, plastic tubes are used to collect fractions during all chromatography steps. Also, to prevent excessive protein degradation, samples are collected into tubes containing protease inhibitors (for a final inhibitor concentration of 1×). (*Caution*: Wear gloves when handling solutions containing protease inhibitors.)

DEAE-CELLULOSE ANION EXCHANGE CHROMATOGRAPHY. The dialyzed P5 fraction is loaded onto the DEAE-cellulose matrix. The column is then washed with buffer B-10 (five column volumes). Bound proteins are eluted from the column with a 120-ml gradient of 0–150 mM NaCl in buffer B-10 at a rate of 0.4 ml/min; 80- × 1.25-ml fractions are collected throughout the gradient. Zyxin elutes between 40 and 60 mM NaCl (i.e., between fractions 40 and 60). Because, at this point in the purification, it is not possible to see distinct zyxin protein bands on silver-stained gels, zyxin-containing fractions are characterized by Western immunoblot analysis using zyxin-specific antibodies. A silver-stained gel illustrating the complexity of a typical DEAE-cellulose column pool is illustrated in Fig. 3B, lane D.

PHENYL-SEPHAROSE CHROMATOGRAPHY. In preparation for phenyl-Sepharose chromatography, saturated ammonium sulfate is added to the pooled material from the DEAE-cellulose column to a final concentration of 12.5%. This sample is applied directly onto a phenyl-Sepharose CL-4B column (preequilibrated in phenyl-Sepharose buffer). After loading the sample, the column is washed with five volumes of phenyl-Sepharose buffer. The first column elution is performed with a 20-ml decreasing gradient of ammonium sulfate (12.5–0%) in buffer B-10; 1.25-ml fractions are collected at a rate of 0.3 ml/min. (Zyxin does not elute during this procedure). This elution is followed immediately by a second 20-ml linear gradient of 0–50% ethylene glycol in buffer B-10. The 1.25-ml fractions are collected at a rate of 0.3 ml/min; zyxin emerges from the column between 25 and 35% ethylene glycol. At this point, the 82-kDa zyxin protein can be seen in peak fractions by silver staining of denaturing gels (Fig. 3B, lane P).

HYDROXYLAPATITE CHROMATOGRAPHY. The ethylene glycol concentration in the fractions pooled from the phenyl-Sepharose column elution is diluted 2-fold in cold HAP buffer. The diluted sample is applied to a 0.5-ml HPLC-hydroxylapatite column, preequilibrated in HAP buffer. To elute zyxin, the column is subjected to a 15-ml linear gradient of 0–75 mM potassium phosphate in HAP buffer. A series of 0.5-ml fractions is collected from the column at a flow rate of 0.5 ml/min. Purified zyxin elutes between 5 and 15 mM potassium phosphate (see silver-stained gels in Fig. 3B and 3C, lanes H).

General Comments. The procedure just described takes 4 days to complete. It has been repeatedly observed that the yield of zyxin from fresh tissue is approximately five times higher than that obtained from frozen tissue. On average 20 μg of purified zyxin can be recovered from 300 g of fresh gizzard tissue; this yield is representative of approximately 3–5% of the total zyxin protein present in the 26% ammonium sulfate precipitate fraction. In some cases, the final purified end product is contaminated with trace amounts of a 55-kDa protein (see asterisk in Fig. 3B, lane H); this contaminant can be removed by selective pooling of fractions at each step in the procedure or by using fresh tissue. Purified zyxin can be stored at 4° for several days. However, because zyxin is extremely labile and sensitive to proteolysis, it is best to use the isolated material immediately after the final step of purification.

Alternative Purification Protocols

Protocols describing the purification of zyxin from human and porcine platelets have also been developed.[9,31] Purification of human zyxin involves a protocol similar to that described above for avian zyxin,[31] only in this case 20 U of recently outdated human platelets serve as the starting material. (*Caution*: All human blood samples should be treated as potentially infectious; waste contaminated by human blood should be autoclaved and appropriately discarded.) Platelet lysates are generated using a previously described protocol involving a low-ionic-strength detergent extraction.[32,33] The resulting lysate is subjected to the chromatographic procedures detailed above for avian tissue extracts. Generally, 16 μg of human zyxin can be recovered from 20 U of platelets. The purified human protein migrates on SDS–PAGE gels at 84 kDa or as an 84/86-kDa doublet. Similarly, porcine zyxin can be isolated from a hypotonic platelet lysate by a series of chromatographic steps, including separation by cation exchange chromatography (using S-Sepharose FF), followed by hydroxylapatite chromatography, and finally by selective elution from a zinc-chelating column.[9] Using this protocol, 10–20 μg of zyxin are typically recovered from 8×10^{12} platelets (from 60 liters of porcine blood).

Finally, zyxin protein that retains all known functions can be generated from recombinant sources. Full-length avian zyxin and its component do-

---

[31] M. E. Hensler and M. C. Beckerle, *Mol. Biol. Cell* **3S,** 266a (1992).
[32] T. O'Halloran, M. C. Beckerle, and K. Burridge, *Nature (Lond.)* **317,** 449 (1985).
[33] M. C. Beckerle, T. O'Halloran, and K. Burridge, *J. Cell. Biochem.* **30,** 259 (1986).

mains have been successfully expressed and purified as fusion proteins using the glutathione S-transferase bacterial expression system (Pharmacia Biotech).[34] Bacterially expressed zyxin fusion proteins are known to retain their ability to interact with specific protein partners (see discussion of functional assays later), as well as to coordinate zinc, suggesting that these proteins exhibit a native fold.

Assays of Zyxin Function

Zyxin has been shown to interact with a variety of proteins via binding sites that are positioned throughout the protein (Fig. 2A). The proline-rich region of zyxin is responsible for mediating interactions with α-actinin,[6,34] the protooncogene product, vav,[12] and members of the Ena/VASP family[9,18,35]; the LIM region of zyxin has been shown to interact with the cysteine-rich protein, CRP.[3,34] One technique that has been used extensively to identify and characterize all known zyxin-binding partners to date is the blot overlay assay, a method that is described next in detail (Fig. 4).

Characterizing Zyxin-Binding Partners Using the Blot Overlay Assay

Reagents and Buffers

FOR RADIOIODINATION

Labeling buffer: 150 mM NaCl, 50 mM Tris-HCl, pH 7.6

IODO-GEN-coated tubes: 40 μl of a 1.0 mg/ml solution of IODO-GEN (1,3,4,6-tetrachloro-3α,6α-diphenylglycouril; Pierce, Rockford, IL) in chloroform are placed in a 1.7-ml microcentrifuge tube; the solvent is evaporated overnight in a properly vented fume hood

Saturated tyrosine solution: 10 mg/ml tyrosine in distilled H$_2$O

Column elution buffer: 150 mM NaCl, 50 mM Tris-HCl, pH 7.5, 0.2% gelatin, 0.1% 2-mercaptoethanol, 1.0 mM EDTA, 0.1% sodium azide. (*Caution*: Wear gloves when handling azide-containing solutions.)

FOR ^{32}P-LABELING

10× Reaction buffer: 200 mM Tris-HCl, pH 7.5, 10 mM (dithiothreitol (DTT), 1.0 M NaCl, 120 mM MgCl$_2$ PKA solution: 6.25 U/μl catalytic subunit of protein kinase A (PKA; Sigma, St. Louis, MO) in 40 mM DTT GST elution buffer: 15 mM reduced glutathione (Sigma) in 50 mM Tris-HCl, pH 8.0

[34] K. L. Schmeichel and M. C. Beckerle, *Cell* **79**, 211 (1994).

[35] F. B. Gertler, K. Niebuhr, M. Reinhard, J. Wehland, and P. Soriano, *Cell* (*Cambridge*) **87**, 227 (1996).

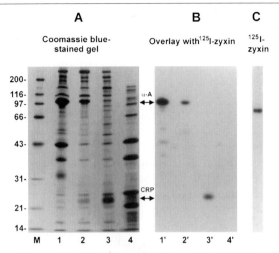

FIG. 4. Detecting zyxin-binding partners using the blot overlay assay. (A) Coomassie blue-stained gel of protein samples from four different fractions derived from an avian smooth muscle extract. M, molecular weight markers; smooth muscle fractions are loaded in lanes 1–4. (B) Autoradiograph of a parallel gel that has been transferred to nitrocellulose and probed with a radioiodinated zyxin probe. The zyxin probe interacts with several proteins on the blot, including α-actinin (α-A, at 100 kDa) and CRP (23 kDa). (C) An autoradiograph of an SDS–PAGE gel loaded with the radioiodinated zyxin used in the blot overlay assay illustrates the quality of the probe. [Reprinted with permission from A. W. Crawford, J. D. Pino, and M. C. Beckerle, *J. Cell. Biol.* **124**, 117 (1994). © 1994 The Rockefeller University Press.]

FOR BLOT OVERLAY ASSAY

Blocking buffer: 150 mM NaCl, 50 mM Tris-HCl, pH 7.6, 0.05% sodium azide, 2.0% bovine serum albumin (BSA) (>96% pure)

Blot overlay buffer: 20 mM HEPES, pH 7.5, 0.5% BSA, 0.25% gelatin, 1.0% Nonidet-P 40, 10 mM NaCl, 1.0 mM EGTA, 0.1% 2-mercapto-ethanol

Washing buffer: 50 mM Tris-HCl, pH 7.6, 150 mM NaCl, 0.05% Tween-20, 0.1% sodium azide, 0.2% gelatin

Making a Radiolabeled Zyxin Probe. Zyxin probes, for use in blot overlay assays, are generated by radioiodinating zyxin purified from avian smooth muscle tissue[3,6] or by labeling glutathione *S*-transferase (GST)-zyxin fusion proteins with ^{32}P.[34,36] (*Caution*: The radionuclides used in these procedures should be handled with extreme care. Proper shielding, protective clothing, and monitoring devices should be utilized during all manipulations. Lead shielding should be used when working with ^{125}I and plexiglass barriers are appropriate for ^{32}P.)

[36] K. L. Schmeichel and M. C. Beckerle, *Mol. Biol. Cell* **8**, 219 (1997).

RADIOIODINATION. An effective method for zyxin radioiodination follows protocols developed by Fraker and Speck[37] that utilize IODO-GEN as a catalyst for coupling ^{125}I to tyrosine residues exposed on the surfaces of target molecules. For this procedure, 100–150 μl of purified zyxin protein (e.g., from the HAP column) are dialyzed into labeling buffer. Dialyzed protein (100 μl) is placed into an IODO-GEN-coated tube. The labeling reaction, once initiated by the addition of 1 mCi Na-^{125}I, is allowed to proceed on ice for 7 min. (*Caution*: Because unconjugated ^{125}I is extremely volatile, these procedures should be preformed in a properly ventilated fume hood.) Saturated tyrosine solution (50 μl) is added to the mixture to react with the remaining unincorporated ^{125}I; the incubation is extended for an additional 5 min. Gel filtration chromatography is used to separate the labeled protein from the radioiodinated tyrosine. In this procedure, the sample is carefully transferred to the surface of a 0.7- $\times$ 15-cm G-50 gel filtration column that has been preequilibrated in column elution buffer and wrapped in a flexible lead shield. After allowing the sample to penetrate into the column matrix, the protein is eluted in column buffer while collecting a series of 500-μl fractions. The radioiodinated zyxin protein elutes from the column at 2500–3000 μl. Fractions are collected to recover the first peak of radioactivity; a second pool of radioiodinated tyrosine molecules should be retained in the column and discarded in an appropriate fashion. A gamma counter is used to determine the cpm/μl in each fraction and the quality of the probe is evaluated via SDS–PAGE followed by autoradiography.

^{32}P-LABELING. ^{32}P-Labeled zyxin probes can be generated from bacterially expressed GST fusion proteins. GST-zyxin fusion proteins expressed from the pGEX-2T-$^{128}/_{129}$vector[34] or the pGEX-2TK vector (available through Pharmacia Biotech) contain engineered protein kinase A (PKA) recognition sites (aa RRASV). Due to this modification, these proteins can be readily ^{32}P-labeled *in vitro* in the presence of [γ-^{32}P]ATP and the catalytic subunit of PKA. In these reactions, GST-zyxin fusion proteins, recovered from a 150-ml culture of isopropyl β-D-thiogalactopyranoside (IPTG)-induced cells, are coupled to 35 μl of glutathione-Sepharose 4B as described elsewhere.[34,38] The protein-coupled beads are then mixed with 5 μl of 10$\times$ reaction buffer, 150 μCi of fresh [γ-^{32}P]ATP (>7000 Ci/mmol stock solution; DuPont–NEN, Boston, MA) and 10 μl of PKA solution (see

[37] P. J. Fraker and J. C. Speck, *Biochem. Biophys. Res. Comm.* **80**, 849 (1978).

[38] F. M. Ausubel, R. Brent, R. E. Kingston, D. D. Moore, J. G. Seidman, J. A. Smith, and K. Struhl, *in* "Current Protocols in Molecular Biology," Vol. 3. John Wiley and Sons, New York, 1997.

[39] J. Devereux, P. Haeberli, and O. Smithies, *Nucleic Acids Res.* **12**, 387 (1984).

recipe above); the mixture is incubated at 25° for 15 min. To terminate the reaction, 1.0 ml phosphate-buffered saline (PBS) is added to the tube and the sample is centrifuged for 5 sec at low speed to recover the matrix; the beads are washed an additional three times with 1.0 ml PBS. ^{32}P-labeled protein is eluted from the affinity matrix by adding 400 μl of GST elution buffer and incubating for 5 min at 25°. Supernatants containing the purified radiolabeled proteins are collected by centrifugation and are transferred to a fresh tube. The cpm/μl of each supernatant is determined by Cerenkov counting and the homogeneity of the probes is evaluated by SDS–PAGE and autoradiographic analysis.

Blot Overlay Assay. For the blot overlay assay, potential zyxin protein partners are resolved on SDS–PAGE gels. The exact amount of protein to be loaded in each lane will vary depending on the interaction; however, one general rule of thumb is to load proteins onto gels at levels that are detectable by Coomassie blue staining. Replicate gels are prepared for parallel analyses by Coomassie blue staining (Fig. 4A) and blot overlay assay (Fig. 4B). Blot overlay gels are transferred to nitrocellulose and the resulting blots are incubated in blocking buffer for a minimum of 4 hr to allow for protein refolding. For the binding assay, nitrocellulose strips are covered with blot overlay buffer in small plastic boxes; for 5- × 10-cm nitrocellulose strips, 10 ml of solution is usually sufficient for the assay as long as the container is not substantially larger than the blot. The radiolabeled zyxin probes (Fig. 4C) are added directly to the overlay buffer: 250,000 cpm/ml for radioiodinated zyxin and 600,000 cpm/ml for a ^{32}P-labeled probe. The blots are covered and incubated with gentle agitation for 4 hr at room temperature. At this point, the overlay solution is discarded and the nitrocellulose strips are washed 6 × 5 min in washing buffer. After wicking off excess buffer, the blots are wrapped in plastic film and analyzed by autoradiography. Generally, overnight exposures should be sufficient to detect relevant interactions (Fig. 4B). Relative protein-binding levels can be determined by performing PhosphorImager analysis of the blots (Molecular Dynamics, Sunnyvale, CA).

General Comments. As a method for detecting protein–protein interactions, the blot overlay assay is quite versatile. Not only does it facilitate the identification of novel zyxin-binding partners, but it is also a useful method for the characterization of protein-binding specificities. Additionally, the blot overlay assay has proven to be useful for mapping the functions of zyxin's component domains.[34,36] In general, this assay is most effective when the zyxin probes are freshly prepared. Positive interactions detected in the blot overlay assay are especially convincing when interactions can be demonstrated in both directions (i.e., when each protein partner can be used as a probe). However it should be noted that, because some proteins behave better than others in terms of their capacity to refold appropriately on the nitrocellulose blots, this is not an absolute requirement. Interactions detected using the blot overlay assay should be confirmed using alternative approaches, such as the solution-binding assays described later.

Alternative Methods for Detecting Zyxin-Dependent Protein–Protein Interactions

Although the blot overlay assay has proven to be a powerful method for detecting and characterizing zyxin-binding partners, a number of additional approaches have been utilized with comparable success. The interaction between zyxin and α-actin has been extensively characterized using both gel filtration and solid phase binding assays.[6] Recent demonstrations that GST-zyxin fusion proteins retain their functional properties have enabled the use of "pull-down" assays for the study of zyxin-binding partners. This type of approach has been used successfully to characterize interactions between zyxin and its partners, CRP,[34] Mena,[35] and vav.[12] Finally, the yeast two-hybrid method has been utilized to demonstrate a specific interaction between zyxin and vav.[12]

Concluding Remarks

Since its initial identification, there has been increasing interest in both the molecular structure and functional properties of the zyxin protein. Zyxin's ability to participate in multiple protein–protein interactions led to the idea that zyxin serves as a molecular scaffold for the assembly of multimeric protein complexes. Due to its subcellular localization at sites of cell adhesion, zyxin and its associated proteins may participate in those cellular events that occur specifically at the focal adhesion, such as the regulation of actin filament assembly or the generation of adhesion-dependent signals.[5] In this short review we have described molecular features that are diagnostic for the zyxin protein. Several methods for the purification of zyxin have been developed that provide sufficient quantities of purified protein for use in biochemical studies as well as in the generation of zyxin-specific antibodies. Assays that have been used to demonstrate interactions between zyxin and various binding partners have also been detailed. Collectively, these diagnostic tools provide a solid basis for future explorations into the structure and function of zyxin and related proteins.

Acknowledgments

This work was supported by grants from the National Institutes of Health. MCB is the recipient of a Faculty Research Award from the American Cancer Society.

[8] Purification and Assays for Paxillin

By Christopher E. Turner and Michael C. Brown

Introduction

Paxillin is a 68-kDa protein that localizes to the cytoplasmic face of sites of cell adhesion to the extracellular matrix.[1] These structures are generally referred to as *focal adhesions* or *focal contacts*.[2] This distribution is shared with a large number of structural and regulatory proteins that serve not only to link physically the extracellular matrix to the actin cytoskeleton but also to transduce extracellular signals.[3] Two important aspects of elucidating the contribution of paxillin to these functions are to determine the identity of other focal adhesion components that interact with paxillin and to evaluate how these interactions may be modulated by posttranslation modifications, such as changes in paxillin phosphorylation. This chapter describes methods for the biochemical and molecular purification of paxillin and paxillin subdomains and assays that can be utilized to study paxillin interactions with other proteins. A summary diagram of the major structural and functional domains of paxillin that will be referred to in this chapter is presented in Fig. 1.

Traditional Biochemical Purification of Paxillin

While the majority of *in vitro* experiments involving paxillin utilize bacterial expression systems (see below), a protocol for purifying paxillin from chicken gizzard smooth muscle tissue has been developed.[4] This protocol incorporates a series of standard column chromatography steps to enrich for this relatively low abundance protein (at least 10 times less abundant than the focal adhesion proteins talin and vinculin), followed by a final immunoaffinity purification. Although not absolutely essential, the steps prior to the affinity isolation enhance considerably the final purity of the paxillin. Paxillin purified by this method retains the ability to interact specifically with vinculin.[4] Other interactions have not been tested.

---

[1] C. E. Turner and J. T. Miller, *J. Cell Sci.* **107**, 1583 (1994).
[2] K. Burridge, K. Fath, T. Kelly, G. Nuckolls, and C. Turner, *Ann. Rev. Cell Biol.* **4**, 487 (1988).
[3] M. A. Schwartz, M. D. Schaller, and M. H. Ginsberg, *Ann. Rev. Cell Biol.* **11**, 549 (1995).
[4] C. E. Turner, J. R. Glenney, and K. Burridge, *J. Cell Biol.* **111**, 1059 (1990).

Domains of Paxillin

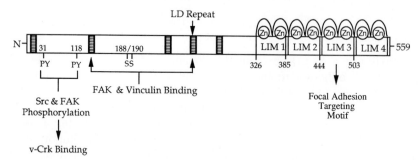

FIG. 1. A schematic representation of the multidomain composition of the focal adhesion protein paxillin. Tyrosine amino acid residues 31 and 118 are targets of the tyrosine kinases FAK and/or src leading to the creation of SH2-binding domains for the SH3- and SH2-domain-containing adapter protein crk.[3,15] Serine residues 188 and 190 are phosphorylated by an unknown serine kinase in response to cell adhesion to fibronectin.[18] The darkened boxes represent 13-amino-acid leucine-rich "LD repeats" denoted as such due to a conserved leucine-aspartate amino acid pair on the amino terminal margin of each repeat.[14] The second and fourth LD repeats are involved in vinculin and FAK binding.[14] The boundaries of each 50-amino-acid LIM domain of this 559-amino-acid protein are demarcated. LIM2 and LIM3 are responsible for localizing paxillin to focal adhesions.[14]

Buffers

Buffer A: 10 mM Tris-HCl, pH 8.0; 2 mM ethylenediaminetetraacetic acid (EDTA); 0.1% 2-mercaptoethanol; 0.5 mM phenylmethylsulfonyl fluoride (PMSF); 5 μM leupeptin

Buffer B: 20 mM Tris-HCl, pH 7.6; 20 mM NaCl; 0.1 mM EDTA; 0.1% 2-mercaptoethanol; 0.5 mM PMSF; 5 μM leupeptin

Buffer C: 10 mM Tris-HCl, pH 7.6; 150 mM NaCl

Other Solutions

15–240 mM Potassium phosphate, pH 7.5
10 mM Sodium phosphate pH 6.8
100 mM Glycine-HCl, pH 2.5

Chromatographic Preparation of Paxillin-Rich Fractions

Fresh, finely chopped chicken gizzard smooth muscle (150 g, transported on ice from the slaughterhouse) is homogenized in 800 ml ice-cold buffer

A with five, 10-sec bursts at top speed in a Waring blender and centrifuged at 16,000g for 10 min in a Sorvall GSA rotor. The supernatant is saved and the pellet resuspended in 800 ml of buffer A, homogenized at medium speed three times for 5 sec each, then centrifuged as above. The two supernatants are combined and filtered through six layers of fine cheese cloth.

The volume of the filtrate is measured and a paxillin-enriched fraction is precipitated by adding solid $(NH)_2SO_4$ (13.4 g/100 ml) with stirring at 4° for 60 min. The precipitate is collected by centrifugation at 12,000g for 10 min, gently resuspended in 20–30 ml of buffer B, and dialyzed overnight against buffer B. Any precipitated protein (mostly myosin contamination) is removed by centrifugation at 100,000g for 30 min (45Ti rotor) and the clarified supernatant is then loaded onto a 100-ml bed-volume column of DEAE-(DE52) cellulose anion-exchange column, preequilibrated in buffer B. After loading, the column is washed extensively with buffer B (300 ml) and the proteins are eluted with a 650 ml NaCl gradient from 0–325 mM in buffer B. Column fractions of 4 ml are assayed for paxillin by SDS–PAGE on 10% gels and Western immunoblotting with paxillin antibody (available commercially). Paxillin migrates as a diffuse band (due to multiple phosphorylation of tyrosine, serine, and threonine residues) of approximately 68 kDa. Proteolytic breakdown products of 44 and 42 kDa may also be observed.[4]

Paxillin-rich fractions are pooled and applied directly to a 25-ml hydroxylapatite column equilibrated in buffer B. Following washing in buffer B (50 ml), proteins are eluted with a 120 ml, 15–240 mM potassium phosphate, pH 7.5, 0.1% 2-mercaptoethanol gradient. Paxillin-enriched fractions (1 ml) are identified by Western immunoblotting, pooled, dialyzed against buffer C, and then loaded batchwise onto a 1-ml paxillin antibody-affinity column. Paxillin antibody can be purchased commercially (the antibody is high affinity and can be reused multiple times thereby reducing the cost) and then coupled to Affi-Gel 10 matrix according to manufacturers' intructions. The matrix is poured into a 1.0- × 5-cm column and washed with 50 ml of buffer C, then briefly with 10 mM sodium phosphate, pH 6.8, before elution of the paxillin protein in 0.5-ml fractions with 100 mM glycine-HCl, pH 2.5, into tubes containing 5–10 μl of 1 M Tris base to neutralize the acid, and 0.1% 2-mercaptoethanol. Paxillin-containing fractions (as determined by SDS–PAGE and Coomassie Blue staining) should be dialyzed immediately into buffer C plus 0.1% 2-mercaptoethanol or buffers appropriate for subsequent assays. This procedure yields approximately 500–800 μg of purified protein. Representative gels and Western blots of the major chromatographic steps used in purification are shown in Fig. 2.

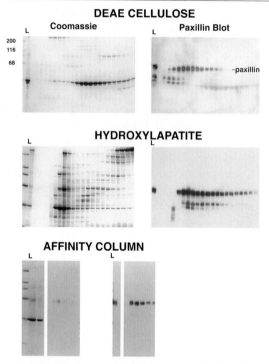

FIG. 2. The three major chromatographic steps in the purification of paxillin from chicken gizzard are presented. On the left is a Coomassie Blue-stained gel of the eluted column fractions (every third fraction is shown). On the right side is an anti-paxillin Western blot of the same fractions. Note that paxillin migrates as a broad band of approximately 68 kDa. Two stable proteolytic fragments of paxillin are seen at approximately 42 and 44 kDa. L, column load.

Use and Preparation of Paxillin Fusion Proteins

The utilization of bacterial expression systems for the generation of large quantities of recombinant proteins/peptides has become an essential component of any comprehensive analysis of protein structure and function. Two strategies employed in the analysis of paxillin domain function are the glutathione S-transferase (GST) and maltose-binding protein (MBP) fusion protein expression cassette systems, which allow for one-step affinity purification of recombinant proteins.[5,6] Paxillin fusion proteins produced

[5] D. B. Smith and K. S. Johnson, Gene 67, 31 (1988).
[6] C. V. Maina, P. D. Riggs, A. G. Grandea III, B. E. Slatko, L. S. Moran, J. A. Tagliamonte, L. A. McReynolds, and C. Guan, Gene 74, 365 (1988).

in such a manner have been used in several precipitation assays as well as in microinjection-based studies.

Media and Buffers

Luria–Bertani (LB): 10 g Bacto-tryptone, 5 g yeast extract, 5 g NaCl (per liter)

Hyperosmotic LB: LB with $0.1 \times$ NaCl, 1 M sorbitol, and 2.5 mM betaine

TBS: 10 mM Tris-HCl, pH 7.6; 150 mM NaCl

Bacterial lysis buffer: TBS, pH 8.0; 1% Triton X-100; 1 mM DTT; 2 mg/ml lysozyme and proteinase inhibitors

Elution buffers:

GST/glutathione: 100 mM Tris-HCl, pH 8.0; 500 mM NaCl; 1 mM DTT; 25 mM glutathione

MBP/maltose: 20 mM Tris-HCl, pH 8.0; 500 mM NaCl; 1 mM DTT; 10 mM maltose

Reagents. Lysozyme (Sigma, St. Louis, MO, L-6876); ampicillin (Sigma, A-9518); isopropyl-β-D-thiogalactopyranose (IPTG) (Promega, Madison, WI); complete protease inhibitor tablets (Boehringer Mannheim, Indianapolis, IN); betaine (Calbiochem, La Jolla, CA); thrombin (Sigma, T-6884); Factor Xa (ICN, Costa Mesa, CA, 191396); glutathione Sepharose 4B (Pharmacia, Piscataway, NJ) or glutathione agarose (Sigma); amylose resin (New England Biolabs, Beverly, MA); glutathione (Sigma, G-4251).

Production of pGEX-2T and pMAL-2p Paxillin Fusion Vectors

With both the GST (Pharmacia) and MBP (NEB) procedures, paxillin coding sequence is subcloned into pGEX-2T or pMAL-2p within the multiple cloning site located at the 3′ region of sequence encoding the carboxyl terminus of either GST or MBP. The primers chosen for subcloning paxillin contain within their nucleotide sequence the enzyme restriction digest sites corresponding to those present in the polycloning site. The sequence of the oligonucleotides also will allow for the insertion of paxillin into the carboxyl terminus such that the encoding sequence of paxillin is in-frame with the carrier, amino-terminal GST or MBP. The GST and MBP systems also have been constructed such that a thrombin-cleavage or Factor Xa-cleavage site, respectively, is present between the carrier protein and the inserted fusion protein.

Expression of Paxillin GST- and MBP-Fusion Proteins

Following construction of the fusion protein vector, the respective plasmids are transformed into bacteria, generally a standard K12 strain (e.g., DH5α), for expression of the engineered fusion protein. Both the GST and

MBP systems allow for the controlled expression of the fusion protein through the use of an IPTG-inducible promoter. Overnight cultures of transformants grown from glycerol stocks in 5-ml standard LB medium with 100 μg/ml ampicillin are diluted 1 : 10 to allow cells to reenter logarithmic phase growth (90 min) followed by induction of protein expression with 0.1 mM IPTG for an additional 180 min at 37° and 250 rpm continuous agitation. Substantial quantities of soluble protein are generally expressed; however, some domains of paxillin (described below) expressed as fusion protein are packaged predominantly into inclusion bodies and are consequently inaccessible for standard purification protocols.

Standard Purification of Soluble Paxillin Fusion Proteins

For soluble fusion proteins, the bacteria are harvested by centrifugation for 10 min at 1500g following induction of protein expression, washed in TBS, and lysed in bacterial lysis buffer for 30 min at room temperature. It is critical that the bacterial suspension be at or above pH 8.0 for optimal lysozyme activity and efficient lysis and in some cases it is helpful to add 1 M Tris base to increase the pH. Triton X-100 is added to 1% and incubated on ice for 10 min. After a brief 10-min 27,000g centrifugation step, the soluble fusion protein-containing supernatant is incubated, end over end, for 90 min at 4° with 100 μl of a 50% slurry of prewashed and equilibrated glutathione-Sepharose 4B beads (Pharmacia) or amylose resin (NEB) for GST or MBP fusion proteins, respectively. Following extensive washing, the quality and quantity of the protein batch is analyzed by SDS–PAGE and Coomassie Brilliant Blue staining of a 5-μl aliquot of a 10% bead slurry.

The beads are utilized directly for precipitation studies or the fusion proteins can be eluted from the solid support by incubation, in batch, with an equal volume of glutathione- or maltose-based elution buffers for 90 min. The eluant is then dialyzed against a buffer appropriate for subsequent assays. Alternatively, the fusion proteins can be cleaved with either thrombin (Sigma) or Factor Xa (ICN) at 1% w/w using standard buffer conditions recommended by the manufacturers, releasing the paxillin portion of the molecule into the buffer. The protein can be recovered and processed as above.

Note that expression of fusion proteins containing double zinc finger LIM domains requires the continuous presence of 10 mM DDT and generally requires a cocktail of proteinase inhibitors for consistent purification of intact, stable fusion proteins with optimal bioactivity. Additionally, although concentration and or dialysis of intact or cleaved paxillin fusion protein can be performed efficiently utilizing Amicon Centricons, "free" LIM domains appear to stick to the membrane and should be concentrated by lyophilization. This will not detrimentally affect LIM domain structure (M. S. Curtis and C. E. Turner, unpublished observation). Fusion proteins are used within 1 week.

Optimization of the Expression of Poorly Soluble Paxillin Fusion Proteins

For most paxillin fusion proteins the above conditions result in the purification of substantial quantities of high-quality protein; however, expression of several forms of recombinant paxillin as GST fusions are recalcitrant to conventional purification schemes including full-length paxillin, the four carboxyl-terminal LIM domains together, combinations of LIM domain pairs, and also LIM3. These proteins are expressed at reduced levels, packaged predominantly within inclusion bodies, and appear to be less stable than fusion proteins derived from the amino terminal 313 amino acids as well as individual fusions of LIM1, LIM2, and LIM4. Interestingly, some paxillin fusion proteins that are poorly soluble as GST fusions are produced much more efficiently as MBP fusions under standard conditions. This is most dramatic with expression of full-length paxillin, the four tandem LIM domains and LIM3. The primary basis for this difference may be the fact that proteins produced as MBP fusion proteins in the pMAL-2p vector are directed through the plasma membrane to the periplasmic space of the bacterium, perhaps avoiding machinery directing these proteins to inclusion bodies and away from intracellular proteinases. Nonetheless, modifications in the culture media formulation and the culture conditions have been made to optimize production of high-quality, soluble fusion protein for those remaining troublesome fusions.

Hyperosmotic Growth to Increase Fusion Protein Solubility

Bacteria containing fusion protein expression vectors are cultured from glycerol stocks overnight in 5 ml LB containing 100 μg/ml ampicillin at 37° with constant agitation of 250 rpm. The overnight culture is transferred into baffled flasks and diluted 1 : 30 into a hyperosmotic LB-based medium; LB containing 0.1× NaCl, 1 M sorbitol, and 2.5 mM betaine.[7] The growth temperature is reduced to room temperature and the cultures are grown for an additional 5 hr prior to induction of protein expression by stimulation with 10 μM IPTG. Protein expression continues for 15 hr at room temperature with agitation. Processing of the fusion proteins then follows standard procedures. When starting with the same amount of bacterial pellet wet-weight material, these alterations in culture conditions result in the genera-

---

[7] J. R. Blackwell and R. Horgan, *FEBS Lett.* **295**, 10 (1991).

tion of soluble, intact fusion protein, with a yield approximately 50% that of freely soluble paxillin fusion proteins prepared by conventional techniques.

In addition to the use of alternative media to increase soluble yields, we have had some success with different *Escherichia coli* K12 strains that are deficient in either the Tsp periplasmic protease, KS1000, or a strain deficient in the DnaJ chaperonin, CAG748. These *E. coli* strains and several others that may be useful in generating intact or soluble fusion proteins are available from NEB. Utilization of these fusion proteins in precipitation assays is described below.

Functional Assays

Paxillin Overlay Assay

Paxillin has been utilized successfully in blot overlay assays either as the soluble probe or the membrane-immobilized target to detect *direct* interactions with vinculin,[4,8] the SH3 domain of c-src,[9] and focal adhesion kinase (FAK).[10,11]

Buffers

Overlay buffer: 20 mM Tris-CHl, pH 7.6; 20 mM NaCl; 0.1 mM EDTA; 0.1% 2-mercaptoethanol; 0.05% Tween 20
Tris-buffered saline: 10 mM Tris-HCl, pH 7.6; 150 mM NaCl

Procedure

The blot overlay procedure is essentially a modification of Western immunoblotting. Tissue or whole-cell lysates can be used to identify binding partners for the protein of interest. However, to enhance the sensitivity of the procedure it is often useful to first fractionate the lysate on an anion-exchange column. This permits the resolution of minor binding proteins away from highly abundant proteins of the same molecular weight that might otherwise interfere with detecting the specific interaction. Details of the fractionation are described elsewhere.[4,12]

Proteins from representative fractions or whole lysate samples are separated by SDS–PAGE and transferred to nitrocellulose. Following transfer,

[8] C. K. Wood, C. E. Turner, P. Jackson, and D. R. Critchley, *J. Cell Sci.* **107**, 709 (1994).
[9] Z. Weng, J. A. Taylor, C. E. Turner, J. S. Brugge, and C. Seidel-Dugan, *J. Biol. Chem.* **268**, 14956 (1993).
[10] M. D. Schaller and J. T. Parsons, *Mol. Cell. Biol.* **15**, 2635 (1995).
[11] K. Tachibana, T. Sato, N. D'Avirro, and C. Morimoto, *J. Exp. Med* **182**, 1089 (1995).
[12] C. E. Turner and K. Burridge, *Eur. J. Cell Biol.* **49**, 202 (1989).

the membrane is subjected to three 10-min washes in overlay buffer to remove residual SDS and methanol. The filter is then blocked for 2 hr at room temperature in overlay buffer plus 3–6% BSA.

Preparation of Iodinated Paxillin

Fifty micrograms of paxillin (50 μl of a 1 mg/ml solution) purified from chicken gizzard smooth muscle, as described above and dialyzed into TBS, is incubated in a 1.5-ml microfuge tube with one iodogen bead (Pierce, Rockford, IL) and 0.5 mCi carrier-free ^{125}iodine (NEN, Rockford, IL) for 20 min at room temperature.[4] Iodinated protein is separated from unincorporated ^{125}I by gel filtration on a Sephadex G-50 column (0.75 × 15 cm) preequilibrated in overlay buffer containing 0.1% BSA. An alternative method of labeling the probe is to express paxillin as a GST fusion protein (see above) using the pGEX-2TK vector (Pharmacia). The resultant fusion contains a consensus phosphorylation site for protein kinase A (PKA) within the amino terminus of the GST component and thus the probe can be labeled with [γ-^{32}P]ATP.[13] It should be noted that paxillin is also an *in vitro* substrate for PKA although it is not known how phosphorylation of paxillin by PKA may modulate its binding to blotted proteins (M. C. Brown and C. E. Turner, unpublished observation).

Blot Overlay

The blocked nitrocellulose filter is placed, protein side up, onto a piece of parafilm that has been laid flat onto a damp glass plate. Excess buffer is removed with a Kimwipe (Kimberly-Clarke, Roswell, CA) and the filter is overlayed with ^{125}I-labeled paxillin diluted to 7.5 × 10^5 cpm/ml in overlay buffer. A humid chamber is created by surrounding the blot with wet paper towels and covering with a glass staining dish. A lead shield should also be employed if using iodinated protein. The blot is incubated at room temperature for 2 hr. No agitation is necessary. The blot is subsequently washed for 60 min in several changes of overlay buffer, air-dried, wrapped in cling film, and exposed to X-ray film overnight at −80°.

Precipitation Assays

Paxillin fusion proteins bound to a solid support have proven to be very useful in the localization and characterization of the various protein–protein interaction motifs present within the primary amino acid sequence of paxillin. This is especially evident in the elucidation of the binding sites on

[13] K. Schmeichel and M. Beckerle, *Cell* **79,** 211 (1994).

paxillin for the focal adhesion proteins vinculin and FAK[14] as well as in the identification of the principal site of FAK phosphorylation of paxillin.[15] We have performed the following two types of precipitation assays using paxillin fusion proteins bound to Sepharose beads and incubated with cell/tissue lysates to examine protein–protein interactions and protein kinase–substrate associations.

Buffers

Lysis buffer: 50 mM Tris-HCl, pH 7.6; 50 mM NaCl; 1 mM EGTA; 2 mM MgCl$_2$; 0.1% mercaptoethanol; 1% Triton X-100; and a cocktail of protease inhibitors
Kinase buffer: 10 mM HEPES, pH 7.5; 3 mM MnCl$_2$

GST-Paxillin Precipitation Binding Assays

A lysate of embryonic day 18–20 chicken gizzard is generally prepared and used for GST–paxillin precipitation binding assays by homogenizing the tissue in 10 volumes of lysis buffer. The lysate is clarified at 15,000g for 15 min at 4°. Aliquots of lysate (1 mg of protein) are incubated with the GST–paxillin fusion protein coupled to the glutathione-Sepharose 4B beads or with GST-Sepharose 4B for 90 min, washed extensively in lysis buffer, and boiled directly in 2× SDS–PAGE sample buffer.

SDS–PAGE is performed using standard procedures utilizing either 7.5–12.5% polyacrylamide gels. Western immunoblotting is executed using monoclonal or polyclonal antibodies to detect the coprecipitation of the protein of interest. Detection is completed using radio-iodinated secondary antibodies, generated as described previously.[4,16] Alternatively, HRP-conjugated secondary antibodies (Sigma) are used in conjunction with the enhanced chemiluminescence system (ECL, Amersham, Arlington Heights, IL).

GST-Paxillin Precipitation Kinase Assays

For precipitation kinase assays, a lysate of day 10–18 embryonic chicken gizzard (or other tissues or cultured cells) is prepared by homogenizing the tissue in 10 volumes of lysis buffer. The lysate is clarified at 15,000g for 15 min at 4°. Aliquots of lysate (1 mg of protein) are incubated with various GST–paxillin fusion proteins coupled to the glutathione-Sepharose 4B beads or with GST-glutathione-Sepharose 4B (as a control) in a total vol-

[14] M. C. Brown, J. A. Perrotta, and C. E. Turner, *J. Cell Biol.* **135,** 1109 (1996).
[15] S. L. Bellis, J. T. Miller, and C. E. Turner, *J. Biol. Chem.* **270,** 17437 (1994).
[16] H. Towbin, T Staehlin, and J. Gordon, *Proc. Natl. Acad. Sci. U.S.A.* **76,** 4350 (1979).

ume of 1 ml lysis buffer for 90 min at 4°, washed extensively in lysis buffer, followed by washing with 1 ml protein kinase buffer. The kinase buffer is aspirated using a Hamilton 700 series syringe (the beads are not aspirated into the syringe) and the pellet is resuspended in 20 μl protein kinase buffer containing 10 μCi of [γ-^{32}P]ATP. The phosphorylation reaction proceeds at room temperature for 20 min with intermittent mixing and is then terminated by boiling directly in 2× SDS–PAGE sample buffer. The samples are processed by SDS–PAGE on 7.5–12.5% gels, stained with Coomassie Blue to confirm equal fusion protein loading, and analyzed by autoradiography following the drying of the gel on Whatman 3M filter paper. Subsequent phosphoamino acid analysis of phosphorylated protein bands can be performed if the gel is dried without prior fixation and staining. To simplify localization of the phophorylated proteins within the unstained gel, spots are placed on the corners of the filter paper using 1 μl of residual ^{35}S-sequencing gel sample prior to autoradiography. Alignment of the resultant autoradiogram signals with the bromophenol blue spots on the filter allows for the definitive localization of phosphorylated proteins. Regions of the dried gel corresponding to the phosphorylated proteins are excised and processed following standard protocols.[17]

Precipitation kinase assays utilizing the full-length paxillin molecule, the amino-terminal 131 amino acids, or other subdomains that include paxillin LD repeats 2 and/or 4[14] will generate an autoradiograph pattern yielding phosphorylation of a 125 kDa (focal adhesion kinase), in addition to the GST–paxillin fusion protein. Phosphoamino acid analysis of the paxillin fusion protein will yield both phosphotyrosine and phosphoserine residues indicating coprecipitation of a serine kinase that also appears to be activated *in vivo* in response to cell adhesion.[18]

Stimulation and Characterization of Paxillin Phosphorylation in Vivo

Interest in paxillin as a mediator of intracellular signaling events is based primarily on the stimulation of paxillin tyrosine phosphorylation in response to adhesion of cells to extracellular matrix[19] or following addition of mitogens to serum-starved cells.[3] While the precise function of paxillin phosphorylation remains to be determined, it is invariably accompanied by the formation of robust focal adhesions and actin-containing stress fibers and also results in direct association with other docking proteins such as

[17] P. van der Geer, K. Luo, B. M. Sefton, and T. Hunter, in "Protein Phosphorylation. A Practical Approach" (D. G. Hardie, ed.) pp. 31–59. Oxford University Press, New York, 1993.
[18] S. L. Bellis, J. A. Perrotta, M. S. Curtis, and C. E. Turner, *Biochem. J.* **325,** 375 (1997).
[19] K. Burridge, C. E. Turner, and L. Romer, *J. Cell Biol.* **119,** 893 (1992).

crk.[20] In this section we describe the basic method used to study adhesion-mediated phosphorylation utilizing rat embryonic fibroblasts (REF 52).[19] Readers interested in studying growth factor stimulated phosphorylation are directed to other reviews on this subject.[21]

Reagents. Poly-L-lysine hydrobromide, MW 70,000–150,000 (Sigma); fibronectin, 0.1% from human plasma (Sigma); phosphate buffered saline (PBS), calcium and magnesium free; trypsin-EDTA, 0.25% trypsin and 1 mM EDTA without calcium or magnesium; serum-containing medium; serum-free culture medium.

Substrate Preparation

Adhesion-dependent paxillin tyrosine phosphorylation results from integrin-mediated adhesion to extracellular matrix proteins such as fibronectin, vitronectin, and collagen. While cells will adhere to poly-L-lysine coated dishes, paxillin phosphorylation is not induced. Therefore, this serves as a useful control substrate. First, 100-mm cell culture dishes are coated with a 2-ml solution of either 0.1 mg/ml poly-L-lysine (in sterile water) or 1–50 μg/ml fibronectin (in serum-free medium), or other matrix protein, for 60 min at 37°. Excess matrix is washed from the dish with PBS, followed by a wash in serum-free medium just prior to plating the cells. If necessary, these plates can be prepared 24 hr ahead of time and stored at 4° under PBS.

Cell Preparation and Assay

Cells should be harvested when approximately 80% confluent. Cells are washed once in PBS and then released from the dish with a brief trypsin-EDTA treatment, or for less strongly adherent cells (e.g., CHO.K1) a brief incubation in a PBS–0.02% EDTA solution. The harvested cells are pooled and washed once in serum-containing medium to inactivate the trypsin and then twice in serum-free medium. The cells are then maintained in suspension in serum-free medium in a tissue culture incubator for 30–60 min. This results in a decrease in tyrosine phosphorylation levels of intracellular proteins to baseline levels. At this time the cells can be plated onto the extracellular matrix (ECM)-coated dishes or poly-L-lysine. As the cells were grown to approximately 80% confluence prior to harvesting, it is recommended that one dish of these cells be used per ECM-coated dish. This will optimize cell number while permitting sufficient room for cell spreading. This formula can be applied to all cell types regardless of their

[20] R. B. Birge, J. E. Fajardo, C. Reichman, S. E. Shoelson, Z. Songyang, L. C. Cantley, and H. Hanafusa, *Mol. Cell Biol.* **13,** 4648 (1993).
[21] E. Rozengurt, *Cancer Surveys* **24,** 81 (1995).

size. As a reference point, the time of maximal phosphorylation following plating is generally 1–3 hr for REF 52 cells. Cells should be harvested in lysis buffer containing 50 μM sodium orthovanadate to inhibit tyrosine phosphatase activity. At this stage, the sample can either be subjected to SDS–PAGE and Western blotted with anti-phosphotyrosine antibodies or first subjected to immunoprecipitation with anti-paxillin antibody. In the latter case, it is suggested that one-third of a dish of cells be used per immunoprecipitation.

Recent reports indicate that substantial increases in serine phosphorylation of paxillin also accompany cell spreading.[18,22] Unfortunately, none of the commercial anti-phosphoserine antibodies recognize this modification on paxillin (our unpublished observations). Therefore, metabolic labeling of cells using inorganic ^{32}P is recommended for these studies. Detailed procedures for labeling of cells in this manner can be found elsewhere.[17]

Acknowledgments

Work in the authors' laboratory was supported by grants from the NIHGMS and AHA.

[22] M. O. De Nichilo and K. M. Yamada, *J. Biol. Chem.* **271**, 11016 (1996).

[9] Assay and Purification of Focal Adhesion Kinase

By JUDITH LACOSTE, AMY MA, and J. THOMAS PARSONS

Introduction

Members of the integrin family are heterodimeric cell surface receptors that fulfill a dual role in mediating cell–cell and cell–extracellular matrix (ECM) interactions.[1–3] Integrin-mediated cellular adhesion and migration are essential to both physiologic (cell proliferation, survival, differentiation[4–12]) and pathologic (tumor growth, metastasis[13–16]) processes. Structur-

[1] B. M. Jockusch, P. Bubeck, K. Giehl, M. Kroemker, J. Moschner, M. Rothkegel, M. Rudiger, K. Schlüter, G. Stanke, and J. Winkler, *Annu. Rev. Cell Dev. Biol.* **11**, 379 (1995).
[2] R. O. Hynes, *Cell* **69**, 11 (1992).
[3] K. Burridge and M. Chrzanowska-Wodnicka, *Annu. Rev. Cell Dev. Biol.* **12**, 463 (1996).
[4] A. S. Clarke, M. M. Lotz, C. Chao, and A. M. Mercurio, *J. Biol. Chem.* **270**, 22673 (1995).
[5] J. Meredith, Jr., Y. Takada, M. Fornaro, L. R. Languino, and M. A. Schwartz, *Science* **269**, 1570 (1995).
[6] F. Fang, G. Orend, N. Watanabe, T. Hunter, and E. Ruoslahti, *Science* **271**, 499 (1996).
[7] X. Zhu, M. Ohtsubo, R. M. Bohmer, J. M. Roberts, and R. K. Assoian, *J. Cell Biol.* **133**, 391 (1996).

ally, integrins provide a link from the external milieu to the intracellular actin cytoskeleton. In adherent cells in culture, integrins are found in actin-rich structures called focal adhesions, which provide attachment to the underlying ECM. In addition to their structural role, integrins function as transducers of molecular signals that are generated from the exterior of the cell. Engagement of integrins by ligands (e.g., fibronectin, laminin, collagen) triggers a number of events, including tyrosine phosphorylation of intracellular proteins.[2,3] One of the major proteins phosphorylated on tyrosine following integrin ligand binding is focal adhesion kinase (FAK), a 125-kDa protein tyrosine kinase localized to focal adhesions.[17–21] Because integrins themselves lack catalytic activity, the activation of FAK on integrin occupancy suggests a role for FAK in the signaling the underlies integrin-mediated cell adhesion and migration. Indeed, the results from several experimental systems provide evidence for such a role for FAK in cell adhesion and migration. Mouse embryos deficient for FAK expression die early in development and cells derived from such early embryos exhibit reduced mobility *in vitro*.[22] The stable overexpression of FAK in Chinese hamster ovary cells stimulates cell migration on fibronectin.[23] Recent studies have shown that overexpression of FRNK (FAK-related nonkinase), the

[8] S. M. Frisch and H. Francis, *J. Cell Biol.* **124**, 619 (1994).

[9] E. Ruoslahti and J. C. Reed, *Cell* **77**, 477 (1994).

[10] A. D. Yurochko, D. Y. Liu, D. Eierman, and S. Haskill, *Proc. Natl. Acad. Sci. U.S.A.* **89**, 9034 (1992).

[11] P. Huhtala, M. J. Humphries, J. B. McCarthy, P. M. Tremble, Z. Werb, and C. H. Damsky, *J. Cell Biol.* **129**, 867 (1995).

[12] C. H. Streuli, G. M. Edwards, M. Delcommenne, C. B. Whitelaw, T. G. Burdon, C. Schindler, and C. J. Watson, *J. Biol. Chem.* **270**, 21639 (1995).

[13] B. M. Chan, N. Matsuura, Y. Takada, B. R. Zetter, and M. E. Hemler, *Science* **251**, 1600 (1991).

[14] M. Agrez, A. Chen, R. I. Cone, R. Pytela, and D. Sheppard, *J. Cell Biol.* **127**, 547 (1994).

[15] P. C. Brooks, A. M. Montgomery, M. Rosenfeld, R. A. Reisfeld, T. Hu, G. Klier, and D. A. Cheresh, *Cell* **79**, 1157 (1994).

[16] J. A. Varner, D. A. Emerson, and R. L. Juliano, *Mol. Biol. Cell* **6**, 725 (1995).

[17] L. Kornberg, H. S. Earp, J. T. Parsons, M. Schaller, and R. L. Juliano, *J. Biol. Chem.* **267**, 23439 (1992).

[18] K. Burridge, C. E. Turner, and L. H. Romer, *J. Cell Biol.* **119**, 893 (1992).

[19] L. Lipfert, B. Haimovich, M. D. Schaller, B. S. Cobb, J. T. Parsons, and J. S. Brugge, *J. Cell Biol.* **119**, 905 (1992).

[20] S. K. Hanks, M. B. Calalb, M. C. Harper, and S. K. Patel, *Proc. Natl. Acad. Sci. U.S.A* **89**, 8487 (1992).

[21] J.-L. Guan and D. Shalloway, *Nature* **358**, 690 (1992).

[22] D. Ilić, Y. Furuta, S. Kanazawa, N. Takeda, K. Sobue, N. Nakatsuji, S. Nomura, J. Fujimoto, M. Okada, T. Yamamoto, and S. Alzawa, *Nature* **377**, 539 (1995).

[23] L. A. Cary, J. F. Chang, and J.-L. Guan, *J. Cell Sci.* **109**, 1787 (1996).

autonomously expressed C-terminal domain of FAK,[24] in chick embryo fibroblasts reduces FAK tyrosine phosphorylation concomitant with a delay in cell spreading on fibronectin.[25] Similar findings were obtained by microinjecting BALB/c 3T3 cells and human umbilical vein endothelial cells with a glutathione S-transferase (GST) fusion protein containing the FAK C-terminal domain.[26] In human cancer cells, invasive tumor cells overexpress FAK relative to normal control cells.[27,28] Human cells expressing high levels of FAK lose attachment and undergo apoptosis when treated with FAK antisense oligonucleotides.[29] Taken together, these results implicate FAK in events necessary for cell spreading and migration in both normal and malignant cells.

Insights as to how FAK functions to mediate integrin signaling come from the analysis of the functions of the domains of FAK. *In vitro*, the N-terminal domain of FAK interacts with peptides corresponding to the cytoplasmic tails of the $\beta 1$, $\beta 2$, and $\beta 3$ integrin subunits.[30] FAK contains a number of phosphoacceptor tyrosine residues. Tyrosine 397 (Tyr-397) is a major autophosphorylation site and is located at the boundary of the catalytic and integrin binding domains.[31] *In vivo* and *in vitro* phosphorylation at Tyr-397 residue promotes SH2-mediated interactions with Src-family kinases (Src, Fyn) possibly generating multienzyme signaling complexes.[31–34] A recent report also implicated Tyr-397 in an SH2-dependent interaction with the regulatory subunit of phosphatidylinositol 3-kinase (PI3K, p85 subunit).[35] Phosphorylation of Tyr-926 leads to Grb2 SH2 domain binding and activation of the Ras-MAP kinase pathway.[36,37] Four other major tyrosine phosphorylation sites have been described (Tyr-407, Tyr-576, Tyr-577, and Tyr-861), two of which (Tyr-576 and Tyr-577) might

[24] M. D. Schaller, C. A. Borgman, and J. T. Parsons, *Mol. Cell. Biol.* **13,** 785 (1993).
[25] A. Richardson and J. T. Parsons, *Nature* **380,** 538 (1996).
[26] A. P. Gilmore and L. H. Romer, *Mol. Biol. Cell* **7,** 1209 (1996).
[27] L. V. Owens, L. Xu, R. J. Craven, G. A. Dent, T. M. Weiner, L. Kornberg, E. T. Liu, and W. G. Cance, *Cancer Res.* **55,** 2752 (1995).
[28] T. M. Weiner, E. T. Liu, R. J. Craven, and W. G. Cance, *Lancet* **342,** 1024 (1993).
[29] L. H. Xu, L. V. Owens, G. C. Sturge, X. H. Yang, E. T. Liu, R. J. Craven, and W. G. Cance, *Cell Growth Differ.* **7,** 413 (1996).
[30] M. D. Schaller, C. A. Otey, J. D. Hildebrand, and J. T. Parsons, *J. Cell Biol.* **130,** 1181 (1995).
[31] M. D. Schaller, J. D. Hildebrand, J. D. Shannon, J. W. Fox, R. R. Vines, and J. T. Parsons, *Mol. Cell. Biol.* **14,** 1680 (1994).
[32] B. S. Cobb, M. D. Schaller, T.-H. Leu, and J. T. Parsons, *Mol. Cell. Biol.* **14,** 147 (1994).
[33] Z. Xing, H.-C. Chen, J. K. Nowlen, S. J. Taylor, D. Shalloway, and J.-L. Guan, *Mol. Biol. Cell* **5,** 413 (1994).
[34] T. R. Polte and S. K. Hanks, *Proc. Natl. Acad. Sci. U.S.A.* **92,** 10678 (1995).
[35] H.-C. Chen, P. A. Appeddu, H. Isoda, and J.-L. Guan, *J. Biol. Chem.* **271,** 26329 (1996).
[36] D. D. Schlaepfer, S. K. Hanks, T. Hunter, and P. Van der Geer, *Nature* **372,** 786 (1994).
[37] L. C. Cantley and Z. Songyang, *J. Cell Sci. Suppl.* **18,** 121 (1994).

play a role in regulation of FAK catalytic activity.[38,39] A structural domain located within the C terminus (amino acids 904–1040) allows targeting of FAK to focal adhesions.[40,41] The C-terminal portion of FAK also directs the interaction of FAK with a number of structural and signaling molecules. The proline-rich motif $P^{712}PKP^{715}SR$ binds the SH3 domains of the Crk-associated substrate (p130^{CAS}),[42,43] and may interact with the SH3 domain of the p85 regulatory subunit of PI3K as well.[44] The proline-rich motif $P^{875}KKP^{878}PR$ has been shown to bind the SH3 domain of a novel GTPase activating protein (GAP) for the Rho family GTPases, Cdc42 and RhoA.[45] Finally, the C-terminal portion of FAK directs the interaction of FAK with the focal adhesion proteins paxillin (residues 904–1053) and talin.[41,46] The interaction of FAK with paxillin and p130^{CAS} recruits other adapter molecules to focal adhesions such as Crk,[47–49] which can also link to the Ras-MAP kinase pathway because of Crk's ability to bind C3G and SOS, two guanine nucleotide exchange proteins for Ras.[50,51] Considering the nature of the proteins recruited/activated by FAK (directly or indirectly), FAK likely plays an important role in integrin-mediated signaling.

The purpose of this article is to review methodology for measuring FAK protein expression, tyrosine phosphorylation, and kinase activity. Tables I and II describe the antibodies available for such assays. We discuss the preparation of cell lysates, assays used to measure FAK protein expression

[38] M. B. Calalb, T. R. Polte, and S. K. Hanks, *Mol. Cell. Biol.* **15,** 954 (1995).

[39] M. B. Calalb, X. Zhang, T. R. Polte, and S. K. Hanks, *Biochem. Biophys. Res. Commun.* **228,** 662 (1996).

[40] J. D. Hildebrand, M. D. Schaller, and J. T. Parsons, *J. Cell Biol.* **123,** 993 (1993).

[41] J. D. Hildebrand, M. D. Schaller, and J. T. Parsons, *Mol. Biol. Cell* **6,** 637 (1995).

[42] M. T. Harte, J. D. Hildebrand, M. R. Burnham, A. H. Bouton, and J. T. Parsons, *J. Biol. Chem.* **271,** 13649 (1996).

[43] T. R. Polte and S. K. Hanks, *Proc. Natl. Acad. Sci. U.S.A.* **92,** 10678 (1995).

[44] C. Guinebault, B. Payrastre, C. Racaud-Sultan, H. Mazarguil, M. Breton, G. Mauco, M. Plantavid, and H. Chap, *J. Cell Biol.* **129,** 831 (1995).

[45] J. D. Hildebrand, J. M. Taylor, and J. T. Parsons, *Mol. Cell. Biol.* **16,** 3169 (1996).

[46] H.-C. Chen, P. A. Appeddu, J. T. Parsons, J. D. Hildebrand, M. D. Schaller, and J.-L. Guan, *J. Biol. Chem.* **270,** 16995 (1995).

[47] R. B. Birge, J. E. Fajardo, C. Reichman, S. E. Shoelson, Z. Songyang, L. C. Cantley, and H. Hanafusa, *Mol. Cell. Biol.* **13,** 4648 (1993).

[48] R. Sakai, A. Iwamatsu, N. Hirano, S. Ogawa, T. Tanaka, H. Mano, Y. Yazaki, and H. Hirai, *EMBO J.* **13,** 3748 (1994).

[49] M. Matsuda, B. J. Mayer, and H. Hanafusa, *Mol. Cell. Biol.* **11,** 1607 (1991).

[50] S. Tanaka, T. Morishita, Y. Hashimoto, S. Hattori, S. Nakamura, M. Shibuya, K. Matuoka, T. Takenawa, K. Kurata, K. Nagashima, and M. Matsuda, *Proc. Natl. Acad. Sci. U.S.A.* **91,** 3443 (1994).

[51] M. Matsuda, Y. Hashimoto, K. Muroya, H. Hasegawa, T. Kurata, S. Tanaka, S. Nakamura, and S. Hattori, *Mol. Cell. Biol.* **14,** 5495 (1994).

TABLE I

ANTIBODIES USED FOR STUDY OF FAK PROTEIN EXPRESSION AND TYROSINE PHOSPHORYLATION

Antibody	Type	Epitope	Species cross-reactivity	Immunoprecipitation	Immunoblotting	Source
2A7	Monoclonal	Residues 904–1040 of chicken FAK	Rat, mouse, hamster, human, dog	10 μg/mg of protein A purified	1–5 μg/mol	J. T. Parsons or Upstate Biotechnology, cat# 05-182[a,b]
BC3	Polyclonal	Residues 651–1028 of chicken FAK	Chicken, rat, mouse, human, bovine	10 μl/mg	1:1000	J. T. Parsons or Upstate Biotechnology, cat# 06-446[c]
BC2	Polyclonal	Residues 311–701 of chicken FAK	Chicken, others not determined	20–30 μl/mg	1:500	J. T. Parsons[c]
HUB3	Polyclonal	Residues 542–880 of human FAK	Human, chicken, bovine	10 μl/mg	1:1000	J. T. Parsons[d]
RC-20:HRPO	Recombinant	Phosphorylated tyrosine residues	Human, dog, rat, mouse, chicken, frog	NA	1:2500	Transduction Laboratories, cat# E120H[e,f]

[a] J. D. Hildebrand, M. D. Schaller, and J. T. Parsons, *Mol. Biol. Cell* **6**, 637 (1995).
[b] S. B. Kanner, A. B. Reynolds, R. R. Vines, and J. T. Parsons, *Proc. Natl. Acad. Sci. U.S.A.* **87**, 3328 (1990).
[c] M. D. Schaller, C. A. Borgman, B. S. Cobb, R. R. Vines, A. B. Reynolds, and J. T. Parsons, *Proc. Natl. Acad. Sci. U.S.A.* **89**, 5192 (1992).
[d] M. D. Schaller and J. T. Parsons, unpublished.
[e] S. Ruff-Jamison, K. Chen, and S. Cohen, *Science* **261**, 1733 (1993).
[f] B. R. Sevetson, X. Kong, and J. C. Lawrence, Jr., *Proc. Natl. Acad. Sci. U.S.A.* **90**, 10305 (1993).

TABLE II
COMMERCIALLY AVAILABLE FAK ANTIBODIES

Antibody	Type	Epitope	Species cross-reactivity	Immunoprecipitation	Immunoblotting	Source
FAK	Monoclonal	Residues 354–533 of chicken FAK	Human, dog, rat, mouse, chicken	+ (denat.)	+	Transduction Laboratories, cat# F15020[a,b,c]
FAK (A-17)	Polyclonal	Residues 2–18 of human FAK	Mouse, rat, human, chicken	+	+	Santa Cruz Biotechnology, cat# sc-557
FAK (C-20)	Polyclonal	Residues 1033–1052 of human FAK	Mouse, rat, human, chicken	+	+	Santa Cruz Biotechnology, cat# sc-558
FAK (C-903)	Polyclonal	Residues 903–1052 of mouse FAK	Mouse, rat, human	+	+	Santa Cruz Biotechnology, cat# sc-932
FAK (H-1)	Monoclonal	Residues 2–18 of human FAK	Human	–	+	Santa Cruz Biotechnology, cat# sc-1688
FAK	Polyclonal	N-D	Human, mouse	+	–	Chemicon, cat# AB1605
FAK	Monoclonal	N-D	Human, mouse, hamster, chicken	+	+	Chemicon, cat# MAB2156
FAK	Polyclonal	Residues 748–1053 of human FAK	Human, mouse, rat, hamster	+	+	Upstate Biotechnology, cat# 06-543

[a] T. Seufferlein, and E. Rozengurt, *J. Biol. Chem.* **269**, 27610 (1994).
[b] S. Greenberg, P. Chang, and S. C. Silverstein, *J. Biol. Chem.* **269**, 3987 (1994).
[c] C. Zhang, M. P. Lambert, C. Bunch, K. Barber, W. S. Wade, G. A. Krafft, and W. L. Klein, *J. Biol. Chem.* **269**, 25247 (1994).

and tyrosine phosphorylation, and protocols for measuring FAK kinase activity. The expression and purification of FAK from baculovirus expression system is described as well as methods for detection of FAK in intact cells using immunofluorescence.

Preparation of Cell Lysates

FAK has been detected in a variety of cell types,[20,52,53] and endogenous levels of FAK are easily detected using the assays described later. Because of its linkage to integrins and cell adhesion, FAK is usually examined in adherent cells. Subconfluent cultures of adherent cells are placed on ice and washed twice in CMF-PBS [calcium, magnesium-free phosphate-buffered saline, 137 mM sodium chloride (NaCl), 2.7 mM potassium chloride (KCl), 8.0 mM disodium phosphate (Na$_2$HPO$_4$ · 7H$_2$O), 1.4 mM potassium phosphate (KH$_2$PO$_4$, pH 7.2)] and lysed by scraping in supplemented radioimmunoprecipitation (S-RIPA) lysis buffer [50 mM N-2-hydroxyethylpiperazine-N'-2-ethanesulfonic acid (HEPES, pH 7.2), 150 mM NaCl, 2 mM ethylenediaminetetraacetic acid (EDTA), 1% (v/v) Nonidet P-40 (NP-40), 0.5% (w/v) sodium deoxycholate (DOC, 5β-cholan-24-oic acid-3α, 12α-diol, C$_{24}$H$_{39}$O$_4$Na), 100 μM leupeptin, 10 μM pepstatin, 0.05 TIU/ml aprotinin, 1 mM phenylmethylsulfonyl fluoride (C$_7$H$_7$FO$_2$S, PMSF), 1 mM benzamidine (C$_7$H$_8$N$_2$ · HCl), 1 mM sodium orthovanadate (Na$_3$VO$_4$), 10 mM sodium pyrophosphate (Na$_4$P$_2$O$_7$ · 10H$_2$O), 40 mM sodium p-nitrophenyl phosphate (C$_6$H$_4$NO$_6$PNa$_2$ · 6H$_2$O), 40 mM sodium fluoride (NaF)].[54,55] Leupeptin, pepstatin, aprotinin, PMSF, and benzamidine are used as protease inhibitors. Sodium orthovanadate, sodium pyrophosphate, sodium p-nitrophenyl phosphate, and sodium fluoride are used to inhibit tyrosine and serine/threonine phosphatases. Usually, 1 ml of S-RIPA lysis buffer is used per 100-mm plate (approximately 10^7 cells). Lysates are cleared by centrifugation in a microfuge (e.g., Eppendorf 5415C) at 4° for 10 min at 16,000g to remove cellular debris and nuclei, and the supernatant is collected in a 1.5-ml microfuge tube. Protein concentrations are determined by the bicinchoninic assay (BCA, Pierce, Rockford, IL). Generally, 2–4 mg of lysate protein is obtained per 10^7 cells.

Cells in suspension (either adherent cells placed in suspension by trypsinization or treatment with 1–10 mM EDTA, or nonadherent cells such as lymphocytes) are collected by centrifugation, washed twice with CMF-

[52] E. André and M. Becker-André, *Biochem. Biophys. Res. Commun.* **190,** 140 (1993).
[53] C. E. Turner, M. D. Schaller, and J. T. Parsons, *J. Cell Sci.* **105,** 637 (1993).
[54] S. B. Kanner, A. B. Reynolds, and J. T. Parsons, *J. Immunol. Methods* **120,** 115 (1989).
[55] A. Richardson, R. K. Malik, J. D. Hildebrand, and J. T. Parsons, *Mol. Cell. Biol.* **17,** 6906 (1997).

PBS, and resuspended in S-RIPA lysis buffer (1 ml per 10^7 cells). Lysates are then cleared by centrifugation as above. For cell adhesion/spreading experiments, cells are first placed in suspension (by trypsinization or treatment with EDTA) in order to inactivate FAK. The cells are then plated on bacterial culture plates (untreated or ECM-coated), and lysed in S-RIPA lysis buffer as described above. These lysates are suitable for immunoprecipitations or direct immunoblots.

For the preparation of whole-cell lysates, cells may be placed directly in 1× SDS sample buffer [62.5 mM tris(hydroxymethyl)aminomethane hydrochloride (Tris-HCl, pH 6.8), 10% (v/v) glycerol, 1% (w/v) sodium dodecyl sulfate (SDS), 1% (v/v) 2-mercaptoethanol, 5 μg/ml bromphenol blue]. For adherent cells, proceed as described above except that cells are scraped in 1 ml of 1× SDS sample buffer and boiled for 10 min. To prepare lysates from suspension cells, proceed as above and resuspend the cell pellets in 1 ml 1× SDS sample buffer and boil for 10 min. These lysates can be used for direct immunoblot analyses. Because of the high concentration of SDS, these lysates are not used for immunoprecipitations.

Detection of FAK Protein Expression and Tyrosine Phosphorylation

Antibodies

Table I provides a summary of antibodies routinely used in our laboratory in the assays described below. Commercially available antibodies to FAK are listed in Table II. Monoclonal antibody 2A7 (Upstate Biotechnology, Lake Placid, NY) was isolated from a screen of monoclonal antibodies derived from mice immunized with partly purified tyrosine phosphorylated proteins from Src-transformed cells.[56] BC3 (Upstate Biotechnology) is a rabbit polyclonal antibody raised against a bacterial TrpE fusion protein containing residues 651–1020 of chicken FAK, which correspond to the C-terminal domain.[57] BC2 antibody is a rabbit polyclonal antisera prepared by immunization with a TrpE fusion protein containing chicken FAK residues 311–701, which correspond to the kinase domain and part of the N-terminal domain.[55] HUB3 is a polyclonal generated against a GST-fusion protein containing residues 542–880 of human FAK. This segment encompasses part of the kinase and C-terminal domains (M. D. Schaller and J. T. Parsons, unpublished).

---

[56] S. B. Kanner, A. B. Reynolds, R. R. Vines, and J. T. Parsons, *Proc. Natl. Acad. Sci. U.S.A.* **87**, 3328 (1990).

[57] M. D. Schaller, C. A. Borgman, B. S. Cobb, R. R. Vines, A. B. Reynolds, and J. T. Parsons, *Proc. Natl. Acad. Sci. U.S.A.* **89**, 5192 (1992).

FAK Immunoprecipitation/Immunoblotting

For immunoprecipitation of FAK, 0.5–1.5 mg of lysate protein is incubated with the FAK-specific antibody for 1 hr at 4° using constant rotation. For monoclonal 2A7, 10 μg of protein A-purified antibody per milligram of lysate is used. For polyclonal sera, 10 μl of BC3, 30 μl of BC2, or 10 μl of HUB3 is used per milligram of cell lysate protein. Immune complexes (IC) are recovered by the addition of 100 μl of protein A-Sepharose beads (1:1 slurry, Pharmacia, Piscataway, NJ) followed by incubation for another hour at 4°, while rotating. When immunoprecipitations with antibody 2A7 are performed, the beads are precoated with 20 μg of donkey anti-mouse immunoglobulin G (IgG, Jackson ImmunoResearch Laboratories, Bar Harbor, MN). IC are collected by centrifugation in a microfuge at 16 000g for 1 min, and then washed twice with 1 ml of ice-cold S-RIPA buffer and twice with 1 ml of ice-cold TBS [Tris-buffered saline, 50 mM Tris-HCl (pH 7.4), 150 mM NaCl]. For each wash, the IC are centrifuged as above and resuspended by flicking the microtube. After the washes, IC are recovered by centrifugation in a microfuge and solubilized from the Sepharose beads by boiling for 10 min in an equal volume (usually 50 μl) of 2× SDS sample buffer.

The solubilized proteins are resolved by electrophoresis on an 8% SDS polyacrylamide gel [0.8% cross-link N, N'-bismethylene acrylamide, 0.375 M Tris-HCl (pH 8.8), 0.1% (w/v) SDS] run at 6 mA for 16 hr in 1× SDS running buffer [25 mM Tris-HCl (pH 8.3), 0.2 M glycine, 0.1% (w/v) SDS]. For direct immunoblot, 5–50 μg of lysates are loaded on the same type of gel and electrophoresis is carried out as above. The amount of lysate used depends on the levels of expression (i.e., endogenous versus overexpressed). Resolved proteins are transferred to a nitrocellulose membrane in transfer buffer [25 mM Tris-HCl (pH 8.3), 0.2 M glycine, 0.1% (w/v) SDS, 20% (v/v) methanol] for 1.2 hr at 22 V at 4°. The membrane is then blocked for 1 hr at room temperature in 5% (w/v) instant nonfat dry milk (we prefer Super G™, the Giant™ supermarket brand)/TBST [50 mM Tris-HCl (pH 7.4), 150 mM NaCl, 0.1% (v/v) polyoxyethylenesorbitan monolaurate (Tween-20)] and immunoblotted for another hour in fresh blotting solution containing the primary anti-FAK antibody. For immunoblotting, antibody 2A7, BC3, or HUB3 is used at a dilution of 1:1000 in 25 ml of blotting solution. Antibody BC2 is diluted 1:500 in 25 ml of blotting solution. Following incubation with the primary antibody, the membrane is then quickly washed twice with TBST and incubated for another hour at room temperature with the secondary antibody (donkey anti-rabbit IgG, Amersham, Arlington Heights, IL, or sheep antimouse IgG, Amersham) linked to horseradish peroxidase (HRP) in 5% (w/v) instant nonfat dry milk/

TBST at a 1:1000 dilution (25 ml total volume). HRP-antibody binding is visualized with enhanced chemiluminescence (ECL) following the manufacturer's instructions (Amersham).

In some tissues and cells, the C terminus of FAK is expressed autonomously as a 41/43-kDA protein named FRNK.[24] FRNK is expressed endogenously in chick embryo fibroblasts,[24] and in rat and mouse embryo fibroblasts (A. Richardson and J. T. Parsons, unpublished observations). FRNK mRNA has also been detected in heart, brain, and lung tissues (A. Richardson and J. T. Parsons, unpublished observations). For optimal detection of FRNK, 5–50 μg of lysates are resolved by SDS–PAGE (SDS–polyacrylamide gel electrophoresis), transferred to nitrocellulose as described above and blotted with BC3 antibody diluted 1:1000 (25 ml total volume) in 5% (w/v) instant nonfat dry milk/TBS in the absence of Tween-20.

Detection of FAK Tyrosine Phosphorylation

To assess FAK tyrosine phosphorylation, cell lysates are first immunoprecipitated with antibodies BC3 or 2A7 described above. IC are recovered and divided equally. Each aliquot of IC is subjected to SDS–PAGE, and transferred to nitrocellulose as described in the previous section. One membrane is blotted for FAK (see above) and serves as a loading control. Detection of tyrosine phosphorylated FAK is carried out by blocking the second membrane in 1% (w/v) bovine serum albumin (BSA)/TBST for at least 20 min at 37°, followed by incubation with the phosphotyrosine antibody RC-20:HRPO (Transduction Laboratories, Lexington, KY) diluted 1:2500 (0.1 μg/ml) in 1% (w/v) BSA/TBST for at least 20 min at 37°. The membrane is washed three times with TBST at room temperature for 10 min, and tyrosine phosphorylated FAK is detected by enhanced chemiluminescence.[58] Note that RC-20:HRPO is already conjugated with horseradish peroxidase and therefore detection does not require incubation with a conjugated secondary antibody.

Detection of FAK Kinase Activity in Vitro

Assay for Autophosphorylation Activity

For the detection of FAK autophosphorylation activity, approximately 0.5–1.5 mg of S-RIPA lysates is immunoprecipitated using antibodies BC3 or 2A7. The resulting IC are recovered with protein A-Sepharose beads, washed twice in S-RIPA lysis buffer and twice in 20 mM piperazine-N,N'-

[58] M. D. Schaller and J. T. Parsons, *Mol. Cell Biol.* **15,** 2635 (1995).

bis(2-ethanesulfonic acid) (PIPES, pH 7.2), 3 mM manganese chloride (MnCl$_2$). The wash buffer is carefully removed and the beads resuspended in 40 μl of 20 mM PIPES (pH 7.2), 3 mM MnCl$_2$, 100 μM ATP, and 10 μCi [γ-^{32}P]ATP (3000 Ci/mmol, DuPont–NEN, Boston, MA) for 15 min at room temperature. An equal volume of 2× SDS sample buffer is added and the samples are boiled for 10 min and labeled FAK is resolved by SDS–PAGE. The gel is dried and signals detected by autoradiography or by PhosphorImager analysis.[21,31,55] In this assay FAK autophosphorylation activity is proportional to the amount of FAK present in the immune complex.

Paxillin and Poly(Glu, Tyr) Substrates

FAK catalytic activity can also be assayed using heterologous substrates. A physiologically relevant substrate is paxillin, an adapter molecule found in focal adhesions. Alternatively, a synthetic substrate may also be used to assay FAK activity. FAK activity toward paxillin is measured *in vitro* by incubating 1 μg of purified GST–paxillin fusion protein with the IC recovered from the FAK immunoprecipitation as described above. The GST–paxillin fusion protein contains the first 1000 N-terminal residues of human paxillin.[59] The kinase reaction conditions are identical to those described for detection of FAK autokinase activity. Phosphorylated GST–paxillin is detected by resolving the reaction samples on SDS–PAGE, followed by autoradiography or PhosphorImager analysis. For measuring FAK kinase activity with a synthetic peptide substrate, FAK immunoprecipitates are incubated with 0.4 μg of poly(Glu, Tyr) 4:1 (Sigma, 1 mg/ml stock solution prepared in kinase buffer) in the kinase buffer described above for 20 min at 30°. Reactions are terminated by adding an equal volume of ice-cold 0.5 M EDTA. The beads are then centrifuged and half the supernatant (75 μl) is spotted to 3MM Whatman paper squares (10 × 10 mm^2). These are washed three times in ice-cold 10% (w/v) trichloroacetic acid, 10 mM disodium phosphate, 10 mM sodium pyrophosphate by swirling, dried, and the radioactivity is quantified by liquid scintillation counting.[31,60]

Production of FAK in the Baculovirus Expression System

Full-length FAK has not been successfully produced in bacterial expression systems. However, the following baculovirus expression system has

[59] R. Salgia, J.-L. Li, S. H. Lo, B. Brunkhorst, G. S. Kansas, E. S. Sobhany, Y. Sun, E. Pisick, M. Hallek, T. Ernst, R. Tantravahi, L. B. Chen, and J. D. Griffin, *J. Biol. Chem.* **270,** 5039 (1995).

[60] J. E. Peterson, T. Jelinek, M. Kaleko, K. Siddle, and M. J. Weber, *J. Biol. Chem.* **269,** 27315 (1994).

been used to produce large quantities of full-length FAK for biochemical studies.

Construction and Expression of Vectors Encoding FAK

Spodoptera frugiperda Sf9 and Sf21 cells (obtained from Pharmingen, San Diego, CA or Invitrogen, Carlsbad, CA) are maintained in Grace's insect cell medium (GIBCO BRL, Gaithersburg, MD) supplemented with 10% fetal bovine serum, lactalbumin hydrolysate, TC yeastolate, and glutamine (complete Grace's medium). High Five™ cells are grown in Excell-400 serum-free medium (JRH Biosciences, Lenexa, KS). Linearized baculoviral DNA (BaculoGold) and the baculovirus transfer vector pVL1393 were obtained from Pharmingen.

The chicken FAK cDNA[57] was subcloned from pBluescript (pBS-FAK) into the baculovirus transfer vector. To facilitate subcloning of the FAK cDNA into the baculovirus transfer vector, a *Bam*HI site was introduced into the FAK cDNA 35 base pairs 5′ to the translation initiation codon. The *Bam*HI site was introduced through polymerase chain reaction (PCR) amplification of the 5′ region of FAK including the initiator ATG extending to nucleotide 525. The PCR amplified fragment, which includes a unique internal *Kpn*I site, was sequenced to confirm fidelity of amplification and subcloned into pBS-FAK using the *Bam*HI site in the multiple cloning site of pBS and the internal *Kpn*I site. The FAK cDNA was then inserted into the *Bam*HI and *Eco*RI sites of the baculovirus transfer vector pVL1393. The resulting recombinant transfer plasmid, pBF5, was cotransfected into Sf9 cells with baculoviral genomic DNA using the BaculoGold transfection kit (Pharmingen). Three days after transfection, viral supernatants were collected and recombinant viruses were subjected to three rounds of plaque purification.[61] Recombinant viruses expressing high levels of FAK were further amplified to generate working viral stocks, which were titered by limiting dilution.

Production of Baculoviral FAK Protein

Stocks of baculovirus containing the FAK cDNA are used to infect Sf9 and High Five cells. FAK expression was found to be about 20-fold more efficient in High Five cells as compared to Sf9 cells. For small-scale expression, High Five cells are cultured and infected on tissue culture plastic as adherent monolayers. For large-scale production of protein, High Five cells are seeded in spinner cultures at a density of 0.5×10^6 cells per ml in

---

[61] D. R. O'Reilly, L. K. Miller, and V. A. Luckow, "Baculovirus Expression Vectors: A Laboratory Manual," pp. 124–132. W. H. Freeman and Company, New York, 1992.

Excell-400 medium and infected with virus 24 hr later at a multiplicity of infection (MOI) of 5 plaque forming units (pfu) per cell. Optimal expression of FAK was observed at 48 hr postinfection, based on Western blot analysis with the polyclonal antiserum BC3.

Partial Purification of Baculoviral FAK

The following method provides partial purification of enzymatically active FAK. Infected cells are collected by centrifugation, washed once in CMF-PBS, and resuspended in NP-40 lysis buffer [20 mM Tris-HCl (pH 7.5), 137 mM NaCl, 0.5 mM EDTA, 0.5 mM ethyleneglycol-bis (β-amino-ethyl ether) N,N,N',N'-tetraacetic acid (EGTA), 10% (v/v) glycerol, 1% (v/v) NP-40, 1 mM Na$_3$VO$_4$, 1 mM PMSF, 0.05 TIU/ml aprotinin, 100 μM leupeptin, 1 mM benzamidine] at 4°. The cells are lysed by Dounce homogenization on ice and the lysate is centrifuged at $\geq$20,000g at 4° for 30 min. The supernatant is filtered through a 0.22-μM filter, and loaded onto a Superdex 200 sizing column (Pharmacia) previously equilibrated with one column volume of deionized/distilled water followed by three column volumes of gel filtration buffer [20 mM Tris-HCl (pH 7.5), 137 mM NaCl, 0.5 mM EDTA, 10% (v/v) glycerol, 1 mM DL-dithiothreitol (DTT), 1 mM benzamidine]. Western blot analysis revealed that the majority of FAK elutes in a broad peak with a molecular weight range of 160,000–670,000.

The fractions collected from the Superdex 200 column containing FAK are pooled and dialyzed against buffer A [50 mM Tris-HCl (pH 7.5), 0.5 mM EDTA, 0.5 mM EGTA, 1 mM DTT, 0.1 mM PMSF, 1 mM benzamidine] and applied to a Mono Q column. Following loading of the sample, the column is washed with 30 column volumes of buffer A and eluted with a 2 column volume salt gradient ranging from buffer A to buffer A with 1 M NaCl. The bulk of FAK elutes in the range of 210–275 mM NaCl as detected by Western blotting. After these two chromatographic steps, FAK can also be visualized by Coomassie Blue staining of the fractions collected from the Mono Q column and appeared to be one of approximately 12–15 major proteins remaining. FAK produced in the baculovirus system is enzymatically active as determined by its ability to autophosphorylate and to phosphorylate exogenous substrates (see above).

FAK Immunofluorescence

This protocol is designed to detect FAK in well spread cells (e.g., fibroblasts or epithelial cells). Cells are seeded on glass coverslips [either uncoated or coated with fibronectin (1–5 μg/cm^2) or with poly-L-Lysine (0.2–

0.5 mg/ml) as a control], and grown overnight or for various periods of time. The coverslips containing adherent cells are washed in CMF-PBS twice and cells are fixed in 3% (w/v) paraformaldehyde/PBS (see below for preparation) for 10 min at room temperature. The coverslips are then rinsed with CMF-PBS and the cells are permeabilized in CMF-PBS containing 0.5% (v/v) Triton X-100 for 5 min at room temperature. Coverslips are then rinsed three times with CMF-PBS and incubated with monoclonal antibody 2A7 (10 μg/ml) for 60 min at room temperature. Coverslips are rinsed and incubated with 5 μg/ml goat anti-mouse IgG (Jackson ImmunoResearch Laboratories) for 60 min at room temperature. Finally, coverslips are incubated with 5 μg/ml fluorescein isothiocyanate (FITC)-conjugated donkey anti-goat IgG (Jackson ImmunoResearch Laboratories) for 60 min at room temperature. Samples are visualized with a fluorescence microscope.[57] Alternatively, the polyclonal antibody BC3 antibody can be used at a 1:500 dilution, followed by an incubation with 5 μg/ml FITC-conjugated goat anti-rabbit IgG (Jackson ImmunoResearch Laboratories). Note that both 2A7 and BC3 recognize phosphorylated and unphosphorylated forms of FAK and therefore no distinction between active and inactive FAK using immunofluorescence can be done.

For the preparation of the paraformaldehyde solution, 70 ml of deionized/distilled water is heated to 60°. In a fume hood, 3 g of paraformaldehyde is added to the warm water and mixed on a stirring plate. To solubilize the paraformaldehyde, one to two drops of 1 N NaOH are added and the solution stirred for 20–30 min. The mixture is removed from the heat and 10 ml of 10× CMF-PBS are added. The pH of the solution is adjusted to pH 7.4 with HCl and deionized/distilled water is added to a final volume of 100 ml. The paraformaldehyde solution can be aliquoted and stored at −20° for about 1 month.

Acknowledgments

JTP is supported by grants from the DHHS-National Cancer Institute, CA29243, CA40042, and the Council for Tobacco Research USA, Inc. JL is a recipient of a Human Frontier Science Program fellowship. We thank R. B. Adams, C. A. Borgman, M. T. Harte, R. K. Malik, and W.-C. Xiong for sharing their versions of the protocols.

[10] Purification and Assays of Vasodilator-Stimulated Phosphoprotein

By Thomas Jarchau, Thomas Mund, Matthias Reinhard, and Ulrich Walter

Introduction

The vasodilator-stimulated phosphoprotein (VASP) was discovered and initially characterized as a 46/50-kDa substrate of cAMP- and cGMP-dependent protein kinases (cAK, cGK) in human platelets.[1-4] VASP is phosphorylated in intact platelets in response to cyclic nucleotide regulating vasodilators such as cAMP-elevating prostaglandins and cGMP-elevating NO donors, and its phosphorylation closely correlates with platelet inhibition.[1,5-7] Three distinct phosphorylation sites were identified in VASP (serine-157, serine-239, and threonine-278), which are phosphorylated *in vitro* and in intact human platelets by both cyclic nucleotide-dependent protein kinases.[4] Phosphorylation of serine-157 leads to a marked shift in apparent molecular mass of VASP in sodium dodecyl sulfate–polyacrylamide gel electrophoresis (SDS–PAGE) from 46 to 50 kDa.[2,4] This mobility shift has been widely used as an indicator of cyclic nucleotide-dependent protein kinase activity in various intact cell systems.[8] Phosphorylation of serine-157 correlates particularly well with inhibition of fibrinogen binding to integrin $\alpha_{IIb}\beta_3$ of human platelets.[7] Molecular cloning of the human and canine VASP cDNAs and of the murine VASP gene predicted highly homologous proteins of 380, 384, and 376 amino acids, respectively, and revealed a proline-rich protein that is organized into three structural segments of locally different amino acid composition and

---

[1] R. Waldmann, M. Nieberding, and U. Walter, *Eur. J. Biochem.* **167,** 441 (1987).

[2] M. Halbrügge and U. Walter, *Eur. J. Biochem.* **185,** 41 (1989).

[3] M. Eigenthaler, C. Nolte, M. Halbrügge, and U. Walter, *Eur. J. Biochem.* **205,** 471 (1992).

[4] E. Butt, K. Abel, M. Krieger, D. Palm, V. Hoppe, J. Hoppe, and U. Walter, *J. Biol. Chem.* **269,** 14509 (1994).

[5] M. Halbrügge, C. Friedrich, M. Eigenthaler, P. Schauzenbächer, and U. Walter, *J. Biol. Chem.* **265,** 3088 (1990).

[6] C. Nolte, M. Eigenthaler, P. Schanzenbächer, and U. Walter, *J. Biol. Chem.* **266,** 14808 (1991).

[7] K. Horstrup, B. Jablonka, P. Hönig-Liedl, M. Just, K. Kochsiek, and U. Walter, *Eur. J. Biochem.* **225,** 21 (1994).

[8] U. Walter, M. Eigenthaler, J. Geiger, and M. Reinhard, *Adv. Exp. Med. Biol.* **344,** 237 (1993).

sequence complexity.[9,10] VASP is the founding member of a new family of proline-rich proteins, which includes Enabled (Ena), a dose-dependent suppressor of *Drosophila Abl-* and *Disabled*-dependent phenotypes,[11] its mammalian homolog Mena, and the Ena-VASP-like protein Evl.[12] These proteins all share a tripartite overall domain organization, comprising highly homologous N-terminal and C-terminal domains, which are separated by low complexity regions including a proline-rich central segment of 60–90 amino acids in length.[12,13] In platelets where the highest VASP levels are found as well as in a wide variety of other cells including vascular smooth muscle cells, endothelial cells, and fibroblasts, VASP has been found to be associated with stress fibers, focal adhesions, and highly dynamic membrane regions.[14] VASP colocalizes with profilin and directly binds to its poly(L-proline)-binding site.[15] VASP also binds to and colocalizes with porcine and human zyxin.[16] ActA, a surface protein of the bacterial pathogen *Listeria monocytogenes*, which exploits an actin polymerization-based system to promote its intracellular motility, also binds to VASP.[17] In cells infected with this bacterium, VASP is found at the interface between the moving bacterium and its actin tail, which is the site where actin polymerization is thought to take place.[17] Functional evidence indicates that VASP is a crucial factor involved in the enhancement of spatially confined actin filament formation (for a review, see Reinhard *et al.*[13]). Furthermore, listerial ActA and zyxin share a proline-

[9] C. Haffner, T. Jarchau, M. Reinhard, J. Hoppe, S. M. Lohmann, and U. Walter, *EMBO J.* **14**, 19 (1995).

[10] M. Zimmer, T. Fink, L. Fischer, W. Hauser, K. Scherer, P. Lichter, and U. Walter, *Genomics* **36**, 227 (1996).

[11] F. B. Gertler, A. R. Comer, J.-L. Juang, S. M. Ahern, M. J. Clark, E. C. Liebl, and F. M. Hoffmann, *Genes Dev.* **9**, 521 (1995).

[12] F. B. Gertler, K. Neibuhr, M. Reinhard, J. Wehland, and P. Soriano, *Cell* **87**, 227 (1996).

[13] M. Reinhard, T. Jarchau, K. Reinhard, and U. Walter, *in* "Guidebook to the Cytoskeletal and Motor Proteins" (T. Kreis, and R. Vale, eds.). Oxford University Press, Oxford and New York, 1998.

[14] M. Reinhard, M. Halbrügge, U. Scheer, C. Wiegand, B. M. Jockusch, and U. Walter, *EMBO J.* **11**, 2063 (1992).

[15] M. Reinhard, K. Giehl, K. Abel, C. Haffner, T. Jarchau, V. Hoppe, B. M. Jockusch, and U. Walter, *EMBO J.* **14**, 1583 (1995a).

[16] M. Reinhard, K. Jouvenal, D. Tripier, and U. Walter, *Proc. Natl. Acad. Sci. U.S.A.* **92**, 7956 (1995b).

[17] T. Chakraborty, F. Ebel, E. Domann, K. Niebuhr, B. Gerstel, S. Pistor, C. J. Temm-Grove, B. M. Jockusch, M. Reinhard, U. Walter, and J. Wehland, *EMBO J.* **14**, 1314 (1995).

rich motif with vinculin,[18,19] which also binds VASP via this motif although with a somewhat different binding characteristic.[20]

The following sections describe methods for the purification and characterization of biologically active VASP from platelets and after heterologous expression from *Spodoptera frugiperda* cells infected with recombinant baculovirus.

Purification of VASP from Human or Porcine Platelets

VASP can be purified[2,21] with a typical yield of several hundred micrograms starting from a 3-liter human platelet concentrate containing 10^9 cells/ml. Such fresh human platelet concentrates may be obtained in medical centers which treat certain patients with pathologic thrombocytosis by therapeutic thrombocytapheresis. All manipulations with intact platelets are done at room temperature using siliconized glass or polypropylene plasticware unless stated otherwise. Platelet-rich plasma (PRP) is prepared by centrifugation at $25°$ for 15 min at $290g$. After addition of EDTA (final concentration, 1 mM) platelets are pelleted from PRP by centrifugation at $25°$ for 15 min at $900g$, resuspended in the original volume of buffer A (4.3 mM KH_2PO_4–K_2HPO_4, 28.7 mM NaH_2PO_4–Na_2HPO_4, pH 6.5, 113 mM NaCl, 5.5 mM glucose, and 1 mM EDTA), and washed twice in this buffer with an incubation for 30 min at room temperature in between. All subsequent manipulations are done at $4°$. Pelleted platelets of the last washing step are resuspended to yield a concentration of 10^{10} cells/ml in ice-cold buffer B, pH 7.6 [20 mM NaH_2PO_4–NaOH, 2 mM EGTA, 2 mM EDTA, 1 phenylmethylsulfonyl fluoride (PMSF), and 100 U/ml aprotinin], incubated on ice for 30 min, and rapidly frozen in liquid nitrogen. The frozen platelet suspension is thawed at room temperature in a water bath, and thawed material is immediately decanted and collected on ice. A particulate fraction from this lysate is then prepared by centrifugation for 60 min at $100,000g$. After resuspension of the particulate material equivalent to a concentration of 2.5×10^{10} platelets/ml in lysis buffer containing 250 mM NaCl, this fraction is extracted for 30 min under constant stirring at $4°$ and then centrifuged at $100,000g$ for 60 min. The supernatant is dialyzed for a total of 4 hr against 3×2 liters of buffer B (pH 7.5). Insoluble protein is removed by centrifugation ($1500g$ for 5 min). The dialyzed extract is incu-

[18] E. Domann, J. Wehland, M. Rohde, S. Pistor, M. Hartl, W. Goebel, M. Leimeister-Wächter, M. Wuenscher, and T. Chakraborty, *EMBO J.* **11,** 1981 (1992).
[19] K. Niebuhr, F. Ebel, R. Frank, M. Reinhard, E. Domann, U. Carl, U. Walter, F. B. Gertler, J. Wehland, and T. Chakraborty, *EMBO J.* **16,** 5433 (1997).
[20] M. Reinhard, M. Rüdiger, B. M. Jockusch, and U. Walter, *FEBS Lett.* **399,** 103 (1996).
[21] M. Halbrügge and U. Walter, *J. Chromatogr.* **521,** 335 (1990).

bated for 30 min in a batch procedure with 140 ml Q-Sepharose FF (Pharmacia Piscataway, NJ) preequilibrated with buffer B (pH 7.5). The anion exchanger is removed by filtration through gauze. The VASP containing filtrate is then applied (flow rate of 3 ml/min) to an Orange A (Amicon, Witten, FRG) column (8 cm × 1.3 cm^2) that has been preequilibrated with buffer B without protease inhibitors. After washing the column in this buffer until the baseline of the recorded absorbance remains constant, bound protein is eluted in a segmented linear gradient of 0–1 M NaCl in buffer B (pH 7.5), without protease inhibitors (from 0 to 100 mM in 10 ml, from 100 to 300 mM in 110 ml, from 300 mM to 1 M in 20 ml followed by 1 M salt) in a total gradient elution volume of 440 ml using a flow rate of 1.5 ml/min. VASP-containing fractions are identified either by Western blot analysis and detection by ECL (Amersham, Arlington Heights, IL) or by cGK-dependent phosphorylation of 5-μl aliquots in the presence of [γ-^{32}P]ATP (see below). Positive fractions are pooled, dialyzed overnight against buffer B (pH 6.3), and are subsequently applied to a Mono S HR 5/5 (Pharmacia) column that has been equilibrated with buffer B (pH 6.3). The column is washed with buffer B (pH 6.3), and elution is performed with a segmented linear gradient of 0–1 M NaCl (from 0 to 80 mM in 5 ml, from 80 to 250 mM in 40 ml, from 250 mM to 1 M in 2 ml followed by 20 ml at 1 M salt and 20 ml salt-free buffer) in buffer B (pH 6.3) at a flow rate of 0.75 ml/min. The purity of the VASP preparation can be assessed by Coomassie Brilliant Blue staining after SDS–PAGE. Depending on the quality of the preparation, VASP-containing fractions may be chromatographed after dialysis (10 mM NaH$_2$PO$_4$–NaOH, pH 7.5, 75 mM NaCl, 4 mM EDTA, 0.5 mM DTT) on a hydroxylapatite column (Bio-Gel HT, Bio-Rad, Richmond, CA; batch characteristics: A, 230.6; B, 260) eluted (0.75 ml/min) with a 55-ml linear gradient from 5% buffer C (10 mM KH$_2$PO$_4$–NaOH, pH 7.5, 75 mM NaCl, 0.5 mM DTT), 95% buffer D (500 mM KH$_2$PO$_4$–NaOH, pH 7.5, 0.5 mM DTT) to 60% buffer C, 40% buffer D. After SDS–PAGE VASP appears as a 46/50-kDa doublet band, typical with a purity of >95%. Finally the preparation is dialyzed against phosphate-buffered saline (PBS) containing 50% glycerol and 1 mM DTT for storage at −20°. Because purified VASP tends to adhere nonspecifically to plastic surfaces, dilute concentrations should be handled with siliconized or bovine serum albumin (BSA)-coated tips and caps.

An alternative protocol has been used for VASP purification from porcine platelets.[16] In this protocol, the 250 mM NaCl extract of the particulate platelet fraction after dialysis is chromatographed at pH 7.0 on an S-Sepharose FF (Pharmacia) column (4 × 33 cm). The column is eluted with a 1-liter gradient from 0 to 500 mM NaCl in 10 mM NaH$_2$PO$_4$–NaOH (pH

7.0), 4 mM EDTA, and 50 mM 2-mercaptoethanol. Solid ammonium sulfate is added to the VASP-containing fractions to a final concentration of 30% saturation. The precipitate is dissolved in 6 ml of 10 mM NaH$_2$PO$_4$–NaOH, pH 7.0, 4 mM EDTA, 50 mM 2-mercaptoethanol, and 75 mM NaCl, dialyzed against the same buffer, and loaded at a flow rate of 0.5 ml/min onto a column of 5–6 ml of hydroxylapatite (Bio-Gel HT, Bio-Rad; batch characteristics: A, 150; B, 230). The column is eluted (1 ml/min) with a 35-ml linear gradient from 10 mM KH$_2$PO$_4$–NaOH, pH 7.5, 50 mM 2-mercaptoethanol, and 75 mM NaCl to 100 mM KH$_2$PO$_4$–NaOH (pH 7.5) and 50 mM 2-mercaptoethanol. VASP-containing fractions are pooled, dialyzed against buffer B (pH 7.5) for Orange A chromatography, and processed as described above.

Purification of Recombinant VASP Expressed in *Spodoptera frugiperda* Cells

Production of VASP Baculovirus

A recombinant baculovirus transfer vector pBacPAK8-VASP, containing the 1140-bp open reading frame of human VASP cDNA flanked by 65 bp of 5′-untranslated and 333 bp of 3′-untranslated region, was constructed by cloning a 1561-bp *Bam*HI/*Xba*I fragment from plasmid[9] p14/19 into the *Bam*HI/*Xba*I sites of baculovirus transfer vector pBacPAK8 (Clontech, Palo Alto, CA). Cotransfection of *Bsu*36I-digested BacPAK6 baculovirus DNA (Clontech) with the transfer plasmid pBacPAK8-VASP into *Spodoptera frugiperda* (Sf21) cells was then performed using lipofectin to generate recombinant virus that contained the full-length cDNA encoding VASP under control of the polyhedrin promoter. After 3 days, virus-containing culture medium was collected and used to infect new Sf21 cells. Expression of VASP in these cells was confirmed by Western blot analysis of whole-cell extracts harvested 2 days postinfection using polyclonal antibodies raised against VASP purified from human platelets (see below). The virus-containing culture medium from these cells was used to isolate clones of recombinant virus by three series of end-point dilution. Purified virus was then repassaged on fresh Sf21 cells twice to amplify the virus titer. Virus-containing medium was collected and cleared of cellular debris by low-speed centrifugation, and viruses were concentrated by centrifugation through a 25% (w/w) sucrose cushion for 75 min at 80,000g in a swinging bucket rotor. After resuspension and filtration through a 0.2-μm filter, the virus preparations were plaque-titered and stored at 4°.

Purification of VASP from Baculovirus-Infected Insect Cells by Immunoaffinity Chromatography

Preparation of an Immunoaffinity Column. Insect cell lysates expressing recombinant human VASP in sufficient high amounts and availability of a panel of well-characterized monoclonal antibodies directed against this protein[22] make immunoaffinity purification of VASP an alternative and time-saving approach for the isolation of this protein that is not dependent on large amounts of rare and variable human biological material. A major disadvantage of an immunoaffinity purification protocol however might be the sensitivity of the isolated protein to the extreme conditions that are often required for the disruption of antigen–antibody complexes. We therefore screened among a panel of anti-VASP monoclonal antibodies for those that release the antigen under mild elution conditions using a modified enzyme-linked immunosorbent assay (ELISA) as described by Thompson *et al.*[23,24] Briefly, after immobilization of VASP on polystyrene microtiter plates and blocking of nonspecific binding sites, antigen–antibody complexes are allowed to form, washed in PBS, and are incubated with different combinations of low molecular weight polyhydroxylated compounds (polyols) and nonchaotropic salts. We tested different concentrations of propylene glycol or ethylene glycol together with sodium chloride or ammonium sulfate. The amount of antibody resistant to elution is then detected by an enzyme-conjugated secondary antibody followed by reaction with its substrate and compared to a control for desorption of the antigen under these conditions. Among seven anti-VASP monoclonal antibodies tested in this way two were found to be polyol responsive. One antibody retained this property after its immobilization and reaction with VASP in solution. This monoclonal antibody (subtype IgG_1), which binds with high specificity to VASP, was prepared by protein A chromatography from tissue culture supernatants of its hybridoma using a high-salt purification protocol.[25] Two milligrams of this antibody were covalently bound to a 1 ml N-hydroxysuccinimide (NHS)-activated HiTrap column (Pharmacia) according to the instructions of the manufacturer and washed under elution conditions (see below). It is stored at 4° in PBS containing 0.02% sodium azide.

Purification of VASP from Baculovirus-Infected Insect Cells. Sf21 cells are grown at 27° with good aeration in 80-ml suspension cultures in 250-

[22] K. Abel, A. Lingnau, K. Niebuhr, J. Wehland, and U. Walter, *Eur. J. Cell Biol.* **69,** Suppl. 42, 39a (1996).

[23] N. E. Thompson, D. B. Aronson, and R. R. Burgess, *J. Biol. Chem.* **265,** 7069 (1990).

[24] N. E. Thompson, D. A. Hager, and R. R. Burgess, *Biochemistry* **31,** 7003 (1992).

[25] E. Harlow and D. Lane, *in* "Antibodies: A Laboratory Manual." Cold Spring Harbor Laboratory Press, New York, 1988.

ml culture flasks in Grace's insect medium containing 10% fetal calf serum, 0.1% (v/v) Pluronic F-68 solution, 100 μg/ml streptomycin and 100 U/ml penicillin. At a cell density of 1×10^6 cells/ml recombinant VASP baculovirus is added at a multiplicity of infection of one to two pfus per cell. Seventy-two hr postinfection, the cells are collected, pelleted by centrifugation at 1000g for 5 min, and washed twice in cold PBS. All subsequent manipulations are performed at 4° unless otherwise stated. Cell pellets from six 80-ml suspension cultures are resuspended in 5 ml of lysis buffer (20 mM NaH$_2$PO$_4$, pH 7.6, 2 mM EDTA, 2 mM EGTA, 1 mM PMSF, and 100 U/ml aprotinin) and are incubated for 10 min on ice. Cells are then lysed by 10 passages through a 0.5- × 25-mm needle (25-gauge), and complete lysis is controlled by phase-contrast light microscopy. The lysate is adjusted to 250 mM NaCl, extracted for 30 min under constant stirring, and centrifuged at 100,000g for 60 min. After resuspension and reextraction of the pellet in an original volume of extraction buffer, the supernatants after centrifugation are combined. A lysate of six 80-ml suspension cultures of Sf21 cells typically yields about 400 mg of protein, of which 1–2 mg is VASP. VASP is then precipitated from this salt extract by the addition of solid ammonium sulfate to a final concentration of 30% saturation. The precipitate is dissolved in 10 ml of PBS (pH 7.4), cleared by low-speed centrifugation and loaded at a flow rate of 0.02 ml/min onto the immunoaffinity column (see above) that has been preequilibrated with 10 ml of PBS. After washing the column, VASP is eluted (flow rate of 0.5 ml/min) at room temperature with 4 ml of 50% propylene glycol (v/v), 1.25 M NaCl in 50 mM Tris-HCl (pH 7.9), and 0.1 mM EDTA. Buffer flow is stopped for 30 min after application of one column volume of elution buffer. VASP containing fractions are identified by SDS–PAGE and Coomassie Brilliant Blue staining and show a 46-kDa band with a purity of >95% (Fig. 1). They are dialyzed against 25 mM HEPES (pH 7.0), 75 mM KCl, and 1 mM DTT, concentrated by dry dialysis using Aquacide II (molecular weight 500,000; Calbiochem, La Jolla, CA) and stored at 4°.

Assays of VASP Activities

Detection of VASP and VASP Phosphorylation

Three phosphorylation sites (serine-157, serine-239, and threonine-278) phosphorylated by both cAMP-dependent protein kinase (cAK) and cGMP-dependent protein kinase I (cGK) have been identified *in vitro* and in experiments with intact platelets by limited proteolysis of phosphorylated VASP followed by microsequencing.[4] cAK first phosphorylates serine-157, followed by serine-239, whereas cGK shows opposite site preference which

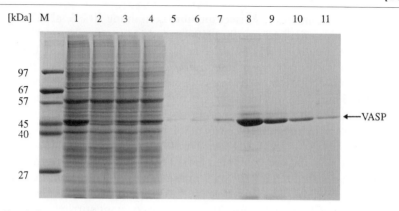

FIG. 1. Immunoaffinity purification of VASP from insect cell lysates. Samples (10 μl) of the dissolved ammonium sulfate precipitate (lane 1) of a salt-extracted and cleared cell lysate, the flow-through material after passage over the immunoaffinity column (lane 2), fractions after washing the column in PBS (lane 3 and 4), and fractions from the final elution in the presence of 50% propylene glycol and 1.25 M NaCl (lane 5 to 11) were analyzed on a 10% SDS–polyacrylamide gel followed by Coomassie Blue staining. Positions of molecular weight markers in kDa, lane M) are indicated.

is particularly evident in the *in vitro* reaction. With both kinases, threonine-278 appears to be the poorest out of the three substrate sites of VASP. Phosphorylated VASP is a substrate for protein phosphatases (PP) 2A, 2B, and 2C *in vitro*, with PP2A being a major phosphatase involved in VASP dephosphorylation in platelets.[26] VASP phosphorylation at serine-157 can be detected by a phosphorylation-induced shift in apparent molecular mass of VASP in SDS–PAGE from 46 to 50 kDa either by autoradiography after ^{32}P incorporation or by Western blot analysis using polyclonal anti-VASP rabbit serum M4[4,5] or a panel of monoclonal VASP antibodies[22] directed against human VASP (immunoGlobe, Grossostheim, Germany). A monoclonal antibody has been developed which recognizes VASP only when serine-239 is phosphorylated.[26a] This antibody will be useful to specifically detect the serine-239 phosphorylation status of VASP. These assays provide convenient tools to monitor the phosphorylation of VASP and thus the activity of cyclic nucleotide-dependent protein kinases *in vitro* and *in vivo* without the need for isotopic labeling of cells and tissues (see, for example, Ref. 7).

Beside its use in Western blot analysis (at a dilution of 1 : 1500–1 : 3000) VASP rabbit serum M4 has also been used for VASP detection by immuno-

[26] K. Abel, G. Mieskes, and U. Walter, *FEBS Lett.* **370**, 184 (1995).
[26a] A. Smolenski, C. Bachmann, K. Reinhard, P. Hönig-Liedl, T. Jarchau, H. Hoschuetzky, and U. Walter, *J. Biol. Chem.* (1998), in press.

VASP rabbit serum M4 has also been used for VASP detection by immuno-fluorescence after formaldehyde fixation (at a dilution of 1:500 in PBS) and for immunoprecipitation (1:125 in a RIPA-type buffer) applications. For Western blot analysis best results are obtained using a hemoglobin containing blocking buffer [PBS containing 0.3% (v/v) Triton X-100, 0.05% (v/v) Tween 20, 1% (w/v) hemoglobin (Sigma, St. Louis, MO), and 0.01% (w/v) NaN$_3$] and ^{125}I-labeled protein A (ICN, Costa Mesa, CA or Amer-sham, Braunschweig, Germany) as the secondary detection reagent. When immunoprecipitated material is subjected to an anti-VASP Western blot, using ^{125}I-labeled protein A also circumvents the problem of masking VASP by an otherwise strong reaction of a secondary antibody with the immuno-globulin heavy chains separated in the gel.

Phosphorylation of VASP in Vitro

In vitro, purified VASP or VASP in total cell lysates can be phosphory-lated both with cGMP-dependent protein kinase (cGK) purified from bo-vine lung[27] and cAMP-dependent protein kinase (cAK) purified from bo-vine heart.[28] Due to its higher degree of specificity cGK is the kinase of choice for VASP phosphorylation in crude preparations or total cell homogenates. When phosphorylation is performed in the presence of [γ-^{32}P]ATP, Laemmli SDS–PAGE (8% separating gel, 3% stacking gel) followed by autoradiography can be used for separation and detection of the phospho-VASP forms. Samples of VASP are incubated either with 2 μg/ml cGK, 20 μM cGMP, or with 2 μg/ml cAK (catalytic subunit) in HEPES buffer (10 mM HEPES, pH 7.4, at room temperature, 5 mM MgCl$_2$, 1 mM dithioerythritol (DTE), 0.2 mM EGTA) for 15 min at 30° in the presence of 4 μM [γ-^{32}P]ATP (370 GBq/mmol). Reactions are stopped by the addition of SDS sample buffer.

Blot Overlays Using VASP as Probe

Blot Overlays. Blot overlays with radiolabeled VASP as a soluble ligand have been used to identify zyxin as a VASP-binding protein[16] and to test for VASP interaction with the *Listeria monocytogenes* protein ActA.[17] A zyxin and ActA derived Phe–Pro–Pro–Pro–Pro (FP$_4$) containing peptide motif involved in these interactions has been characterized in great detail using VASP overlay assays of matrix-bound synthetic peptides.[19] VASP overlays are very sensitive assays for the interactions with proteins con-taining multiple FP$_4$ motifs, such as ActA and zyxin, which are selectively

[27] U. Walter, P. Miller, F. Wilson, D. Menkes, and P. Greengard, *J. Biol. Chem.* **255**, 3757 (1980).
[28] J. A. Beavo, P. J. Bechtel, and E. G. Krebs, *Methods Enzymol.* **38**, 299 (1974).

recognized by VASP even in those complex mixtures of proteins that are present in total cell lysates. However this method fails to reveal VASP interactions with other known VASP-binding proteins that either contain the binding motif in a single copy (e.g., vinculin[20]) or that bind via an unrelated sequence motif (e.g., profilins[15]). Proteins separated by SDS–PAGE are transferred to nitrocellulose (BA85; Schleicher & Schuell, Dassel, Germany) or Nylon membranes (Immobilon P; Millipore, Eschborn, Germany). Membranes are blocked for 30 min at 37° in blocking buffer [10 mM NaPi, pH 7.4 at room temperature, 150 mM NaCl, 0.3% (v/v) Triton X-100, 0.05% (v/v) Tween 20, 0.5 mM DTT, 1% (w/v) hemoglobin (Sigma), 0.01% (w/v) NaN$_3$] and subsequently are incubated for at least 30 min at room temperature with 0.1 μg/ml ^{32}P-labeled VASP (in blocking buffer) prepared as described below. Depending on the sensitivity required, overlays may also perform well using [^{32}P]VASP at 10 ng/ml or even lower concentrations. After washing (three times at 5–10 min) with PBS containing 0.3% (v/v) Triton X-100, 0.05% (v/v) Tween 20, the sheets are dried and exposed to an autoradiographic film.

Phosphorylation of VASP as Probe for Blot Overlays. The following section describes the use of VASP as a probe for blot overlays after ^{32}P-labeling by *in vitro* phosphorylation, which allows for detection of a direct interaction of purified VASP with immobilized and denatured zyxin or ActA protein. Alternatively, if the presence of other proteins in the probe does not perturb the interaction assay, a ^{35}S-labeled VASP probe obtained by an *in vitro* transcription and translation system may be used instead of the purified protein.[19]

For its use in blot overlays VASP is phosphorylated with cGMP-dependent protein kinase (cGK) in the presence of [γ-^{32}P]ATP with high specific radioactivity. However, the following protocol can also be adapted for VASP phosphorylation with the catalytic subunit of cAMP-dependent protein kinase, which will yield similar results in overlay assays. The phosphorylation mixture contains VASP (38 μg/ml) and cGK (15.8 μg/ml) in a buffer containing 10 mM HEPES (pH 7.4 at 20°), 0.2 μg/ml BSA, 5 mM MgCl$_2$, 1 mM DTE, 0.2 mM EGTA, and 20 μM cGMP. The reaction is started by addition of [γ-^{32}P]ATP (5.9 TBq/mmol) to a final concentration of 2.5 μM. After incubation at 30° for 1 hr, unlabeled ATP is added to a final concentration of 200 μM and phosphorylation is allowed to proceed for an additional 30 min. Although quantitative phosphorylation is not a prerequisite for VASP to be active in overlay assays, the 30-min extension step is highly recommended to ensure defined starting conditions. The assay mixture is diluted with an equal volume of double concentrated blocking buffer and the volume is adjusted to 145 μl with blocking buffer. Free [γ-^{32}P]ATP is removed by centrifugation (2 min at 730g) through a

MicroSpin S-200 HR column (Pharmacia) equilibrated in blocking buffer. Due to its quaternary structure,[9] VASP will appear in the flow-through fraction. The flow-through is diluted with blocking buffer to a final VASP concentration of 0.1 μg/ml.

Solid Phase Binding Assays for Interaction of VASP with Purified Binding Proteins

Direct binding of VASP to purified zyxin, ActA protein, profilins, and vinculin has been shown in solid phase binding assays with VASP used as the soluble ligand and its binding partners coated to a plastic surface. In contrast to blot overlays this assay is permissive for structural integrity of both reaction partners, which is a prerequisite for VASP–profilin and VASP–vinculin interactions.

Individually removable microtiter wells (Removawells; Dynatech, Denkendorf, Germany) are coated for at least 1.5 hr at room temperature with purified VASP-binding proteins diluted with PBS (zyxin: 10–80 ng/ml; Act A protein: 30 μg/ml; profilin: 20–25 μg/ml; vinculin: 2.5 μg/ml). All assays are done in triplicate. Unspecific binding sites are blocked with 3% BSA in PBS (1 hr at room temperature). VASP is added directly into the wells and is subsequently diluted with a 3% BSA solution to a final concentration of 0.6–2.2 μg/ml in an incubation volume of 200 or 400 μl. After incubation on a rotary shaker for 2 hr at room temperature, wells are rinsed three times with PBS and bound VASP is detected by subsequent incubations with anti-VASP antiserum M4 (1:1000–1:1500 in 3% BSA in PBS; 1 hr) and ^{125}I-labeled protein A (3.7–4.8 kBq/ml in 3% BSA in PBS; 1 hr) with PBS washings in between. Wells without coating and wells without VASP addition serve as controls. Individual wells are removed from the microtiter plate frame and bound radioactivity is determined in a gamma counter. This assay yields reproducible and reliable results, provided that the proteins are applied directly as a concentrated sample to the microtiter wells and diluent buffer is added thereafter.

Acknowledgment

This work was supported by the Deutsche Forschungsgemeinschaft (SFB 355/C3).

[11] *Listeria Monocytogenes*-Based Assays for Actin Assembly Factors

By Julie A. Theriot and David C. Fung

Introduction

Filamentous actin assembly is required for most forms of cell crawling, including lamellipodial, filopodial, and pseudopodial extension.[1] Actin polymerization is also necessary for elaboration of cell surface protrusions such as microvilli and hair cell cilia.[2] Biochemical analysis aimed at understanding how the cell controls actin assembly and protrusion at its surface has been hampered because of the intimate role of the plasma membrane in these processes, with its attendant biochemical complexity and amorphous structure. The plasma membrane not only defines the boundary of cell surface protrusions, but also provides the signals directing localized actin assembly.

Within the past several years, the actin-based motility of certain pathogenic bacteria has become a widely used model system for analysis of actin assembly-dependent protrusive movement.[3,4] Several different species of pathogenic bacteria (including *Listeria monocytogenes, Shigella flexneri,* and *Rickettsia rickettsii*) and at least one virus (Vaccinia) grow directly in the cytoplasm of infected host cells, and induce host cell actin assembly at their surface. This pathogen-driven relocalization of actin filament polymerization results in the formation of a "comet tail" that propels the infectious agent through the host cell cytoplasm and facilitates cell-to-cell spread.[5] In this type of actin-based motility, the bacteria (or virus) appear to imitate a fragment of the host cell leading edge so that the tail can be thought of as a geometrically and dynamically simplified lamellipodium.[6] Movement of these pathogens can be reconstituted *in vitro* in vertebrate cell cytoplasmic extracts, including extracts from *Xenopus laevis* eggs[7] (Fig. 1) and human

---

[1] J. Condeelis, *Annu. Rev. Cell Biol.* **9,** 411 (1993).
[2] M. B. Heintzelman and M. S. Mooseker, *Curr. Top. Dev. Biol.* **26,** 93 (1992).
[3] J. A. Theriot, *Annu. Rev. Cell Dev. Biol.* **11,** 213 (1995).
[4] I. Lasa and P. Cossart, *Trends Cell Biol.* **6,** 109 (1996).
[5] L. G. Tilney and D. A. Portnoy, *J. Cell Biol.* **109,** 1597 (1989).
[6] J. A. Theriot, T. J. Mitchison, L. G. Tilney, and D. A. Portnoy, *Nature* **357,** 257 (1992).
[7] J. A. Theriot, J. Rosenblatt, D. A. Portnoy, P. J. Goldschmidt-Clermont, and T. J. Mitchison, *Cell* **76,** 505 (1994).

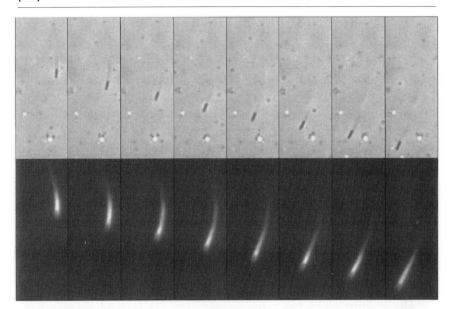

Fig. 1. Reconstitution of *L. monocytogenes* actin-based motility in *Xenopus* egg cytoplasmic extract. Eight frames are shown from a video sequence, separated by 25-sec intervals. Top row, phase-contrast images. Bottom row, fluorescence of rhodamine-actin. Paired phase and fluorescence images were collected with an intensified CCD camera about 1 sec apart. The bacterium is 2.2 μm in length.

platelets.[8] This advance has greatly facilitated the biochemical investigation of the regulation of actin assembly and polymerization-dependent motility.

Listeria monocytogenes-based *in vitro* actin assembly assays have two primary uses. First, they can be used in the biochemical analysis of bacterial and host cell components that regulate actin dynamics.[7–10] Bacterial interaction with host cell filamentous actin occurs in two stages; first, a symmetric "cloud" forms around the bacterium, followed by reorganization to form the polarized "comet tail" associated with movement.[5] Both of these steps require the *L. monocytogenes* ActA protein[11,12] in combination with two

[8] M. D. Welch, A. Iwamatsu, and T. J. Mitchison, *Nature* **385,** 265 (1997).

[9] J. B. Marchand, P. Moreau, A. Paoletti, P. Cossart, M. F. Carlier, and D. Pantaloni, *J. Cell Biol.* **130,** 331 (1995).

[10] J. Rosenblatt, B. J. Agnew, H. Abe, J. R. Bamburg, and T. J. Mitchison, *J. Cell Biol.* **136,** 1323 (1997).

[11] C. Kocks, E. Gouin, M. Tabouret, P. Berche, H. Ohayon, and P. Cossart, *Cell* **68,** 521 (1992).

[12] E. Domann, J. Wehland, M. Rohde, S. Pistor, M. Hartl, W. Goebel, M. Leimeister-Wachter, M. Wuenscher, and T. Chakraborty, *EMBO J.* **11,** 1981 (1992).

different sets of host cell factors.[8] Thus each step can be biochemically dissected independently.

Second, "comet tail" *in vitro* assays are excellent for quantitative analysis of actin dynamics in combination with motion. The tails are geometrically simple and highly similar from one individual to the next, unlike lamellipodia or pseudopodia. Bacterial movement is a consequence of actin filament assembly, and the rate of movement is tightly coupled to the rate of polymerization,[6,13] so the easily quantitated parameter of bacterial speed can be used as a readout for filament polymerization. Actin assembly is coordinated with force generation as well as movement (Fung and Theriot, unpublished manuscript, 1998). Since it is a cell-free assay, labeled actin tracers can be added directly to the cytoplasm, without the need for microinjection. The thin chamber in which the *in vitro* assays are performed allows for a very low and constant fluorescence background, so that quantitation of actin filament density and dynamics is straightforward.

Actin Assembly and Motility Assay

The motility assay has three components: (1) *L. monocytogenes* (or other bacteria, or particles that replace the bacteria), (2) labeled actin for visualization of clouds and tails, and (3) host cell cytoplasm (or a cytoplasmic fraction, or purified protein). Here we briefly outline issues to consider when choosing each of the three components, and give protocols for the preparation of one type of each component.

Listeria Monocytogenes or Alternatives

Considerations. Several strains of *L. monocytogenes* have been reported to move in cytoplasmic extracts, including a "hyperhemolytic" strain, SLCC-5764,[7] and a wild-type strain carrying the *actA* gene on a high-copy-number plasmid.[9] A high surface density of ActA is essential for *in vitro* motility; wild-type strains expressing lower or regulated levels of this virulence factor do not move well in extracts (Theriot, 1994).

Listeria monocytogenes is a human pathogen that can cause serious infections in immunocompromised people and miscarriage in pregnant women. For this reason, many researchers may prefer to use a substitute. *Escherichia coli* expressing the *icsA* gene from *S. flexneri* move well in extracts but are noninfectious.[14,15] They can be fixed with formaldehyde

---

[13] J. M. Sanger, J. W. Sanger, and F. S. Southwick, *Infect. Immun.* **60**, 3609 (1992).
[14] M. B. Goldberg and J. A. Theriot, *Proc. Natl. Acad. Sci. U.S.A.* **92**, 6572 (1995).
[15] C. Kocks, J. B. Marchand, E. Gouin, H. d'Hauteville, P. J. Sansonetti, M. F. Carlier, and P. Cossart, *Mol. Microbiol.* **18**, 413 (1995).

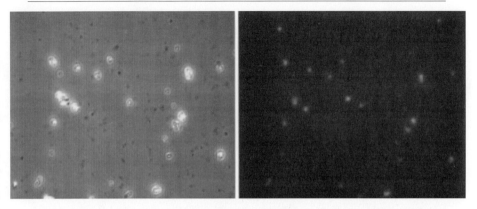

FIG. 2. Nucleation of actin filament assembly by polystyrene beads coated with a 15-amino-acid peptide derived from the *L. monocytogenes* ActA protein. Left panel, phase-contrast. Right panel, fluorescence of rhodamine-actin. A peptide with amino acid sequence EIKKRR KAIASSDSEC was synthesized and attached covalently to 0.5-μm polystyrene beads through the C-terminal cysteine residues. Beads were incubated in *Xenopus* egg extract with rhodamine actin. Filamentous actin is associated with some, but not all, of the beads.

and stored for at least several weeks without losing their motile behavior.[14] Similarly, *Listeria innocua*, a nonpathogenic cousin of *L. monocytogenes*, can form comet tails *in vitro* if it is expressing ActA.[15] *Streptococcus pneumoniae* coated with a purified ActA-derived fusion protein can also move in extracts.[16] Polystyrene beads coated with a peptide derived from ActA can induce the first step of actin assembly, cloud formation, but do not form tails (Fig. 2). Even pure phospholipid vesicles can induce actin tail formation under the appropriate conditions.[16a]

Preparation. Streak *L. monocytogenes* strain SLCC-5764 onto a brain–heart infusion (BHI) agar plate and incubate at 37° until colonies are 1–2 mm in diameter. Put 2 ml BHI broth in a 15-ml sterile disposable test tube. Inoculate the broth with a single colony from the plate. Grow overnight with vigorous aeration at 37°. After the culture has reached stationary phase, transfer 1 ml to an Eppendorf microcentrifuge tube. Pellet the bacteria by spinning 1–2 min at low speed (about 3000 rpm). Resuspend the pellet in 1 ml XB (*Xenopus* extract buffer[17]: 100 mM KCl, 0.1 mM CaCl$_2$, 2 mM MgCl$_2$, 5 mM EGTA, 10 mM K-HEPES, pH 7.7, 50 mM sucrose). Pellet bacteria again and resuspend in 0.1 ml XB. The bacterial suspension can be stored on ice for at least 8 hr with no loss of activity. If it is not practical

[16] G. A. Smith, D. A. Portnoy, and J. A. Theriot, *Mol. Microbiol.* **17,** 945 (1995).
[16a] L. Ma, L. C. Cantley, P. A. Janmey, and M. W. Kirschner, *J. Cell Biol.* **140,** 1125 (1998).
[17] A. W. Murray, *Methods Cell Biol.* **36,** 581 (1991).

or desirable to grow fresh bacteria for each day of experiments, an active killed stock can be prepared by treating the washed bacteria in 1 ml XB with 10 mM iodoacetic acid for 20 min at room temperature. Killed bacteria can be stored in 50% glycerol at $-20°$ for several months with no loss of motile activity.

Labeled Actin

Considerations. Actin can be prepared from a variety of sources, but most easily and in large amounts from vertebrate skeletal muscle.[18] Since skeletal muscle actin readily copolymerizes with other actin isoforms, we and others have routinely used only skeletal actin in these assays. For fluorescent visualization of actin filament distribution, it is best to covalently label the actin itself, since fluorescent phalloidin strongly interferes with actin assembly dynamics. To date, all published reports of this type of assay have used actin covalently labeled with tetramethylrhodamine, a relatively photostable and widely used fluorophore. Actin may be labeled either on the single reactive cysteine (C-374) with tetramethylrhodamine iodoacet-amide (TMR-IA), or on any of several lysine residues with tetramethylrho-damine N-hydroxysuccinimide or tetramethylrhodamine isothiocyanate.[19] One disadvantage of rhodamine is that it is equally fluorescent on G- or F-actin, and so cannot be used in a quantitative spectrophotometric assay to measure the extent of actin assembly induced by *L. monocytogenes.* We have attempted to use pyrene-actin in this manner, but found the signal-to-noise ratio to be too low for meaningful interpretation.

Preparation. Dialyze purified actin in G buffer (5 mM Tris, pH 8.0, 0.2 mM CaCl$_2$, 0.2 mM ATP) with 10 μM dithiothreitol (DTT) at 4° for 4 hr. Measure the actin concentration and dilute to 5 mg/ml in G buffer with 10 μM DTT. Warm solution to room temperature and add 50 μl of 8 mM TMR-IA stock solution per milliliter of actin (dissolve TMR-IA in anhy-drous dimethyl sulfoxide to give stock solution). Mix thoroughly, wrap the tube in aluminum foil, and rotate at 4° overnight. Equilibrate gel filtration column (P-10 or G-25 resin) with G buffer with 0.2 mM DTT. Centrifuge the reaction solution at top speed in a microcentrifuge for 10 min to remove precipitates. Carefully load the column and run with G buffer with 0.2 mM DTT; labeled actin will elute in void volume while unreacted label will remain in resin. Pool the pink actin-containing fractions and measure the total volume. Polymerize actin by adding one-quarter volume 5× G to F conversion buffer (80 mM Tris, pH 8.0, 250 mM KCl, 1 mM ATP, 2 mM MgCl$_2$). Let sit at 4° overnight. Spin out filaments at 100,000g for 1 hr at

[18] J. D. Pardee and J. A. Spudich, *Methods Enzymol.* **85,** 164 (1982).
[19] P. Sheterline, J. Clayton, and J. C. Sparrow, "Actins." Academic Press, London, 1996.

4°. Remove supernatant and overlay pellet with a small amount of G buffer with 0.2 mM DTT and let sit on ice for 1 hr to soften. Very gently resuspend pellet in a minimal volume of G buffer with 0.2 mM DTT, then dialyze against this same buffer overnight. Clarify in a centrifuge, freeze small aliquots in liquid nitrogen, and store at −80°.

Host Cell Cytoplasm or Host Proteins

Considerations. Both extracts from *Xenopus* eggs[7] and human platelets[8] can be prepared at very high protein concentration (including a high concentration of cytoskeletal components), typically 80–100 mg/ml total protein. More dilute cytoplasmic extracts prepared from tissue culture cells do not reconstitute motility. Besides crude cytoplasm, purified proteins or partially purified extract fractions can be used (see article by Welch and Mitchison in this volume).

Preparation. Crude cytostatic factor-arrested *Xenopus* egg extracts are prepared essentially according to the method of Murray[17] with the important modification that cytochalasin is omitted at every step. Dejelly the eggs in 2% cysteine in XB (pH 7.8) and rinse 4 times in XB. Transfer the eggs in a minimum volume of buffer to disposable centrifuge tubes containing 1 ml XB with 10 μg/ml each of the protease inhibitors leupeptin, pepstatin, and chymostatin, then overlay with 1 ml of versilube oil. Pack the eggs with a low-speed spin in a clinical centrifuge, and remove the oil and buffer from the top of the tube. Next, crush the eggs by spinning at 10,000 rpm in a swinging bucket rotor for 10 min at 16°. Collect the orange or straw-colored cytoplasmic layer in the middle of the tube by puncturing the tube with a 16-gauge hypodermic syringe and drawing out the crude extract. Add protease inhibitors to a final concentration of 10 μg/ml each and one-tenth volume sterile 1 M sucrose. An ATP regenerating system (7.5 mM creatine phosphate, 1 mM ATP, 0.1 mM EGTA, 1 mM MgCl$_2$) can also be added from a 20× stock. The extracts can be used immediately or frozen in liquid nitrogen in small (50–200 μl) aliquots and stored at −80° for at least 6 months without loss of activity.

Motility Assay

Preparation. Once each of the three components has been prepared, they need only be mixed together. We typically combine 10 μl of extract, 1 μl of washed bacterial suspension in XB, and 0.5 μl of a 0.5 mg/ml solution of rhodamine-actin in a small Eppendorf tube, pipette gently up and down to mix (never vortex), and incubate on ice for approximately 30 min (up to 2 hr). After the incubation, remove 1 μl of assay mix and spot it on a clean glass slide. Gently overlay the droplet with a clean glass 22- × 22-

mm coverslip and press down until the droplet has spread to the edges. Seal the edges of the coverslip with melted beeswax or VALAP (1:1:1 mixture of vaseline, lanolin, and paraffin). Incubate the sealed slide in the dark at room temperature for about 30 min prior to examination on an upright or inverted microscope equipped with phase-contrast and epifluorescence optics. Actin-rich comet tails associated with moving bacteria will easily be observed in the rhodamine channel using a 40× or higher power objective. Movement can be best appreciated by recording time-lapse video sequences (Fig. 1). Typical movement rates range from about 3 to 30 μm/min.

In the thin chamber, the fluorescence background due to unpolymerized rhodamine-actin is uniform and low. The extract is of essentially uniform viscosity, and the speed of the bacteria is close to constant under most conditions (Fung and Theriot, unpublished manuscript, 1998). These properties make the *L. monocytogenes*-based *in vitro* motility assay particularly attractive for quantitative analysis of the relationship between actin assembly and movement.

Quantitative Analysis

In a quantitative analysis of this type of actin-assembly driven movement, two parameters can be measured: the trajectory (and speed) of the bacterium and the dynamic distribution of actin filaments in the "comet tail." The former is best measured with phase-contrast microscopy, whereas the latter requires fluorescence microscopy. Both types of images may be recorded from the same sample by using a computer to control shutters that switch between transmitted and epifluorescence illumination sources. The computer is also used to acquire the images digitally and for subsequent analysis. While many types of image acquisition and analysis software may be suitable, we use Metamorph from Universal Imaging, which conveniently handles time-lapse sequences as stacks of linked images.

For many experiments, bacterial speeds can be obtained by manually tracking the pixel position of their "nose" from frame to frame. However, much more precise tracking of bacterial trajectories can be accomplished with centroid analysis of the phase-contrast images. A threshold level is chosen that distinguishes the bacteria from the background, and the centroids of the thresholded areas calculated to give the position of the bacteria. This method can easily attain subpixel resolution.

To record both a phase-contrast image of the bacterial cell and a rhodamine fluorescence image of the tail without compromising the quality of either, it is necessary to either use two cameras and split the light path, or to use a single camera that records paired phase-contrast and fluorescence images that are slightly separated in time. The main disadvantage of two cameras is that it is difficult to register the two images perfectly; for subpixel analysis of slow movements this can introduce serious systematic error. The main disadvantage of using one camera is that the phase and fluorescence images are not truly simultaneous (in practice, on our system they are usually separated by about one second).

The reliability of the quantitative analysis of actin filament dynamics depends on the assumption that the camera response is linear with the amount of incident light. Cooled CCD cameras give excellent linearity but are in general too insensitive for use with dim live samples; by the time the camera has integrated a sufficient number of frames for visualization of the tail, the bacterium may have moved a significant distance. Silicon-intensified tube cameras are very sensitive and excellent for real-time recording, but show poor linearity. We use an intensified CCD camera (Dage-MTI GenIISys/CCD-c72, Michigan City, IN) that provides a reasonable compromise; it is sensitive and acquires images quickly, but has a broad linear range. It is important to calibrate the linear range of the camera so that analysis is performed only on experiments that fall within the appropriate parameters.

As with any other *in vitro* reconstitution of a biological activity, it is important to bear in mind that results from the *in vitro* system may not accurately reflect the situation inside living, intact cells. In these actin assembly and movement assays, this caveat is particularly relevant for analysis of the relative contribution of different host factors and bacterial protein domains to the speed of movement, since it is clear that the rate-limiting step is different in different cytoplasmic environments. *Listeria monocytogenes* moves about five times faster in macrophage-like cells than in epithelial cells.[20] A mutant strain of *L. monocytogenes* carrying an in-frame deletion of the proline-rich repeat region of ActA shows a severe motility defect inside of epithelial cells, with a drastic decrease in both the rate and the frequency of movement, though an equivalent strain in *Xenopus* egg extracts moves well with only a modest decrease in average speed.[21] *Listeria monocytogenes* and *S. flexneri* move over an identical range of rates in epithelial cells, but *E. coli* expressing IcsA move about three times faster than *L. monocytogenes* expressing ActA in *Xenopus* egg extracts, even when the two types of bacteria are mixed in a single sample (Theriot, 1995). Thus despite the usefulness of *L. monocytogenes*-based actin assem-

[20] G. A. Dabiri, J. M. Sanger, D. A. Portnoy, and F. S. Southwick, *Proc. Natl. Acad. Sci. U.S.A.* **87**, 6068 (1990).
[21] G. A. Smith, J. A. Theriot, and D. A. Portnoy, *J. Cell Biol.* **135**, 647 (1996).

bly and motility assays for biochemical and quantitative analysis of the coupling between actin polymerization and protrusive movements, conclusions drawn from these types of experiments must always be validated in living cells.

Section II

Analysis of Microtubule-Related Systems

[12] Preparation and Characterization of Caged Fluorescein Tubulin

By ARSHAD DESAI and TIMOTHY J. MITCHISON

Introduction

The microtubule cytoskeleton plays an important role in fundamental cellular processes such as cell division, cytoplasmic organization, and the establishment and maintenance of cellular asymmetry. Understanding the mechanism of these processes requires the development of methods to investigate the dynamics of microtubules in specific subcellular locations. Initially, fluorescence photobleaching was the method of choice for monitoring the local dynamics of microtubules *in vivo*.[1] More recently the technique of photoactivation of fluorescence has provided an alternative approach to analyzing intracellular dynamics of the microtubule cytoskeleton. In the photoactivation approach, tubulin is covalently tagged with a caged fluorochrome that is nonfluorescent until exposed to a brief pulse of ultraviolet (UV) light. Photoactivation is preferable to photobleaching because of reduced photodamage and because photoactivation results in a bright signal against a dark background. More extensive discussion of the advantages of the photoactivation approach and detailed descriptions of the chemical strategies used for caging fluorochromes other than fluorescein have been presented elsewhere.[2,3]

In the initial application of the photoactivation of fluorescence method, a caged fluorescent derivative of carboxyfluorescein (bis-caged carboxyfluorescein, henceforth, referred to as C2CF) was covalently coupled to tubulin, microinjected into cells, and allowed to incorporate into the cellular microtubule cytoskeleton. Subsequent uncaging using UV irradiation localized to a small zone in the mitotic spindle resulted in the discovery of the slow poleward translocation of kinetochore microtubules during metaphase.[2] Since then C2CF-tubulin has been used for a number of other applications: determining the contribution of poleward flux of kinetochore microtubules to anaphase chromosome movement in vertebrate somatic

[1] E. D. Salmon, R. J. Leslie, W. M. Saxton, M. L. Karow, and J. R. McIntosh, *J. Cell Biol.* **99,** 2165 (1984).
[2] T. J. Mitchison, *J. Cell Biol.* **109,** 637 (1989).
[3] T. J. Mitchison, K. E. Sawin, J. A. Theriot, K. Gee, and A. Mallavarapu, *Methods Enzymol.* (1998).

cells,[4,5] demonstrating poleward microtubule flux in spindles assembled *in vitro* in *Xenopus* egg extracts and defining the minimum requirements for flux in this system,[6,7] measuring the differential turnover of spindle microtubule subpopulations,[5] analyzing changes in microtubule dynamics across the cell cycle[8] and during cytoplasmic reorganization,[9] characterizing the mechanism of tubulin transport in neurons,[10,11] and demonstrating the ability of kinetochore microtubule poleward flux to produce a force within the mitotic spindle.[12]

To date, C2CF remains the caged fluorescent probe of choice for analyzing the subcellular dynamics of the microtubule cytoskeleton. C2CF is stably nonfluorescent prior to UV activation both *in vitro* and *in vivo*, and can be covalently coupled to tubulin without inactivating its polymerization competency. The primary limitation of C2CF is the intrinsically poor photostability of fluorescein itself, which tends to photobleach rapidly *in vivo*. A caged version of a more photostable fluorophore such as rhodamine or 2′,7′-difluorofluorescein (Oregon Green; Molecular Probes, Eugene, OR) is necessary to circumvent this limitation. Recently, we have synthesized a useful cage rhodamine and have applied it to photoactivation studies of actin filaments *in vivo*, but not yet microtubules. The merits of different caged fluorochromes and the synthesis of caged rhodamine have been recently discussed.[3]

C2CF is activated for derivatizing proteins via a spacer arm ending in a sulfo-*N*-hydroxysuccinimide (SNHS) ester. Use of a SNHS ester, instead of an NHS ester, increases the water solubility of the hydrophobic caged compound. However, C2CF-SNHS is still too hydrophobic to label many proteins, including actin. At present the only commercial source for C2CF-SNHS is in Japan (Dojindo Laboratories, Kumamoto, Japan). We have only used C2CF-SNHS synthesized in house according to the protocol in the original paper.[2] Below we describe in detail a procedure for coupling C2CF-SNHS ester to bovine brain tubulin. This is a general procedure for coupling moieties with reactive succinimidyl esters to tubulin and we have used it successfully to derivatize tubulin not only with C2CF-SNHS ester but also with succinimidyl esters of biotin, digoxigenin, and a wide range

[4] T. J. Mitchison and E. D. Salmon, *J. Cell Biol.* **119**, 569 (1992).
[5] Y. Zhai, P. J. Kronebusch, and G. G. Borisy, *J. Cell Biol.* **131**, 721 (1995).
[6] K. E. Sawin and T. J. Mitchison, *J. Cell Biol.* **112**, 941 (1991).
[7] K. E. Sawin and T. J. Mitchison, *Mol. Biol. Cell* **5**, 217 (1994).
[8] Y. Zhai, P. J. Kronebusch, P. M. Simon, and G. G. Borisy, *J. Cell Biol.* **135**, 201 (1996).
[9] V. I. Rodionov, S. S. Lim, V. I. Gelfand, and G. G. Borisy, *J. Cell Biol.* **126**, 1455 (1994).
[10] S. S. Reinsch, T. J. Mitchison, and M. Kirschner, *J. Cell Biol.* **115**, 365 (1991).
[11] T. Funakoshi, S. Takeda, and N. Hirokawa, *J. Cell Biol.* **133**, 1347 (1996).
[12] J. C. Waters, T. J. Mitchison, C. L. Rieder, and E. D. Salmon, *Mol. Biol. Cell* **7**, 1547 (1996).

of fluorochromes such as tetramethylrhodamine, X-rhodamine, fluorescein, Oregon Green, Cy3, and Cy5. The rationale of the procedure is well established and involves labeling polymeric tubulin, thereby protecting residues important for microtubule assembly.[13] The labeling is performed at high pH to optimize the reaction with the sulfo-succinimidyl ester, and functional tubulin is selected after the labeling reaction by one or more cycles of polymerization and depolymerization.[13]

Preparation of C2CF-Tubulin

Reagents and Buffers

  5–10 mg/ml Bovine brain tubulin (We store 3-ml aliquots of phospho-cellulose flow-through fractions at −80° after a large-scale bovine brain preparation and use ~50 mg for a large-scale labeling. A detailed protocol for a large-scale bovine brain tubulin prep is available at http://skye.med.harvard.edu.)

  0.1 M C2CF-SNHS ester (made up in anhydrous DMSO and stored in a light-tight container at −80°; C2CF is light sensitive and should only be handled in the dark under a safelight)

  BRB80 (1×): 80 mM PIPES, 1 mM MgCl$_2$, 1 mM EGTA, pH 6.8, with KOH (generally made as a 5× stock and stored at 4°)

  High pH Cushion: 0.1 M sodium HEPES, pH 8.6, 1 mM MgCl$_2$, 1 mM EGTA, 60% (v/v) glycerol

  Labeling buffer: 0.1 M sodium HEPES, pH 8.6, 1 mM MgCl$_2$, 1 mM EGTA, 40% (v/v) glycerol

  Quench: 2× BRB80, 100 mM potassium glutamate, 40% (v/v) glycerol

  Low pH cushion: 60% (v/v) glycerol in 1× BRB80

  10× 1B (injection buffer): 500 mM potassium glutamate, 5 mM MgCl$_2$ (pH of 1× ~ 7.0)

Note

  1 M HEPES; titrate to 8.6 with NaOH and store at −20°.

  2 M potassium glutamate; dissolve glutamic acid to 2 M, carefully titrate with KOH such that 50 mM has a pH ~7.0 and store at −20°.

  (All buffers can be stored indefinitely at −20°.)

Labeling Procedure

  The procedure described below can be scaled down if desired. It is essential to perform all steps involving C2CF-SNHS under a safelight in a

---

[13] A. Hyman, D. Drechsel, D. Kellogg, S. Salser, K. Sawin, P. Steffen, L. Wordeman, and T. Mitchison, *Methods Enzymol.* **196**, 478 (1991).

room well shielded from light. A piece of red acetate sheet taped over a dimly lit lamp is adequate as a safelight for the labeling procedure.

1. Thaw 50 mg tubulin and adjust to a final buffer composition of 1× BRB80, 4 mM MgCl$_2$, 1 mM GTP and transfer to 37° for 5 min. Add DMSO to 10% final in two steps, mixing gently but thoroughly and incubate at 37° for 30 min. In a side-by-side comparison, for reasons that are not clear, using DMSO instead of glycerol for the first polymerization step appears to increase the labeling stoichiometry by ~25% for C2CF-SNHS ester.[14]

2. Layer polymerized tubulin onto 20 ml warm High pH Cushion in two 50.2 Ti tubes. Pellet microtubules in a Beckman (Palo Alto, CA) ultracentrifuge in a 50.2 Ti rotor at 40,000 rpm for 45 min at 35°.

3. Aspirate the supernatant above the cushion and rinse the supernatant–cushion interface twice with warm (37°) labeling buffer. Aspirate the cushion and resuspend the pellet using a cutoff large pipette tip in 1–1.5 ml of warm labeling buffer. Take care to keep the tubulin warm during the resuspension and continue resuspending until no chunks of tubulin are visible.

From this point onward, all subsequent steps should be done under a safelight and any exposure to room light should be avoided.

4. Add 25 μl 0.1 M C2CF-SNHS ester, mix vigorously, and incubate at 37°. Mix intermittently (every 3–5 min) by inversion or gentle vortexing for 30 min. Add an additional 25 μl 0.1 M C2CF-SNHS ester and continue mixing for an additional 30 min. This results in a 10- to 20-fold molar excess of the dye to the tubulin. After a total labeling time of 60 min, add an equal volume of Quench to the labeling reaction and mix well.

5. Layer the quenched labeling reaction onto two TLA100.3 (or TLA100.4) tubes containing 1.5 ml of Low pH Cushion. Spin at 80,000 rpm for 20 min at 35° in a TLA100.3 or TLA100.4 rotor in a Beckman TLA100 ultracentrifuge.

6. Aspirate the supernatant above the cushion and rinse the supernatant–cushion interface twice with 1× BRB80. Aspirate the cushion and resuspend the pellet using a cutoff pipette tip in 1 ml of ice-cold 1× IB. Transfer resuspended chunks of the pellet to a small ice-cold dounce homogenizer (1- or 2-ml volume) in an ice-water bath. Resuspend the pellet by gently douncing till the suspension is uniform. Continue douncing intermittently for a total time of 30 min at 0°.

---

[14] T. J. Mitchison, K. E. Sawin, and J. A. Theriot, in "Cell Biology: A Laboratory Handbook" (J. E. Celis, ed.), Vol. 2, p. 66. Academic Press, San Diego, 1994.

Cold IB seems to promote more rapid depolymerization than BRB80; therefore, we use IB in the depolymerization step for all labeling procedures. For small-scale labelings the pellet can be resuspended directly in the centrifuge tube and sonicated gently using a microtip sonicator to speed depolymerization.

7. Spin the depolymerized tubulin in a TLA100.2 (or TLA100.3) rotor at 80,000 rpm for 10 min at 2°.

8. Recover the supernatant from the cold spin, add BRB80 to 1× (from a 5× stock), $MgCl_2$ to 4 mM, GTP to 1 mM, and add one-half volume of glycerol (33% final). Mix well and polymerize at 37° for 30 min.

9. Layer the polymerization reaction on a 1 ml Low pH Cushion in a TLA100.3 tube and pellet the microtubules at 80,000 rpm in a TLA100.3 rotor for 20 min at 37°.

10. Aspirate the supernatant above the cushion and rinse the supernatant–cushion interface twice with warm IB. Aspirate the cushion and rinse the pellet twice with 1 ml warm IB to remove any residual glycerol. Resuspend the pellet using a cutoff pipette tip in 0.2–0.3 ml of ice-cold IB. This pellet should resuspend easily. Incubate at 0° for 20–30 min.

11. Spin the depolymerized tubulin in a TLA120.1 (or TLA100 or TLA100.2) rotor at 80,000 rpm for 10 min at 2°. Recover the supernatant, quickly estimate the tubulin concentration, adjust with IB if desired, and freeze in 3- to 5-μl aliquots in liquid nitrogen. We generally aim for a final tubulin concentration of 5–10 mg/ml (50–100 μM). Careful determination of tubulin concentration and labeling stoichiometry can be performed as described below, after the tubulin has been aliquoted and frozen. C2CF-tubulin should be stored at −80° in a foil-wrapped box. C2CF-tubulin stored at −80° retains polymerization competency and does not exhibit any significant activation of fluorescence for >1 yr.

Characterization and Use of C2CF-Tubulin

Determination of Concentration, Labeling Stoichiometry, and Polymerization/Activation Competency

Accurate estimations of tubulin concentration can be made by the method of Bradford[15] using tubulin calibrated by A_{280} as a standard. Pure tubulin in the absence of free nucleotide has an extinction coefficient at 280 nm of 115,000 M^{-1} cm^{-1}. This extinction coefficient was calculated from

---

[15] M. M. Bradford, *Anal. Biochem.* **72**, 248 (1976).

A.

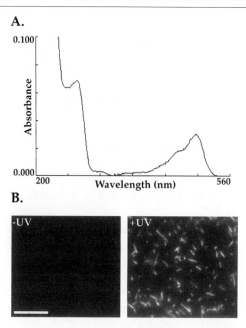

B.

FIG. 1. (A) Absorption spectrum of C2CF-tubulin after uncaging for 30 min. Prior to uncaging there is no absorbance at 495 nm.[2] (B) Assay of polymerization competency and photoactivatibility of C2CF-tubulin. Microtubules polymerized with C2CF-tubulin were observed in the fluorescein channel before (−UV) and after (+UV) exposure to UV for 2 secs. Bar = 10 μm.

tubulin sequences[16] and includes the contribution of the two bound guanine nucleotides to the absorbance at 280 nm. Thus, labeled tubulin concentration can be directly estimated from a wavelength spectrum after subtracting the contribution of the coupled dye to the absorbance at 280 nm. For fluorescein, the absorbance at 280 nm is ~20% that of its peak absorbance at 495 nm.

To determine the concentration and labeling stoichiometry the C2CF must be first uncaged to fluorescein. To do this, dilute the labeled tubulin 1/50 to 1/100 in IB + 2 mM DTT in an Eppendorf tube. Put the Eppendorf tube on a handheld UV lamp, cover with foil (shiny side down), and expose to long-wavelength UV for 30 min. Obtain a wavelength spectrum from 200 to 600 nm after the 30-min activation, using IB + 2 mM DTT exposed to UV in parallel as a blank. An example of such a spectrum is shown in Fig. 1A. Assuming a 100% efficiency for the uncaging reaction, the molar

[16] S. C. Gill and P. H. von Hippel, *Anal. Biochem.* **182,** 319 (1989).

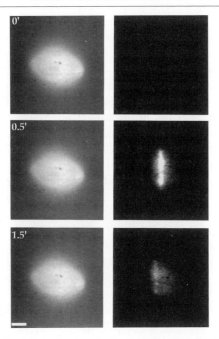

FIG. 2. Poleward microtubule flux in spindles assembled in *Xenopus* egg extracts. Left-hand panels show the X-rhodamine channel and right-hand panels show the fluorescein channel. Prior to activation, no fluorescence is present in the fluorescein channel (top panels). Immediately after a 2-sec activation pulse (using 360-nm light passed through a 100-μm-wide slit placed in a plane conjugate with the specimen[3]), a bright fluorescent mark is evident in the fluorescein channel (middle panels). After 1 min the mark has translocated poleward and diminished in intensity due to rapid spindle microtubule turnover (bottom panels). Images were acquired using a 60×, 1.4 NA Planapo oil immersion objective; exposure was limited using shutters to minimize duration of exposure and neutral density filters to reduce illumination intensity. Bar = 10 μm.

concentration of C2CF and of tubulin can be calculated from the spectrum after activation as follows:

$$\text{Molar concentration of C2CF} = (A_{495} \times \text{dilution factor} \times 1.2)/80,000$$

(The factor of 1.2 corrects for the pH dependence of the absorption spectrum of fluorescein.[17])

$$\text{Molar concentration of tubulin} = [\{A_{280} - (0.2 \times A^{495})\}$$
$$\times \text{dilution factor}]/115,000$$

[17] R. P. Haugland, *in* "Handbook of Fluorescent Probes and Research Chemical." Molecular Probes, Eugene, OR, 1996.

The labeling stoichiometry can be calculated using the ratio of the C2CF concentration to the tubulin concentration. For the example shown in Fig. 1A (diluted 1/100 in IB + 2 mM DTT and activated for 30 min) the concentration of tubulin is calculated to be 56 μM, the concentration of dye is calculated to be 45 μM, resulting in a labeling stoichiometry of ~0.80 mol C2CF/mole tubulin. Using the protocol described above, labeling stoichiometries generally fall in the range of 0.5–0.8.

To assay the polymerization competency of the C2CF-tubulin as well as its ability to be photoactivated, polymerize the C2CF-tubulin (2 mg/ml in 1× BRB80, 1 mM GTP) using either 33% glycerol or 10% DMSO for 30 min at 37° in a foil-wrapped tube. Dilute the polymerized tubulin 1/100 in BRB80 containing 10 μM taxol, and squash 1–2 μl under an 18- × 18-mm coverslip. Using an epifluorescence microscope, compare the same field in the fluorescein channel before and after a 2-sec exposure to UV (performed using a 360-nm DAPI/Hoechst excitation filter). As shown in Fig. 1B, microtubules will be visible in the fluorescein channel after, but not before, UV activation. Microtubules polymerized with pure C2CF tubulin are shorter and more numerous than those polymerized with a similar concentration of unlabeled tubulin, suggesting that the labeled tubulin promotes nucleation and may be slightly altered in its properties relative to unlabeled tubulin.[2]

Use of C2CF-Tubulin for Marking Experiments in Vivo and in Vitro

Microinjection of C2CF-tubulin should be performed in a dimly illuminated room where the lights have been covered with a red acetate sheet. C2CF-tubulin is quite stable *in vivo* (up to 72 hr in *Xenopus* embryos) without significant fluorescence activation.[10] The optical requirements for local activation of C2CF and details on the acquisition of shuttered time lapse images are discussed elsewhere.[3,6] For *in vitro* experiments in *Xenopus* egg extracts, C2CF-tubulin is added to the extract at a final concentration of 100–150 μg/ml. Marking experiments are performed by including X-rhodamine- or Cy5-tubulin to visualize the bulk microtubule distribution, and acquiring paired shuttered time lapse images in the fluorescein and X-rhodamine/Cy 5 channels. Poleward microtubule flux in a metaphase spindle assembled *in vitro* in *Xenopus* egg extracts can be seen in the photoactivation experiment shown in Fig. 2, with the left panels showing the reference X-rhodamine images and the right panels showing the poleward movement and rapid turnover of the fluorescent mark.

[13] Purification of Novel Kinesins from Embryonic Systems

By David Meyer, Daniel R. Rines, Anna Kashina, Douglas G. Cole, and Jonathan M. Scholey

Introduction

Eggs and early embryos of echinoderms and fruit flies have emerged as important model systems for studying structure–function relationships in microtubule-based motor proteins. Here, we briefly describe some of the microtubule-based transport events that are likely to depend on the action of microtubule motors in these early embryos, we outline a general strategy for identifying and purifying kinesin motors from these systems, and, finally, we describe detailed methods for purifying two kinesin holoenzymes, the heterotrimeric kinesin-II motor from sea urchins and the bipolar kinesin, KLP61F (also known as KRP_{130}), from *Drosophila.*

Microtubule-Based Motor Proteins and Intracellular Transport in Eggs and Early Embryos

Intracellular transport systems that move and position subcellular cargoes play critical roles in organizing the cytoplasm of eukaryotic cells, for example, by moving and stationing membrane-bound organelles, driving vesicular transport between these organelles, localizing proteins and RNA molecules, assembling meiotic and mitotic spindles, moving chromosomes, and specifying cleavage planes. Many of these intracellular transport events depend on microtubule-based motor proteins that hydrolyze adenosine triphosphate (ATP) and use the energy released to transport a cargo along microtubule (MT) tracks.[1,2] These motor proteins fall into two families, the kinesins[2] and the dyneins.[3]

In echinoderms and the insect, *Drosophila melanogaster*, MT-based motility plays critical roles in gametogenesis, fertilization, and the polarization of early embryos. Many of these events are likely to depend on the action of MT-based motor proteins, and these early embryonic systems are particularly well suited to the study of such motors because they contain abundant supplies of various motor holoenzymes that can be purified in

[1] L. S. B. Goldstein, *Annu. Rev. Genet.* **27,** 319 (1993).
[2] G. S. Bloom and S. Endow, *Protein Profile* **1,** 1059 (1994).
[3] E. Holzbaur and R. B. Vallee, *Annu. Rev. Cell. Biol.* **10,** 339 (1994).

an active, native multimeric state.[4] Indeed, early embryos of sea urchins and *Drosophila* represent important systems for analyzing the cellular and developmental roles of MT-based transport.[5–9] Each of these two embryonic systems has particular advantages. It is possible to obtain liter amounts of sea urchin eggs from which gram quantities of MTs can be isolated, making this system very well-suited to biochemistry. Unfortunately, sea urchins lack genetics. It is also possible, albeit more difficult, to isolate biochemically useful amounts of MT proteins, including some motor holoenzymes, from fly embryos and this system has the advantage of well-characterized genetics.

Microtubule-Based Motility in Sea Urchin Embryos

Unfertilized sea urchin eggs stockpile depolymerized MT proteins, including tubulin, microtubule-associated proteins (MAPs), and motors. MTs first polymerize on the sperm aster when the fertilizing sperm becomes incorporated into the egg, and they serve as tracks for the migration of the female pronucleus toward the sperm aster at rates of approximately 0.25 μm/sec,[10,11] which may be driven by a minus-end-directed MT motor protein. Nuclear migration determines the position of the nucleus at M phase, and thus postions the mitotic spindle, and the spindle in turn positions cleavage planes.

Following nuclear fusion and centration, MTs associated with the organizing centers are rearranged to form the bipolar mitotic spindle, a self-organizing protein machine that uses MTs and MT-associated motor proteins to assemble, coordinate chromosome motion, and determine the position of the cleavage planes.[7,12,13] As these dynamic MT reorganization events continue, MTs serve as tracks for organizing the endomembrane system and for the delivery of vesicles to specific destinations in the embryo.[14–17] For example, the dynamic, Ca^{2+}-sequestering endoplasmic reticu-

[4] D. G. Cole and J. M. Scholey, *Biophysical J.* **68,** 158s (1995).
[5] K. Dan, *Int. Rev. Cytol.* **9,** 321 (1960).
[6] T. E. Schroeder, *Methods Cell Biol.* **27,** 1 (1986).
[7] B. D. Wright and J. M. Scholey, *Curr. Topics Dev. Biol.* **26,** 71 (1992).
[8] E. A. Fryberg and L. S. B. Goldstein, *Annu. Rev. Cell Biol.* **6,** 559 (1990).
[9] W. Sullivan and W. E. Theurkauf, *Curr. Opin. Cell Biol.* **7,** 18 (1995).
[10] M. S. Hamaguchi and Y. Hiramoto, *Dev. Growth Differ.* **28,** 143 (1986).
[11] T. H. Bestor and G. Schatten, *Dev. Biol.* **88,** 80 (1981).
[12] R. Rappaport, *in* "Cytokinesis in Animal Cells," p. 386. Cambridge University Press, Cambridge, UK, 1996.
[13] V. I. Gelfand and J. M. Scholey, *Nature* **359,** 480 (1992).
[14] J. M. Scholey, *J. Cell Biol.* **133,** 1 (1996).
[15] B. D. Wright, J. H. Henson, K. P. Wedaman, P. J. Willy, J. N. Morand, and J. M. Scholey, *J. Cell Biol.* **113,** 817 (1991).
[16] B. D. Wright, M. Terasaki, and J. M. Scholey, *J. Cell Biol.* **123,** 681 (1993).

lum is reorganized in association with MTs,[18] and the vesicles that are actively transported along MTs are thought to deliver new membrane, extracellular matrix (ECM) material, secretory proteins, and ciliary precursors to the embryonic periphery. This vesicular trafficking may underlie the sorting of ECM components which are stockpiled in the oocyte in several distinct vesicle populations for delivery to the cell surface following fertilization[19]; the insertion of new membrane and ECM derived from cytoplasmic precursor vesicles at the cell surface during cytokinesis; the assembly of cilia on the blastula stage sea urchin embryo[14,20]; and the secretion of the "hatching" enzyme that degrades the fertilization envelope allowing the swimming blastula to emerge.[21] Finally, prior to the fourth division when cells of the 8-cell embryo divide to form the 16-cell embryo, nuclei in the vegetal tier actively migrate toward the vegetal pole, leading to an asymmetric (macromere–micromere) division,[22] and this nuclear transport may depend on the action of a minus-end-directed MT motor protein. MT-based motors are good candidates for driving nuclear migration, vesicular transport, spindle function, and ciliogenesis in early sea urchin embryos.

Microtubule-Based Motility in Early Drosophila Embryos

The early development of *Drosophila* is also characterized by nuclear migrations and multiple mitoses that are likely to depend on MT motors. For example, fruit fly embryogenesis begins with 13 mitotic divisions that occur without cytokinesis and each of these mitoses uses MTs and their associated motors to drive spindle assembly and to coordinate chromosome segregation. Throughout these syncytial divisions, a series of stereotyped nuclear movements occurs, including the MT-dependent transport of nuclei out to the cell cortex during nuclear cycles 7–10, which produces a syncytial blastoderm characterized by a monolayer of cortical nuclei.[9,23,24] Prior to migration, the nuclei are connected by interdigitating MTs, and it has been proposed that an antiparallel MT–MT sliding motor could push the connected nuclei apart by a mechanism analogous to anaphase B, leading to cortical migration.[9] As nuclei reach the embryonic surface during cycle 10,

---

[17] G-Q. Bi, R. L. Morris, G. Liao, J. M. Alderton, J. M. Scholey, and R. A. Steinhardt, *J. Cell Biol.* **138**, 999 (1997).

[18] M. Terasaki and L. A. Jaffe, *J. Cell Biol.* **114**, 929 (1991).

[19] J. C. Matese, S. D. Black, and D. R. McClay, *Dev. Biol.* **186**, 16 (1997).

[20] R. L. Morris and J. M. Scholey, *J. Cell Biol.* **138**, 1009 (1997).

[21] T. Lepage, C. Sardet, and C. Gache, *Dev. Biol.* **150**, 23 (1992).

[22] T. E. Schroeder, *Dev. Biol.* **124**, 9 (1987).

[23] V. E. Foe and B. M. Alberts, *J. Cell Sci.* **61**, 31 (1983).

[24] T. L. Karr and B. M. Alberts, *J. Cell Biol.* **102**, 1494 (1986).

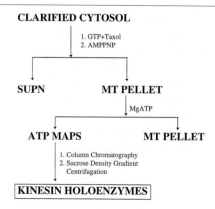

Fig. 1. General scheme for purification of kinesin holoenzymes. Kinesins cosediment with Taxol-stabilized microtubules in the presence of AMPPNP (SUPN = supernatant), and are eluted from the MT pellets using ATP. The resulting ATP MAPs fraction is subjected to conventional fractionation procedures.

there begins a series of saltatory movements of vesicles along microtubules oriented perpendicular to the surface, which likely involves MT motors; vesicles move outward from the central yolk mass to the plasma membrane at 1.7 μm/sec and back again at 0.6 μm/sec.[23] The syncytial divisions are followed by cellularization and the commencement of gastrulation following stage 13. MT-based motor proteins are likely to participate in nuclear migration, spindle assembly, and chromosome segregation in *Drosophila* embryos.

Methodology

General Strategy for Purifying Embryonic Kinesins

Kinesin holoenzymes that participate in some of the aforementioned motile events can be purified from egg/embryo extracts via microtubule affinity precipitation (Fig. 1). In our laboratory, the purification of novel kinesins from embryonic systems has been monitored using "pan-kinesin" antibodies that react with multiple members of the kinesin superfamily. Two types of "pan-kinesin" antibodies have been produced. The first type includes polyclonal anti-peptide antibodies that were raised against conserved peptides located within the motor domain of various members of the kinesin superfamily.[25,26] Procedures for producing such antibodies are

---

[25] D. G. Cole, W. Z. Cande, R. J. Baskin, D. A. Skoufias, C. J. Hogan, and J. M. Scholey, *J. Cell Sci.* **101,** 291 (1992).

[26] K. Sawin, T. J. Mitchison, and L. Wordeman, *J. Cell Sci.* **101,** 303 (1992).

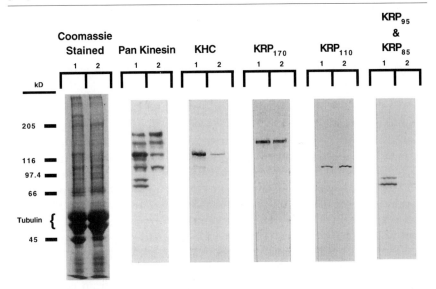

1 = Sea Urchin AMPPNP Microtubules

2 = Sea Urchin ATP Microtubules

FIG. 2. Detection of kinesin polypeptides that copurify with sea urchin egg microtubule precipitates by immunoblotting. Microtubules were prepared from AMPPNP (lane 1) or ATP (lane 2) treated cytosol. The microtubule proteins were run on SDS–PAGE and stained with Coomassie Brilliant Blue (left-hand panel) or transferred to nitrocellulose filters and probed with kinesin antibodies. The pan-kinesin antibodies (panel 2) react with multiple kinesin polypeptides, including kinesin heavy chain (KHC; panel 3), a close relative of Bim C (KRP_{170}; panel 4), a close relative of CHO1 (KRP_{110}; panel 5), and the two motor subunits of the heterotrimeric kinesin-II motor (KRP_{85} and KRP_{95}; panel 6).

detailed elsewhere in this volume by Field *et al.* The second type is a polyclonal antibody, HD, which was raised against recombinant *Drosophila* kinesin motor domain and shown to react with multiple kinesins.[27,28] In our experience, the pan-kinesin peptide antibodies have been most useful for detecting kinesins in samples obtained via AMPPNP-enhanced microtubule affinity binding and biochemical fractionation by immunoblotting.[4] We routinely obtain the pan-kinesin peptide antibodies from the Berkeley Antibody Company (BABCO, Richmond, CA) and use them on immunoblots at 1 : 1000. For example, Fig. 2 shows the spectrum of polypeptides recognized by a pan-kinesin peptide antibody in sea urchin egg MT prepara-

[27] V. Rodionov, F. Gyoeva, and V. Gelfand, *Proc. Natl. Acad. Sci. U.S.A.* **88,** 4956 (1997).
[28] S. L. Rogers, I. S. Tint, P. C. Fanapouz, and V. I. Gelfand, *Proc. Natl. Acad. Sci. U.S.A.* **94,** 3720 (1997).

tions. Some of the polypeptides have been identified as *bona fide* kinesins, using subunit-specific antibodies (Fig. 2).

Purification of the Heterotrimeric Kinesin-II Motor Protein from Sea Urchin Eggs

Sea urchins (*Strongylocentrotus purpuratus*) are collected between December and May from tidepools at Point Arena on the Pacific Coast and housed in 1000-gal holding tanks at the UC Davis marine laboratory in Bodega Bay, approximately 2 hr drive from the main campus. At periodic intervals, members of the laboratory collect packed eggs (approximately 1 liter from 50 female urchins), which are transported back to campus on ice to yield approximately 500 ml cytosolic extract (described in detail by Buster and Scholey[29]). Typically, we prepare various kinesin holoenzymes from 250 ml cytosol, and the remainder is stored at −80° for use when gravid urchins are not available. However, because the Bodega Bay tanks are refrigerated, we are also able to maintain a colony of gravid animals through the summer. Sometimes, animals are ordered from commercial suppliers such as Marinus, Inc. (Long Beach, CA). Kinesins are purified from fresh or frozen cytosol using procedures that are well established in the laboratory.[29-34] Briefly, taxol is used to promote the assembly of MTs in the cytosolic extracts[30] and AMPPNP added to induce strong binding of kinesin holoenzymes to the assembled MTs.[31] MT–kinesin complexes are pelleted and the kinesins desorbed from the MTs by differential centrifugation in ATP and/or high salt for further biochemical fractionation.

Collection of Gametes and Cytosolic Extract Production from Sea Urchin Eggs

1. *Strongylocentrotus purpuratus* are transported as rapidly as possible in Styrofoam boxes, packed in layers with paper well soaked with seawater and cooled with ice packs.

2. Induce sea urchins to shed their gametes by injecting approximately 2 ml of 0.56 *M* KCl solution into the coelomic cavity. The yellow eggs are

[29] D. Buster and J. M. Scholey, *J. Cell Sci.* **14s,** 109 (1991).
[30] J. M. Scholey, B. Neighbors, J. R. McIntosh, and E. D. Salmon, *J. Biol. Chem.* **259,** 6516 (1984).
[31] J. M. Scholey, M. E. Porter, P. M. Grissom, and J. R. McIntosh, *Nature* **318** 483 (1985).
[32] D. Skoufias, D. G. Cole, K. P. Wedaman, and J. M. Scholey, *J. Biol. Chem.* **269,** 1477 (1994).
[33] D. G. Cole, S. W. Chinn, K. P. Wedaman, K. Hall, T. Vuong, and J. M. Scholey, *Nature* **366,** 268 (1993).
[34] K. P. Wedaman, D. W. Meyer, D. J. Rashid, D. G. Cole, and J. M. Scholey, *J. Cell. Biol.* **132,** 371 (1996).

collected by inverting the shedding females onto beakers filled with seawater for 1–2 hr. The white sperm are collected by inverting males onto petri dishes. For production of high-speed supernatant (HSS), perform this and subsequent steps at 0–4°.

3. Decant excess seawater from the egg-containing beakers, pool the eggs, and pass the eggs 10–15 times through a seawater-moistened Nitex screen (150 μm mesh; Tetko Inc., Kansas City, MO) to remove debris and the eggs' jelly coats.

4. To produce HSS, transfer the eggs to 50-ml plastic tubes and gently pellet using a low-speed (18,000g), short-duration (1-min) spin in a clinical centrifuge. Aspirate off the supernatant and gently resuspend the egg pellets in approximately seven volumes of 19 : 1 buffer (Table I; Table II). Repeat the centrifugation and resuspension steps twice more to remove completely the eggs' jelly coats and to minimize the Ca^{2+} concentration. (Many sea urchin proteases are Ca^{2+} activated, and MT polymerization is Ca^{2+} sensitive.) Then resuspend the eggs in sufficient PMEG buffer to reduce NaCl concentration to less than 0.1 M and gently pellet again.

TABLE I
SEA URCHIN—BUFFERS AND SOLUTIONS

Name	Composition
Stock solutions	100 mM ATP in 100 mM Tris-HCl, pH 6.9, 100 mM GP in Tris-HCl, pH 6.9, 100 mM AMPPNP in water, 10 mM Taxol in dimethyl sulfoxide (DMSO), 3 M KCl in water, 1 M MgSO$_4$ in water (Sigma Chemicals, St. Louis, MO)
Protease inhibitors (stock solutions)	1 M Dithiothreitol (DTT) in water, 100 mM phenylmethylsulfonyl fluoride (PMSF) in DMSO, 1 mg/ml pepstatin A in ethanol, 2 mg/ml aprotinin in water, 1 mg/ml leupeptin in water
19 : 1 Buffer	530 mM NaCl, 28 mM KCl, 1 mM ethylenediaminetetraacetic acid (EDTA), 5 mM Tris-HCl in distilled water, pH 7
PMEG buffer	85 mM K$_2$ PIPES, pH 6.9, 15 mM acid PIPES, 2.5 mM MgSO$_4$, 0.5 mM EDTA, 5.0 mM ethylene glycol-bis(β-aminoethyl ether)-N,N,N',N'-tetraacetic acid (EGTA), 0.9 M glycerol, 1.0 mM NaN$_3$, 1 μg/ml pepstatin and leupeptin, 2 μg/ml aprotinin, 100 μg/ml STBI, 0.1 mM PMSF, 1 mg/ml tosyl-L-arginine methyl ester (TAME), 1 mM DTT, 20 μg/ml benzamidine
PEG Mg^{2+}-free buffer	85 mM K$_2$ PIPES, pH 6.9, 15 mM acid PIPES, 15 mM EDTA, 5.0 mM EGTA, 0.9 M glycerol, 1.0 mM NaN$_3$, 1 μg/ml pepstatin and leupeptin, 2 μg/ml aprotinin, 100 μg/ml SBTI, 0.1 mM PMSF, 1 mg/ml TAME, 1 mM DTT, 20 μg/ml benzamidine
PME glycerol-free buffer	85 mM K$_2$ PIPES, pH 6.9, 15 mM acid PIPES, 2.5 mM MgSO$_4$, 0.5 mM EDTA, 5.0 mM EGTA, 1.0 mM NaN$_3$
TME buffer	20 mM Tris-base, pH 8.06 at 4°, 2.5 mM MgSO$_4$, 0.5 mM EDTA, 1.0 mM NaN$_3$

TABLE II

SEA URCHIN—EQUIPMENT FOR EMBRYO COLLECTION AND ENZYME PURIFICATION

Name	Equipment
Nitex screen	150-μm mesh (Tetko Inc.)
Embryo Collection	6 Large plastic trays, 2 L 0.56 M KCl, 18-gauge needles, 60-ml syringes
Gel filtration	Bio-Gel A1.5M gel filtration column (Bio-Rad, Richmond, CA), approx. 100-ml bed volume
Anion exchange	Mono Q column (Pharmacia, Piscataway, NJ), 1-ml volume

5. Resuspend the eggs in two volumes PMEG buffer and homogenize on ice using a prechilled Dounce homogenizer until intact eggs are no longer visible under a stereomicroscope (about 10–20 strokes).

6. Centrifuge the homogenate at 85,000g for 30 min at 4°. Discard the large pellet and the orange, lipid layer; save and centrifuge the clear, intermediate supernatant at 175,000g for 1 hr at 4°. Once again, save the clear, intermediate, HSS. This material is a cytosolic extract and can be stored for extended periods if immediately frozen in liquid nitrogen and then transferred to −80°.

Purification of Polymerized Microtubules and Microtubule-Associated Proteins

1. Thaw *S. purpuratus* HSS by submerging frozen tubes in a 25° water bath while mixing occasionally until thawed. (*Note*: Fresh protease inhibitors can be added to HSS once thawing is complete.)

2. As an *optional* step, you may use 10 units of hexokinase/ml of HSS and 0.9 g glucose/100 ml HSS to partially deplete actomyosin complexes prior to MAPs purification:

   a. Dissolve glucose rapidly in 5 ml PMEG/g glucose to prevent clumping.

   b. Dissolve hexokinase in glucose solution.

   c. Add the hexokinase/glucose mix to the 25° HSS. While stirring slowly, watch for flocculent actomyosin precipitation.

   d. After precipitation begins, wait an additional 15–30 min and centrifuge at 60,000g in 10° for 20–30 min (17,000 rpm in Sorvall SS34 or 25,000 rpm in Beckman Ti50.2 rotor).

   e. Separate the supernatant, which contains the microtubules and MAPs, from the pellet containing the actomyosin complex (the pellet may be discarded).

3. Polymerize microtubules by rewarming the supernatant from the

actomyosin depletion (step 2, if applicable) to 25° and add 1 mM guanosine triphosphate (GTP) and 10–20 μM taxol.

4. Rock the solution gently for 15–20 min at which point MT formation is observed with increased turbidity. (*Note*: If little or no MTs form during this step, 10–20 mg of salt-stripped MTs can be added to the solution.)

5. Binding of MAPs to MTs is accomplished in the presence of 1 mM AMPPNP, and rocking the solution for 20–30 min at 25°.

6. Separation of MTs and MAPs from cytosolic contaminants is accomplished by differential centrifugation through a 15% sucrose cushion with an equal v/v ratio of HSS to sucrose solution:

a. Prepare the 15% sucrose cushion using the PMEG buffer (Table I) with 100 μM GTP, 100 μM AMPPNP, and 5 μM Taxol in open-ended centrifuge tubes.

b. Gently layer the HSS with polymerized MTs over an equal volume of the 15% sucrose cushion in the centrifuge tube with the aid of a graduated pipette (attempt to add the entire volume slowly and evenly).

c. Centrifuge at 50,000g in a 10° swinging bucket rotor for 60 min (13,000 rpm in a Sorvall SS34). At this stage, the pellet contains both the polymerized MTs and rigor bound MAPs while the supernatant contains the 15% sucrose cushion and cytosolic contaminants. On completion, decant the supernatant.

7. Although optional, it is recommended that you wash microtubules and bound proteins with a Mg^{2+}-free buffer.

a. Prepare the wash solution volume with approximately 5 ml/100 ml of initial HSS using the PEG Mg^{2+}-free buffer (Table I) containing 10 μM Taxol, 100 μM GTP, and 100 μM AMPPNP (optional).

b. Gently resuspend the MT/MAPs pellet in the wash solution using a transfer pipette.

c. Quantitatively transfer the MT/MAPs solution to a small glass homogenizer (Wheaton, Millville, NJ) and gently apply five passes with a tight pestle.

d. Transfer the homogenate to an ultracentrifuge tube and centrifuge at 75,000g in a 10° rotor for 20 min (25,000 rpm in a Beckman Ti50.2 rotor). Again, at this stage the pellet will contain both the polymerized MTs and MAPs while the supernatant, with contaminants, can be decanted. Once decanted, place the tubes with pellets on ice.

Elution of Kinesin, Kinesin II, and Other ATP/Salt-Dependent MAPs

1. Using a transfer pipette, gently resuspend the MT/MAPs pellet in 3–6 ml of PMEG buffer containing 150 mM KCl, 10 mM ATP, 1 mM GTP, 10 mM $MgSO_4$, and 10 μM taxol.

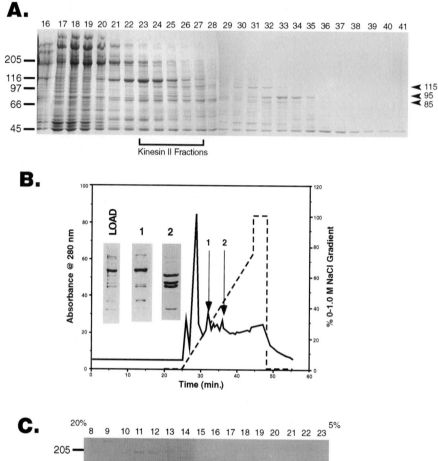

FIG. 3. Purification of the heterotrimeric kinesin-II motor from sea urchin eggs. (A) Coomassie-stained gel of Bio-Gel gel-filtration fractions; pooled fractions 23–27 contain kinesin and kinesin-II proteins; fraction size: 2 ml. (B) Concentration and enrichment of the kinesin-II polypeptides on a Mono Q column. Gel insets show Coomassie-stained gels of the material loaded on the column (load) and the proteins that bound to the resin were eluted using a salt gradient into peak 1 (kinesin) and peak 2 (kinesin-II). Solid line, absorbance at 280 nm; dashed line, salt concentration; fraction size: 0.75 ml. (See Ref. 34 for details.)

2. Quantitatively transfer the resuspended pellet into a prechilled (4°) glass homogenizer (Wheaton) and apply 10 passes with the pestle. Once completed, allow the solution to incubate in homogenizer for 4 hr or more at 4°.

3. For best elution results, add an additional 5 mM MgSO$_4$ and 5 mM ATP before incubating for 30 more minutes at 4°.

4. Centrifuge at 150,000g in a 4° rotor for 20 min (40,000 rpm in a Beckman Ti70.1 rotor). This time the pellet will contain only the polymerized MTs while the supernatant holds the eluted MAPs. (*Important*: do not throw out the supernatant, it has the proteins of interest.)

5. Carefully remove the supernatant from the centrifuge tubes while avoiding the cloudy portions on the top and at the bottom of the tube.

6. If the volume of the supernatant is larger than 3 ml, concentrate proteins by spinning in a concentration tube (Centriprep-30, Amicon, Danvers, MA) at 1500g until the volume is approximately 3 ml.

Gel Filtration Chromatography of Kinesin-II (Fig. 3A). Once the MAPs have been isolated from the MTs and concentrated down into 3 ml, the proteins can be further purified by gel filtration and anion-exchange chromatography:

1. Equilibrate gel filtration column (Bio-Gel A1.5M; Table II) with PME buffer containing 100 μM ATP and 1.0 mM DTT. Flow rates of about 0.2–0.3 ml/min are appropriate.

2. Load the concentrated ATP/KCl eluate onto the column.

3. Collect fractions of approximately 2 ml in sterile polystyrene tubes (do not use glass tubes because kinesin proteins bind directly to glass).

4. Monitor elution by spectrophotometer at A$_{280}$. Kinesin-II will elute between the void volume peak and the large ATP peak at the end. It routinely elutes from our gel filtration column in about 15–20 ml.

5. Pool the peak kinesin-II fractions (which will also contain some kinesin) and dialyze into low ionic strength Tris (TME buffer, pH 8.06) without protease inhibitors for approximately 1–2 hr.

Ion-Exchange Chromatography of Kinesin-II (Fig. 3B)

1. Tris is a basic buffer, whereas PIPES interacts with the positively charged Mono Q anion-exchange column. Lower ionic strength will minimize charge competition between buffer and protein for affinity

---

(C) Coomassie-stained gel of 5–20% sucrose gradient fractions; pooled fractions 12–14 contain highest concentration of kinesin-II; fraction size: 0.2 ml. (To scan the whole gradient, two identical gels were run, with the junction lying between fractions 14 and 15.)

to the resin, and by eliminating most protease inhibitors we get rid of other proteins that might compete for affinity to the resin. At pH 8.06, kinesin-II is sufficiently charged to allow for selective affinity to the Mono Q resin.

2. Two changes of the 1 liter TME containing 100 μM ATP after approximately 45-min intervals are recommended.

3. Equilibrate the anion-exchange (Pharmacia Mono Q) column with the TME buffer.

4. Load the dialyzed material onto the column using a superloop apparatus or other concentrated loading mechanism.

5. After the entire sample is loaded, wash the column with TME equilibration buffer until the A_{280} profile returns to baseline.

6. Elute proteins from the column with a gradient of 0–750 mM NaCl in TME buffer for 20 min using a flow rate of 1 ml/min (approximately 20 column volumes) while collecting 0.75-ml fractions. A gradient from 0–750 mM NaCl (0–75% pump B) will resolve kinesin and kinesin-II as they are eluted in two distinct peaks. In addition, the gradient is steep enough to keep kinesin-II as concentrated as possible for addition to a sucrose gradient in the next step.

7. Kinesin will elute first at about 200 mM NaCl, then kinesin-II will follow at about 350 mM NaCl.

5–20% Sucrose Density Gradient Centrifugation of Kinesin-II (Fig. 3C)

1. Prepare 5 and 20% sucrose solutions by mixing 5 and 20% w/v sucrose solutions in PMEG buffer (Table I), containing fresh 0.1 mM ATP on a gradient mixer in a 5-ml centrifuge tube for use in a swinging bucket rotor (Beckman SW55 Ti). Be careful not to disrupt the finished gradient before and especially after centrifugation.

2. Carefully load 200–250 μl of the peak kinesin-II anion-exchange fractions onto the gradients.

3. Centrifuge gradients for a minimum of 9 hr at 300,000g in a 4° swinging bucket rotor (55,000 rpm in a Beckman SW55 ultracentrifuge rotor).

4. The gradient is fractionated by carefully inserting a 10-μl capillary tube through the center of the gradient and drawing it from bottom to top using a peristaltic pump leading to a fraction collector. Collect 200-μl fractions.

Yield: Fractions obtained during the purification of kinesin-II are shown in Fig. 4. Please note that all fractions shown in Figs. 3 and 4 come from a single, representative kinesin-II purification run. Typically, we obtain 0.2–1.0 mg highly purified kinesin-II from 200 ml cytosol.

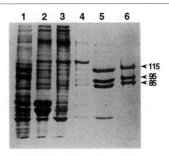

Fig. 4. Coomassie-stained gel of (1) cytosol, (2) AMPPNP microtubules, (3) ATP eluate from AMPPNP microtubules, (4) pooled fractions from gel filtration, (5) peak fraction from Mono Q, anion-exchange column, and (6) pooled fractions from sucrose density gradient containing purified kinesin-II.

Properties of the Heterotrimeric Kinesin-II Motor

The heterotrimeric kinesin-II complex is the founding member of a family of kinesin motors named the "heteromeric kinesins"[35] and consists of three subunits, all of which we have now cloned and sequenced.[33,34,36] Two of the subunits are kinesin-related motor polypeptides named SpKRP85[33] and SpKRP95.[36] Sequence analysis and *in vitro* assembly assays reveal that these two subunits form a heterodimeric coiled-coil with hetero-dimerization being favored over homodimerization; moreover, homodimer-ization is thought to be suppressed by electrostatic repulsion between resi-dues of like charge.

The sequence of the third subunit, SpKAP115, reveals that it is a "pio-neer" protein, unrelated to any known protein in the sequence databases, that may regulate the conformation or cargo-binding properties of the motor subunits.[34] Subsequently, homologs of SpKAP115 have been se-quenced and characterized from mouse[37] and human.[38] Interestingly, the human and urchin homologs have been found to contain armadillo repeats that are protein–protein interaction motifs that may be involved in the attachment of kinesin-II to its intracellular cargo.[38,39] In addition, Shimizu *et al.*[38] showed that the human homolog is a substrate for tyrosine kinases *in vitro*, as we had predicted based on sequence analysis,[34] and is consistent

[35] R. D. Vale and R. Fletterick, *Ann. Rev. Cell Dev. Biol.* **13**, 745 (1997).

[36] D. J. Rashid, K. P. Wedaman, and J. M. Scholey, *J. Molec. Biol.* **252**, 157 (1995).

[37] H. Yamazaki, T. Nakata, Y. Okada, and N. Hirokawa, *Proc. Natl. Acad. Sci. U.S.A.* **93**, 8443 (1996).

[38] K. Shimizu, H. Kawabe, S. Minami, T. Honda, K. Takaishi, H. Shirataki, and Y. Takai, *J. Biol. Chem.* **271**, 27013 (1996).

[39] J. G. Gindhart and L. S. B. Goldstein, *Trends Cell Biol.* **6**, 415 (1996).

with the hypothesis that phosphorylation of this subunit may modulate kinesin-II-driven motility or its attachment to cargo.

The native heterotrimeric kinesin-II motor transports particles toward the plus ends of MT tracks at approximately 0.4 μm/sec in an *in vitro* motility assay.[33] Kinesin-II is associated with punctate, vesicle-like particles that accumulate in the metaphase half spindles and anaphase interzone of dividing sea urchin embryonic cells[40] and with the midpiece and axoneme of spermatozoa.[41] In addition, close relatives of the motor subunits of sea urchin kinesin-II have been identified in *Chlamydomonas, C. elegans, Drosophila,* and mouse (reviewed by Scholey[14]). Analysis of organisms carrying mutations in the genes encoding close relatives of SpKRP85 and SpKRP95 are consistent with the hypothesis that kinesin-II functions as a membrane traffic motor in axons, axonemes, and spindles. Thus, we hypothesized that kinesin-II in sea urchin embryos drives vesicle transport along spindle MTs and may deliver new membrane to the developing cleavage furrow or deliver new material required for ciliogenesis.[14] This hypothesis was recently tested using antibody microinjection, which revealed that kinesin-II is clearly important for ciliogenesis, but these experiments did not reveal any role in cell division.[20]

Purification of the Bipolar Kinesin, KLP61F from Drosophila melanogaster Embryos

Drosophila embryos are grown in population cages in the laboratory and used to prepare embryonic kinesin holoenzymes via microtubule affinity precipitation. In the case of the bipolar kinesin KLP61F, good yields can be obtained from 200 ml frozen cytosol. This requires obtaining approximately 200 g of *Drosophila* embryos (Oregon, R), which in turn requires several embryo collections. For this reason, the batches of cytosol obtained from each collection are rapidly frozen in liquid nitrogen and stored at $-80°$ until enough cytosol is accumulated for a KLP61F preparation. The basic protocol for collecting embryos and preparing microtubules is derived from procedures developed by Saxton *et al.*[42]

Drosophila "Farm" Maintenance. Drosophila colonies are maintained with one or two collection plates taken from cages (prepared as described in next section) early in the adult life cycle as follows:

[40] J. H. Henson, D. G. Cole, M. Terasaki, D. Rashid, and J. M. Scholey, *Dev. Biol.* **171,** 182 (1995).
[41] J. H. Henson, D. G. Cole, C. D. Roesener, S. Capuano, R. J. Mendola, and J. M. Scholey, *Cell Motil. Cytoskel.* **38,** 29 (1997).
[42] W. M. Saxon, M. E. Porter, S. A. Cohn, J. M. Scholey, E. C. Raff, and J. R. McInosh, *Proc. Natl. Acad. Sci. U.S.A.* **85,** 1109 (1988).

1. Remove the embryos from a collection plate by gently rinsing them off into a set of graded copper screens (40-, 60-, and 120-μm mesh) with room temperature water.
2. Rinse the embryos three times with the embryo wash solution (Table III), and submerge filter unit with embryos into a 50% commercial bleach/embryo wash solution for 90 sec–2 minutes.
3. Remove filter with embryos from the 50% bleach solution and rinse in the wash solution until the odor of the bleach is gone.
4. Blot the screen dry with paper towels and transfer 25 g of damp embryos into 300 ml of a sucrose solution (Table III). Stir the sucrose solution to homogeneity and use a 25-ml glass pipette to aliquot 5 ml of the solution into embryo jars (Table IV).

TABLE III
DROSOPHILA—BUFFERS AND SOLUTIONS

Name	Composition
Stock solutions	100 mM ATP in 100 mM Tris-HCl, pH 6.9, 100 mM GTP in 100 mM Tris-HCl, pH 6.9, 100 mM AMPPNP in water, 10 mM Taxol in DMSO, 3 M KCl in water, 1 M MgSO$_4$ in water (Sigma Chemicals)
Protease solutions (stock solutions)	1 M DTT in water, 100 mM PMSF in DMSO, 1 mg/ml pepstatin A in ethanol, 2 mg/ml aprotinin in water, 1 mg/ml leupeptin in water
Buffer A	1 Liter of 100 mM PIPES-KOH, 0.5 mM EDTA, 5 mM EGTA, 2.5 mM MgSO$_4$, 150 mM KCl, 1 mM NaN$_3$, pH 6.9, Buffer is supplemented right before use with 1 mM DTT, 0.1 mM PMSF, 1 μg/ml pepstatin A, 2 μg/ml aprotinin, 1 μg/ml leupeptin, 100 μg/ml soybean trypsin inhibitor (SBTI), and 0.1 mM ATP, and filtered through 0.22-μm membranes
Buffer B	100 ml of 100 mM PIPES-KOH, 0.5 mM EGTA, 2.5 mM MgSO$_4$, 0.9 M glycerol, pH 6.9. Buffer is supplemented right before use with 1 mM DTT, 0.1 mM PMSF, 1 μg/ml pepstatin A, 2 μg/ml aprotinin, 1 μg/ml leupeptin, 100 μg/ml SBTI, and 1 mg/ml p-tosyl-L-arginine methyl ester (TAME)
Embryo wash solution	3 Liters of 0.4% NaCl, 0.03% Triton X-100 in water
50% Bleach solution	1 Part bleach/1 part embryo wash solution
Sucrose solution	90 g Sucrose/450 ml water, 50 ml embryo wash solution
Tubulin	0.5 ml Frozen bovine brain phosphocellulose-purified tubulin at a concentration of at least 2 mg/ml
Fly embryo food	9 g agar, 40 g nutritional yeast, 100 g glucose, 1 liter boiling H$_2$O, 3 ml propionic acid, 20 ml n-butyl p-hydroxybenzoate (10 g/100 ml 70% ethanol); cook dry ingredients and H$_2$O on low heat (low bubble) for approx. 5 min, then add acid/n-butyl p-hydroxybenzoate solutions

TABLE IV

DROSOPHILA—EQUIPMENT FOR EMBRYO COLLECTION AND ENZYME PURIFICATION

Name	Equipment
Fly cages	Plexiglas rectangular 9 × 10 × 20 in.
Embryo jars	1-Liter jars with screw cap lids and screen inserts (Nalgene, Rochester, NY) filled with approx. 100 ml fly embryo food
Gel-filtration column	Superose 6 FPLC column (bed volume 120 ml) equilibrated with buffer A, flow rate of 0.5 ml/min, for 4 hr or more, just prior to or during the preparation
Alternate gel-filtration	Bio-Gel A1.5M gel filtration column (Bio-Rad), approx. 100-ml bed volume

5. Add a paper towel to the side of the jar to avoid excess condensation, cap and place jars into the 25° incubator.
6. Feed the embryos every 4 days with a very light layer of dry baker's yeast until adult flies begin to appear. Adult flies will hatch 12–15 days after initial seeding.
7. Once adults are available, use a CO_2 canister to anesthetize the flies and quantitatively transfer 80–100 ml of immobile flies into a clean 0.5 m^3 cage. Add fresh yeast trays to the cage and begin collecting embryos 24 hr later.

Drosophila Embryo Collection

1. From 2 to 12 population cages (consisting of approximately 0.5-m^3 Plexiglas containers with Nylon screens) are used to store and feed *Drosophila* colonies. The cages are stored in large incubators (Percival Scientific) at a temperature of 25° with a relative humidity of 60–80%. To collect fresh embryos, place fresh tray (consisting of styrofoam meat tray containing molasses/agar) into each cage and allow egg-laying and embryonic development to proceed for desired time course (e.g. 24 hr for 0–24 hr embryo collection).
2. Remove the embryos from trays by rinsing them off into a set of graded copper screens (40-, 60-, and 120-μm mesh) with room temperature water.
3. Rinse the embryos trapped in the bottom screen three times with the embryo wash solution, and submerge for 2 min in 50% commercial bleach to remove the chorions.
4. Rinse them in water until the odor of the bleach is gone, rinse once in buffer A, then blot them with a paper towel to absorb excess buffer. Note that the embryos at this stage are still alive and since

buffer A is toxic to them, it is critical to homogenize the embryos as quickly as possible after this wash.

Drosophila Cytosol Production

1. Weigh the embryos, transfer them into a prechilled Dounce glass homogenizer on ice, and add an equal volume of cold buffer A.
2. Homogenize embryos with five strokes of the loose pestle ("A") followed by seven strokes with the tight pestle ("B"). Alternatively, a motor-driven Teflon pestle in a Potter–Elvehjem-type homogenizer tube can be used.
3. Place the homogenate into prechilled centrifuge tubes and centrifuge at 15,000g for 40 min at 4°.
4. Pick the hardened lipid layer off the top with a cotton tip applicator, transfer supernatant to another set of centrifuge tubes, and centrifuge a second time in a fixed-angle rotor at 150,000g for 30 min at 4° (40,000 rpm in a Beckman 50.2Ti rotor).
5. Collect clear supernatant from this spin into 50-ml plastic polystyrene tubes, freeze rapidly in liquid nitrogen, and store at −80°.
6. Thaw 200 ml of *D. melanogaster* embryonic cytosol (see previous step) in a 15–20° water bath and immediately place on ice. Add 1 mM DTT, 0.1 mM PMSF, 1 μg/ml pepstatin A, 2 μg/ml aprotinin, 1 μg/ml leupeptin, and 100 μg/ml soybean trypsin inhibitor (SBTI) from 1000× stock solutions.
7. To clarify the cytosol of aggregated proteins, pour the cytosol into prechilled centrifuge tubes and centrifuge at 150,000g in a 4° rotor for 40 min (40,000 rpm in a Beckman 50.2Ti rotor).
8. Carefully remove the clear supernatant from between the small lipid layer on top and the loose pellet on the bottom of the tube. This procedure should yield about 150–160 ml of HSS.

Purification of Polymerized Microtubules and Microtubule-Associated Proteins

1. Warm the supernatant to room temperature by briefly putting it into a 37° water bath.
2. Add 1 mM GTP, and 20 μM Taxol from 100× and 400× stock solutions, respectively, and incubate at room temperature for 20 min with occasional stirring (about once very 5 min). The supernatant should become turbid due to microtubule formation.

A

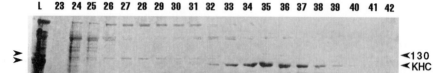

B

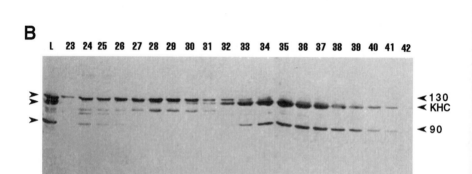

C

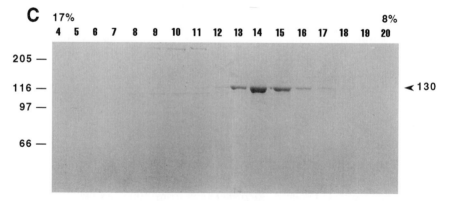

FIG. 5. Purification of the bipolar kinesin KLP61F. (A, B) Gels and blots of fractions from the gel filtration fractionation of *Drosophila* MAPs. (A) is Coomassie-stained and (B) is a corresponding immunoblot probed with a pan-kinesin antibody. *L* refers to the MAP fraction that was loaded onto the column. Numbers above indicate the gel filtration fractions. Arrow-

3. After 20 min, add 100 mM AMPPNP to a final concentration of 1 mM, and incubate the mixture for an additional 20 min with occasional stirring. During this incubation, prepare a 15% (w/v) solution of sucrose in buffer A (Table III).

4. After the incubation, carefully layer approximately 10 ml of the supernatant over an equal amount of 15% sucrose solution in several centrifuge tubes and spin at 35,000g for 60 min at 10° to pellet microtubule-motor complexes.

5. Wash the microtubules by resuspending all pellets in approximately 15 ml of buffer B containing 10 mM EDTA in place of MgSO$_4$ plus 1 mM GTP and 20 μM Taxol in a prechilled glass (Wheaton) homogenizer. Centrifuge for 30 min at 150,000g in a 4° rotor (40,000 rpm in a Beckman 50.2Ti rotor). Decant the supernatant and transfer the centrifuge tubes with the pellets onto ice.

6. To elute the proteins from the polymerized microtubules, gently resuspend the pellets in approximately 5 ml (1/3 volume) of ice-cold buffer A that has been previously supplemented with 250 mM KCl, 1 mM GTP, 20 μM Taxol, 10 mM MgSO$_4$, and 10 mM ATP.

7. Incubate the resuspended pellets for at least 4 hr on ice.

8. After incubation, rewarm the solution to room temperature, incubate for an additional 20 min, and centrifuge at 200,000g for 20–30 min in a 10° rotor (45,000 rpm in Beckman SW55 Ti, or 55,000 rpm in Beckman TLA 55). The supernatant (ATP eluate) should now contain a mixture of microtubule motor proteins.

Gel Filtration Chromatography of KLP61F

1. Add the ATP eluate to a concentrator tube (Centriprep 30) and centrifuge at 1500g until its volume is reduced to approximately 3 ml. Filter through a 0.22-μm membrane.

2. Load onto a preequilibrated Bio-Gel A1.5M (1- $\times$ 90-cm) open-bed column (Fig. 5) or a Superose 6 FPLC gel filtration column. For FPLC use a flow rate of 0.5 ml/min and collect 3-ml fractions. The presence of KLP61F is monitored by sodium dodecyl sulfate–polyacrylamide gel electrophoresis

---

heads on the sides indicate the 130-kDa subunit of KLP61F (130), kinesin heavy chain (KHC), and a 90-kDa band that may represent ncd (90). (C) Coomassie-stained gel of sucrose density gradient fractions. Arrowheads indicate the peak fraction containing purified KLP61F. Top and bottom percentages of sucrose are indicated above the gel, and molecular weight standards are indicated on the left. See Ref. 43 and 44 for details.

(SDS–PAGE) and immunoblotting, where the 130-kDa band should be clearly visible in fractions immediately following the void volume.

3. Pool the fractions containing a clearly visible band of KLP61F and load into a concentrator tube (Centriprep 30), pretreated for 10–15 min with a solution of 1 mg/ml SBTI in buffer A to prevent the nonspecific binding of proteins to the concentrator membrane.

4. Centrifuge the concentrator tube at 1500g until the volume of the pooled KLP61F fractions is reduced to approximately 3–5 ml. This may require several centrifugation steps of 10–15 min each, discarding the excess liquid after each step.

5. In parallel with the previous step, polymerize 0.5 ml of bovine brain microtubules from phosphocellulose-purified tubulin by adding 1 mM GTP and 20 μM Taxol and incubate at 37° for 30–45 min.

6. Mix the pooled KLP61F fractions with Taxol-stabilized PC microtubules in 2 mM AMPPNP and buffer A and pellet them at 100,000g for 20 min.

7. Decant the supernatant from the previous step and resuspend the pellet in 200 μl buffer A containing 250 mM KCl, 20 μM Taxol, 10 mM MgSO$_4$, and 10 mM ATP by gentle pipetting up and down.

8. Incubate for 1 hr at room temperature with occasional pipetting, then centrifuge at 100,000g for 20 min at 10° in a fixed-angle rotor (TLA 55 rotor in a Beckman TL100 centrifuge).

5–20% Sucrose Density Gradient Centrifugation of KLP61F

1. Prepare 5 to 20% (w/v) sucrose gradient in buffer A (Table III), containing fresh 0.1 mM ATP on a gradient mixer (Jule, Inc., New Haven, CT) into a 5-ml centrifuge tube to be used in a swinging bucket rotor (Beckman SW55 Ti). Be careful not to disrupt the finished gradient before and especially after centrifugation.

2. Gently load the supernatant directly onto the freshly prepared sucrose gradient and centrifuge for 9 hr at 300,000g in a swinging bucket rotor (55,000 rpm in a Beckman SW55 ultracentrifuge rotor).

3. Fractionate the gradient by carefully inserting a 10-μl capillary tube through the center of the gradient, all the way to the bottom of the tube, and draw the contents from bottom to top using a peristaltic pump connected to a fraction collector. Collect 200-μl fractions. The presence of KLP61F is monitored by SDS–PAGE. A major 130-kDa band should appear approximately in the middle of the gradient.

Yield: Fractions obtained during the purification of KLP61F are shown in Fig. 5. The final yield is estimated to be 150 μg of KLP61F from 200 ml cytosol in an average preparation.

Properties of KLP61F

Pan-kinesin peptide antibodies react with several putative kinesins in extracts of fruit fly embryos[43] (Fig. 5) one of which is KLP61F.[43–49] This protein is closely related to Eg5, a member of the Bim C family of kinesins. Members of the Bim C family of kinesins, first identified genetically in several laboratories, are motor proteins that play vital roles in the assembly and functioning of a normal bipolar mitotic spindle and thus play important roles in chromosome segregation (see Ref. 48 for a review). *Drosophila* embryonic KLP61F is the first member of this family to be purified from its natural host cell as a holoenzyme.[43–45] We showed that the biochemically purified protein is the product of the KLP61F gene by peptide mapping and microsequencing of the purified protein.[45] In addition, mutations in the KLP61F gene lead to defects in centrosome separation and spindle assembly.[47] Hydrodynamic studies suggested that the native KLP61F holoenzyme is a 490-kDa homotetramer of four identical 130-kDa kinesin-related motor subunits, leading us to hypothesize that the individual motor subunits might assemble into a structure analogous to a miniature myosin filament that could drive spindle pole separation by crosslinking and sliding apart anti-parallel microtubules, using a sort of "sliding filament" mechanism.[43] This hypothesis was tested by using rotary shadow electron microscopy to image KLP61F holoenzymes alone or following their decoration with antibody raised against the recombinant motor domain.[44] The results supported our hypothesis for the bipolar "minifilamentous" nature of the molecule, and suggests that this bipolar kinesin may drive spindle assembly by crosslinking microtubules and inducing MT–MT sliding.

Summary

Several kinesin holoenzymes, including the heterotrimeric kinesin-II and bipolar KLP61F complexes described here, are being purified in our laboratory using microtubule affinity precipitation and conventional biochemical fractionation procedures.[4] These protocols have been optimized

[43] D. G. Cole, W. M. Saxton, K. B. Sheehan, and J. M. Scholey, *J. Biol. Chem.* **269,** 22913 (1994).
[44] A. S. Kashina, R. J. Baskin, D. G. Cole, K. P. Wedaman, W. M. Saxton, and J. M. Scholey, *Nature* **379,** 270 (1996).
[45] A. S. Kashina, J. M. Scholey, J. D. Leszyk, and W. M. Saxton, *Nature* **384,** 225 (1996).
[46] N. R. Barton, A. J. Pereira, and L. S. B. Goldstein, *Mol. Biol. Cell* **6,** 1563 (1995).
[47] M. M. Heck, A. Perreira, P. Pesavento, Y. Yannoni, A. C. Spradling, and L. S. Goldstein, *J. Cell. Biol.* **123,** 665 (1993).
[48] A. S. Kashina, G. C. Rogers, and J. M. Scholey, *Biochim. Biophys. Acta* **1357,** 257 (1997).
[49] D. G. Cole and J. M. Scholey, *Trends Cell Biol.* **5,** 259 (1995).

by using pan-kinesin peptide antibodies and subunit-specific antibodies to monitor the enrichment of kinesin-related polypeptides in particular fractions by immunoblotting. Protein purification represents the most direct route available for determining the oligomeric state and subunit composition of a kinesin holoenzyme, for identifying tightly associated accessory subunits such as SpKAP115, and for determining the molecular architecture and functional properties of native kinesin motors.[49] Protein purification methods therefore represent an important complementary approach to molecular genetic approaches that are being pursued in many other laboratories.

[14] Assaying Processive Movement of Kinesin by Fluorescence Microscopy

By Daniel W. Pierce and Ronald D. Vale

Introduction

A single molecule of conventional kinesin[1] has the remarkable capability of moving over many tubulin subunits without detaching and diffusing away from the microtubule. Such continuous single-molecule motility has also been called processivity, which is a term adopted from the polymerase field to describe the long-range movements of these enzymes along a DNA polymer.[2] Processivity is an uncommon property for cytoskeletal motor proteins. For example, all myosin motors tested thus far spend the majority of their ATPase cycle dissociated or weakly bound to the actin filament.[3] As a result, these motors execute a single mechanical step and then become easily separated from the actin filament. This is an advantage for motors that work in large arrays such as muscle myosin, since motors that are attached but not producing force may exert an unwanted drag. However, for motors that operate in small numbers, such as conventional kinesin which powers organelle transport, processivity is an advantage because it allows efficient motion with minimal time spent searching for and rebinding to a filament. Whether other motors in the kinesin superfamily are also processive is an issue of active study, although preliminary work from our

---

[1] R. D. Vale, B. J. Schnapp, T. S. Reese, and M. P. Sheetz, *J. Cell Biol.* **101,** 37a (1985).
[2] T. L. Capson, S. J. Benkovic, and N. G. Nossal, *Cell* **65,** 249 (1991).
[3] J. T. Finer, R. M. Simmons, and J. A. Spudich, *Nature* **368,** 113 (1994).

laboratory indicates that unc104 (a member of the monomeric family of kinesin motors[4]) and Ncd (a member of the C-terminal family of kinesin motors[5,6]) are not highly processive like conventional kinesin*.

The mechanism by which kinesin generates processive movement has been the focus of considerable research. The current view is that the two heads of conventional kinesin operate by a hand-over-hand mechanism that involves coordination of the two enzymatic cycles.[7] Consistent with this general idea, a truncated, monomeric motor domain of kinesin is a functional motor, but is no longer capable of processive motion.[8,9] However, there is still considerable uncertainty as to the structural basis of processivity and the precise sequence of events that occur during the enzymatic cycle.

To study the mechanism of kinesin, it is important to have a reliable motility assay for measuring processivity. Single-molecule motility of kinesin was first described using a microtubule gliding assay in which the kinesin is adsorbed to the surface of a coverslip.[10] Microtubules are captured from solution and pulled along the surface while they are imaged by differential interference contrast or fluorescence using labeled tubulin. However, demonstration of single-molecule motility using this assay required substantial amounts of both luck and hard work. To absorb kinesin to the glass surface at the very low density while retaining activity, surfaces were pre-coated with tubulin and the kinesin co-adsorbed with cytochrome c. This strategy cannot be expected to succeed with any particular motor, and native conventional kinesin holoenzyme is thus far unique in remaining active when surface adsorbed at very low density. Furthermore, even if the experimental barriers to obtaining a low-density assay can be overcome, it is necessary to show that there is a linear relationship between the rate of spontaneous capture of microtubules from solution and the motor surface density. Only then can the desired measurement—the distribution of run lengths—be made, and there is still no certainty that any particular event is indeed powered by a single kinesin. When microtubules are attached by a single kinesin to a surface, they often pivot about this attachment point. While such pivoting motion is often quoted as evidence for single-molecule

[4] A. J. Otsuka, A. Jeyaprakash, A. J. Garcia, L. Z. Tang, G. Fisk, T. Hartshorne, R. Franco, and T. Born, *Neuron* **6**, 113 (1991).

[5] H. B. McDonald and L. S. Goldstein, *Cell* **61**, 991 (1990).

[6] R. A. Walker, E. D. Salmon, and S. A. Endow, *Nature* **347**, 780 (1990).

[7] D. D. Hackney, *Proc. Natl. Acad. Sci. U.S.A.* **91**, 6865 (1994).

[8] D. D. Hackney, *J. Biol. Chem.* **269**, 16508 (1994).

[9] R. D. Vale, T. Funatsu, D. W. Pierce, L. Romberg, Y. Harada, and T. Yanagida, *Nature* **380**, 451 (1996).

[10] J. Howard, A. J. Hudspeth, and R. D. Vale, *Nature* **342**, 154 (1989).

motility, it is not a reliable measure because aggregates can also give rise to this behavior.

Single-molecule processive motion of conventional kinesin has also been demonstrated in microsphere assays in which the kinesin is adsorbed to microspheres, and the microspheres are observed as they travel along a surface-bound microtubule or axoneme.[11] The key argument that single kinesin molecules can give rise to motility is again a statistical one: As the kinesin/microsphere stoichiometry is varied, the fraction of motile microspheres is correctly predicted by assuming a Poisson distribution of kinesins per microsphere and allowing for some minor loss of kinesin due to nonfunctional adsorption and adsorption to the walls of the container. Again, this argument depends on the unique ability of conventional kinesin holoenzyme to adsorb to a surface at very low densities with very little loss of activity, and the number of motors present on any particular microsphere is not known.

In this article, we describe an alternative method for measuring processivity by directly visualizing single motor proteins using fluorescence microscopy (Fig. 1). This assay involves preparing a fluorescent derivative of kinesin and observing single-molecule motility along axonemal microtubules using a microscope that can detect individual fluorescent dye molecules. From observing fluorescently labeled kinesin molecules, travel distance, association time, and spot intensity can be measured, allowing assessment of velocity, processivity (here defined as the mean distance traveled before dissociation), and the number of fluorophores present in the moving spot. Hundreds of moving spots can be scored in 10 min of recorded videotape. This assay does not require surface adsorption of the kinesin, which can lead to inactivation, and it provides an intrinsic readout of the number of motors present in the motile unit.[9] The assay is rather insensitive to the presence of inactive motors in the preparation (a 2% active fraction is readily detected).

When a motor traveling along a filament becomes momentarily detached, it may either bind anew and resume motion or be separated from the filament by diffusion. The kinetic partitioning between these outcomes depends on the rate of rebinding and the diffusion constant acting to separate the motor from the filament. The magnitude of this diffusion constant depends in turn on the geometry of the assay. In the microtubule gliding assay, the motor position is fixed and the relevant diffusion constant is $\sim 1 \times 10^{-9}$ cm^2/sec, for a 2-μm-long microtubule perpendicular to the long axis[11] in water at 25°. The value for a 200-nm microsphere is 1.1×10^{-8} cm^2/sec. In the single-molecule fluorescence assay, the motor protein

[11] S. M. Block, L. S. Goldstein, and B. J. Schnapp, *Nature* **348**, 348 (1990).

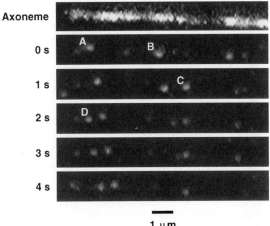

Axoneme

0 s

1 s

2 s

3 s

4 s

1 μm

FIG. 1. Movement of kinesin green fluorescent protein (GFP) on axonemal microtubules. The topmost panel shows an image of a Cy5-labeled axoneme, recorded with a Cy5 filter set and 632-nm excitation. Subsequent panels show the same area imaged with a GFP filter set and 488-nm, 10-mW excitation. Each panel shows a single video frame acquired with a four-frame rolling average. The numbers on the left indicate time in seconds. The letters A, B, C, and D indicate kinesin-GFP molecules. Molecules A and B are bound to and moving along the axoneme over the entire 4-sec interval displayed. Molecule B becomes dimmer between 1 and 2 sec, probably due to photobleaching of one of the two GFP molecules present in the kinesin-GFP dimer. Molecule C binds to the axoneme or adjacent glass surface between 0 and 1 sec, but does not move. Molecule D binds just behind molecule A between 1 and 2 sec and maintains its position relative to molecule A.

itself is the diffusing object and the diffusion constant will depend on its size and shape, but a value of 6.4×10^{-7} cm^2/sec (for a hydrated, 100-kDa spherical protein) is representative. The exact dependence of the measured processivity on the magnitude of this diffusion constant is model dependent, but the values measured for kinesin of 5 μm in a microtubule gliding assay,[10] 1.4 μm in a microsphere assay with 200-nm microspheres,[11] and 0.6 μm in the single-molecule fluorescence assay[9] follow the expected trend. It is possible that motor proteins that do not exhibit detectable processive movement as unattached, single molecules in buffer might be capable of moving processively in their *in vivo* context (i.e., when attached to much larger cargoes and working in viscous cytoplasm). Nevertheless, conventional kinesin shows robust movement, indicating that at least for this motor high viscosity and cargo attachment are not prerequisites for high processivity.

This assay also has several limitations that the reader should be aware of from the outset. First, the single-molecule fluorescence assay has a spatial resolution restricted by the diffraction limit of light (~200 nm). In principle, this could be improved by centroid analysis of the spot size, but because the single-molecule fluorescence images are noisy, the assay will never be capable of nanometer precision motion measurements as have been made using optical traps. As a practical consequence of the spatial limitations, unless a motor is capable of long-distance (100-nm) movement, no motility will be observed. Other important limitations are set by the photobleaching lifetime of the fluorophore employed ($\sim$0.12 sec^{-1} for green fluorescent protein at an excitation power of 10 mW and $\sim$0.06 sec^{-1} for Cy3 at an excitation power of 5 mW), which limits the observation time of a single protein.

The basic requirements for the assay are fluorescent kinesin and a microscope capable of imaging single-molecule fluorescence.[12] Fluorescent kinesin can be obtained by chemical labeling on specific introduced cysteine residues with organic fluorophores such as Cy3, or by generation of green fluorescent protein (GFP) fusion proteins. In the case of cysteine labeling, it is desirable to introduce the cysteine adjacent to positively charged amino acids to enhance its reactivity and therefore the specificity of the labeling. We routinely prepare GFP fusions instead of cysteine-containing constructs for Cy3 labeling for the following reasons:

1. Some proteins are easily inactivated by chemical modification. While we obtained satisfactory results with Cy3 labeling of kinesin, we have found that similar dye labeling of Ncd causes inactivation due to the presence of reactive cysteine groups in the motor domain. This scenario is not uncommon, and any new protein under study has to be carefully studied with respect to possible adverse effects of chemical labeling.

2. Chemical labeling requires the availability of milligram quantities of highly purified protein, and often leads to large losses ($\sim$90% of the starting material in the case of kinesin). In contrast, GFP fusion proteins can be tested in crude extracts. Dye labeling of a specific protein is impossible under these conditions. Kinesin-GFP motility can be observed after *in vitro* translation in rabbit reticulocyte or wheat germ lysates by simply diluting the lysate into the appropriate motility buffer.

3. Chemical labeling gives a statistical mixture of proteins with differing numbers of fluorophores. This distribution must be characterized and taken into account when analyzing data. In addition, the labeling reactions give variable average stoichiometries. GFP fusions provide a consistent and nonstatistical stoichiometry of fluorophore to protein.

4. The GFP methodology is easier to use than Cy3 labeling. Proteins can be tested immediately after column purification.

[12] T. Funatsu, Y. Harada, M. Tokunaga, K. Saito, and Y. Yanagida, *Nature* **374,** 555 (1995).

In this article we focus on practical aspects of making the necessary reagents and performing the assays. The microscopy is briefly described. The methodology has been published, and further details on the microscope built in our laboratory are given elsewhere.[13]

Microscopy and Single Molecule Fluorophores

Although commercial intensified video cameras are capable of detecting single-molecule fluorescence with laser excitation, conventional fluorescence microscopes are not capable of single-molecule imaging because background light levels are too high.[12] Reduction of this background to levels compatible with single-molecule imaging can be accomplished by changing the illumination mode, optimizing the filter sets employed, and minimizing the autofluorescence of all the optics in the light path. Although other designs are possible, the microscope we have used employs total internal reflection (TIR) illumination of the sample. The excitation laser source is coupled at a low angle by a prism into a quartz microscope slide, and totally internally reflects at the interface between the slide and the aqueous sample. This gives rise to an evanescent wave within the sample that decays with an exponential falloff constant of ~150 nm. Thus, in contrast to epi- or trans-illumination of the sample, only the portion of the sample very near this surface is illuminated and background due to out-of-focus fluorescence in the sample is correspondingly reduced. In addition, since the laser light does not traverse the objective lens, background due to autofluorescence of the glass lenses and cements in the objective is also eliminated. To eliminate background due to laser light scattered at the interface and other background luminescence, we employed custom dichroic mirror/barrier filter sets that pass as much of the fluorescence of the dye as possible while attenuating the corresponding laser illuminating wavelength by at least 8 orders of magnitude. Finally, the microscope slides and relay lenses used to form the image at the camera were made of synthetic fused silica due to the low fluorescence of this material. The microscope is equipped with a 35-mW, 632-nm HeNe laser for imaging Cy5 fluorescence, and a 50-mW tunable argon-ion laser used at 488 nm for GFP (the red-shifted S65T variant[14]) or at 514 nm for Cy3.

To date, single molecules of a number of fluorophores have been successfully imaged, including Cy3, Cy5,[12] texas red, fluorescein isothiocyanate

[13] D. W. Pierce and R. D. Vale, *Meth. Cell Biol.* in press (1998).
[14] R. Heim, A. B. Cubitt, and R. Y. Tsien, *Nature* **373,** 663 (1995).

(FITC),[15] tetramethylrhodamine,[16] and GFP.[17] The most important parameter is the number of photons available from the fluorophore before photobleaching. This is the ratio of the fluorescence quantum yield (Φf) to the photobleaching quantum yield (Φb). Unfortunately, neither of these quantities is commonly known with great accuracy, and both may vary by orders of magnitude for any particular fluorophore depending on solution conditions and the microenvironment of the fluorophore. Fluorophore selection is therefore largely empirical. Of the available GFP variants, the red-shifted mutants are clearly preferable to wild-type due to the increased absorption at 488 nm, which allows excitation at lower laser powers. GFP S65T,[14] GFPmut1 (F64L, S65T), and GFPmut2 (S65A, V68L, S72A)[18] have all been imaged successfully using our system. The spectral properties of GFP S65T and GFPmut1 are similar, but GFPmut1 folds more efficiently at higher expression temperatures. GFPmut2 also expresses very well but has a broader absorption spectrum than the other variants, which is suggestive of greater disorder in the protein surrounding the chromophore. For these reasons, GFPmut1 is at present the variant of choice.

Preparation of Fluorescent Kinesin

Chemical Labeling of Kinesin with Cy3

Human conventional kinesin heavy chain contains a ~340-amino-acid (aa) N-terminal motor domain, a long α-helical coiled-coil stalk domain that mediates dimerization, and a small globular C-terminal domain that is thought to interact with cargo.[19,20] The stalk domain is interrupted in the middle by a proline- and glycine-rich hinge, which may allow the tail domain to fold back on and inhibit the motor domain when not bound to cargo.[21] For *in vitro* motility studies, it is desirable to eliminate the tail domain, and aa 560 was chosen as the truncation point because it lies just before the proline–glycine-rich hinge but leaves ~200 aa of the coiled-coil stalk to mediate dimerization. Monomeric constructs do not yield move-

[15] T. Enderle, T. Ha, D. F. Ogletree, D. S. Chemla, C. Magowan, and S. Weiss, *Proc. Natl. Acad. Sci. U.S.A.* **94,** 520 (1997).
[16] T. Ha, T. Enderle, D. F. Ogletree, D. S. Chemla, P. R. Selvin, and S. Weiss, *Proc. Natl. Acad. Sci. U.S.A.* **93,** 6264 (1996).
[17] D. W. Pierce and R. D. Vale, *Nature* **338,** 388 (1997).
[18] B. P. Cormack, R. H. Valdivia, and S. Falkow, *Gene* **173,** 33 (1996).
[19] J. T. Yang, R. A. Laymon, and L. S. Goldstein, *Cell* **56,** 879 (1989).
[20] N. Hirokawa, K. K. Pfister, H. Yorifuji, M. C. Wagner, S. T. Brady, and G. S. Bloom, *Cell* **56,** 867 (1989).
[21] D. D. Hackney, J. D. Levitt, and J. Suhan, *J. Biol. Chem.* **267,** 8696 (1992).

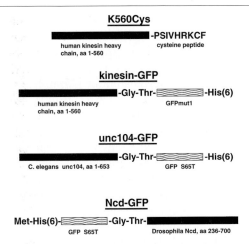

FIG. 2. Constructs described in this chapter. In K560Cys, the positively charged amino acids (arginine and lysine) preceding the introduced cysteine increase its reactivity. The Gly-Thr sequence in kinesin-GFP, unc104-GFP, and Ncd-GFP is introduced due to the *Kpn*1 restriction site used to generate the fusion proteins. The methionine residue preceding the 6xHis tag in Ncd-GFP is introduced to provide a translation start site.

ment,[9,22,23] and we are currently investigating the minimal dimer size necessary for normal processive motion.

The construct K560Cys (Fig. 2) additionally contains a C-terminal, 9-aa, cysteine-containing peptide (PSIVHRKCF) derived from the C terminus of actin.[24] The important aspect of this sequence is the positively charged residues preceding the cysteine that decrease its pK_a and increase its reactivity. In principle, the RKC tripeptide is probably sufficient to add to introduce a reactive cysteine. Bacterial expression of K560Cys is accomplished using the vector pHB40p, a PBR322-derived plasmid that drives expression from a T7 promoter under control of the inducible *lac* repressor and carries an ampicillin resistance gene.

Protein Preparation

1. Inoculate 0.5 ml of LB containing 50 μg/ml ampicillin with one colony of *Escherichia coli* strain BL21(DE3) carrying pHB40p and grow at 37° for 4–8 hr. Inoculate 2.0 liters of LB-amp with 10 μl of this culture and grow at 24° overnight. When the culture reaches an OD_{600} of ~0.5,

[22] D. D. Hackney, *Proc. Natl. Acad. Sci. U.S.A.* **91,** 6865 (1994).

[23] Y.-Z. Ma and E. W. Taylor, *J. Biol. Chem.* **272,** 724 (1997).

[24] S. Itakura, H. Yamakawa, Y. Y. Toyoshima, A. Ishijima, T. Kojima, Y. Harada, T. Yanagida, T. Wakabayashi, and K. Sutoh, *Biochem. Biophys. Res. Comm.* **196,** 1504 (1993).

add isopropyl-β-D-thiogalactopyranoside (IPTG) to a final concentration of 0.1 mM. Continue growth for at least 8 hr.

2. Harvest cells by centrifugation, freeze in liquid nitrogen, and store at $-80°$.

3. Resuspend pellets in 5 ml lysis buffer [25 mM K$^+$ piperazine-N,N'-bis(2-ethanesulfonic acid) (PIPES), pH 6.8, 50 mM NaCl, 1 mM ethyleneglycol-bis(β-aminoethyl)-N,N,N',N'-tetraacetic acid (EGTA), 1 mM dithiothreitol (DTT), 1 mM MgCl$_2$, and 25 μM adenosine triphosphate (ATP)] per gram wet weight. Add phenylmethylsulfonyl fluoride (PMSF) to 1 mM and lyse cells using a French press. Centrifuge the lysate at 40,000g for 30 min at $4°$.

4. Prepare a 5-ml bed volume phosphocellulose column (Whatman P11, Maidstone, England) and equilibrate with lysis buffer. Load the supernatant onto the column at 1 ml/min and wash with 30 ml lysis buffer. Elute the K560Cys with 40 ml 0.2–1.0 M NaCl gradient in column buffer (lysis buffer without NaCl). The K560Cys will elute in a broad peak between 0.3 and 0.5 M NaCl. Identify peak fractions by sodium dodecyl sulfate–polyacrylamide gel electrophoresis (SDS–PAGE).

5. Combine the peak fractions and dilute with column buffer to give a final salt concentration of 0.2 M. Load this solution onto a POROS HS column (PerSeptive Biosystems, Inc., Framingham, MA) and collect the flow-through. Dilute the flow-through 1:1 with column buffer.

6. Load this solution at 1 ml/min onto a 1-ml Mono Q column (Pharmacia, Piscataway, NJ) and wash with 20 ml of column buffer with 50 mM NaCl. Elute the K560Cys with 20 ml 50 mM–1 M NaCl gradient, collecting 0.5-ml fractions. The K560Cys should be largely in two fractions at 0.25–0.3 M NaCl. Pool peak fractions, add 10% w/v sucrose, and freeze and store small aliquots in liquid nitrogen. Typical yields are 0.5–1.5 mg K560Cys.

Labeling K560Cys with Cy3-maleimide. Monofunctional Cy3 (Amersham, Arlington Heights, IL) has one N-hydroxysuccinimide group for labeling primary amines. To convert this functional group to one that will label the sulfhydryl group of a cysteine residue, the Cy3 is first reacted with N-(2-(1-piperazinyl)ethyl)maleimide (PEM; Dojindo Corp., Japan).

1. Dissolve 2.0 mg of PEM (MW 245) in 2 ml of dry dimethylformamide (DMF). Add 24.5 μl (100 nmol) of PEM to 1 tube of monofunctional Cy3 and dissolve by vortexing. Wrap the tube in foil to protect from light and incubate at $40°$ for ~12 hr.

2. Dialyze an aliquot of K560Cys in a microdialyzer versus BRB80 (80 mM K$^+$ PIPES, pH 6.8, 1 mM MgCl$_2$, 1 mM EGTA) for 1 hr at $4°$. Measure the protein concentration using a Bradford assay with bovine serum albumin

(BSA) standard. Calculate the molarity and total moles of the protein solution. A minimum of 0.5 mg of K560Cys is generally required for a successful reaction.

3. Measure the A_{552} nm of a 1:1000 dilution of the Cy3 solution in BRB80 to determine its concentration ($\varepsilon = 130,000\ M^{-1}\ cm^{-1}$), and calculate the number of moles of Cy3 present. If there are more than 100 nmol of Cy3, PEM was the limiting reagent in the Cy3–PEM reaction and the concentration of Cy3–maleimide should be calculated based on the concentration of PEM (4.1 mM). If there are fewer than 100 nmol of Cy3, calculate the amount of Cy3-maleimide based on the measured Cy3 absorption.

4. Dilute the Cy3–maleimide 10-fold in 50 mM glycine, pH 7.0, on ice. This will inactivate any unreacted Cy3. After 60 sec. add a 3-fold molar excess of Cy3-maleimide to the K560Cys solution on ice, keeping the final proportion of DMF to less than 5% (v/v). Incubate on ice in the dark for 60 min.

5. Prepare an NAP-5 gel filtration column (Pharmacia). Equilibrate with 10 ml of 5 mg/ml BSA in BRB80 at 4°, and allow to stand for 20 min to block nonspecific protein binding sites on the column. Wash the column with 20 ml of BRB80.

6. Stop the labeling reaction by adding DTT to 10 mM. Centrifuge at 350,000g for 10 min at 4° to remove aggregates. Dilute the supernatant to 0.2 ml with BRB80 and load into the NAP-5 column. Collect 0.2-ml fractions, and measure the absorbance at 280 and 552 nm to determine the labeling stoichiometry. Cy3 absorbs at 280 nm; the absorbance is ~8% of that at 552 nm. The extinction coefficient at 280 nm of K560Cys is ~44,000, including the contribution of one bound ADP. The labeling stoichiometry is therefore $44,000 \times A_{552}/[130,000 \times (A_{280}-0.08 \times A_{552})]$. Add 25 μM ATP and 1 mM DDT to the fractions that contain K560-Cy3. Motor activity will be retained on ice for several days, or aliquots may be frozen and stored in liquid nitrogen after adding 0.5 mg/ml casein and 10% (w/v) sucrose.

Expression and Purification of Kinesin-GFP

Kinesin-GFP contains the same truncated human kinesin as K560Cys, but the sequence Gly-Thr-GFPmut1-His-His-His-His-His-His is substituted for the 9-aa cysteine-containing peptide (Fig. 2). The Gly-Thr sequence is introduced due to the *Kpn*1 restriction site used to generate the fusion protein, and the 6xHis sequence is introduced to allow purification on metal-chelate affinity media. The bacterial expression plasmid is pET17b (Qiagen, Valencia, CA).

1. Inoculate 1.0 liter TPM with 1 colony of *E. coli* strain BL21(DE3) carrying the pET17b-kinesin-GFP expression vector. Grow at 37° for ~8 hr, or until the OD_{600} of the culture is 1–2. Cool the culture to 23°, add

IPTG to 0.2 mM, and grow an additional 12 hr at 23°. Harvest bacteria by centrifugation, freeze in liquid nitrogen, and store at −80°. TPM medium contains 20 g tryptone, 15 g yeast extract, 4 g NaCl, 2 g Na$_2$HPO$_4$, and 1 g KH$_2$PO$_4$ per liter. Ten milliliters of 20% (w/v) glucose and 2 ml of 50 mg/ml ampicillin are added just prior to inoculation.

2. Resuspend the bacteria in 100 ml cold lysis buffer (50 mM NaPO$_4$, pH 8.0, 250 mM NaCl, 1 mM MgCl$_2$) supplemented with 0.5 mM ATP, 0.1% Tween 20, 0.24 mg/ml of the protease inhibitor Pefablock (AEBSF, Boehringer, Indianapolis, IN), and 2 μg/ml each of aprotinin, leupeptin, and pepstatin. Add 2 mg/ml lysozyme (Sigma, St. Louis, MO) and stir at 4° for 20 min. Add 0.5 μg/ml DNase 1 (Sigma). Sonicate at high power for 1 min on ice, and centrifuge at 40,000g for 30 min.

3. Equilibrate 4 ml of Ni^{2+} NTA agarose (Qiagen) in batch mode with 4 × 15 ml lysis buffer, allowing 10 min for equilibration with each change of buffer. Add 20 mM imidazole : Cl$^-$ (from a 1 M pH 8.0 stock) to the supernatant, and then add the equilibrated resin. Rotate at 4° for 45 min. Pour into a column and wash with 50 ml of wash buffer (50 mM NaPO$_4$, pH 6.0, 250 mM NaCl, 1 mM MgCl$_2$, 0.1 mM ATP). Elute with elution buffer (50 mM NaPO$_4$, pH 7.2, 500 mM imidazole, 250 mM NaCl, 1 mM MgCl$_2$, 0.1 mM ATP), collecting 0.5-ml fractions. Pool the green fractions and dilute 10-fold with column buffer (25 mM K$^+$ PIPES, pH 6.8, 2 mM MgCl$_2$, 1 mM EGTA, 25 μM ATP). Load onto a 1-ml Mono Q column (Pharmacia). Wash with 5 ml of column buffer, 7 ml of column buffer with 0.2 M NaCl, and another 5 ml of column buffer. Elute with 20 ml 0–1 M NaCl gradient in column buffer. Several 1-ml green fractions should elute at ∼0.3 M NaCl. Add 10% (w/v) sucrose and freeze and store aliquots in liquid nitrogen.

Preparation of GFP Fusion Proteins of Other Motors in the Kinesin Superfamily

Preparation of Unc104-GFP. Unc104-GFP contains the first 653 aa of the *Caenorhabditis elegans* monomeric kinesin unc104,[4] with the same Gly-Thr-GFP S65T-His-His-His-His-His-His sequence fused to the C terminus as for kinesin-GFP. Expression and purification procedures are similar to those for kinesin-GFP, except as noted:

1. Before sonication of the lysozyme-treated bacteria, add 10 mM 2-mercaptoethanol (2ME). Also include 10 mM 2ME in the wash and elution buffers.
2. After elution from the Ni^{2+} NTA resin, pool green fractions and dilute 10-fold in column buffer (25 mM K$^+$ PIPES, pH 6.8 2 mM MgCl$_2$, 1 mM EGTA, 1 mM DTT, 0.1 mM ATP). Load onto a 1-ml Mono S column (Pharmacia) and wash with 10 ml of column buffer

with 0.1 M NaCl. Elute the unc104-GFP with 20 ml 0.1–1 M NaCl gradient in column buffer. Unc104-GFP should elute as an intense green band at ~0.25 M NaCl.

Preparation of Ncd-GFP. Because Ncd has a C-terminal motor domain,[6,25] the GFP fusion is made N terminal to the motor domain in this case and the 6xHis tag is located at the N terminus of the GFP. Residues 236–700 of Ncd are included. The complete polypeptide is then Met-His-His-His-His-His-His-GFP S65T-Gly-Thr-Ncd (aa 236–700). Differences relative to the kinesin-GFP expression and purification protocol are noted below.

1. The TPM medium should contain 8 g NaCl per liter.
2. Before sonication, add 2ME to 10 mM. Also include 10 mM 2ME in the wash and elution buffers for the Ni^{2+} NTA column.
3. Dilute the peak fractions from the Ni^{2+} NTA column 10-fold in 50 mM NaCl, 2 mM MgCl$_2$, 1 mM EGTA, 1 mM DTT, and 0.1 mM ATP. Load onto a 1-ml Mono S column equilibrated with column buffer (10 mM NaPO$_4$, pH 7.2, 100 mM NaCl, 2 mM MgCl$_2$, 1 mM EGTA, 1 mM DTT, 0.1 mM ATP) and wash with 10 ml of this buffer. Elute the Ncd-GFP with 20 ml 0–1 M NaCl gradient in column buffer (0.1–1.1 M total NaCl). Ncd-GFP should elute at ~0.25 M total NaCl. Fractions eluting above ~0.35 M NaCl are generally not active.

Expression of Kinesin-GFP by *in Vitro* Translation

Active kinesin-GFP in sufficient quantity for the single-molecule fluorescence assay may be obtained by *in vitro* translation in what germ extract or rabbit reticulocyte lysate cell-free translation systems. mRNA may be prepared in advance using commercial kits such as the mMessage mMachine kit from Ambion Corp. (Austin, TX), or coupled transcription–translation may be performed using commercially available kits from Promega (Madison, WI). At a minimum, however, the *in vitro* translation must yield a final concentration of about 100 nM kinesin-GFP. Some component(s) of the translation cocktail appear to be inhibitory in the assay and must be diluted 20-fold in the final assay mixture. A 100-nM concentration thus allows 20-fold dilution with a final kinesin-GFP concentration of 5 nM. The GFP concentration in the *in vitro* translation reaction may be conveniently determined by measurement of the fluorescence spectrum of a 50- to 200-fold dilution on a sufficiently sensitive fluorimeter and comparison with

[25] H. B. McDonald, R. J. Stewart, and L. S. Goldstein, *Cell* **63**, 1159 (1990).

known concentrations of purified GFP. The *in vitro* translation cocktail alone gives rise to background fluorescence that must be taken into account.

Preparation of Other Reagents for the Motility Assay

Fuorescent Sea Urchin Sperm Flagellar Axonemes

This assay relies on the fact that axonemes will adhere to glass surfaces in the presence of 7.5 mg/ml BSA, while kinesin-GFP largely remains in solution. Presumably, adhesion is mediated by axonemal proteins other than tubulin since microtubules are not retained by the surface under these conditions. It would clearly be desirable to establish a microtubule assay. The difficulty has been that since the surface of the slide (not the coverslip) is imaged, it has been necessary to use very thin samples that preclude the use of perfusion chambers (flow cells). Since all components had to be introduced in a single solution, and buffer conditions under which microtubules adhere to glass generally lead to quantitative removal of kinesin from solution, microtubule assays have not been possible. However, the recent availability of longer working distance, 1.4-NA objectives (Nikon CF∞ series) allows thin flow cells to be used, and we are currently preparing a microtubule assay.

Axoneme Preparation. Strongylocentrotus purpuratus can be obtained along the northern California coast from January to March. The following protocol is based on that of Gibbons and Fronk.[26]

1. Inject live sea urchins (*S. purpuratus*) with a few ml of 0.5 M KCl. Set them on Styrofoam cups at 4° so that semen or eggs will drip into the cup.
2. Collect the semen and dilute 3-fold with seawater (or Instant Ocean, sold at aquarium stores) at 20°. Yield may be increased by rinsing remaining semen off the urchins. All solutions should be maintained at 4° for the remainder of the preparation.
3. Centrifuge at 2000g for 5 min to pellet sperm.
4. Resuspend by douncing the pellet in the same volume of buffer 1 (5 mM imidazole : Cl⁻, pH 7.0, 100 mM NaCl, 4 mM MgSO$_4$, 1 mM CaCl$_2$, 0.1 mM EDTA, 0.1 mM ATP, 7 mM 2ME) with 1% Triton X-100. Centrifuge at 1500g for 5 min (to pellet sperm heads). Centrifuge the supernatant at 12,000g for 5 min.
5. Repeat step 4.
6. Resuspend the pellet by gently douncing in buffer 1. Centrifuge at 12,000g for 5 min. Repeat.

---

[26] I. R. Gibbons and E. Fronk, *J. Biol. Chem.* **254,** 187 (1979).

7. Resuspend the pellet by gently douncing in buffer 2 (5 mM imidazole: Cl$^-$, pH 7.0, 600 mM NaCl, 4 mM MgSO$_4$, 1 mM CaCl$_2$, 0.1 mM EDTA, 7 mM 2ME, 1 mM DTT). Incubate for 10 min, then centrifuge at 12,000g for 5 min.
8. Repeat step 7, but add 1% Triton X-100 to buffer 2 and raise the pH to 8.0.
9. Repeat step 6 with one added resuspension and centrifugation (total of three).
10. Resuspend pellet in one-fifth volume of buffer 1 containing 50% v/v glycerol. Store at $-20°$.

Cy5-Labeling of Sea Urchin Sperm Flagellar Axonemes. This protocol yields axonemes that are lightly labeled but easily visible when using a single-molecule fluorescence microscope.

1. Add 0.25 ml dry dimethyl sulfoxide (DMSO) to one tube of monofunctional Cy5. Measure the A_{650} of a 1:200 dilution in HEPES buffer (50 mM N-2-hydroxyethylpiperazine-N'-2-ethanesulfonic acid (HEPES), 1 mM MgCl$_2$, 1 mM EGTA). The absorbance is equal to the concentration of the stock solution in mM (ε_{650} for Cy5 is 2 $\times$ 10^5). Dilute an aliquot of the stock to 10 μM in DMSO.
2. Add four volumes of HEPES buffer to one volume of axoneme stock. Centrifuge at 13,000g for 5 min at 4°. Resuspend axonemes in one volume HEPES buffer. Keep on ice.
3. Add 1/100 volume of 10 μM Cy5 to the axoneme stock. Incubate on ice for 20 min in the dark. Centrifuge at 13,000g for 5 min at 4°.
4. Resuspend in five volumes BRB80 and centrifuge. Repeat.
5. Resuspend in one-fifth volume of 1:1 BRB80:glycerol. Aliquot and store at $-20°$.

Other Reagents

Oxygen Scavenger Stocks. This oxygen depletion system is based on that of Harada *et al.*[27] All three parts may be aliquoted and stored at $-20°$.
Part A: Prepare a solution of 43 mg/ml glucose oxidase (Sigma G 2133) in 2$\times$ assay buffer (see part C), and add an equal volume of glycerol.
Part B: Catalase. Dilute catalase (Sigma C 3155, supplied as a stabilized solution) with 2$\times$ assay buffer to a concentration of 7.2 mg/ml and add an equal volume of glycerol.
Part C: Glucose. Prepare a 450 mg/ml solution of glucose in 1$\times$ assay buffer (12 mM K$^+$ PIPES, pH 6.8, 2 mM MgCl$_2$, 1 mM EGTA).

[27] Y. Harada, K. Sakurada, T. Aoki, D. D. Thomas, and T. Yanagida, *J. Mol. Biol.* **216,** 49 (1990).

BSA. Prepare a solution of 30 mg/ml lipid-free BSA (Sigma A 0281, less pure grades contain fluorescent impurities) in 1× assay buffer. Centrifuge at 350,000g for 15 min at 4°. Freeze aliquots of the supernatant in liquid nitrogen and store at −80°.

Cleaned Quartz Slides. Quartz slides may be obtained from Matsunami Trading Company (Japan). If the slides have been previously used, soak in a detergent solution for >2 hr to remove coverslips and sealant. Rinse thoroughly with deionized water, and place the slides in a bath sonicator in acetone. Sonicate for 15 min, allow to stand >1 hr, and repeat sonication. Rinse thoroughly with water. Immerse slides in 0.1 M KOH and sonicate for 15 min. Allow to stand for >4 hr and repeat sonication. Rinse thoroughly with water. Immerse in ethanol and sonicate for 15 min. Transfer to a clean ethanol solution and store submerged in ethanol.

Assay and Analysis for Processive Movement of Fuorescent Kinesin

In the assay described below, a fluorescently labeled kinesin motor is combined with ATP, BSA, and Cy5-labeled sperm flagellar axonemes. This mixture is spotted onto a quartz slide, covered with a coverslip, sealed with rubber cement, and imaged. The BSA reduces nonspecific surface adsorption of the labeled kinesin to the glass surfaces but does not prevent the axonemes from sticking. Individual fluorescent spots containing one or more fluorophores can then be observed as they bind to and move along a surface-bound axoneme. Travel distance, association time, and spot intensity can be measured, allowing assessment of velocity, processivity (here defined as the mean distance traveled before dissociation), and the number of fluorophores present in the moving spot.

Assay Procedure

All dilution and mixing steps should be performed on ice.

1. Dilute the fluorescent motor in assay buffer (12 mM K$^+$ PIPES, pH 6.8, 2 mM MgCl$_2$, 1 mM EGTA) containing 30 mg/ml BSA. The final concentration should be between 0.5 and 50 nM.

2. Dilute a small aliquot of Cy5-labeled axonemes at least 10-fold with assay buffer. The exact dilution will depend on the concentration of the axoneme stock.

3. Prepare a solution containing 4 mM ATP, 2% v/v 2ME, and 1:25 diluted glucose solution (from the oxygen scavenger 100× stock) in assay buffer.

4. Prepare a solution containing a 1:25 dilution of the glucose oxidase and catalase 100× stocks in assay buffer. This solution should be used within 30 min of mixing.

5. Remove a cleaned quartz slide from ethanol and rinse thoroughly in ≥18 MΩ/cm deionized water that has been filtered with a 0.2-μm filter. Remove water with compressed air or other gas.

6. Mix 1.1 μl each of the diluted motor, diluted axonemes, ATP solution, and oxygen scavenger enzymes on ice. Remove 4 μl and spot onto the slide. Cover with an 18-mm^2 coverslip and seal with rubber cement (nail polish has been found to interfere with GFP fluorescence[28]).

7. Image the sample. It is generally useful to record an image of the axonemes using a HeNe laser to excite Cy5 fluorescence, and then change filter sets and image GFP using 488-nm argon laser illumination of the same field. For optimal detection of movement, relatively high (10–50 nM) concentrations of fluorescent motor should be used. For quantitative analysis of movements, the concentration should be kept low (0.5–2 nM) to reduce the background fluorescence in the images and to reduce the frequency of movements so that there is generally only one motor moving on a given axoneme at any time. Frame averaging may be useful to reduce noise if full video rate data are not required.

Analysis of Motility Data

If motility is observed, the velocity and distance of movements may be measured, as well as the association rate of fluorescent motors with the axoneme. For kinesin-GFP at 1 nM concentration, a field containing 20 μm of axonemes will give rise to ~600 movements in 10 min of observation. Scoring motility data requires frame-by-frame analysis of many minutes of video data to determine exactly when and where a movement started and stopped, and to detect short associations that may be missed when the images are viewed in real time. At present, it is still most convenient to record data in analog format on sVHS videotape as opposed to digital capture and storage of images due to the limitations of available digital storage devices. Analysis therefore requires a computer-controlled sVHS video cassette player that is capable of displaying images frame by frame. Mouse input of the screen positions of spots and automated storage and calibration of the resulting position and time data are also highly desirable features. The processivity is determined by fitting the measured run-length distribution to an exponential. Since the first bin (corresponding to events of 300 nm or less) may contain random association events with the glass

[28] M. Chalfie, Y. Tu, G. Euskirchen, W. W. Ward, and D. C. Prasher, *Science* **263**, 802 (1994).

surface near the axoneme, it is advisable to exclude it when fitting the data. For exponentially distributed run lengths, a 100-nm processivity corresponds to 5% of the events exceeding the 300-nm threshold, while a 50-nm processivity corresponds to only 0.25%. At this latter low level of detectable events, issues of possible trace contamination of the preparation and misinterpretation of random movements or noise events become troublesome, and little can be said with confidence. However, even if association events do not give rise to measurable movements, the association time can still be measured down to the shortest frame rate with which acceptable signal-to-noise images can be obtained (1/30 sec on our apparatus). If the assumption that undetectable movements are occuring can be justified, and the velocity is known, this time can be converted into a distance that is not constrained by the spatial resolution of the microscope. In this situation, the ATP-concentration dependence of the measured distribution of association times would be helpful in corroborating the ATP-dependent nature of the association events.

It is also highly desirable to measure the intensity distribution of moving versus still (glass-adsorbed) spots in order to determine the association state of the moving molecules. The association state of the glass-adsorbed molecules (taken to be representative of the preparation) can be determined by hydrodynamic methods, and confirmed by the photobleaching behavior of the spots. Spots containing one fluorophore generally photobleach in abrupt, quantal events. Moving versus still spot intensity distributions may be measured using the area integration function of an Argus 20 image processor (Hamamatsu, Bridgewater, NJ). Spot intensity records for analysis of photobleaching behavior are difficult to obtain from analog images, and for this purpose it is desirable to digitize segments of video data. Analysis can then be performed using software packages such as NIH Image or LABView Concept VI (we use the latter). When spot intensities are compared between different images or different samples, it is important that the background intensity be measured in each image and subtracted because the background may vary between samples or even between different regions of the same image.

Conclusion

In summary, this assay offers a direct and unambiguous measurement of the ability of single molecules of motor proteins to undergo directed movements of several hundred nanometers or more. There is no requirement that all or even most of the motor preparation be active, because a 2% active fraction of conventional kinesin can be readily detected. If aggregated proteins are present, it is obvious in the raw data. In contrast to microtubule gliding and microsphere motility assays, interpretation of the data is straightforward. Because the motor proteins are directly imaged, the number of molecules present in a moving spot can be determined by simply measuring the intensity of the spot, whereas other types of assays require careful statistical analysis to be confident that single-molecule motility is being observed.

* Note added in proof: Ncd has been shown to be non-processive using this assay (Case et al., Cell 90(5) 959–966, 1997).

[15] Purification of Dynactin and Dynein from Brain Tissue

By JAMES B. BINGHAM, STEPHEN J. KING, and TRINA A. SCHROER

Introduction

Dynactin is a large, multiprotein complex that is required for cytoplasmic dynein function, both *in vivo* and *in vitro*.[1] Although the precise details of the way in which dynein and dynactin interact to form a productive motor complex have yet to be determined, it is generally assumed that dynactin serves as a linker molecule that targets and tethers dynein to different intracellular structures.[2] These cargoes are thought to include membranes of the endocytic pathway, Golgi membranes, the ER–Golgi intermediate compartment (ERGIC), chromosomes, and microtubules. The complexity and unusual structure of the dynactin molecule may reflect its ability to bind these diverse targets. For this reason, a number of laboratories have become interested in understanding the function of the individual subunits that comprise the dynactin molecule and their organization within the protein complex. In this article, we describe a novel method for purifying dynactin in milligram quantities from bovine brain. Recent modifications to the established dynactin purification method are also discussed. Because cytoplasmic dynein copurifies with dynactin until the very last step of both methods, its purification is also described.

Comparisons of Dynactin Purification Protocols

Our laboratory uses two different protocols to purify dynactin from brain tissue. The first protocol, hereafter referred to as the *microtubule*

[1] T. A. Schroer, *J. Cell Biol.* **127**, 1 (1994).
[2] V. Allan, *Curr. Biol.* **6**, 630 (1996).

affinity protocol, was developed by Schroer and Sheetz[3] and allows microgram quantities of dynactin and dynein to be purified based on their affinity for microtubules in the presence or absence of adenosine triphosphate (ATP). The second "large-scale" protocol relies on successive rounds of ion-exchange chromatography and sucrose density centrifugation and provides milligram quantities of dynactin. Before choosing one protocol over the other, three points should be considered: (1) the total amount of dynactin desired, (2) the cost of setting up and performing the purification, and (3) the percentage of cellular dynactin recovered. For many applications the amount of dynactin desired will be the deciding factor. The microtubule affinity protocol routinely purifies 20 μg of dynactin starting from 20 g of chick embryo brains, whereas the large-scale protocol purifies approximately 20 mg of dynactin starting from 2 kg of bovine brain tissue. A second important consideration is the cost and/or availability of the equipment and materials required for the large-scale protocol. Multiple ultracentrifuges and several large chromatography columns are required for the large-scale preparation whereas a single ultracentrifuge and a Mono Q column are required for the microtubule affinity protocol. Both protocols use expensive reagents, most notably guanosine triphosphate (GTP), adenylimidodiphosphate (AMP-PNP), and Taxol for the microtubule affinity protocol and large amounts of ATP, dithiothreitol (DTT), and PIPES for the large-scale protocol. The third consideration is the population of cellular dynactin that is purified in each protocol. The microtubule affinity protocol only allows recovery of that subclass of dynactin that is able to directly or indirectly bind to microtubules in the absence of ATP and be released in the presence of ATP; this corresponds to less than 2% of the initial cellular dynactin pool.[4] The large-scale protocol allows recovery of approximately 25% of the initial cellular dynactin pool from start to finish. In the large-scale protocol, dynactin is purified based on the overall ionic character and size of the complex and not on a functional property (microtubule affinity) that may be highly regulated both during the cell cycle and locally throughout the cell.

Large-Scale Bovine Dynactin/Dynein Purification

Protocol Overview

A flowchart of the overall purification scheme is shown in Fig. 1. Briefly, five bovine brains are homogenized and a cleared, high-speed supernatant is generated. This supernatant is loaded onto an SP-Sepharose Fast Flow

[3] T. A. Schroer and M. P. Sheetz, *J. Cell Biol.* **115,** 1309 (1991).
[4] S. R. Gill, T. A. Schroer, I. Szilak, E. R. Steuer, M. P. Sheetz, and D. W. Cleveland, *J. Cell Biol.* **115,** 1639 (1991).

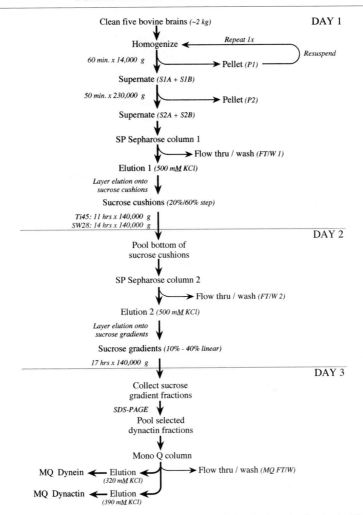

Clean five bovine brains *(~2 kg)*   DAY 1

Homogenize ← *Repeat 1x*

60 *min. x 14,000 g*   → Pellet *(P1)* —— *Resuspend*

Supernate *(S1A + S1B)*

50 *min. x 230,000 g*   → Pellet *(P2)*

Supernate *(S2A + S2B)*

SP Sepharose column 1

→ Flow thru / wash *(FT/W 1)*

Elution 1 *(500 mM KCl)*

Layer elution onto sucrose cushions

Sucrose cushions *(20%/60% step)*

Ti45: 11 hrs x 140,000 g
SW28: 14 hrs x 140,000 g   DAY 2

Pool bottom of sucrose cushions

SP Sepharose column 2

→ Flow thru / wash *(FT/W 2)*

Elution 2 *(500 mM KCl)*

Layer elution onto sucrose gradients

Sucrose gradients *(10% - 40% linear)*

17 hrs x 140,000 g   DAY 3

Collect sucrose gradient fractions

SDS-PAGE

Pool selected dynactin fractions

Mono Q column

MQ Dynein ← Elution ←   → Flow thru / wash *(MQ FT/W)*
(320 mM KCl)

MQ Dynactin ← Elution ←
(390 mM KCl)

Fig. 1. Flowchart of the purification of dynactin and dynein from bovine brain. The steps are divided to show those performed on each of 3 consecutive days.

chromatography column and the 0.5 *M* KCl elution peak is layered onto sucrose cushions and centrifuged overnight. The next day, the sucrose cushions are recovered, loaded onto a second SP-Sepharose Fast Flow column, and the 0.5 *M* KCl elution peak is layered onto sucrose gradients and centrifuged overnight. On the third day, the sucrose gradients are fractionated and the fractions are analyzed by sodium dodecyl sulfate–polyacrylamide gel electrophoresis (SDS–PAGE) to identify those fractions that contain predominantly dynactin polypeptides and a small number of contaminant polypeptides. These fractions are pooled and loaded onto a Mono Q column and the protein peaks are separated by ion-exchange chromatography. The enrichment, yield, and purity of each purification step from a typical preparation are shown in Table I and Fig. 2. Further purification (additional sucrose density centrifugation and Mono Q chromatography) steps can be performed depending on the purity of dynactin desired.

Required Equipment

The following equipment or their equivalents are necessary:

Tissue homogenizer (Tissumizer from Tekmar Cincinnati, OH) with macro generator probe (20 × 195 mm)

Two Beckman (Fullerton, CA) low-speed centrifuges with two JA-10 rotors

Six Beckman ultracentrifuges with three Ti45 rotors and three SW28 rotors

Linear gradient maker with appropriate mixing device and peristaltic pump

740-ml bed volume XK 50/60 chromatography column (5-cm inner diameter, 60-cm length) with flow adaptors (Pharmacia, Piscataway, NJ)

347-ml bed volume Econo-Column chromatography column (2.5-cm inner diameter, 75-cm length) with flow adaptors (Bio-Rad, Richmond, CA)

1.1 Liters SP-Sepharose Fast Flow resin (Pharmacia)

Masterflex peristaltic pump, pump head and #14 and #16 tubing with connectors (Cole Parmer, Vernon Hills, IL)

A_{280} Absorbance detector and flow cell (ISCO Lincoln, NE UA5 or comparable)

LCC-500 Plus Core FPLC system, Mono Q HR 10/10 column (~8-ml bed volume), and 50- or 150-ml Superloop (Pharmacia)

Materials

The following reagents are required for the protocol. Approximate volumes are indicated. ATP, DTT, and protease inhibitors should be made fresh and added to buffers immediately prior to use. Unless otherwise noted, most of the chemicals can be purchased from Sigma, Baker, or any other supplier.

Five bovine brains (~2 kg total) on ice, fresh from a local slaughterhouse

PMEE buffer (500 ml): 35 mM PIPES [piperazine-N,N'-bis(2-ethane-

TABLE I

Dynactin and Dynein Purification Profile

Sample	Total volume (ml)	Total protein (mg)	Step enrichment	Total enrichment	Dynactin Total (mg)[a]	Dynactin Yield (%)[b]	Dynactin Purity (%)[c]	Dynein Total (mg)[a]	Dynein Yield (%)[b]	Dynein Purity (%)[c]
Homogenate	3800	98581.5	—	—	102.2 ± 17	100	0.1	N.D.[d]	N.D.	N.D.
S2 pool	2840	19358.5	5.1	5.1	51.0 ± 22	50.0	0.3	83.8 ± 33	100	0.4
Elution 1	755	3118.7	6.2	31.6	57.4 ± 19	56.2	1.8	72.9 ± 8	87.0	2.3
SC pool	919	883.1	3.5	111.6	56.6 ± 6	55.4	6.4	40.5 ± 10	48.3	4.6
Elution 2	256	494.6	1.8	199.3	35.1 ± 11	34.3	7.1	38.7 ± 24	46.2	7.8
SG pool	106	68.1	7.3	1448.4	27.9 ± 4	27.3	40.9	19.2 ± 4	22.9	28.2
MQ dynactin	3.0	24.1	2.8	3010.3	23.3 ± 1	22.8	96.8	N.A.	N.A.	N.A.
MQ dynein	4.2	16.9	4.0	5841.5	N.A.[d]	N.A.	N.A.	15.9 ± 1	19.0	94.3

[a] The amounts of dynactin and dynein in the MQ dynactin and MQ dynein samples were determined as the total milligrams of protein in the respective sample multiplied by the percent purity of the sample (see below). The amounts of dynactin and dynein in the other samples were determined by quantitative immunoblotting and comparisons to parallel immunoblots of a dilution series of known amounts of MQ dynactin or MQ dynein. The 45A and 70.1 monoclonal antibodies were used to recognize Arp1 and dynein intermediate chain polypeptides, respectively. Extrapolations to the total milligrams of dynactin and dynein were performed assuming that all Arp1 and dynein intermediate chain polypeptides present in the sample were assembled into the dynactin and dynein complexes. Values are ± standard deviations.

[b] The yield of dynactin was determined as the amount of dynactin in the relevant sample divided by the amount of dynactin present in the homogenate. For dynein, the yield was determined as the amount of dynein compared to the amount present in the S2 pool.

[c] MQ dynactin was determined to be 96.8% pure from scans of Coomassie-stained gels. The purity of dynactin in the other samples was determined as the total milligrams of dynactin (see above) compared to the total milligrams of protein in the sample. MQ dynein was determined to be 94.3% pure from scans of Coomassie-stained gels. The purity of dynein in the other samples was determined as the milligrams of dynein compared to the total milligrams of protein in the sample.

[d] N.A., not applicable; N.D. not determined.

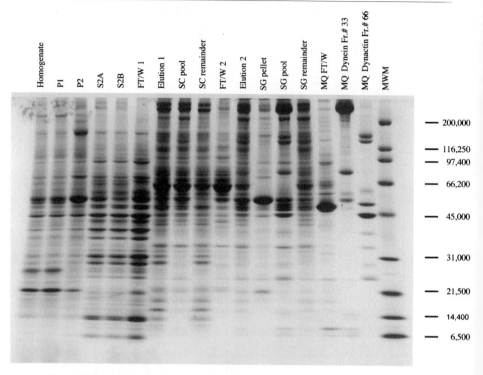

Fɪɢ. 2. Coomassie Brilliant Blue-stained gel showing the polypeptide compositions of relevant fractions across the purification protocol. Ten micrograms of protein was loaded in each lane. The lanes are labeled as described in the text and in Fig. 1. Positions of molecular weight markers are shown to the right of the gel.

sulfonic acid)], 35 mM KOH, 5 mM MgSO$_4$, 1 mM EGTA [ethylene glycol-bis(β-aminoethyl ether) N,N,N′,N′-tetraacetic acid], 0.5 mM EDTA [(ethylenedinitrilo)tetraacetic acid], pH 7.2

ATP (37.5 ml): 200 mM adenosine 5′-triphosphate disodium salt (Sigma), pH 7.0, with NaOH

DTT (37.5 ml): 400 mM 1,4-dithio-ᴅʟ-threitol (Fluka Biochemicals, Milwaukee, WI) in 0.1 mM EDTA, pH 8.0

1000× Protease inhibitors (PIs; 5.5 ml each): 1 mg/ml leupeptin in water, 1 mg/ml pepstatin A in methanol, 40 mg/ml PMSF (phenylmethylsulfonyl fluoride) in ethanol, 10 mg/ml TAME (Nα-p-tosyl-ʟ-arginine methyl ester) in ethanol, 10 mg/ml BAME (Nα-benzoyl-ʟ-arginine methyl ester) in ethanol, 10 mg/ml TPCK (N-tosyl-ʟ-phenylalanine chloromethyl ketone) in ethanol, 10 mg/ml TLCK (Nα-p-tosyl-ʟ-lysine chloromethyl ketone) in ethanol

Homogenization buffer (4 liters): PMEE buffer containing 1 mM DTT, 0.5 mM ATP, and 1× PIs

SP-Sepharose equilibration buffer (6 liters): PMEE buffer containing 1 mM DTT and 0.5 mM ATP

SP-Sepharose elution buffer (3 liters): SP-Sepharose equilibration buffer containing 0.5 M KCl

Sucrose solutions: 10% (300 ml), 20% (1.1 liters), 40% (300 ml), and 60% (250 ml) sucrose (Ultra Pure, ICN Biochemicals, Aurora, OH) in PMEE buffer containing 1 mM DTT, 0.5 mM ATP, with or without 1× PIs as indicated in the text

Mono Q equilibration buffer (500 ml): 35 mM Tris (Ultra Pure, ICN Biochemicals), 5 mM MgSO₄, pH 7.2

Mono Q elution buffer (250 ml): Mono Q equilibration buffer containing 1 M KCl

Cryostorage medium (10 ml): 2.5 M sucrose, 35 mM Tris, 5 mM MgSO₄, pH 7.2

Day 1: Homogenization and Production of High-Speed Supernatant

Obtain five bovine brains (~2 kg) fresh from slaughterhouse and transport them on ice to the laboratory as quickly as possible. Unless otherwise noted the remainder of the purification is performed at 4°. Rinse brains (which include cerebrum, cerebellum, and brain stem tissue) with PMEE buffer to remove excess blood and bone splinters. Remove meninges and blood with Kimwipes. Pinch off small pieces (1–2 cm^3) of the entire brain tissue and drop into a 5-liter beaker containing 2 liters of homogenization buffer. Homogenize using a Tekmar Tissumizer (maximum setting). Move the large probe around the beaker to homogenize the chunks of tissue. When all the large chunks are suspended, bring the probe to the bottom of the beaker and slowly raise it until vigorous mixing begins without the formation of air bubbles. Time homogenization for 3 min from this point, let the homogenate sit for 5 min to depolymerize microtubules, and homogenize again for 3 min. Centrifuge the homogenate (~4 liter final volume) in two Beckman JA10 rotors (14,000g) for 60 min at 4°. Carefully decant the supernatant (S1A) and save. Resuspend the pellets from the first spin in 2 liters of homogenization buffer and repeat the homogenization and centrifugation as described above. Carefully decant the supernatant (S1B). Spin S1A and S1B in ultracentrifuges (Beckman Ti45 rotors) at 230,000g for 50 min at 4°. It will take six to seven total Ti45 runs. Carefully decant the high-speed supernatants (S2A and S2B) and pool (approximately 2.8 liters). Do not allow the S2 pool to become contaminated with material from the pellet.

SP-Sepharose Chromatography and Sucrose Cushion Sedimentation.
Load the S2 pool onto a 740-ml bed volume column of SP-Sepharose Fast
Flow resin (flow rate: 36 ml/min) that has been preequilibrated in SP-
Sepharose equilibration buffer. Wash with SP-Sepharose equilibration
buffer at 75 ml/min until the A_{280} returns to baseline (approximately 2
liters). Elute the column with 2 liters of SP-Sepharose elution buffer at 36
ml/min and collect the material corresponding to the A_{280} absorbance peak
(elution 1; approximately 750 ml). In SW28 and Ti45 tubes (18 each),
prepare 36 two-step sucrose cushions (60 and 20% sucrose with 1× PIs).
SW28: 4 ml of 60% sucrose overlayered with 20 ml of 20% sucrose. *Ti45*:
8 ml of 60% sucrose overlayered with 33 ml of 20% sucrose. Carefully layer
elution 1 onto the sucrose cushions. Centrifuge the SW28 tubes for 14 hr
at 140,000g and the Ti45 tubes for 11 hr at 140,000g (4°). Other combinations
of SW28 and Ti45 rotors can be used as long as the entire elution 1 is
accommodated.

Day 2: Sucrose Cushion Harvest and SP-Sepharose Chromatography

Collect the bottom 33 ml from the Ti45 tubes and the bottom 17 ml
from the SW28 tubes (sucrose cushion pool; approximately 900 ml) and
gently mix end over end in a graduated cylinder (overmixing causes air
bubble formation that will result in a floccular precipitate of protein). Load
the sucrose cushion pool onto a 347-ml bed volume column of SP-Sepharose
Fast Flow resin (flow rate: 4 ml/min) that has been preequilibrated in SP-
Sepharose equilibration buffer. Wash with 1 liter SP-Sepharose equilibration buffer at 6 ml/min. Elute the column with 1 liter SP-Sepharose elution
buffer at 6 ml/min and collect the material corresponding to the A_{280}
absorbance peak (elution 2; approximately 260 ml).

Sucrose Gradient Sedimentation. Pour linear 10–40% sucrose gradients
(24 ml each, without PIs) in 18 SW28 ultraclear tubes. Carefully layer the
gently mixed elution 2 onto the gradients. Spin for 17 hr at 140,000g at 4°.

Day 3: Sucrose Gradient Harvest and Determination of Peak Fractions

Avoiding the pellet, collect fifteen 2.2-ml fractions from the bottom of
the sucrose gradients. Combine fractions collected from the same part of
the 18 gradients (i.e., fractions 1, 2, 3, etc.) into 50-ml conical tubes (each
will contain approximately 36 ml). If desired, resuspend pellets in a total
of 36 ml PMEE buffer. Run an SDS–PAGE gel on ~7.5 μl of each of the
fractions. Examine the Coomassie-stained gel to determine which fractions
to pool, taking care to avoid fractions that contain contaminating polypeptides (a sample gel and detailed instructions on which fractions to pool

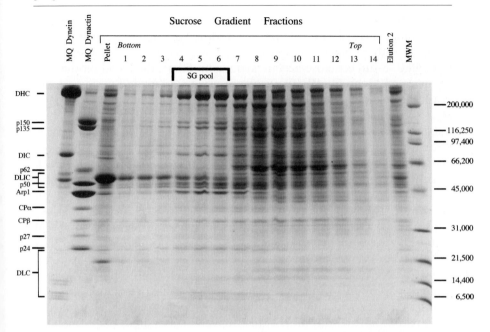

FIG. 3. Coomassie Brilliant Blue-stained gel of fractions from the 10–40% linear sucrose gradients (day 2). Equal volumes of individual sucrose gradient fractions from a representative preparation were analyzed by SDS–PAGE to determine which fractions to pool before loading onto the Mono Q column. The gel contains (left to right) purified Mono Q dynein (MQ Dynein) and Mono Q dynactin (MQ Dynactin) (if possible, these should be included to facilitate identification of appropriate polypeptides), the sucrose gradient pellet (Pellet), consecutive fractions (1–14) of the sucrose gradients, the material loaded onto the gradients (Elution 2), and molecular weight markers (MWM). In this preparation, dynactin and cytoplasmic dynein polypeptides were predominantly found in fractions 2–10. However, only fractions 4, 5, and 6 were pooled together (SG pool) and loaded onto the Mono Q column. Fractions 2 and 3 were not included because they contained only a relatively small percentage of the total dynactin and dynein and they were not significantly enriched for either protein complex. Fractions 7–10 were not included because they contained large amounts of contaminating polypeptides, especially those at approximately 65, 100–110, and 220 kDa. The positions of dynactin subunits [p150, p135, p62, dynamitin (p50), Arp1, actin capping protein α subunit (CPα), actin capping protein β subunit (CPβ), p27, and p24] are shown by the long dashes and positions of cytoplasmic dynein subunits [dynein heavy chains (DHC), intermediate chains (DIC), light intermediate chains (DLIC), and light chains (DLC)] are indicated by short dashes to the left of the gel. Positions of molecular weight markers are shown to the right of the gel.

are given in Fig. 3). Pool the selected fractions (sucrose gradient pool; approximately 100–150 ml) and store on ice.

Mono Q Chromatography. The FPLC system can be used at room temperature as long as peak fractions are placed on ice *immediately* after separation. The A_{280} sensitivity of the chart recorder should be set to the 2 AU range. Load FPLC pump A with Mono Q equilibration buffer and pump B with Mono Q elution buffer. Filter the sucrose gradient pool through 0.45-μm cellulose acetate syringe-tip filters, then load onto a pre-equilibrated Mono Q HR 10/10 column using a 50- or 150-ml Superloop. Wash the column with 20 ml of Mono Q equilibration buffer to return the A_{280} to baseline. Elute the column using the elution program described in Table II and collect 1.4-ml fractions. The first major peak, which contains cytoplasmic dynein (MQ dynein), elutes from the column at 320 mM KCl and corresponds to a little over three fractions. The second major peak usually elutes at 390 mM KCl in two fractions and contains dynactin (MQ dynactin). Immediately add 0.5 mM ATP and 1.0 mM DTT to each fraction of interest. The peak fractions can be analyzed by SDS–PAGE to assess purity (Fig. 4). Quantitative scans of dynactin subunits from such gels reveal that the bovine brain dynactin subunits are present in apparent molar ratios of $2:1:4:10:1:1:1:3$, respectively, for the (p150 + p135), p62, dynamitin, (Arp1 + actin), actin capping protein α subunit, actin capping protein β subunit, p27, and p24 subunits. Densitometry of high-resolution gels has shown that Arp1 and actin are present in the purified dynactin complex in an apparent molar ratio of $9:1$, respectively. As desired, peak fractions can be assayed for protein concentration by a Bradford assay using BSA as a standard. We typically obtain approximately 20 mg each of cytoplasmic dynein and dynactin.

Further Purification and Cryostorage

As seen in Fig. 4, at this point in the purification some contaminating dynein is present in the dynactin peak and some contaminating dynactin

TABLE II
LARGE-SCALE PROTOCOL MONO Q COLUMN
ELUTION PROGRAM

Salt concentration	Volume (ml)	Flow rate (ml/min)
0 mM KCl	14.9	2.0
0–280 mM KCl	25.2	2.0
280–320 mM KCl	Step	
320–340 mM KCl	46.8	1.0
340–390 mM KCl	Step	
390–500 mM KCl	28.0	0.5
500–1000 mM KCl	2.0	2.0
1000 mM KCl	12.0	2.0
1000–0 mM KCl	2.0	2.0

is present in the dynein peak. Depending on the desired use of the protein, these two complexes can be further purified by sedimentation through a second set of sucrose gradients followed by a second round of Mono Q chromatography. The conditions used are the same as described above except that the sucrose gradients are larger (32-ml total volume) and are spun for a longer time (20 hr). The highly purified dynein and dynactin can be cryoprotected and subject to long-term storage at $-80°$ before use in experimental assays. For cryostorage, the protein peaks are diluted $1:1$ in cryostorage medium, mixed gently but completely by pipetting, aliquoted, flash frozen in liquid nitrogen, and transferred to a $-80°$ freezer.

Scaling Down the Bovine Dynactin/Dynein Purification Protocol

The scale of the purification protocol can be altered as necessary to suit the desired dynactin yield or to accommodate the available equipment. The SP-Sepharose Fast Flow resin has a capacity, per milliliter of packed resin bed, of 5 mg protein in the S2 A + B pool or 1.5 mg protein in a sucrose cushion pool. The volume of elution 1 from the first SP-Sepharose column will be 1.0–1.2 times the bed volume of that column. The volume of elution 2 from the second SP-Sepharose column is usually 70% the bed volume of that column. We have determined the linear flow rates that optimize these column's capacities for dynactin and dynein. The first SP-Sepharose column should be loaded and eluted at 110 cm/hr (for a 5-cm internal diameter column this translates to $\pi \times (2.5 \text{ cm})^2 \times 110 \text{ cm/hr} \times 1 \text{ hr}/60 \text{ min} = 36 \text{ ml/min}$) and washed at 230 cm/hr. The second SP-Sepharose column should be loaded at 50 cm/hr and washed and eluted at 75 cm/hr. The capacity of Mono Q resin for the sucrose gradient pool is approximately 10 mg protein per milliliter of resin. If a smaller Mono Q column is used (HR 5/5), the elution program described in Table I can still be used but all flow rates and volumes should be reduced two-fold.

Yields of 3–5 mg of dynactin and dynein can be obtained from one bovine brain. Follow the general large-scale protocol. The volumes obtained for all steps will be one-fifth of the full-scale protocol. The bed volume of the first SP-Sepharose column should be 150 ml; the elution volume (elution 1) will be approximately 165 ml. Elution 1 can be sedimented through sucrose cushions using a single Ti45 rotor. The second SP-Sepharose column bed volume should be 80 ml; the elution volume (elution 2) will be approximately 60 ml. Elution 2 can be sedimented through sucrose cushions using a single SW28 rotor. Either an HR10/10 or an HR5/5 Mono Q column can be used for the final step.

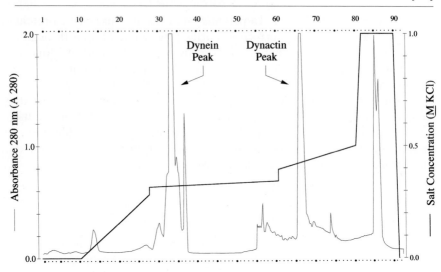

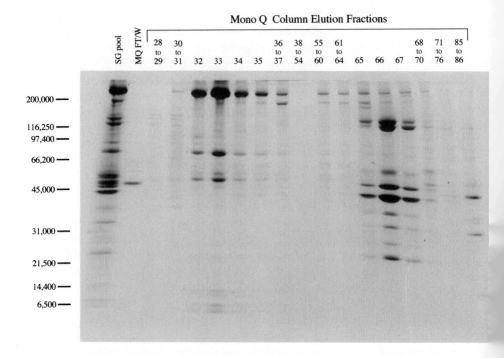

Microtubule Affinity Purification Protocol

The procedure originally developed to purify dynactin was based on the ATP-dependent binding and release of cytoplasmic dynein and dynactin from microtubules.[3] This procedure also allows isolation of highly purified cytoplasmic dynein. Briefly, proteins in a high-speed supernatant are induced to bind to and copellet with polymerized microtubules by inclusion of the nonhydrolyzable ATP analog AMP-PNP. A subset of the microtubule binding proteins is then released from the microtubules by the addition of excess ATP. The released proteins are then subject to sucrose density centrifugation and the 20S pool is fractionated by Mono Q ion-exchange chromatography. Using 20 g of chick embryo brains, approximately 20 μg of dynactin could be purified. Our laboratory uses the protocol described in detail by Schroer and Sheetz[3] with the following minor modifications that increase the yield and purity of dynein and dynactin while decreasing the amount of AMP-PNP used.

The homogenization buffer includes 10% glycerol.

Endogenous polymerized microtubules in the high-speed supernatant are *not* removed by centrifugation prior to AMP-PNP-induced binding but are instead supplemented with 0.25 mg/ml of Taxol-stabilized bovine microtubules.

The binding of polypeptides to microtubules is induced by the addition of 1 mM AMP-PNP, 10 mg/ml glucose, and 1 unit hexokinase per milliliter of high-speed supernatant for 30 min.

The microtubule pellet is washed by resuspending in homogenization buffer (without glycerol) that is supplemented with 1 mM GTP and 20 μM Taxol, then centrifuged at 188,000g for 30 min at room temperature prior to adding the ATP release buffer.

The ATP release buffer additionally contains 200 μM sodium ortho-vanadate, no supplemental MgSO$_4$, and only 2 mM ATP. After centrifugation and collection of the dynein- and dynactin-containing

---

FIG. 4. Mono Q elution profile and corresponding Coomassie Brilliant Blue-stained gel of cytoplasmic dynein and dynactin complexes. The upper panel shows a representative column profile depicting A$_{280}$ nm (thin line) as a function of KCl concentration (thick line). The numbers on the top indicate individual column fractions. The two major protein peaks were in fractions 33 and 66 and are composed of cytoplasmic dynein and dynactin, respectively. In the lower panel, equal volumes of relevant fractions were analyzed by SDS–PAGE. The left lane contains the sucrose gradient pool that was loaded onto the column (SG Pool, see Fig. 2). The next lane contains the flow-through and wash from the Mono Q column (MQ FT/W). The numbers above the remaining lanes correspond to single or pooled fractions from the Mono Q elution. Positions of molecular weight markers are shown to the left of the gel.

TABLE III
MICROTUBULE AFFINITY PROTOCOL MONO Q
COLUMN ELUTION PROGRAM

Salt concentration	Volume (ml)	Flow rate (ml/min)
0 mM KCl	10.0	1.0
0–210 mM KCl	12.6	1.0
210–250 mM KCl	Step	
250 mM KCl	2.0	0.5
250–280 mM KCl	6.4	1.0
280–310 mM KCl	Step	
310–400 mM KCl	21.0	0.5
400–1000 mM KCl	1.0	2.0
1000 mM KCl	4.0	2.0
1000–0 mM KCl	2.0	2.0

supernatant, a second ATP release step is performed identical to the first.

The Mono Q HR5/5 elution program is as described in Table III.

Assays for Dynactin Activity

The two assays that exist for the determination of the activity of purified dynactin make use of the requirement of dynactin for cytoplasmic dynein function in organelle motility assays. The first of these is an *in vitro* vesicle motility assay described in detail by Schroer and Sheetz.[3] Mono Q purified dynein is unable to translocate vesicles along stationary microtubules unless dynactin is included in the assay sample. A second assay that can be used to measure dynactin function is an *in vitro* aster formation assay.[5–7] In this assay, extracts from mitotic HeLa cells are induced to form microtubule asters; aster formation is blocked by the immunodepletion of endogenous dynactin and can be rescued by the addition of exogenous dynactin preparations.[6] Both of these assays are labor intensive and require meticulous adherence to the protocols described. Due to the scope of this article, we are unable to provide details of these assays beyond the descriptions provided in the original Schroer and Sheetz[3] paper and in Chapter [27] in this volume.[7]

[5] T. Gaglio, A. Saredi, and D. A. Compton, *J. Cell Biol.* **131,** 693 (1995).
[6] T. Gaglio, A. Saredi, J. B. Bingham, M. J. Hasbani, S. R. Gill, T. A. Schroer, and D. A. Compton, *J. Cell Biol.* **135,** 399 (1996).
[7] D. A. Compton, *Methods Enzymol.* **298,** [27], 1998 (this volume).

[16] Isolation and Characterization of Kinectin

By JANARDAN KUMAR, ITARU TOYOSHIMA, and MICHAEL P. SHEETZ

Introduction

Current models of microtubule-dependent vesicular organelle motility predict that a motor protein binds a vesicle cargo through a specific interaction with a vesicle membrane protein. Many motor proteins have been identified and a number of these are associated with membranous vesicle motility (see reviews by Vallee and Sheetz[1] and Hirokawa[2]). It thus becomes important to devise strategies to identify and purify the membrane receptors for the motors. This article describes a strategy that has worked for purifying a major kinesin-binding membrane protein; similar strategies may work for purifying other motor binding proteins.

Microtubule-dependent transport is involved in a variety of intracellular trafficking events. While microtubules act as tracks, the anterograde and retrograde motors, kinesin and cytoplasmic dynein, respectively, power the vesicle movements. Kinesin has been implicated in the transport of intracellular organelles, lysosomes,[3] endoplasmic reticulum-like network formation,[4] melanophores,[5] axonal vesicles,[6] Golgi,[7] and mitochondria.[8] However, the choice of the motor protein that interacts with the vesicle surface determines the directionality of vesicular transport.

Although considerable progress has been made in defining the enzymatic activity of motor proteins, their structure and composition, and their interaction with microtubules, the basis of anchorage to membrane-bound organelles is only beginning to be understood. Presumably, motor proteins interact with integral proteins on a vesicle surface that contain the vesicle signals for the directionality of movement on microtubules. Recent studies identified kinectin, a coiled-coil protein of M_r 160 kDa from chicken embryo

---

[1] R. B. Vallee and M. P. Sheetz, *Science* **271**, 1539 (1996).

[2] N. Hirokawa, *Prog. Clin. Biol. Res.* **390**, 117 (1994).

[3] P. J. Hollenbeck and J. A. Swanson, *Nature* **346**, 864 (1990).

[4] S. L. Dabora and M. P. Sheetz, *Cell* **54**, 27 (1988).

[5] V. I. Rodionov, F. K. Gyoeva, A. S. Kashina, S. A. Kuznetsov, and V. I. Gelfand, *J. Biol. Chem.* **265**, 5702 (1990).

[6] S. T. Brady, K. K. Pfister, and G. S. Bloom, *Proc. Natl. Acad. Sci. U.S.A.* **87**, 1061 (1990).

[7] J. Lippincott-Schwartz, N. B. Cole, A. Marota, P. A. Conard, and G. S. Bloom, *J. Cell Biol.* **128**, 293 (1995).

[8] M. Nagaku, R. Sato-Yoshitake, Y. Okada, Y. Noda, R. Takemura, H. Yamazaki, and N. Hirokawa, *Cell* **79**, 1209 (1994).

brain vesicles, as a kinesin-binding protein involved in organelle motility and have implicated it in cytoplasmic dynein-dependent motility as well.[9–11] We describe the method by which an integral membrane protein, kinectin of M_r 160 kDa, was identified and the development of assays of its involvement in vesicle motility.

Strategy for Isolation

We have devised two methods for isolating kinectin as a solubilized membrane receptor for kinesin: either by use of a kinesin affinity column or a specific antibody affinity column. In the initial purification using a kinesin affinity column, it was important to (1) present the motor with its nonmotor domain exposed and (2) label the cytoplasmic-facing vesicle membrane proteins so that they could be identified in the presence of luminal or soluble proteins. Kinesin was first identified as an anterograde mechanochemical enzyme that produced unidirectional movement along microtubules.[12] Kinesin is composed of two heavy chains and two light chains with distinct globular domains. The amino-terminal domain of each heavy chain forms a large globular head that contains the adenosine triphosphate (ATP) and microtubule-binding domains. The tail contains a long coiled-coil domain that connects the head to the small globular carboxy terminus of each chain, with which kinesin light chain forms a fan-shaped tail.[13–15] The amino-terminal motor domain is highly conserved, and a monoclonal antibody (mAb) to that domain, SUK-4,[16] binds to kinesin from most species. Earlier studies showed that extraction of membrane-bound proteins by hypertonic salt treatment did not block vesicle motility, whereas protease treatment of the remaining vesicle proteins did.[17] This indicates that integral membrane proteins on the cytoplasmic surface of the vesicles are required for motor protein-dependent vesicle motility. Therefore, a reasonable approach to identify the interacting proteins on the membrane vesicles is to prepare a column of the exposed tail region of kinesin by making a sandwich of mAb SUK-4 and kinesin. Furthermore, to aid in identifying the vesicle surface proteins that bind to the column, the

---

[9] I. Toyoshima, H. Yu, E. R. Steuer, and M. P. Sheetz, *J. Cell Biol.* **118**, 1121 (1992).

[10] H. Yu, I. Toyoshima, E. R. Steuer, and M. P. Sheetz, *J. Biol. Chem.* **267**, 20457 (1992).

[11] J. Kumar, H. Yu, and M. P. Sheetz, *Science* **267**, 1834 (1995).

[12] R. D. Vale, T. S. Reese, and M. P. Sheetz, *Cell* **42**, 39 (1985).

[13] J. L. Cyr, G. S. Bloom, C. A. Slaughter, and S. T. Brady, *Proc. Natl. Acad. Sci. U.S.A.* **88**, 10114 (1991).

[14] C. S. Johnson, D. Buster, and J. M. Scholey, *Cell Motil. Cytoskelet.* **16**, 204 (1990).

[15] K. P. Wedman, A. E. Knight, J. J. Kendrick, and J. M. Scholey, *J. Mol. Biol* **231**, 155 (1993).

[16] A. L. Ingold, S. A. Cohn, and J. M. Scholey, *J. Cell Biol.* **107**, 2857 (1988).

[17] T. A. Schroer, E. R. Steuer, and M. P. Sheetz, *Cell* **56**, 937 (1989).

vesicle surface proteins were biotinylated with the membrane-impermeable reagent, sulfo-NHS biotin (Pierce Chemical Co., Rockford, IL).

Isolation of Kinectin

Kinectin was first reported as an integral membrane protein from chick embryo brain vesicles that bound to kinesin.[9] Further characterization of kinectin showed that it was essential for kinesin driven vesicle motility. Here we present a brief procedure for isolation of kinectin from chick embryo brain vesicles.

Solutions

PMEE′ buffer: 35 mM PIPES/KOH, pH 7.4, 5 mM MgCl$_2$, 1 mM ethylene glycol-bis(β-aminoethyl ether)-N,N,N',N'-tetraacetic acid (EGTA), and 0.5 mM ethylenediaminetetraacetic acid (EDTA) with 1 mM dithiothreitol (DTT)

Protease inhibitor cocktail: 1 mM (phenylmethylsulfonyl fluoride (PMSF), 1 μg/ml pepstatin A, 10 μg/ml $N\alpha$-P-tosyl-L-arginine methylester (TAME), 10 μg/ml L-1-chloro-3-(4-tosyl amido)-4-phenyl-2-butanone (TPCK), 1 μg/ml leupeptin, and 10 μg/ml soybean trypsin inhibitor

Sucrose step gradient: 15, 20, 40, and 60% sucrose (w/v) were dissolved in homogenization buffer (PMEE′ with protease inhibitor cocktail)

Salt wash buffer: PMEE′ with 0.15 M sodium chloride and 0.05% Triton X-100

Acid elution buffer: 0.2 M glycine/HCl, pH 2.8, and 0.05% Triton X-100

Procedure

Collection of Chick Embryo Brains (CEB) and Preparation of Heavy Microsomal Fraction. Chick eggs are incubated at 38.5° for 12 days (circulated air incubator, Model 1202, G.Q.F. Manufacture Co., Savannah, GA). The CEB is dissected using sterilized forceps and immediately transferred to PMEE′ buffer kept on ice. Approximately 40 g of CEBs is typically collected from 120 eggs (~0.3 g/brain) and washed several times with PMEE′ to remove blood and debris. CEB (~40 g) is mixed with an equal volume of homogenization buffer and transferred to a Dounce glass homogenizer. The homogenization is performed at 4° using 10 slow strokes to avoid foaming. The homogenate is centrifuged at 15,000g for 15 min at 4° in a SA600 rotor (Sorval Instrument, Wilmington, DE). The supernatant is collected and layered onto a stepwise gradient made up of 2.0 ml 60% sucrose, 2.0 ml 40% sucrose, and 1.0 ml 20% sucrose in homogenization buffer. The gradient is centrifuged at 180,000g for 2 hr using an SW40 rotor at 4° (Beckman Instruments, Palo Alto, CA). The 40/60 interface vesicles are collected and subjected to dialysis overnight against PMEE′ buffer to remove sucrose. Alternatively, we remove the peripheral proteins from the vesicles by treatment with 0.1 M sodium carbonate. For carbonate washing of the vesicles, the 40/60 interface vesicles are incubated on ice for 5 min after mixing with a final concentration of 0.1 M sodium carbonate, pH 11. The vesicles are washed by centrifugation through a 15% sucrose cushion in homogenization buffer at 180,000g for 90 min in the SW40 rotor.

Biotinylation of Vesicle Surface Proteins. After carbonate washing of the microsomes, the membrane proteins of the cytoplasmic surface are biotinylated. The microsomes are resuspended with homogenization buffer and reacted with sulfo-NHS biotin. The sulfo-NHS biotin is used at 1.0 mg/ml in homogenization buffer and incubated for 20 min at room temperature. The reaction mixture is quenched with 20 mM of glutamine and the labeled membranes are washed immediately by centrifugation through a 15% sucrose cushion in homogenization buffer at 180,000g for 1 hr at 4°. The final pellet is resuspended in homogenization buffer.

Antibody Affinity Columns. For making immunoaffinity columns, the CNBr-activated Sepharose 4B (Sigma Co., St. Louis, MO) is treated according to manufacturer's instructions. The kinesin heavy-chain monoclonal antibody, SUK-4 (Sigma Co.)[18] and control antibodies, cytoplasmic dynein heavy-chain mAb 440.1 (Sigma Co.)[19] or nonspecific mouse IgG, are used to conjugate ~5–10 mg of IgG per milliliter of CNBr-activated Sepharose 4B beads. The mAb SUK-4, 440.1 or mouse IgG antibody-conjugated bead columns (1.0 ml each) are prepared simultaneously and approximately 10 ml of a high-speed supernatant of CEBs is loaded onto a bead column to load the antibody column with the specific antigen. The washing of beads is performed with 10 ml of PMEE′ buffer to remove unbound proteins from the column. Now, the biotin-labeled, solubilized vesicle membranes in 0.05% Triton X-100 are incubated with kinesin-loaded beads for 1 hr at 4°. The beads are packed in a column and washed with PMEE′ containing 0.05% Triton X-100. Furthermore, the column is washed with the salt wash buffer (0.15 M sodium chloride with PMEE′ and 0.05% Triton X-100) to remove the proteins interacting with kinesin due to nonspecific interactions. Finally, the bound proteins are eluted with the acid elution buffer and the peak fractions are analyzed by SDS–PAGE and subjected to electroblot. The electroblots are incubated with avidin-conjugated horse radish peroxidase and visualized by peroxidase reaction to identify the interacting

[18] J. M. Scholey, J. Heuser, J. T. Yang, and L. S. B. Goldstein, *Nature* (*Lond.*) **338,** 355 (1989).
[19] E. R. Steuer, T. A. Schroer, L. Wordeman, and M. P. Sheetz, *Nature* (*Lond.*) **345,** 266 (1990).

proteins from the cytoplasmic surface of the vesicles.[20] Remarkably, a 160-kDa band was found on SDS–PAGE by Coomassie Blue stain as well as with HRP-avidin on electroblot and appeared specifically in the eluate from column prepared with the SUK-4 mAb.

The rationale for this strategy is that the membrane-binding domain of kinesin is exposed on the affinity column. Many supernatant proteins in addition to kinesin bind to the SUK-4 column in a kinesin-dependent manner, and it remains to be determined if they are kinesin-binding proteins and if they assist in the binding of kinectin. The membranes are mildly solubilized and a set of the biotinylated proteins bound to the beads in a 0.15 M NaCl in a PMEE'-resistant but kinesin-dependent manner.

Several minor biotinylated proteins were observed inconsistently. On the other hand, we were unable to get enrichment of 160-kDa protein in the eluate of cytoplasmic dynein mAb 440.1 affinity column or with other anti-kinesin antibodies that bound to other domains of the molecule. For a summary of the procedure, see the flowchart in Fig. 1.

Preparation of Monoclonal Antibodies against the 160-kDa Protein. A partially purified kinectin is used to produce monoclonal antibodies. We inject the CEB vesicle fraction after mixing with Ribi adjuvant (Ribi Immunochem. Res., Hamilton, MT) for the primary immunization and the first boost in BALB/c mice. For the second boost, the crude kinectin fraction is mixed with streptavidin agarose washed and injected with Ribi adjuvant. The final boost is given with the same kinectin fraction adsorbed on streptavidin without adjuvant. Spleen cells of the BALB/c mice are isolated and fused with myeloma cell line X63Ag8.653. Hybridomas are cloned twice and screening of the antibodies is performed by ELISA (enzyme linked immunosorbent assay).[20] Immunoblots are incubated with the hybridoma cell supernatant and visualized with secondary antibody of goat anti-mouse IgG conjugated with alkaline-phosphatase. A second series of mAbs is also prepared by immunization of BALB/c mice with a crude fraction of 160-kDa protein and boosted twice with pure 160-kDa protein. The final boost is intraperitoneal without Ribi adjuvant. Thus, we were able to make one mAb from the first series (KR160.1B9), and 10 mAbs were obtained from a second series (KR160.2 to KR 160.11). Interestingly, KR 160.10 and KR160.6 recognize only the 160-kDa band on vesicles, whereas others recognized the 160-kDa as well as bands in 120-kDa range.[9]

A third series of mAbs was prepared by immunization of BALB/c mice with carbonate washed 40/60% sucrose interface vesicles from CEB as described earlier. Here, the major objective was to find a series of mAbs

---

[20] E. Harlow and D. Lane, *in* "Antibodies: A Laboratory Manual." Cold Spring Harbor Laboratory Press, New York, 1988.

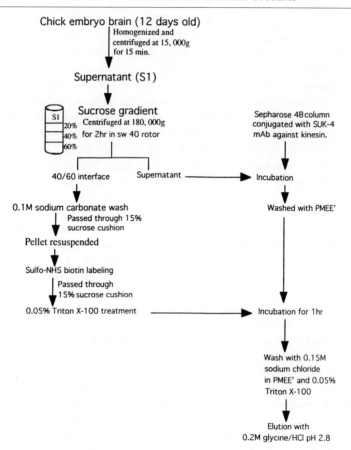

Fig. 1. Schematic representation of protocol for affinity purification of brain microsomal kinesin binding protein(s). The procedure for identification of kinesin binding proteins on the cytoplasmic face of brain microsomes is presented. [Adopted from Toyoshima *et al.*, *J. Cell Biol.* **118**, 1121 (1992).]

against cytoplasmic surface proteins of the vesicle. Remarkably, one mAb VSP4D recognized kinectin 160 kDa in the native form and showed a dramatic inhibition of microtubule-dependent vesicle motility.[11]

Identification of Soluble Kinesin–Membrane Protein Complex. The mAbs against kinectin provided a means to analyze the characteristic behavior of the 160-kDa protein. Further, we characterized the kinesin–membrane protein complex by using a SUK-4 immunoaffinity column. Fresh 40/60 interface vesicles (7.0 mg/ml) are solubilized without carbonate wash in homogenization buffer containing 0.1% Triton X-100 and 0.5 *M* sodium chloride. Following incubation for 1 hr at 4°, the supernatant is collected by centrifugation at 239,000g for 30 min in a TL100.3 rotor (Beckman Instruments). The supernatant is diluted with homogenization buffer (final sodium chloride concentration was 0.15 M) and incubated with 100 μl of SUK-4 conjugated Sepharose beads for 1 hr at 4°. Subsequently, the beads are washed with approximately 10 ml of suspension buffer and the kinesin–membrane protein complex is eluted with 100 μl of acid elution buffer containing 0.1% Triton X-100.

In the eluant, we analyzed the kinesin and kinectin on Western blots using kinesin mAb CKHC1 and KR 160.11 hybridoma supernatant, respectively. Kinesin and kinectin bands on blots were quantitated using video densitometer. With SDS–PAGE destained gels or immunoblots, images were taken with a video camera (Type 67; Dage-MTI, Inc., Michigan City, IN) and processed with a microcomputer-based image analysis (Image 1.29, developed by Wayne Rasband, National Institutes of Health Research Service Branch, National Institute of Mental Health). To define a standard curve, we ran known amounts of the purified kinesin and 160-kDa protein on the same gel. The quantitation of bound kinesin on the 40/60 interface vesicles was 0.9 pmol/mg of vesicle proteins. Furthermore, the detergent extraction study suggested that in presence of 0.5 M sodium chloride >80% of the kinesin was solubilized and after dilution to 0.15 M, >20% of the solubilized kinesin binds to a SUK-4 column and elutes with an acidic solution. Several minor bands appeared in the eluate of solubilized vesicle extract, but the major bands were kinesin and 160-kDa and 47- to 50-kDa proteins (see Fig. 2).[9]

Cosedimentation of Solubilized Kinectin and Kinesin. The 40/60 interface vesicles of CEB from sucrose gradient are solubilized as described earlier except that the vesicles are incubated with 0.1% $C_{12}E_8$ (octaethyleneglycoldodecyl ether; Calbiochem Corp., La Jolla, CA) and 0.5 M sodium chloride for 30 min at 24°. Further, the sample is centrifuged at 190,000g for 30 min at 24° by using a TLA 100.2 rotor to remove the insoluble material. The supernatant is collected and dialyzed for 2.5 hr against PMEE′ at 20° with or without addition to kinesin. A linear gradient of 5–20% sucrose (11.0 ml) in PMEE′ was prepared by using a linear gradient maker. After dialysis, the supernatants are loaded on linear gradients and centrifuged at 180,000g in an SW40 rotor at 20° for 11 hr. After collecting each fraction of 1.0 ml from the tube, the sedimentation of the complex was analyzed by using immunoblot with antibodies: 9E5.KHC2 for kinesin and KR 160.11 for kinectin. To determine the sedimentation of kinesin and cytoplasmic dynein, parallel centrifugations were performed separately under similar conditions as standards. The kinesin and kinectin (160 kDa)

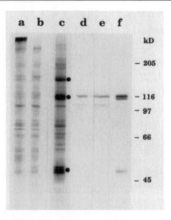

FIG. 2. Soluble kinesin–membrane protein complex. These silver-stained gels (lanes a–c) and Western blots with CKHC.10 mAb (lanes d–f) show that kinesin is present on the heavy microsomes (lanes a and d) and that after solubilization it is coisolated with a 160-kDa band. For this experiment a solubilized fraction (lanes b and e) of native heavy microsomes (lanes a and d) was applied to a SUK-4 antibody affinity column and eluted with acid (lanes c and f). Kinesin heavy chain in each fraction was detected with CKHC.10 mAb (lanes d–f) after SDS–PAGE and electroblot. The major components of the extract are noted (dots): 160-kDa protein, kinesin heavy chain, and 47- to 50-kDa component. [Photographs taken from Toyoshima *et al.*, *J. Cell Biol.* **118**, 1121 (1992).]

cosedimented with a broad range of sedimentation coefficients but peaked at 13S from fresh chick embryo brains whereas kinesin has an apparent sedimentation coefficient of 5S and kinectin has 9S from frozen brains as given in Fig. 3.

Assay for Kinectin-Dependent Vesicle Motility. Although no direct assay of the activity of kinectin has been developed yet, we have developed an assay to find the role of kinectin in vesicle motility by using mAb against the cytoplasmic surface of the vesicle.

In vitro vesicle motility assay. Primary cultures of chick embryo fibroblasts (CEFs) are prepared in our laboratory from 11-day-old embryos by following a procedure described earlier[21] and are maintained in MEM media (Earle's minimal essential medium (EMEM), supplemented with 5% fetal calf serum, penicillin, and streptomycin, all from Life Technologies). After growing the cells in roller bottles for 2 days, the aliquots of the primary cultures are frozen in 60% Iscove's modified Dulbecco's modified medium, 30% fetal calf serum, 10% dimethyl sulfoxide (Life Technologies, Inc.) overnight at −70° in cryogenic vials by using Nalgene cryogenic con-

[21] P. M. Kelly and M. J. Schlesinger, *Cell* **15**, 1277 (1978).

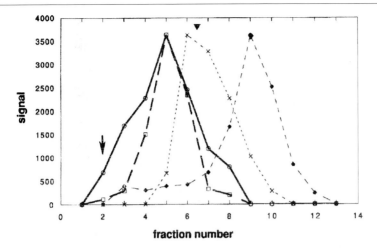

FIG. 3. Cosedimentation of solubilized 160-kDa protein and kinesin. The plots of the normalized immunostaining intensity of kinesin and 160-kDa protein versus gradient fraction show the distribution of both kinesin and 160-kDa protein after velocity sedimentation in a 5–20% sucrose gradient. Cosedimentation of kinesin (□) and 160-kDa protein (○) at 13S is seen in extracts of 40/60% sucrose interface microsomes prepared from fresh brain. In corresponding extracts from frozen brain, kinesin (×) sedimented with its normal sedimentation coefficient (9S) and 160-kDa protein (◆) at 5S. In these experiments, the 40/60% microsomes were solubilized and dialyzed with added kinesin (1 μg of kinesin with 1 mg of solubilized microsome protein). The position of cytoplasmic dynein (long arrow) and kinesin (solid inverted triangle) under such conditions are noted. [Data adopted from Toyoshima *et al.*, *J. Cell Biol.* **118**, 1121 (1992).]

trolled-rate freezing containers (Fisher Scientific, Pittsburgh, PA) and maintained under liquid nitrogen.

Secondary cultures of CEFs are grown in MEM media from primary cultures at 37° supplemented with CO_2 for 2 days in roller bottles after plating. For preparation of the cytosolic fraction and vesicles, the CEFs are treated with 1 mM dibutyryl cyclic AMP in MEM media for 1 hr at 37° and cells are collected by trypsinization. The CEFs are pelleted at 1500g for 15 min at 4° and washed with PMEE′ buffer and resuspended with an equal volume of ice-cold PMEE′ containing 1 mM DTT and protease inhibitor cocktail. The CEFs are homogenized by using a prechilled ball bearing homogenizer. Unbroken cells and nuclei are pelleted after centrifugation at 1000g for 15 min at 4°. The supernatant was centrifuged at 100,000g in a TLS-55 rotor (Beckman) for 30 min at 4°. The cytosolic supernatant and the pellets are recovered in Eppendorf tubes gently to avoid mixing of vesicles in cytosol. Here, the pellet is resuspended and treated with 0.1 M sodium carbonate at pH 11 for 10 min and passed through a 15% sucrose cushion in homogenization buffer to remove peripheral proteins of the vesicles by centrifugation under the same conditions. Furthermore, the vesicle pellet is washed in PMEE' to remove sucrose contamination, which interferes with the vesicle motility. On the other hand, the resulting cytosolic supernatant is adjusted to 1 mM GTP (Sigma Co.) and 20 μM Taxol (Calbiochem Co, La Jolla, CA) and incubated at 37° for 15 min to polymerize endogenous microtubules. The polymerized microtubules are removed by centrifugation at 100,000g for 5 min in a Beckman Airfuge at room temperature. Finally, the supernatant (S3 cytosol) is collected and kept on ice to use as the source of cytosolic motors.

The vesicle motility assay is performed on polarized microtubules using a flow chamber made up of acid-washed coverslips (20% nitric acid overnight, rinsed with deionized distilled water, and stored in 70% ethanol) overlaid on two strips of double-stick tape on a standard microscope slide. The tapes are spaced 2–3 mm apart to form a chamber of 5- to 10-μl volume.

Porcine brain tubulin is purified as previously described[12,22] and stored at −80° until use. Axonemes (a kind gift from Ted Salmon, University of North Carolina, Chapel Hill) are washed in PMEE' buffer, and 10 μl axonemes (~1.0 mg/ml) are added to a flow chamber and the chamber inverted for 15 min to settle the axonemes on the surface. The N-ethylmaleimide (NEM)-treated porcine tubulin is prepared as described earlier.[11] One volume of NEM-treated tubulin (3.0 mg/ml) is mixed with two volumes of porcine brain tubulin (3.0 mg/ml), and 10 μl of tubulin solution is flowed through the flow chamber and incubated for 15 min at 37° after keeping the flow chamber in a humidified chamber. Then, PMEE' buffer containing 1 mM guanosine triphosphate (GTP) and 20 μM Taxol is passed through the flow chamber. The sample is analyzed by video-enhanced differential interference contrast microscope (VEDIC) for the proper density of polarized microtubules. If the density of the polarized microtubules is too high, then a serial dilution of tubulin or axonemes is performed in PMEE' buffer to get about two to three polarized microtubules per video field. To analyze the vesicle motility, a serial dilution of vesicles was performed in PMEE' buffer to optimize for observation of significant number of vesicle movements. Further, the carbonate washed (CW) vesicle protein concentration is measured by using a Bio-Rad (Richmond, CA) protein assay kit and the vesicles were mixed with 150 μg/ml alpha-casein. CW vesicles (0.2 mg/ml) are incubated with monoclonal antibodies 0.1 mg/ml for 2–3 hr with rocking at 4°. The CW vesicle concentration and mAb concentration for dilutions are kept exactly the same for each analysis in a flow chamber. A sample of 15 μl is mixed with 10 μl of cytosolic fraction (S3), 2 μl of vesicles with final concentration of 5 mM Mg-ATP. The resulting mixture is gently added

[22] R. C. Williams and J. C. Lee, *Methods Enzymol.* **85,** 376 (1982).

to the flow chamber. Samples are observed on a heated Zeiss Axiovert-100 microscope at 35–37°. The polarized microtubules on an axoneme and a kinesin-dependent vesicle movement (plus-end directed) are shown in Fig. 4.

The number of vesicle movements were normalized to the concentration of vesicles in the medium and the length of microtubules in the field. The data were expressed as shown in Table I as movements per minute per 36 μm of length of microtubules per 70 vesicles in 4 μm^2 of buffer per minute. Moving vesicles were counted only once. Vesicles that stopped and restarted movement were not counted as a second movement. Means and standard error of means were calculated and weighted by the number of minutes observed in each field. Here, we found that a monoclonal antibody (mAb) VSP4D, which recognized a kinectin epitope on the cytoplasmic surface of the vesicles,[11] inhibited ~90% of vesicle motility toward the plus-end, considered to be kinesin-dependent vesicle motility. As a control, the mAb

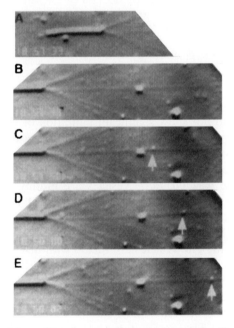

FIG. 4. *In vitro* vesicle motility assay on polarized microtubules. To assess the effects of mAbs on either plus-end (kinesin-driven) or minus-end (cytoplasmic dynein-driven)-directed vesicle motility in the presence of a high-speed cytoplasmic supernatant of the homogenized CEF cells, we quantified movements using an axoneme-based motility assay. A typical plus-end-directed vesicle movement (arrows indicate the positions of the moving vesicle) is shown. (A and B) indicate the video frame before attachment of a vesicle onto a polarized microtubule and (C–E) represent a vesicle attaching and moving toward the plus-end direction. [Photograph adopted from Kumar *et al.*, *Science* **267**, 1834 (1995).]

TABLE I

MONOCLONAL ANTIBODY INHIBITION OF MICROTUBULE-BASED VESICLE MOTILITY[a]

mAb	Vesicle movements (no. of movements per minute ± SEM)		No. of vesicles attached (n)
	Plus-end directed	Minus-end directed	
Control (60)	2.32 ± 0.35	13.98 ± 1.12	25.12 ± 2.30 (60)
VSP2B whole IgG (24)	2.87 ± 0.75	11.48 ± 0.95	23.54 ± 2.89 (24)
VSP2B Fab (24)	2.30 ± 0.87	10.04 ± 1.20	17.59 ± 3.21 (24)
VSP4D whole IgG	0.23 ± 0.08	4.37 ± 0.36	7.72 ± 0.97 (35)
VSP4D Fab	0.15 ± 0.06	6.61 ± 1.10	10.8 ± 2.75 (33)

[a] Anti-kinectin mAb VSP4D was incubated with chick embryo fibroblast vesicles (0.1 mg of mAb per 0.2 mg of vesicle proteins), and the vesicles were used in a directionality motility assay. The vesicle movements were compared with controls (no mAb and VSP2B). The number of movements per minute is the normalized number of moving vesicles per 36-μm microtubule length per (70 vesicles in 4 μm^2 of buffer per minute). n is the number of fields analyzed, and each field was measured from 1 to 3 min so that the time measured ranged from 32 to 92 min. The normalized number of vesicles attached is the sum of the number of stationary and moving vesicles attached to 36 μm of microtubule length.
Adapted from Kumar *et al.*, *Science* **267**, 1834 (1995).

VSP2B recognizes a 96- to 98-kDa antigen on the cytoplasmic surface of the vesicles and did not show significant inhibition of vesicle motility.

Monoclonal Antibody Affinity Purification of Kinectin. We have utilized monoclonal antibodies, VSP4D and 160.9, on protein A-Sepharose CL-4B (Schleicher and Schuell, Keene, NH) to identify antigen by antibody affinity. The carbonate washed 40/60 vesicles are biotinylated and collected at a 15/60% sucrose interface after centrifugation at 192,000g for 30 min in a SW 55Ti rotor (Beckman Instruments). The collected vesicles are dialyzed against PMEE' before incubation with 3% bovine serum albumen (BSA) at 4° for 1 hr. The vesicles are mixed with mAbs, VSP4D or 160.9 separately, and incubated for 5 hr at 4° and collected at a 15/60% sucrose interface after centrifugation to remove unbound antibodies from the vesicles. The vesicles are solubilized with 1.0% Triton X-100 and 0.5 M NaCl in PMEE' and insoluble material is removed by sedimentation at 150,000g for 30 min. The supernatants are incubated with the protein A-Sepharose CL-4B (prewashed with 0.2 M bicarbonate, 0.1 M sodium chloride buffer pH 8.0 and subsequently with PMEE' buffer) for 4–5 hr at 4° and washed with NET buffer (150 mM NaCl, 50 mM Tris-HCl, 5 mM EDTA, pH 8.0). We analyzed the antigen bound with antibodies after mixing the protein A-Sepharose CL-4B with Laemmli sample buffer by

SDS–PAGE and electroblot.[20,23] Kinectin is found as the major protein bound with the mAb 160.9 or VSP4D.

Conclusion

We have purified kinectin both on the basis of its affinity for the kinesin–SUK4 complex and utilizing the monoclonal antibody 160.9 after detergent solubilization of vesicles. In addition, a second, soluble form of kinectin was isolated using the antibody affinity column. Reconstitution of kinectin into vesicles and stimulation of motile activity has not been achieved; however, *in vitro* translation studies do show that kinectin is posttranslationally inserted into membranes. Purified kinectin does retain motor binding activity and has been found to bind to both kinesin and cytoplasmic dynein. In terms of the critical aspects of the binding interaction, the presentation of the motor molecule was critical in that kinesin bound to other anti-kinesin antibodies would not bind kinectin and we have not found an anti-dynein antibody that would present it in a manner capable of isolating kinectin. Biotinylation of the vesicles was critical for identifying which vesicle-specific components were binding. For future studies of motor-binding sites on vesicles, it is recommended that motor domain antibodies, vesicle labeling, and a variety of detergent conditions be utilized.

[23] U. K. Laemmli, *Nature (Lond.)* **277,** 680 (1970).

[17] Purification and Assay of CLIP-170

By Georgios S. Diamantopoulos, Jochen Scheel, Thomas E. Kreis, and Janet E. Rickard

Introduction

One of the most striking features of intracellular organization is the specific arrangement of organelles within the cytoplasm. The microtubule network plays a crucial role in this process as it defines the movement and positioning of many intracellular organelles.[1,2] These activities are regulated by microtubule-binding proteins that can mediate interactions between microtubules and membranes. The microtubule-based motors kinesin and

[1] M. Kirschner and T. Mitchison, *Cell* **45,** 329 (1986).
[2] T. E. Kreis, *Cell Motil. Cytoskel.* **15,** 67 (1990).

dynein, which are required for directed organelle movement, are the best characterized of these proteins.[3] Recently, a new class of cytoplasmic linker proteins (CLIPs), which can participate in the interactions between microtubules and organelles, has been described.[4] CLIP-170, the best characterized member of this class, was identified in HeLa cells where it localizes to the plus ends of microtubules.[5] *In vitro* binding assays,[6] as well as *in vivo* data,[7] strongly suggest that CLIP-170 acts as a linker protein between microtubules and endosomes. Different domains of CLIP-170 that have been characterized *in vivo*[8] have also been expressed and purified in bacteria.[9] However, it is also useful to analyze the native protein in purified form. In this article, we describe a protocol for purifing native CLIP-170 from human placenta. The activity of the purified protein can be tested using an *in vitro* assay in which tubulin is polymerized from centrosomes and the association of placenta CLIP-170 with the microtubules is examined by immunofluorescence.

Purification

CLIP-170 was initially isolated as a component of an extract of microtubule-binding proteins from HeLa cells.[5] Because this preparation contains many other microtubule-binding proteins, we developed a specific purification method using a monoclonal antibody (mAb). The high efficiency of this antibody affinity column as a purification step justifies the time involved in producing the antibody. This method was initially used for a one-step purification of CLIP-170 from HeLa cell high-speed supernatant.[10] This preparation was not of high purity, and the low yield obtained from convenient quantities of cultured cells made further purification steps impractical. We therefore developed a protocol using a tissue source (human placenta) that increases the yield of protein and allows a further purification step. A three-step procedure is followed: (1) preparation of high-speed supernatant (HSS) from human placenta, (2) immunoisolation of CLIP-170 with an antibody affinity chromatography column, and (3) enrichment and concentration of CLIP-170 by binding to microtubules.

[3] H. V. Goodson, C. Valetti, and T. E. Kreis, *Curr. Opin. Cell Biol.* **9,** 18 (1997).
[4] J. E. Rickard and T. E. Kreis, *Trends Cell Biol.* **6,** 178 (1996).
[5] J. E. Rickard and T. E. Kreis, *J. Cell Biol.* **110,** 1623 (1990).
[6] P. Pierre, J. Scheel, J. E. Rickard, and T. E. Kreis, *Cell* **70,** 887 (1992).
[7] O. Rosorius and T. E. Kreis, unpublished data.
[8] P. Pierre, R. Pepperkok, and T. E. Kreis, *J. Cell Sci.* **107,** 1909 (1994).
[9] J. Scheel, P. Pierre, J. E. Rickard, G. S. Diamantopoulos, C. Valetti, F. G. Van der Goot, M. Haeuer, U. Aebi, and T. E. Kreis, in preparation.
[10] J. E. Rickard and T. E. Kreis, *J. Biol. Chem.* **266,** 17597 (1991).

Preparation of High-Speed Supernatant from Human Placenta

Because human tissue is a potential source of human pathogens (hepatitis virus, human immunodeficiency virus, etc.), gloves and laboratory coats are used throughout the purification. We always disinfect all equipment, centrifuge tubes and glassware used at the initial steps (including HSS) with lysoformin (Lysoform) overnight. The tissue should be taken from the hospital as soon as possible, usually 30–40 min after birth (kept at 4° until collection). The tissue is placed in 1 liter of PE_5M (100 mM K PIPES, 5 mM ethylene glycol-*bis*(β-aminoethyl ether)-N,N,N′,N′-tetraacetic acid [EGTA], 1 mM $MgSO_4$, pH 6.7) adjusted at room temperature for transport to the lab. All subsequent manipulations are performed in the cold room. The umbilical cord and most of the connective tissue are removed since they can cause problems at the homogenization step. The tissue is weighed in a cooled beaker; the weight of one human placenta is around 500 g. The tissue is placed on a metal plate cooled on ice and cut into small pieces with a pair of scissors. The pieces are immediately washed in ice-cold PE_5M. The tissue is then resuspended in an equal volume of PE_5M including 1 mM dithiothreitol (DTT) and a mixture of protease inhibitors. Two stock solutions of protease inhibitors consisting of (A) 50 mM phenylmethylsulfonyl fluoride (PMSF), 2 mg/ml chymostatin, 1 mg/ml pepstatin in dimethyl sulfoxide (DMSO) and (B) 0.2 mg/ml aprotinin, 1 mg/ml leupeptin, 0.4 mM *trans*-epoxysuccinyl L-leucylamido(4-guanidino) butane in water are stored at $-70°$ and used at a final dilution of 1 : 200 for each. The tissue is homogenized in a Waring blender at maximum speed, three bursts of 20 sec with a 30-sec pause in between to avoid heating. The homogenate is then spun for 15 min at 40,000g (19,000 rpm) at 4° in a TFA 20.500 rotor (Kontron Instruments, Zürich, Switzerland). The resulting supernatant is then centrifuged for 1 hr at 150,000g (36,000 rpm) at 4° using two TFA 45.94 rotors (Kontron). The HSS ($\sim$700 ml from a 500-g placenta) can be either frozen in liquid N_2 or used immediately for the next step of purification. We prefer to use the HSS immediately for purification of CLIP-170. This makes the procedure longer (20 hr) but provides higher yields of the protein.

Affinity Purification of CLIP-170 Using a Monoclonal Antibody

A monoclonal antibody (clone 3A3) raised against CLIP-170[10] is used to affinity purify CLIP-170 from the placenta HSS. We first briefly describe the preparation of the antibody and then describe how this antibody is used to purify CLIP-170 using an affinity chromatography column.

Preparation of the Antibody Column. The production of the hybridoma cell line producing the 3A3 antibody has been described.[10] Ascitic fluid from the 3A3 hybridoma was used for purification of the IgG by DE52 ion-exchange chromatography,[11] giving yields of IgG between 5 and 30 mg/ml of ascites. Purified 3A3 IgG was conjugated to CNBr-activated Sepharose 4B (Pharmacia, Piscataway, NJ) at a coupling ratio of 10 mg of IgG/ml of swollen beads in 0.1 M NaHCO$_3$, 0.5 M NaCl, pH 8.9, overnight at 4° followed by high- and low-pH washes as recommended by Pharmacia. The beads can be reused many times for the purification of CLIP-170 and are stored at 4° in PBS (137 mM NaCl, 2.7 mM KCl, 4.3 mM Na$_2$HPO$_4$, and 1.4 mM KH$_2$PO$_4$) with 0.1% (v/v) Triton X-100, 0.2 mM PMSF, and 0.02% (w/v) NaN$_3$.

Immunoisolation of CLIP-170 from High-Speed Supernatant. The HSS is made to 0.1% (v/v) Triton X-100, which reduces nonspecific binding, and is then mixed with the 3A3 beads. The volume of beads used is ~1/8 of the HSS volume (80 ml for one placenta). Beads and HSS are incubated with rotation at 4° for 2.5–3 hr. This is long enough for binding of most of the CLIP-170 to the beads (Fig. 1b, lanes 1 and 2). The mixture is loaded into a column (2.60 cm in diameter), packed by gravity, and the flow-through is collected for analysis of CLIP-170 content. The column is washed with two column volumes of PE$_5$MT [PE$_5$M with 0.1% (v/v) Triton X-100], two column volumes of PE$_5$MT plus 1 M NaCl, and two column volumes of PE$_5$MT, using a pump at a speed of 5 ml/min. Bound CLIP-170 on the column is eluted with 50 mM diethylamine, 2 mM EGTA, 1 mM MgSO$_4$, 0.1% (v/v) Triton X-100, pH 11.5, at 3 ml/min. Eluted fractions at high pH are neutralized with K PIPES, pH 6.8, at a final concentration of 60 mM. We add this buffer to the collection tubes before starting the elution and then mix each fraction well by inverting the tube three times in order to neutralize the pH very quickly as the protein elutes. Fractions can be checked for CLIP-170 content by SDS–PAGE. However since the method is reproducible, it is quicker to locate the protein-containing fractions by protein assay. We always check where the pH changes during the elution because this is an indication of the peak fractions of CLIP-170. The peak fractions identified by gel electrophoresis or protein assay are pooled and further purified and concentrated in the final step. The antibody column is regenerated by using a total of one column volume of elution buffer, then two column volumes of 0.1 M sodium acetate, 0.5 M NaCl, pH 4, and then at least three column volumes of PEM (same as PE$_5$M but with 1 mM EGTA). The beads are retrieved with a glass pipette and are kept in the buffer described earlier. We have used the same beads with no loss of activity for more than 20 purifications.

Affinity Purification and Concentration by Microtubule Binding

CLIP-170 eluted from the affinity chromatography column contains a significant amount of contaminating protein (Fig. 1a, lane 3). To remove

---

[11] P. Parham, *Methods Enzymol.* **92,** 110 (1983).

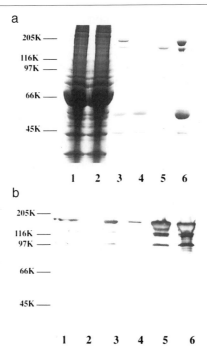

FIG. 1. Analysis by gel electrophoresis of protein samples taken at different stages during purification of human placenta CLIP-170. Samples were separated on sodium dodecyl sulfate (SDS)–polyacrylamide gels and stained with Coomassie Blue (a) or transferred to nitrocellulose and probed with a polyclonal antibody against CLIP-170 (b). Lane 1, High-speed supernatant; 2, 3A3 column flow-through; 3, 3A3 column eluate; 4, supernatant after paclitaxel microtubules (MTs) sedimentation; 5, supernatant after salt elution of paclitaxel MTs pellet; 6, MTs pellet after salt elution. Numbers at the left indicate molecular weights.

these proteins, we use a second purification step which takes advantage of the ability of CLIP-170 to bind to microtubules. This procedure also results in concentration of the protein. We pool the most concentrated fractions determined from the protein gel (10 fractions above 30 ng/μl CLIP-170) and add to them 20 μM paclitaxel, 1 mM DTT, the same protease inhibitors as before, and paclitaxel-stabilized pig brain microtubules. About 0.5–1 mg of tubulin is added per tube used (usually five), which is enough to bind all microtubule-binding proteins. The pooled fractions are then incubated at 37° for 30 min. The incubation mixture is then transferred to TST 41.14 tubes (Kontron). Using a Pasteur pipette, we underlay the samples with one-third of their volume of 10% (w/v) sucrose in PEM (filtered through a 0.22-μm filter and prewarmed to 37°) containing 20 μM paclitaxel. Tubes are spun in a prewarmed TST 41.14 swingout rotor for 30 min at 30° and 40,000g (17,000 rpm). The supernatant is aspirated and the sucrose interface is washed with warm PEM. The cushion is then carefully aspirated and the pellets are rinsed in a small volume of prewarmed PEM containing 20 μM paclitaxel. The tubes are subsequently left in an inverted position for a few minutes to drain all liquid. The pellet is then resuspended in a small volume (~300 μl total for all the tubes) of prewarmed 0.8 M NaCl in PEM containing 20 μM paclitaxel. We use a cut blue tip to resuspend the pellet by pipetting up and down several times. All the resuspended pellets are pooled and left for 10 min at room temperature. The sample is loaded in a centrifuge tube and spun for 45 min at 90,000g (43,000 rpm) at 30° in a prewarmed fixed-angle TFT 80.4 rotor (Kontron). The supernatant contains CLIP-170 in high purity revealed by Coomassie staining and immunoblotting (Figs. 1a and 1b, lane 5). We find the final step of resuspension very important, especially the amount of the resuspension buffer that is used. A large volume of resuspension buffer will result in low concentration of the recovered protein, whereas too little is insufficient for solubilizing and detaching the protein from the pellet. Because CLIP-170 is a rather sticky protein, we avoid extracting it in a large volume since subsequent concentration steps result in significant losses of the protein. We have found 300 μl to be the optimum volume. The affinity chromatography step and the microtubule binding step are very efficient since they provide a high enrichment for CLIP-170 (Fig. 1b, lanes 3 and 5; see also Ref. 9). However a significant amount of the protein is lost during the microtubule-binding step since ~10% of the pooled fractions does not bind to paclitaxel microtubules and almost 50% remains attached to the pellet after the salt elution step (Figs. 1a and 1b, lanes 4 and 6). We usually obtain high-purity CLIP-170 (Figs. 1a and 1b, lane 5) at a final concentration of 0.4–0.7 mg/ml and a yield of 125–250 μg.

Assays for CLIP-170

The distribution of CLIP-170 in cells can be assayed by immunofluorescence using monoclonal and polyclonal antibodies specific for CLIP-170.[5,10] Immunoblotting can be used to estimate the CLIP-170 content of extracts and is used as discussed next to evaluate the efficiency of the purification. Immunoprecipitation of CLIP-170 can be used to study the regulation of its phosphorylation state.[10] Finally, an *in vitro* assay has been used to test binding of purified placenta CLIP-170 to polymerized tubulin.

Immunoblots for CLIP-170. The efficiency of the purification is tested by immunoblots using a polyclonal antibody (α55, Ref. 6) against CLIP-170. Samples from different steps of the purification were separated on polyacrylamide gels according to Laemmli[12] with an 8% running gel and a

[12] U. K. Laemmli, *Nature (Lond.)* **227**, 680 (1970).

3% stacking gel, using 0.75-mm-thick mini gels (Bio-Rad, Hercules, CA) run at 25 mA. Proteins were stained with Coomassie Brilliant Blue (Fig. 1a) or subsequently transferred to nitrocellulose papers (0.2-mm pore size, Schleicher and Schuell, Keene, NH) using a Genie electroblotter (Idea Scientific Co., Corvallis, OR) with 25 mM Tris(hydroxymethyl) methylamine, 192 mM glycine. Excess binding sites on the filters were blocked with PBS containing 5% (w/v) dried nonfat milk powder and 0.2% (v/v) Triton X-100 before incubation with antibody, diluted in the same buffer, for 16–24 hr at 4°. Bound antibody was visualized using alkaline phosphatase-conjugated second antibody followed by NBT/BCIP (Fig. 1b). CLIP-170 is efficiently depleted from HSS and bound to the beads of the column (compare lanes 2 and 3 of Fig. 1b). A significant amount of CLIP-170 does not bind to the paclitaxel microtubules (Fig. 1b, lane 4). It is possible that a fraction of this amount of CLIP-170 is hyperphosphorylated and thus inhibited from binding to microtubules.[10] As seen in lanes 5 and 6 (Fig. 1a) only 50% of the CLIP-170 that is bound to microtubules can be eluted after salt wash of the pellet. Finally the protein is obtained in high purity (lane 5) with some degradation products.

Immunofluorescence Localization of CLIP-170 in HeLa Cells. HeLa cells were grown on coverslips in MEM supplemented with 1% nonessential amino acids, 1% glutamine, and 10% fetal calf serum in humidified incubators at 37°. Cells were fixed in methanol at −20° for 4 min. After fixation, cells were labeled with a mixture of two monoclonal antibodies against CLIP-170 (2D6 and 4D3)[5] and with a rabbit polyclonal antibody specific for tyrosinated tubulin[13] followed by rhodamine or fluorescein-labeled secondary antibodies. Epifluorescence microscopy was performed using a Zeiss 100× Planapo, 1.4 oil immersion objective on a Zeiss inverted microscope (Axiovert 10; Zeiss, Zürich, Switzerland). Images were recorded with a cooled CCD camera, controlled by a Power Macintosh and processed with the software package IPLab Spectrum V 3.0. CLIP-170 exhibits a patchy distribution along microtubules with a distinct localization at their plus ends. This can be better observed at the periphery of the cell where subsets of individual microtubules with well defined plus ends can be seen. In Fig. 2 many of these microtubules contain CLIP-170 staining at their plus ends (arrows), whereas there are some microtubules that are not labeled (arrowheads).

In Vitro Assay for Binding of CLIP-170 to Microtubules. We test the binding of the purified CLIP-170 to microtubules using an *in vitro* assay where pig brain tubulin is polymerized from centrosomes and the centroso-

[13] T. E. Kreis, *EMBO J.* **6**, 2597 (1987).

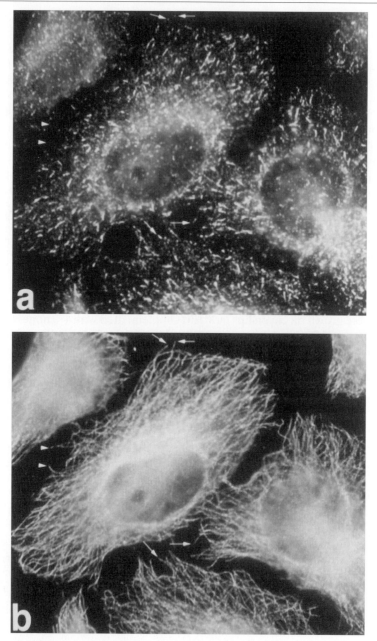

FIG. 2. Immunofluorescence localization of CLIP-170. HeLa cells were fixed in methanol and CLIP-170 and tubulin localized by double immunofluorescence labeling with monoclonal anti-CLIP-170 (a) and polyclonal anti-tubulin (b). Arrows and arrowheads indicate microtubule ends, labeled or unlabeled, respectively, by the antibody to CLIP-170.

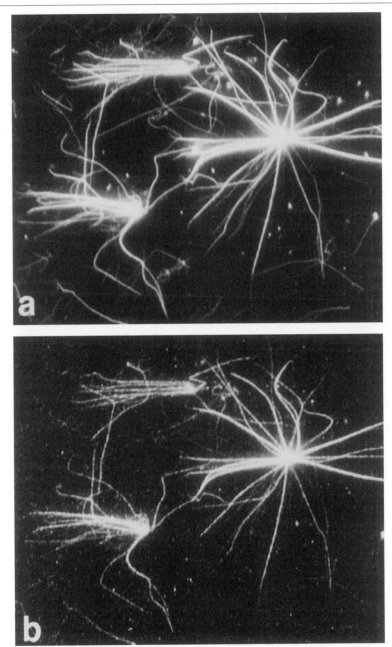

Fig. 3. Centrosomal microtubules observed after a 15-min incubation of purified centrosomes with 1 mg/ml tubulin and 36 μg/ml purified placenta CLIP-170. (a) Tubulin labeling with monoclonal antibody and (b) CLIP-170 labeling with an anti-peptide antibody.

mal microtubules are sedimented on coverslips.[14] Briefly, 1 mg/ml purified tubulin is incubated with centrosomes in the presence of 36 μg/ml CLIP-170 in 80 mM K PIPES, 1 mM MgCl$_2$, 1 mM K EGTA, pH 6.8. The samples are incubated at 37° for 15 min before fixation in glutaraldehyde. The fixed asters are sedimented onto round glass coverslips through a 25% (w/v) sucrose cushion in a HSA 13.94 rotor (Hereaus, Zürich, Switzerland) for 15 min at 25,000g (13,000 rpm) at 20°. They are then postfixed in methanol for 5 min at $-20°$ and processed for immunofluorescence labeling with a rabbit antipeptide antibody against CLIP-170 (anti-KRKV; Ref. 6) and a monoclonal antibody against tyrosinated tubulin.[13] CLIP-170 (Fig. 3a) colocalizes with the centrosomal microtubules (Fig. 3b) and binding of the protein to microtubules changes their phenotype since they are longer and appear more bundled compared to the control without the protein (data not shown). This assay is being used to study further the mechanism by which CLIP-170 localizes to microtubule ends.[15]

Acknowledgments

We are grateful to Mr. Collaud of the Lausanne Abattoir for supplying pig brains and to the nurses of the Maternity Hospital, HUG, Geneva, for their help in obtaining placentas. We thank the National Cancer Institute, Bethesda, MD, for a gift of paclitaxel. We thank Dr. F. Perez for providing Fig. 2. Work performed in the authors' laboratories is supported by the Fonds National Suisse (TEK, JER) and the Canton de Genève (TEK).

[14] T. Mitchison and M. Kirschner, *Nature* (*Lond.*) **312**, 232 (1984).
[15] G. S. Diamantopoulos, F. Perez, H. V. Goodson, G. Batelier, R. Melki, T. E. Kreis, and J. E. Rickard, submitted for publication.

[18] Purification and Assay of the Microtubule-Severing Protein Katanin

By Frank McNally

Introduction

Katanin is an adenosine triphosphate (ATP)-dependent microtubule-severing protein that has been isolated from sea urchin egg cytosol based on its microtubule-severing activity. In the presence of Mg-ATP, katanin generates internal breaks within Taxol-stabilized microtubules.[1] During this reaction, tubulin heterodimers are released from the microtubule polymer

[1] F. J. McNally and R. D. Vale, *Cell* **75**, 419 (1993).

without apparent modification or denaturation of the tubulin. Katanin appears to disrupt protein–protein interactions within a microtubule, thus releasing tubulin and generating breaks in the microtubule.

Katanin also hydrolyzes ATP, and the rate of this hydrolysis is increased in the presence of microtubules.[1] It has been proposed that ATP hydrolysis is required for the microtubule-severing reaction because high concentrations of ATPγS or adenosine diphosphate (ADP) will inhibit both the ATPase activity and the microtubule-severing activity of katanin in the presence of ATP. In addition, ATPγS cannot be substituted for ATP in a microtubule-severing reaction. Katanin is composed of two polypeptides termed p81 and p60.[1] Results of hydrodynamic studies indicate that katanin is composed of one subunit each of p81 and p60.

Immunolocalization studies of katanin in sea urchin embryos have shown that katanin is concentrated around the centrosomes.[2] This localization has led to speculation that katanin may be involved in severing microtubules from their centrosomal attachment points *in vivo*. Immunolocalization in unfertilized sea urchin eggs indicates that katanin is dispersed throughout the cytoplasm and katanin is purified from this cytoplasmic pool.

In this paper a method for isolating katanin is presented that allows simultaneous isolation of tubulin from the same starting material.

Materials

Materials for Microtubule-Severing Assays

3- × 1-in. precleaned microscope slides
18 × 18-mm No. 1 coverslips
3/4-in. Double-stick Scotch tape cut into 3-mm × 3/4-in. strips
Whatman (Maidstone, England) 3MM paper cut into triangles (3 cm/side)

NEM-Treated Xenopus Extract. A *Xenopus laevis* egg extract prepared as described by Murray[3] is treated with 5 mM N-ethylmaleimide (NEM) at 22° for 10 min followed by treatment with 25 mM dithiothreitol (DTT) for 10 min. This preparation can be stored at −80° in small aliquots. NEM-treated *Xenopus* extract is diluted 1 : 10 in BRB80 (see below) immediately before use.

Rhodamine-Labeled Microtubules. Tubulin is purified from bovine brain by cycled assembly and phosphocellulose chromatography and labeled with

[2] F. J. McNally, K. Okawa, A. Iwamatsu, and R. D. Vale, *J. Cell Sci.* **109,** 561 (1996).
[3] A. W. Murray, *in* "Methods in Cell Biology" (B. K. Kay and H. B. Peng, eds.), p. 581. Academic Press, San Diego, 1991.

tetramethylrhodamine as described by Hyman *et al.*[4] Microtubules are assembled by incubation of 10 mg/ml unlabeled tubulin and 1 mg/ml rhodamine-labeled tubulin in BRB80 containing 7% (v/v) dimethyl sulfoxide (DMSO) and 1 mM guanosine triphosphate (GTP) for 45 min at 37°. An equal volume of BRB80 containing 40 μM Taxol is then added. Taxol-stabilized microtubules remain stable for up to 1 week at 20°. During the first 24 hr after Taxol addition, microtubule length increases, making the microtubules more useful for severing assays.

Taxol is purchased from Sigma (St. Louis, MO).

Materials for Katanin Purification

Sea Urchins. Gamete-bearing sea urchins (*Strongylocentrotus purpuratus*) can usually be obtained from Marinus Inc. (Venice, CA) from December through March. The yield of eggs and length of the season can vary dramatically due to environmental conditions.

Buffers

Seawater: natural seawater or Instant Ocean dissolved in deionized water to a specific gravity of 1.025 at 4°

19:1 buffer: 530 mM NaCl, 28 mM KCl, 1 mM Na-EDTA, 5 mM Tris-Cl, pH 7.0

PMEG 7.3: 100 mM K-PIPES, 1 mM MgCl$_2$, 5 mM Na-EGTA, 10% (v/v) glycerol, pH 7.3

BRB80: 80 mM K-PIPES, 1 mM MgCl$_2$, 1 mM Na-EGTA, pH 6.8

Assay buffer: 20 mM K-HEPES, 1 mM MgCl$_2$, 0.1 mM Na-EGTA, 1 mM ATP, 20 μM Taxol, pH 7.5

50 mM N-2-hydroxyethyl piperazine-N'-propanesulfonic acid (K-EPPS), pH 8.0

50 mM K-EPPS, 400 mM KCl, pH 8.0

Hydroxyapatite buffer A: 15 mM potassium phosphate, pH 7.2

Hydroxyapatite buffer B: 1 M potassium phosphate, pH 6.8 is made up from 1 M monobasic and dibasic stock solutions

Mono Q buffer A: 20 mM Tris-Cl, 10% (v/v) glycerol, pH 8.0

Mono Q buffer B: 20 mM Tris-Cl, 10% (v/v) glycerol, 1 M NaCl, pH 8.0

Protease inhibitors: 40 μg/ml tosyl arginine methyl ester, 40 μg/ml tosyl phenylalanyl chloromethyl ketone, 20 μg/ml benzamidine, 2 μg/ml aprotinin, 100 μg/ml soybean trypsin inhibitor

---

[4] A. Hyman, D. Drechsel, D. Kellog, S. Salser, K. Sawin, P. Steffen, L. Wordeman, and T. Mitchison, *Methods Enzymol.* **196**, 478 (1991).

Columns

SP Sepharose Fast Flow is obtained from Pharmacia (Piscataway, NJ). Macroprep High S Support (Bio-Rad, Richmond CA) has also been used successfully.

HYDROXYAPATITE COLUMNS. Three types of hydroxyapatite columns have been used successfully. High-performance crystalline hydroxyapatite prepacked in 5-ml columns (Bio-Rad) provides good separation and yield but backpressure increases with repeated use, making these columns suboptimal. Spherical ceramic hydroxyapatite beads have proven to be more durable. Ceramic hydroxyapatite (10-μm beads) prepacked in 5-ml columns can be obtained from Bio-Rad and bulk ceramic hydroxyapatite (20-μm beads) can be obtained from American International Chemical (Natick, MA) and packed into a Pharmacia HR 10/10 column with the corresponding Pharmacia packing equipment and a high-pressure chromatography system. Ordinary crystalline hydroxyapatite does not have the capacity or resolution required for katanin purification.

ANION-EXCHANGE COLUMNS. Katanin has been successfully purified with a 1-ml Mono Q column (Pharmacia) and with a 1-ml Resource Q column (Pharmacia).

Procedure

Assay of Microtubule-Severing Activity

The most reliable assay of microtubule-severing activity involves immobilization of fluorescently labeled, Taxol-stabilized microtubules on a glass coverslip and observation of these microtubules by fluorescence microscopy after exposure to a solution containing katanin and ATP.

A flow cell is first constructed by placing two strips of double-stick Scotch tape on a standard microscope slide. The strips are placed parallel to each other and approximately 3 mm apart. An 18-mm coverslip is placed on top of the tape strips and sealed by applying gentle pressure on the coverslip over the tape. The volume of the resulting flow cell is approximately 20 μl. Liquid is added through one open end with a micropipettor and withdrawn from the opposite end by absorption into a small piece of Whatman 3MM paper.

Assay Procedure

Assays are conducted at room temperature (20–25°).

1. A dry flow cell is filled with NEM-treated *Xenopus* egg extract that has been diluted 10-fold in BRB80 and incubated for 1 min. Microtu-

bule-binding proteins in the extract bind to the coverslip surface during this incubation.

2. BRB80 (40 μl) is flowed through to wash out unbound extract proteins.

3. Rhodamine-labeled microtubules (20 μl; 20–50 μg/ml tubulin) diluted in BRB80 containing 20 μM Taxol are flowed into the flow cell. A 1-mm incubation allows microtubules to bind to the glass.

4. BRB80 (40 μl) is flowed through to remove unbound microtubules.

5. The test sample diluted in assay buffer with 1 mM ATP (20 μl) is flowed through and incubated for 30 sec to 8 min.

Monitoring the Reaction. Reactions can be monitored in one of three ways.

1. Reactions can be stopped at a defined time after addition of sample by flowing through 20 μl of 1% glutaraldehyde in BRB80. The microtubules are now fixed and can be observed by fluorescence microscopy at any time. This method allows simultaneous assay of a large number of samples (e.g., column fractions) in separate flow cells. For example, reactions can be started sequentially in 20 flow cells at 10-sec intervals. At 3 min after the start of the first reaction, glutaraldehyde can be flowed sequentially through each flow cell at 10-sec intervals. The disadvantage to this method is that the additional flow-through step can dislodge microtubules and wash them out of the flow cell.

Microtubule severing is indicated by a reduced average length of microtubules (compared with controls) and especially by the "dotted line" appearance shown in Figs. 1b and 1c. The generation of these "dotted line" structures requires that some katanin-generated microtubule fragments either rotate on or dissociate from the coverslip. Reactions carried out with purified katanin usually have a more disorganized appearance because the resulting microtubule fragments are not colinear on the coverslip surface. Average microtubule length can be measured by capturing images of microtubule-coated coverslips with a SIT (silicon intensified target) camera and a frame grabber (e.g., a Scion AG5 board) and measuring the lengths of all the microtubules in a field using image processing software (e.g., Signal Analytics IP Lab Spectrum).

2. Reactions can be directly viewed at a specified time point (e.g., 5 min) after sample addition without fixation. This requires continuous movement of the stage during observation so that the same field of view is not illuminated for more than 3 sec. Longer illumination results in photodamage to microtubules that appears very similar to microtubule severing. As with the first method, 20 or more reactions can be carried out simultaneously by starting reactions at 20-sec intervals and observing successive

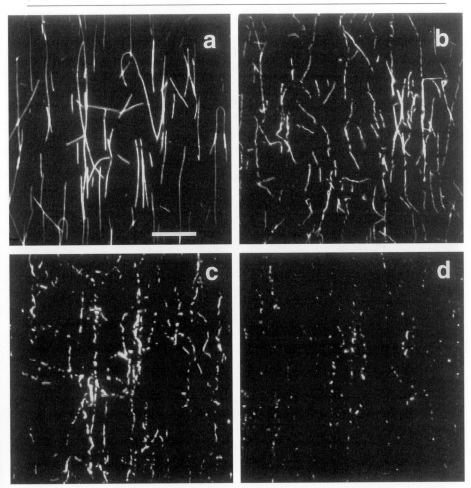

FIG. 1. A microtubule-severing assay. Fluorescence images of rhodamine-labeled microtubules before and after exposure to a sea urchin egg high-speed supernatant (HSS) are shown.
Microtubules before addition of the HSS have an average length that is greater than 10 μm
(a). Microtubules become progressively shorter at 30 sec (b), 1 min (c), and 2 min (d) after
exposure to the HSS. The linear arrays of microtubule fragments seen in (b) and (c) and the
fluorescent spots left on the coverslip in (d) are characteristic of a katanin-mediated microtubule-severing reaction. Bar = 10 μm.

flow cells at 20-sec intervals. Scanning the entire flow cell is recommended
because the extent of microtubule-severing can vary dramatically between
the center and edge of the flow cell and between the two ends of the flow
cell. This method requires some practice.

3. The first two methods described are essentially fixed time point

assays. A microtubule-severing reaction can also be viewed continuously as it occurs. This type of assay is not useful for assaying large numbers of samples because of time constraints. However, direct, continuous observation of a break occurring in a microtubule is the most definitive assay for microtubule severing. Continuous observation of a severing reaction requires the use of neutral density filters to reduce the intensity of illumination and the addition of an "oxygen-scavenging" mixture[5] to reduce the extent of oxygen-mediated photodamage to the microtubules. Photodamage results in the rapid fragmentation of rhodamine-labeled microtubules even in the absence of katanin. Control reactions without katanin or without ATP should be carried out before and after an experiment to ensure that photodamage is not occurring.

Quantitation of Microtubule-Severing Activity

Activity can be roughly quantitated by limiting dilution. Activity in fractions from a column elution can be compared by determining the dilution of each fraction that yields microtubules of the same average length in the same period of time. A sea urchin egg high-speed supernatant (30 mg/ml protein) can be expected to reduce the average microtubule length from greater than 10 μm (Fig. 1a) to less than 1 μm (Fig. 1d) in 5 min when diluted 10-fold in assay buffer. The peak fractions eluting from a hydroxyapatite column (Fig. 2) have the same effect at a dilution of 1:500. The peak of microtubule-severing activity as determined by limiting dilution corresponds exactly with the peak of katanin protein eluting from a hydroxyapatite column (Fig. 2) or a Resource Q column (Fig. 3).

Salt concentrations greater than 50 mM inhibit the microtubule-severing reaction. Column fractions should be diluted sufficiently to reduce any salt concentration to below 50 mM.

Other Assays

Microtubule severing can be assayed in solution by mixing a test sample with microtubules and ATP in a microfuge tube and observing the reduced length of microtubules by mounting the mixture between a slide and coverslip after fixation.[6] This method has two major limitations. First, many types of protein solutions mediate the bundling of microtubules. Bundling inhibits katanin-mediated severing and also makes it difficult to observe breaks in individual microtubules within the bundle. Second, microtubules are

---

[5] A. Kishiino and T. Yanagida, *Nature* **334,** 74 (1988).
[6] R. D. Vale, *Cell* **64,** 827 (1991).

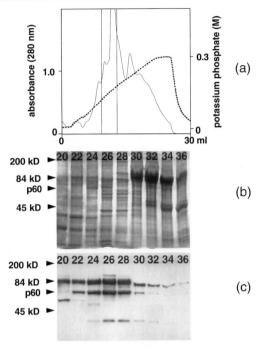

FIG. 2. Hydroxyapatite chromatography of katanin. (a) Elution profile from a 10- × 100-mm hydroxyapatite (20-μm ceramic bead) column. Absorbance at 280 nm is indicated by the solid line with the scale on the left y axis. Concentration of potassium phosphate is indicated by the dotted line with the scale on the right y axis. The elution volume is indicated on the x axis. The vertical solid lines indicate the region of the chromatogram represented in (b) and (c). (b) Coomassie Brilliant Blue-stained SDS–PAGE analysis of fractions (0.5 ml) eluted from the hydroxyapatite column. The fraction numbers are indicated at the top of each lane. Katanin polypeptides cannot yet be distinguished. (c) Immunoblot of the same fractions shown in (b) probed with a mixture of p81 and p60 katanin-specific antibodies. The highest concentration of katanin is seen in fractions 26–28. The highest concentration of microtubule-severing activity is also found in fractions 26–28.

susceptible to breakage by shear forces generated while pipetting or mounting the microtubules. Shear-mediated breakage can give the false impression of microtubule severing. Some proteins may modulate the stiffness of microtubules, making them more susceptible to breakage by shear forces. Thus appropriate negative controls are difficult to devise. One way to minimize shear forces is to use micropipette tips from which the ends have been cut off to slowly pipette the reaction into a flow cell that has been previously coated with a solution of polylysine. Another solution assay that minimizes the effects of shear forces is the fluorescence energy transfer assay described by McNally and Vale.[1] Immobilization of microtubules on

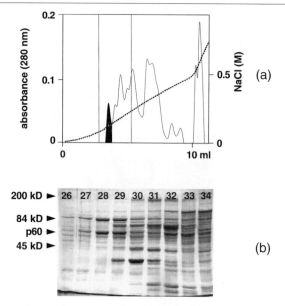

FIG. 3. Anion-exchange chromatography of katanin. (a) Elution profile from a 5- × 50-mm Resource Q (Pharmacia) column. Absorbance at 280 nm is indicated by the solid line with the scale on the left y axis. Concentration of NaCl is indicated by the dotted line with the scale on the right y axis. The elution volume is indicated on the x axis. The vertical solid lines indicate the region of the chromatogram represented in (b). The shaded area of the chromatogram indicates the peak of katanin protein (fractions 28 and 29). (b) Coomassie Brilliant Blue-stained SDS–PAGE analysis of fractions (0.25 ml) eluted from the Resource Q column. The fraction numbers are indicated at the top of each lane. The two subunits of katanin, p81 and p60, are clearly visible in fractions 28 and 29. The highest concentration of microtubule-severing activity is also found in fractions 28 and 29.

a coverslip, however, is the only way to eliminate completely the problem of microtubule bundling.

Purification of Katanin

Katanin does not cosediment with microtubules in sea urchin egg extracts. Thus tubulin can be purified from a sea urchin egg lysate by DMSO-induced assembly and sedimentation and katanin can be isolated from the resulting supernatant. Katanin is purified by successive binding and batch elution from SP-Sepharose, binding and gradient elution from high-performance hydroxyapatite, and binding and gradient elution from a high-performance quaternary amine anion-exchange column. All steps are carried out at 0–4°.

Preparation of High-Speed Supernatants

Sea urchin eggs are dejellied and washed successfully in 19:1 buffer and in PMEG buffer as previously described[7] except that the pH of the PMEG is 7.3 to allow DMSO-induced tubulin assembly.[8] After the final sedimentation, a 100-ml packed volume of eggs is resuspended in 200 ml of PMEG 7.3 with 1 mM DTT and protease inhibitors. Eggs are lysed with a "Polytron"-type homogenizer. Exact duration and settings depend on the model of homogenizer. Lysis is complete when no eggs are visible by light microscopy. Lysates are centrifuged at 25,000g (Beckman JA14, Sorvall GSA) for 45 min at 4°. Supernatants are withdrawn and centrifuged at 45,000g (Beckman JA20, Sorval SS34) for 45 min at 4°. Supernatants are withdrawn and GTP is added to 0.5 mM and DMSO is added to 8% (v/v) final concentration. Lysates are incubated at 25° for 30 min to allow microtubule assembly. Assembled microtubules are then sedimented at 45,000g for 45 min at 25°. A minimal volume of BRB80, 1 mM GTP is added to the microtubule pellets, which are flash frozen in liquid nitrogen and stored at −80°. Tubulin can be purified from these frozen pellets as previously described.[8] Supernatants withdrawn from the microtubule-pelleting spin are then centrifuged at 250,000g (Beckman Type 50.2Ti) for 45 min at 4° and the supernatant is carefully withdrawn. This high-speed supernatant (HSS) can be flash frozen in liquid nitrogen and stored at −80°. If lysates are frozen before complete clearing by centrifugation, katanin loses its microtubule-severing activity.

S-Sepharose Chromatography

HSS (200 ml; approximately 30 mg/ml protein) is quickly thawed in a 37° water bath and immediately cooled to 0°. Two volumes of cold 50 mM K-EPPS, pH 8.0, is added to the HSS and 20 ml (packed volume) of SP-Sepharose Fast Flow (Pharmacia) that has been equilibrated with 50 mM K-EPPS, pH 8.0, is added. This suspension is incubated at 0° with gentle agitation for 45 min. The SP-Sepharose is allowed to settle and the supernatant is removed. The SP-Sepharose is transferred to 2- × 50-ml Corning tubes and washed three times in 50 mM K-EPPS by centrifugation at 1000g for 5 min at 4°, followed by removal of the supernatant and gentle resuspension with 40 ml buffer. Protein is eluted by addition of 40 ml 50 mM K-EPPS, pH 8.0, and 400 mM KCl to 15 ml of packed SP-Sepharose and gentle agitation for 20 min at 0°. After centrifugation at 1000g the superna-

[7] D. Buster and J. M. Scholey, *J. Cell Sci.* **Suppl. 14,** 109 (1991).
[8] K. Suprenant and J. C. Marsh, *J. Cell Sci.* **87,** 71 (1987).

tant is removed. Microtubule-severing activity is typically detectable in this eluate in 5 min at a dilution of 1 : 10.

Hydroxyapatite Chromatography

The SP-Sepharose eluate is brought to 1 mM CaCl$_2$ by the addition of 1 M CaCl$_2$ to inactivate residual EGTA, which can interfere with hydroxy-apatite chromatography. Immediately before loading onto the hydroxyapatite column, the SP-Sepharose eluate is centrifuged at 400,000g (Beckman Type 70Ti or TLA100.4) for 10 min at 4° to remove precipitated protein. The supernatant is loaded onto an 8-ml high-performance hydroxyapatite column (see materials) preequilibrated with 15 mM potassium phosphate, pH 7.2, using a 50-ml Superloop (Pharmacia) and a high-pressure chromatography system at a flow rate of 1 ml/min. The column is washed with 8 ml of 15 mM potassium phosphate and eluted with a 30-ml linear gradient from 15 to 300 mM potassium phosphate at a flow rate of 0.5 to 1 ml/min. Then 0.5-ml fractions are collected. The UV absorbance profile of the elution is extremely reproducible (Fig. 2). The peak of microtubule-severing activity is determined by limiting dilution. Peak fractions typically yield significant microtubule-severing activity at a dilution of 1 : 500 to 1 : 1000. The four peak fractions (fractions 25–28 in Fig. 2) are pooled for further purification. Katanin polypeptides are not yet apparent by Coomassie staining at this stage (see Fig. 2). Note that if fractions are assayed in the absence of ATP, a broad peak of microtubule-severing activity can be detected that elutes after the katanin peak. E-MAP is responsible for this discrete activity.[1,9]

Anion-Exchange Chromatography

The pooled fractions from hydroxyapatite chromatography (2 ml) are diluted in 10 volumes of Mono Q buffer A. This material is loaded onto a 1-ml Mono Q or Resource Q column using a 50-ml Superloop (Pharmacia) at a flow rate of 0.25–1 ml/min. Katanin is eluted at 0.25–1 ml/min with a 10-ml linear gradient of 0–500 mM NaCl (50% Mono Q buffer B) and 0.25-ml fractions are collected. The UV absorbance profile of the elution is extremely reproducible and katanin elutes in the first UV absorbing peak that elutes from the column (shaded area in Fig. 3a; fractions 28 and 29 in Fig. 3b). Peak fractions are pooled, aliquoted, flash frozen in liquid nitrogen, and stored at −80°. The yield of katanin from 100 ml of packed eggs is typically 25–100 μg, which represents a 1% yield of microtubule-severing activity. Purified katanin is extremely susceptible to nonspecific surface

[9] Q. Li and K. A. Suprenant, *J. Biol. Chem.* **269**, 31777 (1994).

adsorption and precipitation due to changing buffer conditions. As a result, further chromatography of katanin results in low yields (10–20%) and broad elution profiles (even if rechromatographed on the same Mono Q column). Katanin can be concentrated and exchanged into assay buffer by successive dilution and concentration with a Centricon 30 (Amicon, Danvers, MA) after first adding soybean trypsin inhibitor to 250 μg/ml final concentration as a carrier protein. Yields from this concentration and buffer exchange are low.

Discussion

Microtubule-Severing Assays

The greatest technical difficulty in the assay of microtubule-severing activity is the reproducible immobilization of microtubules on the coverslip surface. The most satisfactory images of a microtubule-severing reaction are obtained when most of the resulting microtubule fragments remain immobilized in their original positions on the glass surface. Two reagents that have been used successfully in microtubule-severing assays are NEM-treated *Xenopus* egg extracts and streptavidin. Kinesin-like proteins are thought to be responsible for the microtubule–glass adhesion properties of *Xenopus* egg extracts because of the increased binding in the presence of adenyl-imidodiphosphate.[6] NEM treatment of these extracts is necessary to inactivate the endogenous microtubule-severing activity. However, microtubules immobilized on a coverslip with NEM-treated *Xenopus* extract sometimes release from the surface during flow-through of ATP-containing buffers. Assembly of microtubules from a mixture of biotin-labeled and rhodamine-labeled tubulin and immobilization on streptavidin-coated coverslips has been used successfully but has similar reproducibility problems. Polylysine is an example of a reagent that mediates tight binding of microtubules to glass but which inhibits katanin-mediated microtubule severing.

Photodamage to microtubules is a second major problem with microtubule-severing assays because the results of photodamage are nearly identical in appearance with the results of a severing reaction. Fixed time point experiments conducted in the absence of illumination avoid this problem.

Purification of Katanin

Obtaining large quantities of sea urchin cytosolic extracts is the major difficulty in purification of katanin. Costs are minimized and yields are maximized when sea urchins can be obtained directly from a commercial fisherman or by collection in the intertidal zone. These options are obviously

increases the purity of katanin eluting from the anion-exchange column. The use of a Mono Q 5/5 (Pharmacia) column in the final step and utilization of only the first half of the katanin peak (fraction 28 in Fig. 3) results in the purest katanin preparations.

[19] Purification and Assay of γ Tubulin Ring Complex

By Yixian Zheng, Mei Lie Wong, Bruce Alberts,
and Tim Mitchison

Introduction

Genetic and cell biological studies have indicated that γ tubulin is involved in microtubule nucleation from microtubule organizing centers[1] (MTOCs). Sucrose gradient sedimentation analysis of the *Xenopus* and *Drosophila* γ tubulin showed that γ tubulin exists as large complexes in *Xenopus* egg and *Drosophila* embryo extracts.[2,3] To understand mechanistically how γ tubulin is involved in microtubule nucleation, it is necessary to study its function biochemically. Unfortunately, it has not yet been possible to produce large quantities of γ tubulin using heterologous protein expression systems. Furthermore, since γ tubulin is a minor cellular protein, purification of the endogenous γ tubulin using conventional fractionation approaches is particularly challenging. Here we describe a purification procedure that combines conventional chromatography steps with antibody affinity columns to purify *Xenopus* γ tubulin complex. We will also present some assays that we used to begin to understand the biochemical functions of the purified γ tubulin complex.

Preparation of *Xenopus* Egg Extracts

We typically use 15 female *Xenopus laevis* for each extract preparation. The method described by Murray[4] is used to make highly concentrated egg cytosolic extract from cytostatic factor (CSF)-arrested *Xenopus* eggs. After the crushing spin in an SW28 rotor at 15,000 rpm for 20 min at 4°, we collect the cytosol, which is spun again in an SS34 rotor at 15,000 rpm for 20 min before storage. This clarified cytosol is separated into 3-ml

---

[1] B. R. Oakley, C. E. Oakley, Y. Yoon, and M. K. Jung, *Cell* **61,** 1289 (1990).
[2] J. W. Raff, D. R. Kellogg, and B. M. Alberts, *J. Cell Biol.* **121,** 823 (1993).
[3] T. Stearns and M. Kirschner, *Cell* **76,** 623 (1994).
[4] A. W. Murray, *Methods Cell Biol.* **36,** 581 (1991).

aliquots, quickly frozen in liquid nitrogen, and stored at −80°. A volume of 4–40 ml of clarified extract can be expected from 15 frogs. We found that we can store the extract for at least 2 months without noticeable effects on the γ tubulin complex.

Purification of the γ Tubulin Complex

We first attempted to use the following series of conventional steps to purify the γ tubulin complex: ammonium sulfate precipitation, gel filtration column, S-Sepharose Fast Flow column, and Mono Q column. We found that not enough purification can be achieved and the yield was too low to add additional purification steps. We then developed a purification scheme that combines both conventional and antibody affinity chromatography to obtain sufficiently pure native γ tubulin complex. We first describe the columns and then the purification procedure.

Columns

Gel Filtration Column. A ~300-ml Sephacryl S-300 HR (Pharmacia, Piscataway, NJ) gel filtration column is poured in an XK 26/70 column housing (Pharmacia) and equilibrated in HEPES 100 plus 0.1 mM guanosine triphosphate (GTP).

Antibody Column. A 1-ml antibody column is made by coupling approximately 1–2 mg of affinity purified, rabbit anti-*Xenopus* γ tubulin C-terminal peptide (AATRPDYISWGTQDK) antibodies to protein A agarose beads following the procedure described by Harlow and Lane.[5] (See article by Chris Field [41] in this volume for more details on the peptide antibody affinity chromatography.) To affinity purify the antibodies, we typically perform sequential elutions with 1.4 M MgCl$_2$, 4 M MgCl$_2$, 0.1 M glycine, pH 2.5, and 0.1 M glycine, pH 2.0. The 1.4 M MgCl$_2$ and glycine, pH 2.5, eluted fractions are pooled separated for the affinity purification. We pour the column in a 1-ml Bio-Spin disposable column (Bio-Rad, Hercules, CA). We found that the antibody column can be reused one to two times. Before each subsequent use, the column is stripped with 5–10 ml of 0.1 mM glycine, pH 2.5, and equilibrated with 10 ml of HEPES 100.

SP-Sepharose Fast Flow Column. A 100-μl SP-Sepharose Fast Flow column is made in a 200-μl micropipette tip. To make this column, we first cut off a small piece from the sharp end of the micropipette tip to make the end wider. Then a piece of Nitex nylon fiber (TETKO, Briarcliff Manor,

---

[5] E. Harlow and D. Lane, *in* "Antibodies: A Laboratory Manual." Cold Spring Harbor Laboratory Press, New York, 1988.

NY) that has a mesh opening of 45 μm is fixed to this end using a suitable piece of Tygon tubing (see Fig. 1).

Purification Procedure

Each step described below is carried out at 4° unless otherwise specified.

1. Thaw 30 ml of frozen *Xenopus* extract. Clarify this extract by centrifuging in an SW55 rotor at 40,000 rpm for 1.5 hr. The supernatant is recovered and an appropriate volume of 100% ammonium sulfate is added to bring the final concentration to 15%. Incubate on ice for 10 min.

2. Centrifuge the mixture in an SS34 rotor at 9600 rpm for 15 min. The supernatant is recovered and the concentration of ammonium sulfate adjusted to 25%. Incubate on ice for 10 min and repeat the centrifugation in the SS34 rotor.

3. Decant the supernatant and resuspend the pellet in 10 ml HEPES 100 containing 0.1 mM GTP. Clarify the mixture by centrifuging in an SW55 rotor at 30,000 rpm for 1 hr. Figure 2a shows an analysis of the ammonium sulfate fractionated supernatants and pellets. About 80–90% of the γ tubulin is recovered in the 25% ammonium pellet fraction.

4. Load the supernatant (~100 mg total protein) onto the Sephacryl S-300 HR gel filtration column that has been preequilibrated in HEPES 100 containing 0.1 mM GTP. The column is run on the fast performance liquid chromatography (FPLC) system at 3 ml/min and 5-ml fractions are collected.

5. One microliter from each fraction is spotted on a nitrocellulose filter and the filter is blotted with anti-γ tubulin antibodies to locate the γ tubulin peak. This step is estimated to give approximately 70% recovery of γ tubulin as quantified by Western blotting.

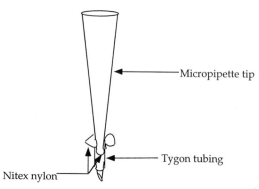

FIG. 1. A diagram of the construction of a 100-μl column housing for the SP-Sepharose Fast Flow column.

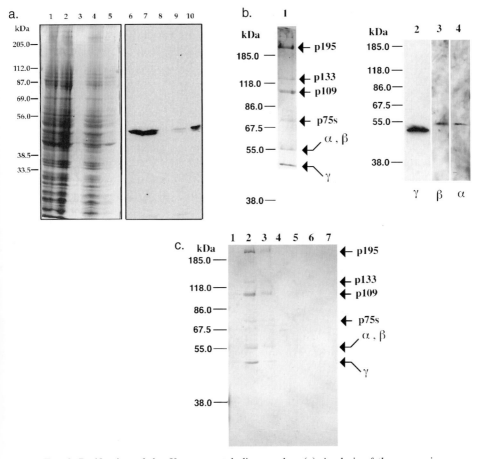

Fɪɢ. 2. Purification of the *Xenopus* γ tubulin complex. (a) Analysis of the ammonium sulfate fractionated *Xenopus* egg extract by sodium dodecyl sulfate–polyacrylamide gel electrophoresis (SDS–PAGE) (lanes 1–5) and Western blotting with an anti-γ tubulin antibody raised against the C-terminal peptide AATRPDYISWGTQDK of *Xenopus* γ tubulin (lanes 6–10). Lanes: 1 and 6, concentrated crude *Xenopus* egg extract; 2 and 7, clarified *Xenopus* egg extract; 3 and 8, 15% ammonium sulfate pellet fraction; 4 and 9, 25% ammonium sulfate supernatant fraction; 5 and 10, 25% ammonium sulfate pellet fraction. The ammonium sulfate pellets were brought to the same volumes as the starting crude extract. One microliter from each of the fractions was loaded on the gel for both Western blotting and Coomassie staining. (b) Analysis by SDS–PAGE of the γ tubulin complex purified through the antibody-column step. Lanes: 1, gel stained with Coomassie Blue to detect total protein; 2, Western blot probed with the anti-γ tubulin antibody; 3, Western blot probed with a monoclonal antibody (tub 2-28-33; Sigma) against β tubulin; 4, Western blot probed with a monoclonal antibody (DM1α; Sigma) against α tubulin. (c) Coomassie Blue-stained gel of the proteins in Part (b) as they eluted from a subsequent SP-Sepharose Fast Flow column. Lanes 1–7 are the salt step-gradient fractions. [Reprinted with permission from *Nature* **378,** 578 (1995) Macmillan Magazines Limited.]

6. Pool the fractions containing γ tubulin (~30 mg protein can be expected at this step) and load onto the 1-ml antibody column at a flow rate of 25 ml/hr. Wash the column sequentially with 25 ml HEPES 100 containing 0.1 mM GTP, 25 ml HEPES 250 containing 0.1 mM GTP (HEPES 250 is the same as HEPES 100 except it contains 250 mM rather than 100 mM NaCl), and 25 ml HEPES 100 containing 0.1 mM GTP.

7. Elute the γ tubulin complex by loading 0.4 mg/ml of the peptide (AATRPDYISWGTQDK) dissolved in 3–5 ml HEPES 100 containing 0.1 mM GTP onto the antibody column, allowing the peptide to enter the column. Then stop the flow. Wait for 2–12 hr to allow the peptide to displace the γ tubulin complex from the antibodies. Continue the elution and collect 0.5-ml fractions. Based on a dot blot, γ tubulin peaks in the first 0.5-ml fraction, although a smaller amount can be found in the second and third fraction (approximate recovery: 50%). Figure 2b shows the analysis of proteins eluted from the antibody column.

8. Pool the first three fractions and load manually onto the 100-μl SP-Sepharose Fast Flow column. Wash the column with 1 ml HEPES 100 containing 0.1 mM GTP. (The peptides do not bind to the column.) Elute the γ tubulin complex by sequentially loading 200 μl of the same HEPES buffer but containing an increasing concentration of NaCl (200, 300, 400, 500, 600, and 700 mM) and collect the flow-through after each loading. As shown in Fig. 2c, most proteins are eluted in the second fraction. Using bovine serum albumin (BSA) as a standard and based on densitometry scanning, the final yield of total protein is 1–3 μg (γ tubulin final yield: 200–800 ng) in a total volume of 400 μl. The recovery at this step is approximately 25–40%. The peak of this purified γ tubulin complex can be further analyzed by sucrose gradient sedimentation to confirm that it has the same S value as the γ tubulin complex present in the extract. The following proteins consistently copurify with γ tubulin: p195, p133, p109 (doublet), p75s, and α and β tubulin.[6]

Additional Purification Steps Used to Characterize the γTuRC

We used the following fractionation methods to characterize the γ tubulin ring complex (γTuRC) while we were developing the purification procedure. These steps may be helpful for developing methods to purify γ tubulin complexes from organisms other than *Xenopus*.

Sucrose Gradient Sedimentation

A 5-ml, 5–40% linear sucrose gradient is made by layering 1 ml each (w/v) 40, 30, 20, 10, and 5% sucrose in HEPES 100 buffer (50 mM HEPES,

[6] Y. Zheng, M. Wong, B. Alberts, and T. Mitchison, *Nature* **378,** 578 (1995).

pH 8.0, 1 mM MgCl$_2$, 1 mM ethylene glycol-bis(β-aminoethyl ether)-N,N,N′,N′-tetraacetic acid (EGTA), 100 mM NaCl) sequentially in an SW55 centrifuge tube and allowing to diffuse overnight at 4°. The extract (200–300 μl) is layered on this gradient and the centrifugation is carried out in an SW55 rotor at 50,000 rpm for 4.5 hr at 4°. Then 300-μl fractions are taken by hand from the top of the gradient and analyzed by SDS–PAGE followed by Western blotting and detection with anti-γ tubulin antibodies. Previously we[6] and others[3] estimated that the S value of the γ tubulin complex is approximately 25S. Initially the largest marker we used was thyroglobulin (19S) and we have since included the *Escherichia coli* 30S ribosome as one of the markers. We found that the γ tubulin complex is heavier than the 30S ribosome. By extrapolating the standard curve obtained from the S values of ovalbumin, catalase, thyroglobulin, and 30S ribosome, we now estimate that the γ tubulin complex is approximately 36S.

Gel Filtration Chromatography

Frozen egg extract (100 μl) is diluted with an equal volume of HEPES 100, and 100 μl of this diluted extract is then passed sequentially through two Bio-Spin 6 columns (Bio-Rad; bed volume of 1 ml) that have been equilibrated in the same buffer. This treatment helps to clean up the extract and prolong the lifetime of the gel filtration column. Then 100 μl of the resulting extract is loaded onto a 24-ml prepacked Superose 6 HR 10/30 column (Pharmacia) in the above buffer at 4°. Using a flow rate of 0.3 ml/min, 0.5-ml fractions are collected, portions of which are subjected to SDS–PAGE and analyzed by Western blotting with antibodies against γ tubulin. As shown in Fig. 3, most γ tubulin exists as a large complex with a Stokes radius greater than 16.2 nm.

Ionic-Exchange Chromatography

We found that the γ tubulin complex binds to both cation- and anion-exchange columns. In our pilot runs using HEPES buffer (buffer A: 50 mM HEPES, pH 8.0, 1 mM MgCl$_2$, 1 mM EGTA; buffer B: buffer A containing 1 M NaCl) for the Mono S column and Bis–Tris propane buffer (buffer A: 20 mM Bis–Tris propane, pH 6.5, 1 mM MgCl$_2$, 1 mM EGTA; buffer B: buffer A containing 1 M NaCl) or Bis–Tris buffer (buffer A: 20 mM Bis–Tris, pH 6.0, 1 mM MgCl$_2$, 1 mM EGTA; buffer B: buffer A containing 1 M NaCl) for the Mono Q column. The Mono S column gave better purification.

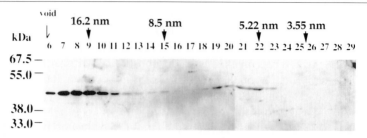

FIG. 3. The γ tubulin complex in *Xenopus* eggs. Extracts made from *Xenopus* eggs were fractionated on Superose 6 gel filtration columns and fractions were analyzed by Western blotting, probing with anti-γ tubulin antibodies. The numbers 6–29 are fraction numbers. The Stokes radii of the standards: ovalbumin (3.55 nm), catalase (5.22 nm), thyroglobulin (8.5 nm), and kinesin (KRP130, 16.2 nm) are indicated by arrows. [Reprinted with permission from *Nature* **378,** 578 (1995) Macmillan Magazines Limited.]

Assay of the Purified γ Tubulin Complex

Electron Microscopy Analysis of the Purified γ Tubulin Complex

Five microliters is taken from each of the SP-Sepharose Fast Flow column fractions and applied onto carbon-coated EM grids. After 2–5 min, the grids are rinsed in water, stained with 1% uranyl acetate in 50% ethanol for 30 sec, and examined in a Philips (Mahwah, NJ) EM400 microscope.

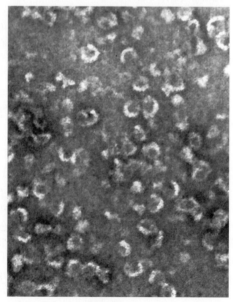

FIG. 4. Electron microscopy of a field of negatively stained γ tubulin complexes.

As shown in Fig. 4, ring structures are present in the γ tubulin complex peak fractions. The number of rings correlates with the amount of protein present in each fraction.

Microtubule Nucleation Assay

The tubulin used in this and the following experiments is prepared from bovine brain by two cycles of temperature-dependent polymerization, followed by phosphocellulose chromatography.[7] This tubulin is cycled one more time (we adjust the final tubulin concentration to 15–20 mg/ml) and frozen in small aliquots using liquid nitrogen, and stored at −80°.[8] Rhodamine-labeled tubulin is made and stored according to Hyman *et al.*[8]

1. Thaw the frozen labeled and unlabeled tubulin. On ice, mix the labeled and unlabeled tubulin at a 1:5 ratio, and adjust the total tubulin concentrations using BRB80 (80 mM K-PIPES, 1 mM MgCl$_2$, 1 mM EGTA, pH 6.8) containing 1 mM GTP to 1, 2, 4, 10, and 12 mg/ml.
2. To 10 μl of the above tubulin solution add 10 μl of the peak γTuRC containing fraction from the SP-Sepharose Fast Flow column or a buffer control.
3. Incubate the mixtures in a 37° water bath for 5 min. After incubation, add to each mixture 100 μl BRB80 containing 1% glutaraldehyde, mix by inverting the tubes a few times, and incubate at room temperature for 3 min. Add 300 μl of BRB80 containing 80% glycerol to each mixture and mix by inverting the tubes.
4. Mount 5 μl of each mixture to observe under the fluorescence microscope. Cut off the ends of the micropipette tips to avoid shearing microtubules. Count the number of microtubules per 50 random fields and plot this number against the tubulin concentration. A typical microtubule nucleation result is shown in Fig. 5.

Microtubule Cosedimentation

The following procedure is used to address whether the purified γTuRC can bind to preformed microtubules.

1. Thaw tubulin and adjust the final concentration to 1.5 mg/ml (final volume 190 μl) with BRB80 containing 1 mM GTP. Incubate at 37° for 20 min. Add 20 μl Taxol at 200 μM and continue incubation for another 10 min.

[7] T. J. Mitchison and M. W. Kirschner, *Nature* **312,** 237 (1984).
[8] A. Hyman, D. Drechsel, D. Kellogg, S. Salser, K. Sawin, P. Steffen, L. Wordeman, and T. Mitchison, *Methods Enzymol.* **196,** 478 (1991).

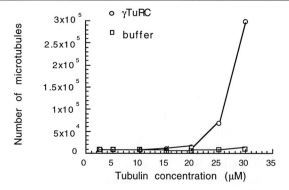

Fɪɢ. 5. Nucleation of microtubules by purified γTuRC. Microtubules were incubated with the indicated concentrations of tubulin in the presence of either γTuRC (circles) or buffer (squares), and the number of microtubules observed after a fixed time was plotted against the tubulin concentration. [Reprinted with permission from *Nature* **378,** 578 (1995) Macmillan Magazines Limited.]

2. Centrifuge the microtubules at 12,000*g* for 5 min at room temperature to remove aggregates.
3. Add increasing volumes of this microtubule mixture to the γTuRC and incubate at room temperature for 10 min. Layer onto a 150-μl cushion (30% v/v glycerol in BRB80) and spin in a TLA100 rotor (Beckman, Fullerton, CA) at 50,000 rpm for 10 min. The pellet is analyzed by Western blotting using anti-γ tubulin and anti-α tubulin antibodies. γ Tubulin cosediments with microtubules and more γ tubulin is sedimented in the presence of more microtubules.

Microtubule End Binding

To address whether γTuRC binds along the sides or the ends of microtubules, we used the following method.

1. Thaw tubulin and adjust the concentration to 5 mg/ml. Mix an equal volume of this tubulin with the purified γTuRC or buffer control. Carry out microtubule nucleation at 37° for 30–40 sec.
2. Add to the above mixture BRB80 containing glutaraldehyde so that the final concentration of the glutaraldehyde is 1% and incubate at room temperature for 2–3 min. Add five volumes of 80% v/v glycerol in BRB80 and mix by inverting the tube.
3. Layer this onto a 3-ml cushion (25% v/v glycerol in BRB80) in a 15-ml modified Corex tube (see Evans *et al.*[9] for the construction of

---

[9] L. Evans, T. Mitchison, and M. Kirschner, *J. Cell Biol.* **100,** 1185 (1985).

the modified Corex tube). At the bottom of the Corex tube, place a round coverslip having two 400-mesh EM grids fixed onto it with a thin layer of Formvar.
4. Centrifuge at 13,000 rpm in an HB-4 rotor for 2 hr at 25° to pellet the microtubules onto the EM grids. Rinse the grids with 20 ml deionized H_2O containing 0.2% Triton X-100 in a petri dish. The grids are then processed for negative staining and examined as described earlier.

Minus-End Blocking Assay

1. Thaw cycled unlabeled tubulin and rhodamine-labeled tubulin. On ice, mix the labeled and unlabeled tubulin at approximately a 1:30 ratio and adjust the final concentration to 2.5 mg/ml with BRB80 containing 1 mM GTP. (The ratio of labeled and unlabeled tubulin needs to be determined for each batch of labeled tubulin by mixing the labeled and unlabeled tubulin at various ratios and polymerizing microtubules. Choose the ratio that gives faintly labeled microtubules.)
2. Add to the above tubulin mix the purified γTuRC (or buffer control) so that the final concentration of the γTuRC is approximately 6 μg/ml. Incubate at 37° for 1 min.

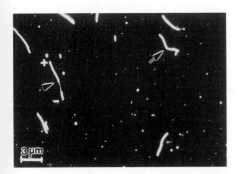 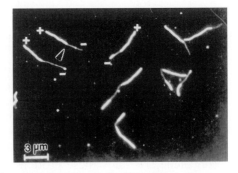

FIG. 6. γTuRC blocks elongation of the minus end of a microtubule. The micrograph on the left is a field of polarity-marked microtubules produced without γTuRC. The bright segment at the plus end of the dimly labeled microtubule is about three times longer than that of the minus end (see the microtubule pointed to by an arrow). On the right is a field of polarity-marked microtubules produced with γTuRC. Many dimly labeled microtubules have no or a very short bright segment at their minus ends, suggesting that γTuRC can block or inhibit the minus-end microtubule growth. In the presence of γTuRC, 33.7% of the total microtubules exhibited no minus-end growth, compared to only 4% in the absence of γTuRC.

3. Add rhodamine-labeled tubulin to the above mix to bring the final ratio of labeled to unlabeled tubulin to 1:3 and the final tubulin concentration to 1.1 mg/ml. Continue incubation for 4 min.

4. Fix microtubules by adding glutaraldehyde to a final concentration of 1% and leave at room temperature for 3 min. Dilute the mixture with five volumes of BRB80 containing 60% glycerol (v/v).

5. Centrifuge the microtubules through a glycerol cushion (25% v/v glycerol in BRB80) onto a round coverslip at the bottom of a modified 15-ml Corex tube.[9] The coverslip is mounted with antifade (18 mM Tris-Cl, pH 8.2, with 80% v/v glycerol and 9.1% p-diaminobenzene). Random fields of microtubules were photographed (Fig. 6) at a final magnification of 600× and the lengths of the bright microtubule segments were measured from the negatives.

Acknowledgments

The authors thank Ona Martin for critically reading the manuscript. MLW was supported by HHMI.

[20] Rapid Isolation of Centrosomes

By MAUREEN BLOMBERG-WIRSCHELL and STEPHEN J. DOXSEY

Introduction

Centrosomes are involved in nucleation and organization of microtubules in interphase cells and during spindle formation in mitosis.[1] Little is known about the molecular components that perform and regulate these important functions. Centrosome components are usually present in low quantities and often go undetected in morphologic and biochemical assays due to the small size and singular nature of the centrosome in most cells. However, a rapidly increasing number of proteins, including many that regulate important cellular events, have been shown to associate with this organelle. To study these proteins, it has become increasingly important to have a fast and simple method to isolate centrosomes.

We have developed procedures for the rapid isolation of centrosomes from a variety of cell types and species. The first method is a modified and abbreviated version of that described by Mitchison and Kirschner.[2] It offers the advantage that nucleation-competent centrosomes can be obtained in

---

[1] D. Kellogg, M. Moritz, and B. Alberts, *Ann. Rev. Biochem.* **63**, 639 (1994).
[2] T. Mitchison and M. Kirschner, *Methods Enzymol.* **134**, 261 (1986).

a short period of time from little starting material. The entire procedure can be completed in 1–1.5 hr, a significantly shorter time than that of conventional procedures. Although centrosome fractions are less pure than those obtained by other methods, they are adequate for most functional, morphologic and biochemical analyses. The second method involves isolation of nuclei with attached centrosomes. It can be completed in 30 min and is primarily useful for morphologic analyses. Both procedures require minimal equipment and are easily adapted to most laboratory settings. Only those steps significantly different from previously published procedures are expanded on here. For details on other methods for isolating centrosomes and other microtubule organizing centers, see Refs. 2–4.

Reagents and Supplies

Solutions

1× PBS: 130 mM NaCl, 2 mM KCl, 8 mM Na_2HPO_4, 2 mM KH_2PO_4

PB: 80 mM PIPES, 1 mM Ethylene glycol-bis(β-aminoethyl ether)-N,N,N′,N′-tetraacetic acid (EGTA), 1 mM $MgCl_2$, pH 6.8 with KOH

PE: 10 mM PIPES, 1 mM Ethylenediamine tetraacetic acid (EDTA), 0.1% 2-mercaptoethanol (2-ME), pH to 7.2 with potassium hydroxide (KOH)

10× Protease inhibitors: 10 mg/ml aprotinin, 10 mg/ml leupeptin, 10 mg/ml pepstatin

Lysis buffer: 1 mM Tris-HCl (pH 8.0), 0.1% 2-ME, 0.5% Triton X-100, 1× protease inhibitors [Optional: 1 mM phenylmethylsulfonyl fluoride (PMSF)]

Nocodozole: 10 mg/ml in dimethyl sulfoxide (DMSO)

Cytochalasin B: 20 mg/ml in DMSO

Supplies

All chemicals and reagents should be the highest grade available and were obtained from Sigma (St. Louis, MO) except Ficoll Type 400 from Pharmacia (Piscataway, NJ), palloidin from Molecular Probes (Eugene, OR), and secondary antibodies from Jackson Labs (West Grove, PA). Tubulin and rhodamine-labeled tubulin were prepared from calf brain as described in Ref. 5. All solutions are made fresh and kept at 4°.

[3] M. Moudjou and M. Bornens, *in* "Cell Biology: A Laboratory Handbook," (J. E. Celis, ed.), Vol. 1. p. 595. Academic Press, San Diego, 1994.

[4] J. V. Kilmartin, *Curr. Opin. Cell Biol.* **6(1)**, 50 (1994).

[5] A. Hyman, D. Drechsel, D. Kellogg, S. Salser, K. Sawin, P. Steffen, L. Wordeman, and T. Mitchison, *Methods Enzymol.* **196**, 478 (1991).

Centrosome Isolation Procedures: Method 1

Cell Plating and Treatments

Many cell types have been used to isolate centrosomes by this method. They are generally plated on two 150-mm plates in the appropriate medium 2 days prior to use. Best results are obtained when cells are confluent on the day of the preparation.

The first step in this method involves depolymerization of cytoskeletal elements to release centrosomes from nuclei and the cytoplasm. Depolymerization is accomplished by adding complete media containing nocodazole and cytochalasin B (final concentrations, 100 μg/ml) to cells and incubating for 15–20 min at 37° in a CO_2 incubator. At these drug concentrations, microtubules and actin filaments are rapidly and effectively depolymerized in most cells without adversely affecting centrosome structure or function (see Figs. 2 and 3) in a later section. For other cells, it may be necessary to use lower concentrations of drugs for longer times (see below). Depolymerization of cytoskeletal elements can be monitored by fluorescence labeling with anti-α tubulin antibodies and fluorescent phalloidin.

Cell Washing and Lysis

Washes are performed at the sink using solutions chilled on wet ice. Cells are washed consecutively with 1× PBS, then 0.1× PBS with 8% w/v sucrose and finally with 8% w/v sucrose in H_2O. Each wash is performed quickly with 30 ml solution per 150-mm plate. The last wash is aspirated carefully to remove the majority of the sucrose solution. Lysis buffer is added (7 ml) and plates are agitated for 10 min at 4° on a rotary shaker. Plates can be manually rocked once or twice during lysis to ensure that all cells are covered with lysis buffer. For suspension cells, washes are done by pelleting and resuspending in consecutive buffers[2] and cells are lysed in 5 ml lysis buffer/10^7 cells. The use of low ionic strength lysis buffer facilitates liberation of centrosomes from nuclei.[2]

Centrosome Isolation

Lysates containing centrosomes are collected into a 15-ml plastic conical tube on ice and components added (from stocks) to achieve a final concentration of 10 mM PIPES, 1 mM EDTA, and 1 μg/ml protease inhibitors. Tubes are centrifuged at 1500g for 5 min to pellet residual nuclei and other cellular debris. For many adherent cells, nuclei are not released from the culture plate, so the nuclear centrifugation step can be omitted. Lysates are added to 15-ml Corex tubes (Corning G, Corning, NY) by pouring through a funnel lined with 40-μm nytex membrane (Small Parts Inc., Miami Lakes, FL), which retains residual chromatin and cellular debris. Addition of DNase II (10 units/ml, Sigma) can also be used to reduce DNA contamination,[3] although this step has not been tested in this protocol. The efficiency of centrosome release from cells is monitored in parallel samples by immunostaining cells plated on coverslips with antibodies to centrosome markers.

Lysates are underlain with 1–1.5 ml 20% (w/v) Ficoll MW 400,000 in PE with 0.1% Triton X-100 and tubes are spun at 25,000g in a swinging bucket rotor (Beckman JS-13, JS-13.1, Sorvall HB4) for 20 min at 4°. The lysate is aspirated to within 2 ml above the Ficoll cushion. Centrosomes will be layered on top of the cushion and are usually not visible. They are collected using a Pasteur pipette (5.75 in.) with a clean-cut tip. The interface layer is drawn into a pipette positioned ~2 mm above the cushion using gentle sweeping motions over the entire layer; the Ficoll appears as an inverted cone when visualized against a dark background. The collection step can be repeated to maximize yields as it is the least reproducible step of the procedure. The centrosome fraction (0.5–0.9 ml) is placed in a plastic tube on ice and immediately used for centrosome assays. The procedure can be scaled up using more than two 150-mm plates of cells, 50-ml plastic tubes to collect lysates and 30-ml Corex tubes to concentrate centrosomes.

Centrosome Isolation Procedures: Method 2

This isolation method involves preparation of a crude fraction of nuclei under conditions that preserve the nuclear–centrosome association.

Solutions

1× PBS: 130 mM NaCl, 2 mM KCl, 8 mM Na$_2$HPO$_4$, 2 mM KH$_2$PO$_4$
Nuclear isolation buffer: 50 mM Tris, 80 mM NaCl, 25 mM EDTA, 1% Triton X-100, pH 7.6
TMNP buffer: 13 mM Tris, 10 mM MgCl$_2$, 10 mM NaCl, pH 7.4

Procedure

Cells are grown to confluency in one 150-mm tissue culture dish and collected by trypsinization. They are washed once with 1× PBS to remove residual traces of media, resuspended in 3 ml of nuclear isolation buffer, and incubated on ice for 5 min to lyse cells. Nuclei are collected by low-speed centrifugation (~200g), washed once in TMNP buffer, and resuspended in 1 ml of TMNP buffer.

General Features of Rapidly Isolated Centrosomes

Method 1

Centrosome fractions that can be prepared easily and rapidly, while not biochemically pure, are a reliable source of material for many functional, morphologic and biochemical studies. For most assays (Figs. 2–4), we find little difference between centrosomes prepared by this and other methods. In fact, this procedure may better preserve microtubule nucleation activity and retain proteins that would otherwise be stripped from centrosomes by shearing forces created during sucrose gradient centrifugation.

The yield from this procedure is about 30–75% depending on the cell type used. Sufficient material has easily been generated from many cell types to perform the assays described. For example, centrosomes from CHO and COS cells are routinely visualized using only 2×35-mm plates ($\sim 1.3 \times 10^5$ cells). As many as 3×10^7 centrosomes have been generated from 3×150-mm plates of CHO or COS cells using a single 15-ml Corex tube.

The short incubation of cells at high drug concentrations to depolymerize microtubules and actin does not alter centrosome structure or function in most cell types tested (Figs. 2 and 3). However, it is possible that this treatment could result in centrosome splitting (only one centriole/centrosome), release of cells from tissue culture plates, or incomplete depolymerization of cytoskeletal elements. These problems can be easily overcome by decreasing drug concentrations (nocodazole, 0.1–10 μg/ml; cytochalasin, 5–20 μg/ml) and extending treatment times 90–120 min) as described.[2,3]

This method has been used to reproducibly isolate centrosomes from cell lines derived from a number of different sources and species. These include adherent cells from mammals (HeLa, COS, NIH 3T3, Chinese hamster ovary (CHO), Rat 1), frogs (XTC), and *Drosophila* (Schneider) (American Type Culture Collection, Rockville, MD). The method is likely to be adaptable to most cultured cells, including suspension cultures, and perhaps other sources such as embryonic systems and tissues.

Method 2

Method 2 has been used to isolate nuclei with attached centrosomes from COS and CHO cells and is likely to work with other cell types. We have used this method primarily for mophologic applications (immunofluorescence, immunoelectron microscopy). It may be useful for microtubule nucleation assays as most microtubules are depolymerized during the cold washes, but this has not been tested. Given the impurity of the fractions, biochemical analyses have not been attempted.

Centrosome Assays

Coverslip Supports and Coverslips

Many centrosome assays involve centrifugation of material onto coverslips. We have designed a simple single-piece Plexiglas coverslip support for these purposes. It is rounded on one end to fit the curvature of the bottom of 15-ml Corex tubes and flat on the other end to accommodate 12-mm coverslips (Fig. 1). This design obviates the need to modify tubes permanently,[2] and allows any 15-ml Corex tube or equivalent to be used. Supports can easily be made in most machine shops as described below. Start with a bar of clear acrylic (Plexigas) 9/16 in. in diameter. Cut pieces 0.6 in. in length with parting tool on a lathe using depth stop in collar. Round one end of the piece with a 9/32-in. radius cutter. Cut slot and hole by end milling with a 1/8-in. bit. The slot and hole are designed to retrieve the coverslip support with a wire tool. A heavy-gauge paper clip is easily modified for this purpose.

Coverslips (12 mm) are acid washed to facilitate adherence of centrosome fractions. They are agitated in 40% HCl for 1 hr in a large beaker, washed 15–20 times in double distilled water, followed by an acetone wash and air-drying on Whatman paper.

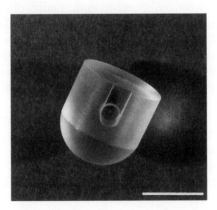

FIG. 1. Plexiglas coverslip support. Rounded end fits curvature of a 15-ml Corex tube and the flat end supports the coverslip. Slot and hole are used to retrieve the support from Corex tubes. See text for details. Bar = 1 cm.

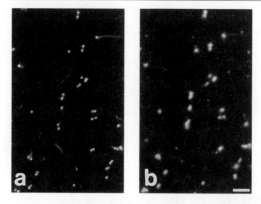

FIG. 2. Immunofluorescence images of centrosomes. Centrosome fractions were prepared in 90 min from one 150-mm plate of CHO cells (~10^7 cells). Five percent of the preparation was centrifuged onto coverslips and processed for immunofluorescence using antibodies to α tubulin to visualize centrioles (a) and pericentrin antibodies to stain the pericentriolar matrix[6] (b). Micrograph shows typical field when viewed with a 100× objective lens. Note doublet staining pattern typical of intact centrosomes. Bar = 5 μm.

Method 1

Immunofluorescence and Immunogold Localization of Centrosome Proteins. Centrosomes prepared by this method can be used to localize centrosome proteins by immunofluorescence microscopy.[2,6] Typically, centrosomes appear as double dots when well preserved, but they can also appear as single dots or aggregates (Fig. 2). The pattern of staining often indicates whether the protein is a component of the centrioles or the surrounding centrosome matrix.[6] Proteins in the matrix often appear as hollow spheres or rings while centriolar antigens appear as smaller densely staining dots. Immunogold electron microscopy on centrosome fractions can be used to confirm and extend these observations.[6] Some antibodies occasionally label noncentrosomal particulate material present in these preparations, so centrosome localization should be confirmed by costaining with antibodies to known centrosome proteins or with antibodies to α or β tubulin, which label centrioles. It is possible that some centrosome antigens are lost during the preparation. An alternative method (although less convincing) to demonstrate centrosome-specific localization involves microtubule-independent centrosome localization *in vivo.*

To prepare centrosomes for immunofluorescence analysis, 50–100 μl of the centrosome preparation is gently mixed with 3–5 ml of PE in 15-ml Corex tubes with acid-washed coverslips on plastic supports (see above).

[6] S. J. Doxsey, P. Stein, L. Evans, P. Calarco, and M. Kirschner, *Cell* **76,** 639 (1994).

Centrosomes are spun at 25,000g for 20 min at 4° in a swinging bucket rotor (see above). The buffer is aspirated to 1 ml above the coverslip and 50 μl of 20% Nonidet P-40 (NP-40) is added to the tube to prevent desiccation of centrosomes during coverslip retrieval (see above). Centrosomes on coverslips are fixed in −20° MeOH (or other fixatives) and processed for immunofluorescence or immunoelectron microscopy.

Many proteins have been localized to centrosomes by immunofluorescence and immunoelectron microscopy using this method.[7–9] In addition, green fluorescent centrosomes produced by transient transfection of cells with a pericentrin-green fluorescent protein construct[10] were isolated by this technique.[10,11]

Microtubule Nucleation from Isolated Centrosomes. Microtubule nucleation from centrosomes prepared fresh by this method is generally more reliable than that from centrosomes stored as frozen sucrose fractions following purification by other methods.[2] Nucleating activity is stable for 1–2 hr on ice but is compromised after freezing. This assay may be useful for examining the effects of chemicals, proteins, antibodies, and other reagents on microtubule nucleation and to reconstitute nucleating activity from chemically treated centrosomes using cellular fractions and isolated proteins. Moreover, centrosomes rapidly isolated from a diversity of species have been used as templates for microtubule nucleation in extracts prepared from *Xenopus* eggs.[6]

For the nucleation assay, microtubules are visualized using tetramethylrhodamine-labeled tubulin.[5] In this way, nucleation can be directly monitored (Fig. 3) obviating the need for immunofluorescence protocols. Centrosomes spun onto coverslips are incubated with 100 μl of tubulin (20 μM), rhodamine tubulin (5 μM), 1 mM guanosine 5′-triphosphate (GTP), in PB at 37° for 10 min. The tubulin solution is aspirated and coverslips are washed gently in PB containing 1 mM GTP. Asters are fixed in 1% glutaraldehyde in PB for 5 min at 37°, rinsed in PB, and postfixed in MeOH at −20° as described.[2] Nucleation from centrosomes prepared from some cells may be inconsistent but can usually be improved by using slightly higher GTP concentrations (1–5 mM) and by including 10–20% glycerol during the nucleation reaction. Nucleation can also be performed in solution and centrosomes with asters centrifuged onto coverslips.[2] The solution assay is less reproducible since rapidly isolated centrosome frac-

---

[7] C. A. Sparks, E. Fey, C. Vidair, and S. J. Doxsey, *J. Cell Sci.* **108,** 3389 (1995).

[8] R. Brown, S. J. Doxsey, R. L. Martin, and W. Welsh, *J. Biol. Chem.* **271,** 824 (1996).

[9] S. M. Pockwinse, G. Krockmalnic, S. J. Doxsey, J. Nickerson, J. Lian, A. van Wijnen, J. Stein, G. Stein, and S. Penman, *Proc. Natl. Acad. Sci. U.S.A.* **94,** 3022 (1997).

[10] R. Heim, A. B. Cubitt, and R. Y. Tsien, *Nature* **373,** 663 (1995).

[11] A. Young, R. Tuft, W. Carrington, and S. J. Doxsey, *Methods Cell Biol.* in press, (1998).

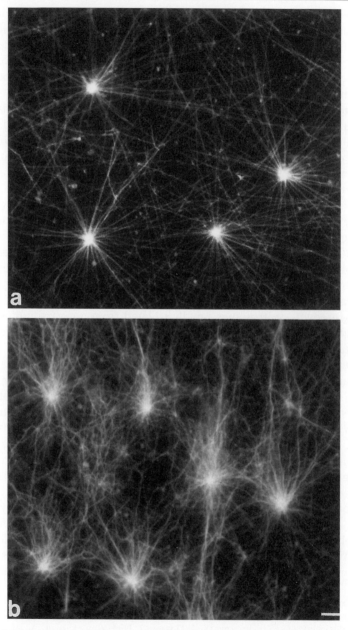

FIG. 3. Microtubule nucleation from rapidly isolated centrosomes. Centrosomes prepared from Rat 1 cells (a) or CHO cells (b) were spun onto coverslips, incubated with rhodamine-labeled tubulin to promote regrowth of microtubules and fixed. See text for details. Bar = 5 μm.

tions are dilute, contain variable amounts of Ficoll, and are subjected to shearing forces of centrifugation that may perturb aster integrity. Using either assay, centrosome proteins can be localized to the center of fixed asters by indirect immunofluorescence.

Biochemical Detection of Centrosome Proteins. Centrosome fractions isolated by this technique are substantially enriched in some centrosome antigens. We have found that several centrosome proteins, while undetectable in whole-cell lysates, give strong, clean signals in Western blots of centrosome fractions (Figs. 4a and 4b). However, other proteins, while present at the centrosome, are also found at other sites in the cell.[8] Since these proteins may be present in material other than centrosomes in these preparations, it is important to confirm that they localize to centrosomes by immunofluorescence before using the biochemical assay. This assay can be used to compare the electrophoretic mobility, posttranslational modifications, and other properties of centrosome proteins from different species and after different treatments. The assay is of limited utility in the biochemical identification of new centrosome proteins due to the relative impurity of the fractions and small amount of material that is generated.

For biochemical analysis, centrosome fractions are diluted at least 1:4

2004

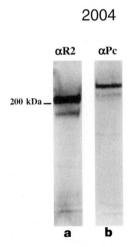

FIG. 4. Biochemical detection of centrosome antigens. Centrosomes prepared from CHO cells were exposed to sodium dodecyl sulfate–polyacrylamide gel electrophoresis (SDS–PAGE) (6% gel) and immunoblotted using antibodies to pericentrin (b) and R2, a novel centrosome protein identified using scleroderma autoimmune sera. (a; M. Blomberg-Wirschell and S. Doxsey, unpublished observations.) Approximately 2×10^6 centrosomes were loaded per lane. Neither protein was detectable by Western blot of whole-cell lysates.

in PE, centrifuged at 25,000g for 20 min at 4° in a 15-ml polypropylene tube (Falcon 2059) and processed for SDS–PAGE[12] and immunoblotting.[13]

Method 2

Immunolocalization of Proteins. Nuclei prepared by this method have been used to localize centrosomal proteins by immunofluorescence and immunogold techniques. For immunofluorescence, 100–200 μl of the nuclear fraction is spun onto acid-washed coverslips and processed as described.[6] For immunoelectron microscopy, nuclei can either be collected on coverslips (100–500 μl) or in a pellet (100–500 μl) and processed as described.[6] Centrosomes from both COS (monkey kidney) and CHO cells have been successfully labeled using these methods.

Acknowledgments

S. J. Doxsey is supported by NIH (RO1 GM5 1994), American Heart Association (Established Investigator) and American Cancer Society.

We wish to thank Aaron Young for assistance with experimental procedures and figure assembly and Jeanette Landrie for photographic assistance.

[12] U. K. Laemelli, *Nature (Lond.)* **227,** 680 (1970).
[13] J. Otto, *in* "Methods in Cell Biology" (D. Asai, ed.), p. 105. Academic Press, San Diego, 1993.

[21] Photoaffinity Labeling Approach to Map the Taxol-Binding Site on the Microtubule

By GEORGE A. ORR, SRINIVASA RAO, CHARLES S. SWINDELL, DAVID G. I. KINGSTON, and SUSAN BAND HORWITZ

Introduction

Taxol is an antitumor drug used in the treatment of advanced ovarian and breast carcinomas[1,2] and is demonstrating encouraging activity in a variety of other human malignancies.[3] The drug originally was isolated

[1] W. P. McGuire, E. K. Rowinsky, N. B. Rosenshein, F. C. Grumbine, D. S. Ettinger, D. K. Armstrong, and R. C. Donehower, *Ann. Intern. Med.* **111,** 273 (1989).
[2] F. A. Holmes, R. S. Walters, R. L. Theriault, A. D. Forman, L. K. Newton, M. N. Raber, A. U. Buzdar, D. K. Frye, and G. N. OHortobagyi, *J. Natl. Cancer. Inst.* **83,** 1797 (1991).
[3] E. K. Rowinsky and R. C. Donehower, *N. Engl. J. Med.* **332,** 1004 (1995).

from the bark of the western yew, *Taxus brevifolia,*[4] and shown to be an antimitotic agent and a potent inhibitor of cell replication.[5] The structure of this novel diterpenoid was first described in 1971[4] and subsequently its unique mechanism of action in the tubulin/microtubule system was identified.[6] *In vitro,* Taxol polymerizes tubulin into stable microtubules in the absence of guanosine triphosphate (GTP), although under normal conditions, the guanine nucleotide is an absolute requirement for tubulin polymerization.[7,8] Taxol-stabilized microtubules are resistant to depolymerization by cold temperature, dilution, and Ca^{2+}. In contrast to other antimicrotubule agents such as colchicine and vinblastine, which bind to the tubulin dimer, the Taxol binding site is present only on the microtubule polymer. Taxol binds to microtubules specifically and reversibly with a stoichiometry, relative to the α/β-tubulin heterodimer, approaching one.[9,10] In cells, incubation with Taxol results in the formation of stable bundles of microtubules.[7]

Photoaffinity labeling is a powerful method to address the nature of the interaction between Taxol and its target protein. Our earlier studies demonstrated that the direct photolabeling of tubulin with ^{3}H-Taxol resulted in the preferential labeling of β tubulin.[11] However, the low extent of photoincorporation of drug precluded the use of ^{3}H-Taxol to map the drug binding site. To overcome this problem, we have made use of Taxol analogs with an arylazide substituent attached to either the C-3' of the A ring side chain or the C-2 of the B ring of the taxoid nucleus (Fig. 1). With both analogs, tritium was incorporated close to the photoreactive group to enable the site(s) of incorporation to be identified and characterized. Recent studies by Nogales and collaborators[12,13] have presented an atomic model of the $\alpha\beta$-tubulin dimer fitted to a 3.7-Å density map obtained by electron crystallography of zinc-induced sheets. Our photolabeling studies indicating that Taxol binds preferentially to the β subunit proved helpful in projection studies that identified the β monomer. The availability of an atomic model

[4] M. C. Wani, H. L. Taylor, M. E. Wall, P. Coggon, and A. T. McPhail, *J. Am. Chem. Soc.* **93,** 2325 (1971).

[5] P. B. Schiff, J. Fant, and S. B. Horwitz, *Nature* **277,** 665 (1979).

[6] P. B. Schiff and S. B. Horwitz, *Proc. Natl. Acad. Sci. U.S.A.* **77,** 1561 (1980).

[7] P. B. Schiff and S. B. Horwitz, *Biochemistry* **20,** 3247 (1981).

[8] N. Kumar, *J. Biol. Chem.* **256,** 10435 (1981).

[9] J. Parness and S. B. Horwitz, *J. Cell. Biol.* **91,** 479 (1981).

[10] J. F. Diaz and J. M. Andreu, *Biochemistry* **32,** 2747 (1993).

[11] S. Rao, S. B. Horwitz, and I. Ringel, *J. Natl. Cancer Inst.* **84,** 785 (1992).

[12] E. Nogales, S. G. Wolf, I. A. Khan, R. F. Luduena, and K. H. Downing, *Nature* **375,** 424 (1995).

[13] E. Nogales, S. G. Wolf, and K. H. Downing, *Nature* **391,** 199 (1998).

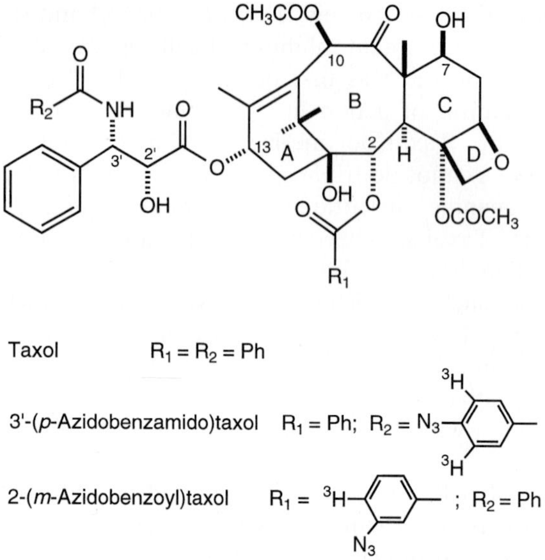

Fig. 1. Molecular structures of Taxol, ^{3}H-2-(m-azidobenzoyl)taxol, and ^{3}H-3′-(p-azido-benzamido)taxol.

of the tubulin heterodimer will play a definitive role in ascertaining the binding site for Taxol in the microtubule.

Experimental Procedures

Microtubule Assembly

Microtubule protein (MTP) is purified from calf brain by two cycles of temperature-dependent assembly–disassembly.[14] The concentration of tubulin in MTP is based on a tubulin content of 85%. For microtubule assembly experiments, MTP (final concentration 1.0–1.5 mg/ml) is suspended in assembly buffer consisting of 0.1 M 2(N-morpholino)ethanesulfonic acid (MES), 1 mM ethylene glycol bis(aminoethyl)-N,N′-tetraacetic acid (EGTA), 0.5 mM MgCl$_2$ and 3 M glycerol, pH 6.6. Assembly at 35° is monitored spectrophotometrically at 350 nm by following changes in turbidity that are representative of polymer mass.[15]

---

[14] M. L. Shelanski, F. Gaskin, and C. R. Cantor, *Proc. Natl. Acad. Sci. U.S.A.* **70,** 765 (1973).
[15] F. Gaskins, C. R. Cantor, and M. L. Shelanski, *J. Mol. Biol.* **89,** 737 (1974).

Labeling of Microtubules with Photoreactive Taxol Analogs

^{3}H-3'-(p-azidobenzamido)taxol (10 μM, 1.7–2.8 Ci/mmol) or ^{3}H-2-(m-azidobenzoyl)taxol (10 μM, 0.12 Ci/mmol) is added to MTP (10 μM tubulin) in assembly buffer and incubated at 37° for 30 min. Aliquots (250 μl) are placed in a multiwell plate (1.7 cm in diameter) which is kept at 4° and irradiated for 30 min at 254 nm with a Mineralight lamp (model R52G, UVP Inc., San Gabriel, CA) at a distance of 7 cm. The extent of photoincorporation is calculated, based on one drug binding site per dimer, by a filter binding assay after precipitation of photolabeled tubulin with cold acetone. The photolabeled samples are analyzed on 9% sodium dodecyl sulfate–polyacrylamide gel electrophoresis (SDS–PAGE) gels.[16] For fluorography, the analytical gels are stained with Coomassie R-250, destained, treated with EN^{3}HANCE, and exposed to Kodak X-Omat AR film at −70°C.

Isolation of β-Tubulin and Formic Acid Cleavage

For the isolation of β tubulin, preparative SDS–PAGE gels (9%, 3 mm) are immersed in ice-cold 20 mM KCl for 5 min to visualize the α- and β-tubulin subunits. The β-tubulin band is excised, washed six to eight times with H$_2$O, and electroeluted for 16 hr with an electroeluter (model 422, Bio-Rad, Richmond, CA) at 10 mA per electroelution tube. The recovery of β tubulin is between 60 and 80%. SDS is removed from electroeluted β tubulin using Extracti-gel D (Pierce, Rockford, IL) and the eluate is treated with trichloroacetic acid (final concentration 12% w/v) to precipitate β tubulin and remove SDS. The precipitate is washed twice with ice-cold acetone to remove residual trichloroacetic acid. The ^{3}H-2-(m-azidobenzyl) taxol-labeled β tubulin is reduced and carboxymethylated prior to formic acid cleavage.[17] Photolabeled β tubulins (10–15 nmol) are dissolved in 400 μl of 75% formic acid and incubated at 37°. After 72 hr, formic acid is removed by evaporation in a Speed-Vac (Savant, Holbrook, NY), the residue washed twice with H$_2$O and dried. The formic acid digestion products are separated on 17.5% SDS–PAGE gels with 0.1 M Tris, 0.1 M Tricine, and 0.1% SDS as the cathode buffer[18] and visualized by fluorography as described earlier.

High-Performance Electrophoresis Chromatography (HPEC)

HPEC (Applied Biosystems, San Jose, CA, model 230A) analysis is performed to fractionate the undigested and formic acid-digested ^{3}H-3'-

[16] U. K. Laemmli, *Nature* **227,** 680 (1970).
[17] F. R. N. Gurd, *Methods Enzymol.* **11,** 532 (1967).
[18] H. Schagger and G. Jagow, *Anal. Biochem.* **166,** 368 (1987).

(*p*-azidobenzamido)taxol-labeled β tubulin. The upper electrode buffer contains 75 m*M* Tris-phosphate, pH 7.5, and 0.1% SDS. The lower electrode buffer, which is also used for elution, consists of 75 m*M* Tris-Cl, pH 7.5. The sample is resuspended in 1× HPEC loading buffer (7.5 m*M* Tris-phosphate, pH 7.5, 0.25% SDS, 0.2% 2-mercaptoethanol, and 15% glycerol) by boiling for 3 min and centrifuging at 13,000 rpm in a microfuge for 10 min. The supernatant (25 μl) is loaded onto a 2.5- × 50-mm gel. A 7% polyacrylamide gel is used to resolve undigested photolabeled β tubulin and a 12% polyacrylamide gel plus 6 *M* urea is used for digested β tubulin. The elution flow rate is 15 μl/min. Fractions (3 min) are collected and counted by liquid scintillation spectrometry.

Determination of Peptide Mass and Sequence of the Radiolabeled Peptide from Formic-Acid-Digested ³H-2-(m-azidobenzoyl)taxol-Labeled β Tubulin

Following formic acid digestion of ³H-3'-(*p*-azidobenzamido)taxol-labeled β tubulin, the cysteine residues are modified by either pyridylethylation[19] or carboxymethylation.[17] After reduction/alkylation, the sample is centrifuged through a Centricon-10 (Amicon, Danvers, MA) microconcentrator and the filtrate is purified by reversed-phase high-performance liquid chromatography (RP-HPLC) on an Aquapore RP-300 (Applied Biosystem, San Jose, CA) (2.1 × 220 mm) C_8 column using an HP1090 liquid chromatograph. The peptides are eluted with a linear gradient (1%/min) of H_2O/ 0.1% trifluoroacetic acid and acetonitrile/0.1% trifluoroacetic acid at a flow rate of 200 μl/min. The fraction containing the major UV absorbing (214-nm) material, which is also the major peak of radioactivity, is collected and concentrated in a Speed-Vac. The sample is ionized by electrospray on a PE-Sciex API-III (Ontario, Canada) mass analyzer and the monoisotopic mass of the sample measured. The measured mass of the sample is obtained from the different charge states. Peptide sequencing was performed on an Applied Biosystems 477A sequencer.

CNBr and Tryptic Digestion of ³H-2-(m-azidobenzoyl)taxol-Labeled β Tubulin

Photolabeled carboxymethylated β tubulin (25–30 nmol) is dissolved in 400 μl of 70% formic acid containing CNBr (Sigma, St. Louis, MO) (20 mg) and incubated at 37°. After 48 hr, the formic acid is evaporated in a Speed-Vac and the sample is washed twice with H_2O and dried. The material is analyzed by SDS–PAGE (15%) and fluorography. For mapping studies, CNBr-digested β tubulin is resuspended in 200 μl of 6 *M* guanidine-HCl

---

[19] A. S. Mak and B. L. Jones, *Anal. Biochem.* **84,** 432 (1978).

containing 2 mM dithiothreitol (DTT), and chromatographed on a pre-packed Hitrap desalting column (Pharmacia, Piscataway, NJ) in the presence of 6 M guanidine-HCl. The major radioactive fractions (0.5 ml) are pooled and subjected to RP-HPLC on an HP1090 liquid chromatograph using an Aquapore RP-300 C-8 column (2.1 × 220 mm). The peptides are eluted using a linear acetonitrile/0.1% trifluoroacetic acid gradient (0–55% over 55 min). The flow rate is 200 μl/min and 1-min fractions were collected. The recovery of the applied radioactivity is approximately 55%. The radioactive fractions isolated from C$_8$ reversed-phase columns are dried, dissolved in 50 mM NH$_4$HCO$_3$, 1 mM CaCl$_2$ and 1 M urea, and treated with trypsin for 48 hr at 25°. The total reaction volume is 400 μl and trypsin is added to the sample in three separate additions (at 0, 16, and 25 hr) to a final ratio of 1 : 40 (w/w). Then 6 M guanidine-HCl containing 2 mM DTT is added to the tryptic digest prior to C$_8$ reverse-phase chromatography. Tryptic peptides are eluted from the column with a linear acetonitrile/0.1% trifluoroacetic acid gradient (20–40% over 40 min). Individual peaks are collected and the radioactivity determined. The radioactive peaks are rechromatographed using an extended acetonitrile gradient (20–30% over 40 min). The purified peptide fractions are sequenced on an Applied Biosystems 477A sequencer.

Photolabeling of Microtubules with ^{3}H-3′-(p-azidobenzamido)-taxol and ^{3}H-2-(m-azidobenzoyl)Taxol

Both ^{3}H-3′-(p-azidobenzamido)taxol[20,21] and ^{3}H-2-(m-azidobenzoyl) taxol,[22] like Taxol, have the ability to (1) enhance *in vitro* assembly of tubulin into cold and Ca^{2+} stable microtubules in the absence of GTP and (2) form stable microtubule bundles in cells (data not shown). Irradiation at 254 nm of microtubules formed in the presence of either Taxol analog specifically photolabels β tubulin (Fig. 2). Under the conditions employed, no labeling of α tubulin was observed. The efficiency of photoincorporation of ^{3}H-3′-(p-azidobenzamido)taxol into β tubulin ranged from 2 to 6% after 30 min of irradiation.[20,21] Under the same conditions the extent of photoincorporation of ^{3}H-2-(m-azidobenzoyl)taxol into β tubulin varied from 10 to 15%.[22] Increasing the time of irradiation did not improve the efficiency

[20] S. Rao, N. E. Krauss, J. M. Heerding, C. S. Swindell, I. Ringel, G. A. Orr, and S. B. Horwitz, *J. Biol. Chem.* **269,** 3132 (1994).

[21] S. B. Horwitz, S. Rao, N. E. Krauss, J. M. Heerding, C. S. Swindell, I. Ringel, and G. A. Orr, in "Taxane Anticancer Agents: Basic Science and Current Status" (G. I. George, T. T. Chen, I. Ojima, and D. M. Vyas, eds.), ACS Symp. Series Vol 583, Ch. 11, pp. 154–161, Maple Press, York, Pennsylvania, 1995.

[22] S. Rao, G. A. Orr, A. G. Chaudhary, D. G. I. Kingston, and S. B. Horwitz, *J. Biol. Chem.* **270,** 20235 (1995).

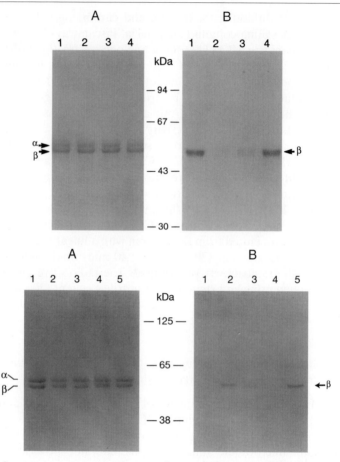

FIG. 2. ^{3}H-3′-(p-azidobenzamido)taxol and ^{3}H-2-(m-azidobenzoyl)taxol specifically photo-incorporate into β tubulin. Upper panel, ^{3}H-3′-(p-azidobenzamido)taxol; lower panel, ^{3}H-2-(m-azidobenzoyl)taxol. In each panel, (A) SDS–PAGE followed by Coomassie staining. (B) Fluorography of part (A). No additions (lane 1), addition of 50 μM Taxol (lane 2), 50 μM unlabeled analog (lane 3) or 50 μM baccatin III (lane 4). [From S. Rao, N. E. Krauss, J. M. Heerding, C. S. Swindell, I. Ringel, G. A. Orr, and S. B. Horwitz, *J. Biol. Chem.* **269,** 3132 (1994).]

of photolabeling with either analog (data not shown). The addition of a 50-fold molar excess of Taxol or the appropriate unlabeled analog to the incubation mixture prior to photolabeling inhibited photoincorporation into β tubulin (Fig. 2). In the case of ^{3}H-2-(m-azidobenzoyl)taxol, the data indicated that Taxol was not as good a competitor of photoincorporation as the unlabeled photoaffinity analog. This observation correlates with the fact that 2-(m-azidobenzoyl)taxol is more potent than Taxol in enhancing

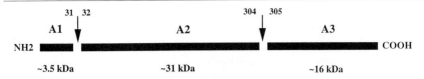

Fig. 3. Schematic representation of formic acid cleavage of β tubulin. [From S. Rao, G. A. Orr, A. G. Chaudhary, D. G. I. Kingston, and S. B. Horwitz, *J. Biol. Chem.* **270**, 20235 (1995).]

tubulin polymerization.[23,24] Due to the extreme hydrophobicity of Taxol, it is not possible to use more than a 50-fold molar excess of Taxol in this experiment. A tenfold excess of baccatin III, a compound that does not have the same biological activities as Taxol,[25] did not influence the photoincorporation of either ^{3}H-3'-(p-azidobenzamido)taxol or ^{3}H-2-(m-azidobenzoyl)taxol into β tubulin. These competition studies suggest that the photoaffinity analogs and Taxol are binding at the same or overlapping sites on the microtubule. Although our studies indicate that both Taxol-derived photoaffinity probes label β tubulin exclusively, other studies have reported some incorporation into α tubulin.[26–28]

Identification of the Site of Photoincorporation of ^{3}H-3'-(p-azidobenzamido)taxol

To locate the site(s) of photoincorporation for ^{3}H-3'-(p-azidobenzamido)taxol in β tubulin, we made use of the fact that formic acid is known to cleave preferentially Asp-Pro bonds.[29] Because β tubulin contains two such linkages at positions 31–32 and 304–305,[30,31] complete formic acid digestion of β tubulin will result in three distinct peptide fragments consisting of amino acids 1–31 (A_1, $M_r = \sim 3{,}500$), 32–304 (A_2, $M_r = \sim 31{,}000$) and 305–445 (A_3, $M_r = \sim 16{,}000$) (see Fig. 3).[20,21] SDS–PAGE/Coomassie

[23] A. G. Chaudhary, M. M. Charpure, J. M. Rimoldi, M. D. Chordia, A. A. L. Gunatilaka, and D. G. I. Kingston, *J. Am. Chem. Soc.* **116**, 4097 (1994).

[24] S. Grover, J. M. Rimoldi, A. A. Molinero, A. G. Chaudhary, D. G. I. Kingston, and E. Hamel, *Biochemistry* **34**, 3927 (1995).

[25] J. Parness, D. G. I. Kingston, R. G. Powell, C. Harracksingh, and S. B. Horwitz, *Biochem. Biophys. Res. Commun.* **105**, 1082 (1982).

[26] D. Dasgupta, H. Park, G. C. D. Harriman, G. I. Georg, and R. H. Himes, *J. Med. Chem.* **37**, 2976 (1994).

[27] C. Combeau, A. Commerçon, C. Mioskowski, B. Rousseau, F. Aubert, and M. Goeldner, *Biochemistry* **33**, 6676 (1994).

[28] C. Loeb, C. Combeau, L. Ehret-Sabatier, A. Breton-Gilet, D. Faucher, B. Rousseau, A. Commerçon, and M. Goeldner, *Biochemistry* **36**, 3820 (1997).

[29] P. Sonderegger, R. Jaussi, H. Gehring, K. Brunschweiler, and P. Christen, *Anal. Biochem.* **122**, 298 (1982).

[30] J. L. Hall, L. Dudley, P. R. Dobner, S. A. Lewis, and N. J. Cowan, *Mol. Cell Biol.* **3**, 854 (1983).

[31] M. G.-S. Lee, C. Loomis, and N. J. Cowan, *Nucleic Acids Res.* **12**, 5823 (1984).

staining of photolabeled β tubulin digested with formic acid for various times up to 120 hr showed the expected partially and totally cleaved fragments with the exception of the N-terminal fragment A_1, which is poorly stained. Fluorography of the same gel demonstrated that radioactivity was associated primarily with fragments A_1 and $A_1 + A_2$. As the digestion time is increased, the amount of radioactivity associated with $A_1 + A_2$ decreases and a concomitant increase occurs in label associated with A_1.[20,21]

The labeled β tubulin and its formic acid digestion products were analyzed further by HPEC. The radiolabel in undigested β tubulin eluted as a single peak (Fig. 4A) as did the A_1 fragment after a 120-hr formic acid digestion (Fig. 4B). The recovery of radiolabel from the HPEC analysis was ~50% of the total radioactivity applied to the gel. The A_1 fragment consisted of 76% of the recovered radioactivity. The remaining radioactivity was not eluted. To confirm the identity of the N-terminal domain, the A_1 fragment from a 48-hr formic acid digest was sequenced and the following partial sequence was identified: Met, Arg, Glu, Ile, Val, His, Ile, Gln, Ala, Gly, Gln, X, Gly, Asn, Gln, Ile, Gly, Ala, Lys, Phe, (X = not identified). This sequence was identical to the N-terminal sequence of human β_2 tubulin and is highly conserved in essentially all (except yeast) of the β tubulins that have been sequenced.[32] Because it was not possible to sequence as far as Asp-31 and the putative A_1 fragment migrated abnormally on HPEC, it was considered necessary to obtain independent proof that formic acid cleavage was occurring at Asp-31–Pro-32. The mass of the N-terminal peptide (A_1) was determined by electrospray mass spectrometry after modification of Cys-12 by either pyridylethylation or carboxymethylation. The mass of the peptide corresponded very closely with the theoretical values.[20,21] This observation confirms that one of the formic acid cleavages takes place specifically at Asp-31–Pro-32 in β tubulin and defines residues 1–31 of β tubulin as the site of photoincorporation of the 3'-p-azidobenzamido group.

Identification of the Site of Photoincorporation of ^{3}H-2-(M-azidobenzoyl)taxol

To locate the photolabeled domain within β tubulin, ^{3}H-2-(m-azidobenzoyl)taxol photolabeled β tubulin was electroeluted from gels, reduced, carboxymethylated, and subjected to either formic acid or CNBr cleavage. The peptides resulting from formic acid digestion of ^{3}H-2-(m-azidobenzoyl)taxol photolabeled β tubulin were separated by SDS–PAGE. The fluorograph of the gel demonstrated that the radiolabel was associated with

[32] D. W. Cleveland and K. F. Sullivan, *Ann. Rev. Biochem.* **54**, 331 (1985).

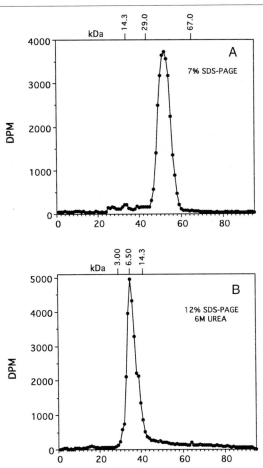

FIG. 4. High-performance electrophoresis chromatographic analysis of ^{3}H-3'-(p-azido-benzamido)taxol-labeled β tubulin and its formic acid digestion products. Intact β-tubulin was resolved on a 7% polyacrylamide gel (panel A) and digested β tubulin on a 12% polyacrylamide gel containing 6 M urea (panel B). In each case, 75,000 dpm in a total volume of 25 μl was loaded. Electrophoresis was conducted either at 0.3 mA for 30 min followed by 1.5 mA for 5 hr (panel A), or at 0.4 mA for 30 min followed by 1.3 mA for 5 hr (panel B). Fractions were counted by liquid scintillation spectrometry. Insulin (3 kDa, A/B chains), aprotinin (6.5 kDa), lysozyme (14.3 kDa), carbonic anhydrase (29 kDa), and bovine serum albumin (67 kDa) were used as markers. [From S. Rao, N. E. Krauss, J. M. Heerding, C. S. Swindell, I. Ringel, G. A. Orr, and S. B. Horwitz, *J. Biol. Chem.* **296**, 3132 (1994).]

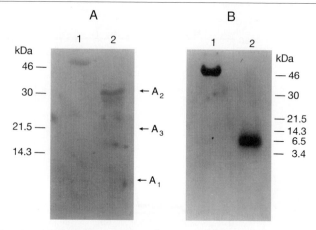

FIG. 5. Formic acid and CNBr cleavage of ^{3}H-2-(m-azidobenzoyl) taxol-labeled β tubulin. The photolabeled β tubulin, after reductive alkylation, was treated with either formic acid (panel A) or CNBr (panel B) as described in the text. Digests were subjected to SDS–PAGE analysis and fluorography. Lanes 1, undigested ^{3}H-2-(m-azidobenzoyl)taxol-labeled β tubulin; lanes 2, ^{3}H-2-(m-azidobenzoyl)taxol-labeled β tubulin digested with formic acid for 72 hr (panel A) or with CNBr for 48 hr (panel B). [From S. Rao, G. A. Orr, A. G. Chaudhary, D. G. I. Kingston, and S. B. Horwitz, *J. Biol. Chem.* **270**, 20235 (1995).]

the A$_2$ fragment (Fig. 5A). A$_1$ and A$_3$ fragments were not labeled under our experimental conditions. The CNBr digestion of photolabeled β tubulin produced a single radiolabeled peptide of approximately 6.5 kDa (Fig. 5B).

To define more precisely the site of photoincorporation, ^{3}H-2-(m-azido-benzoyl)taxol photolabeled β tubulin was isolated, digested with CNBr, and fractionated on a Hightrap desalting column to remove the lower molecular weight CNBr fragments. The major radioactive peak, which eluted at the void volume of the column, was subjected to RP-HPLC on a C$_8$ column. A major peak of radioactivity was eluted between 54 and 60 min (data not shown). The peak was pooled, dried, resuspended in NH$_4$HCO$_3$ buffer, and subjected to trypsin digestion. The resulting tryptic peptides were separated by RP-HPLC. The majority of radioactivity was associated with two closely eluting peaks identified as A and B in Fig. 6. Each peak was rechromatographed on a C$_8$ reversed-phase column, with an extended acetonitrile gradient (Figs. 7A and 7B). The resulting peptides were subjected to N-terminal amino acid sequencing. The following se-quence was identified: L-T/A-T-P-T-Y-G-D-L-N-H-L-V-S-A. This se-quence is identical to the amino acid residues 217–231 of vertebrate β tubulin.[33] In addition to this major peptide, the peak seen in Fig. 7A also

[33] K. F. Sullivan and D. W. Cleveland, *Proc. Natl. Acad. Sci. U.S.A.* **83**, 4327 (1986).

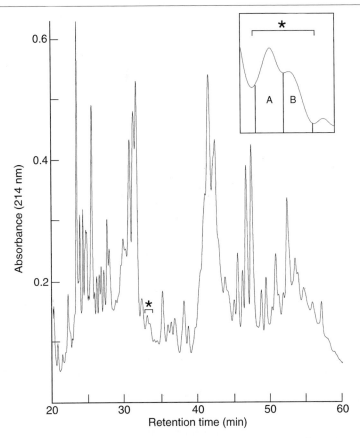

FIG. 6. Reverse-phase HPLC purification of tryptic peptides. The major ^{3}H-2-(m-azidoben-zoyl)taxol-photolabeled peptides obtained after CNBr cleavage and RP-HPLC were pooled, digested with trypsin and analyzed by RP-HPLC. The peptides were resolved with a linear gradient of acetonitrile/0.1% trifluoroacetic acid (20–40% over 40 min). Peaks were collected by hand and an aliquot of each was assayed for radioactivity. The inset is an enlargement of the area that contained the majority of the radioactivity and is identified with an asterisk (*). [From S. Rao, G. A. Orr, A. G. Chaudhary, D. G. I. Kingston, and S. B. Horwitz, *J. Biol. Chem.* **270**, 20235 (1995).]

contained a minor sequence corresponding to amino acid residues 337–350. This contaminating peptide is located within the A$_3$ fragment of formic-acid-digested β tubulin that was never labeled under our experimental conditions (see Fig. 5A). The initial yields of the Edman degradation reaction for each radiolabeled peptide sequenced were in close agreement to the concentrations determined from the specific activity of the incorporated Taxol analog. Therefore residues 217–231 of β tubulin are the site(s) of photoincorporation of the 2-m-azidobenzoyl group. The loss of tritium

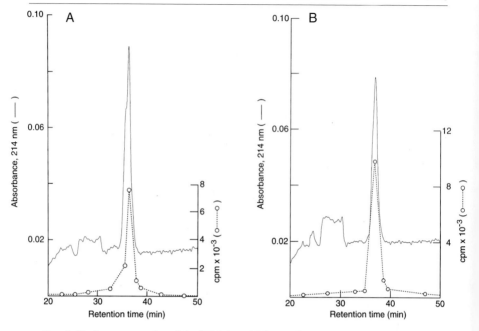

FIG. 7. Rechromatography of the ^{3}H-2-(m-azidobenzoyl)taxol-photolabeled tryptic peptides. After fractionation (Fig. 6) of the major radioactive tryptic peptides (A and B), they were repurified by subjecting them individually to RP-HPLC on a C_8 column and eluting with a linear gradient of acetonitrile/0.1% trifluoroacetic acid (20–30% over 40 min). (solid line) Absorbance at 214 nm; (dashed line) Radioactivity profile.

during the sequencing procedure did not allow us to identify the modified residue(s). This difficulty has been reported in other photoaffinity studies.[34]

Conclusions

The two domains of β tubulin so far identified as forming molecular contacts with Taxol consist of amino acids 1–31[20,21] and 217–231.[22] These domains interact with the 3′-benzamido and the 2-benzoyl groups of the Taxol analogs, respectively. If the Taxol binding site is located within a single α/β heterodimer, these domains, although far apart in the primary sequence, must be located close enough in the microtubule to interact with the bound Taxol. It is interesting to note that in the absence of GTP, a bifunctional thiol specific reagent, N,N'-ethylenebis(iodoacetamide), which contains a 9-Å spacer arm, cross-links the cysteine at position 12 with one

---

[34] B. Jayaram and B. E. Haley, *J. Biol. Chem.* **269**, 3233 (1994).

of the cysteines in either position 201 or 211 in β tubulin.[35] Alternatively, the Taxol binding site could be formed within the contact regions between adjacent α/β heterodimers in the microtubule. It is known that the N-terminal domain of β tubulin and the C-terminal domain of α tubulin form the interdimer contacts along protofilaments.[36] The interaction of the 3'-benzamido group of Taxol with the N-terminal domain of β tubulin may result in a conformational change in the protein that could alter the stability of microtubules. The Cys-12 of β-tubulin has been implicated as an important component of the exchangeable GTP binding site.[37] The *in vitro* assembly of tubulin under normal conditions requires GTP. Taxol can substitute for GTP and assemble microtubules in the absence of this nucleotide. Our results with 3'-(p-azidobenzamido)taxol indicate that it binds in the same region of tubulin as GTP and again implicates the N-terminal domain of β tubulin as an important regulatory site for microtubule assembly. Although these data suggest that both Taxol and GTP are interacting with the same region of β tubulin, earlier experiments have indicated that Taxol does not prevent either GTP binding or hydrolysis.[6,38] Note that other domains that do not include Cys-12 have been implicated as direct photoaffinity binding sites for GTP, such as residues 63–77[39] and 155–174.[40]

The two identified Taxol contact domains on β tubulin are highly conserved during evolution,[33] highlighting the functional importance of these domains in the tubulin assembly–disassembly reactions. Both the N-terminal and central regions of β tubulin have also been implicated in binding other drugs, including colchicine, vinblastine, and benzimidazoles. The direct photoincorporation of ^{3}H-colchicine into α/β-tubulin heterodimers resulted in labeling of residues 1–46 and residues 214–241 of β tubulin.[41] Although colchicine and Taxol have, in some respects, opposite effects in the tubulin/microtubule system, both are antimitotic agents and at low concentrations both affect microtubule dynamics.[42,43] The binding sites for these two antimitotic agents are, however, present in different forms of tubulin. Colchicine interacts with the α/β-tubulin heterodimer and binding

[35] M. Little and R. F. Luduena, *Biochim. Biophys. Acta* **912**, 28 (1987).

[36] K. Kirchner and E.-M. Mandelkow, *EMBO J.* **4**, 2397 (1985).

[37] B. D. Shivanna, M. R. Mejillano, T. D. Williams, and R. H. Himes, *J. Biol. Chem.* **268**, 127 (1993).

[38] M.-F. Carlier and D. Pantaloni, *Biochemistry* **22**, 4814 (1983).

[39] K. Linse and E.-M. Mandelkow, *J. Biol. Chem.* **263**, 15205 (1988).

[40] J. Hesse, M. Thierauf, and H. Ponstingl, *J. Biol. Chem.* **262**, 15472 (1987).

[41] S. Uppuluri, L. Knipling, D. L. Sackett, and J. Wolff, *Proc. Natl. Acad. Sci. U.S.A.* **90**, 11598 (1993).

[42] L. Wilson and M. A. Jordan, *in* "Microtubules" (J. S. Hyams and C. W. Lloyd, eds.), pp. 59–83. John Wiley and Sons, New York, 1994.

[43] W. B. Derry, L. Wilson, and M. A. Jordan, *Biochemistry* **34**, 2203 (1995).

of the drug is not inhibited by Taxol.[44] Conversely, Taxol binding is exclusively to the microtubule and is unaffected by colchicine.[9,45] A fluorescent vinblastine analog photolabels residues 175–213,[46] whereas a photoreactive benzimidazole photolabels residues 63–103 of a nematode β tubulin.[47]

[44] L. Wilson and I. Meza, *J. Cell Biol.* **58,** 709 (1973).
[45] S. B. Horwitz, J. Parness, P. B. Schiff, and J. J. Manfredi, *Cold Spring Harbor Symp. Quant. Biol.* **46,** pp. 219 (1982).
[46] S. S. Rai and J. Wolf, *J. Biol. Chem.* **271,** 14707 (1996).
[47] B. Nare, G. Lubega, R. K. Prichard, and E. Georges, *J. Biol. Chem.* **271,** 8575 (1996).

[22] Use of Drugs to Study Role of Microtubule Assembly Dynamics in Living Cells

By MARY ANN JORDAN and LESLIE WILSON

Introduction

Microtubules are dynamic cytoskeletal components that function in the development and maintenance of cell shape and polarity, in cell division, and in cellular movement. Antimitotic drugs of three diverse classes, the vinca alkaloids, the colchicine-nocodazole-podophyllotoxin class, and the taxanes, have recently been found to suppress dynamics of microtubules resulting in inhibition of specific cellular processes. At very low concentrations that alter microtubule dynamics without altering the mass of microtubule polymer, the drugs block cell division at mitotic metaphase,[1–4] suppress chromosome oscillations and attachment of mitotic chromosomes to microtubules,[1–3,5,6] suppress neurite growth cone morphogenesis,[7] and inhibit fibroblast locomotion.[8] The use of these drugs as chemical probes to alter microtubule assembly dynamics sensitively in well-characterized molecular ways can help elucidate the role of dynamic microtubules in specifying or regulating cell function. These drugs have a range of effects

[1] M. A. Jordan, D. Thrower, and L. Wilson, *Cancer Res.* **51,** 2212 (1991).
[2] M. A. Jordan, D. Thrower, and L. Wilson, *J. Cell Sci.* **102,** 401 (1992).
[3] M. A. Jordan, R. J. Toso, D. Thrower, and L. Wilson, *Proc. Natl. Acad. Sci. U.S.A.* **90,** 9552 (1993).
[4] L. Wilson, R. J. Toso, and M. A. Jordan, *J. Cell. Pharmacol.* **1 (Suppl. 1),** 35 (1993).
[5] K. L. Wendell, L. Wilson, and M. A. Jordan, *J. Cell Sci.* **104,** 261 (1993).
[6] C. Rieder, A. Schultz, R. Cole, and G. Sluder, *J. Cell Biol.* **127,** 1301 (1994).
[7] E. Tanaka, T. Ho, and M. W. Kirschner, *J. Cell Biol.* **128,** 139 (1995).
[8] G. Liao, T. Nagasaki, and G. G. Gundersen, *J. Cell Sci.* **108,** 3473 (1995).

on microtubules *in vitro* and in cells. The effects depend on the particular drug, the drug concentration, and other variables including the rate of uptake into cells. Their rational use as powerful probes of the role of microtubule assembly dynamics in cells requires consideration of these parameters.

Section I contains a brief overview of microtubule assembly dynamics. To use the drugs appropriately and to interpret the results obtained, it is important to understand their mechanisms of action in some detail. Thus Section II describes the current state of knowledge regarding (1) the binding of five major microtubule-interactive drugs, vinblastine, colchicine, podophyllotoxin, nocodazole, and taxol, to tubulin and microtubules, (2) the effects of the drugs on the dynamics of tubulin addition and loss at opposite microtubule ends *in vitro* and in cells, (3) the levels and kinetics of cellular uptake and loss of drug, (4) the drug-concentration-dependence for effects on microtubules and other organelles in cells, and (5) the advantages and disadvantages for the use of a particular drug for the study of microtubule assembly dynamics. Sections III and IV contain methods for measuring drug uptake into cultured cells and microtubule polymer mass, respectively. Section V concerns some general considerations for the use of drugs to study microtubule assembly dynamics in cells.

I. Brief Overview of Microtubule Assembly Dynamics

Microtubules are polymers composed of the protein tubulin (MW 100,000), a heterodimer of two related polypeptides, α and β. Microtubules are intrinsically dynamic polymers, and their dynamic properties are critically involved in their cellular functions.[7–12] Microtubules are not simple equilibrium polymers. They polymerize and depolymerize by the reversible addition and loss of tubulin dimers at the ends of the microtubules. Guanosine triphosphate (GTP) bound to β tubulin is hydrolyzed after addition of tubulin dimers to the microtubule ends, which gives rise to two dynamic behaviors. One dynamic behavior, called *treadmilling* or *flux*, is the net addition of tubulin at one microtubule end and the net loss of tubulin at

[9] L. Wordeman and T. J. Mitchison, *in* "Microtubules" (J. S. Hyams and C. W. Lloyd, eds.), pp. 287–302. Wiley-Liss, New York, 1994.

[10] J. R. McIntosh, *in* "Microtubules" (J. S. Hyams and C. W. Lloyd, eds.), pp. 413–434. Wiley-Liss, New York, 1994.

[11] L. Wilson and M. A. Jordan, *in* "Microtubules" (J. Hyams and C. Lloyd, eds.), pp. 59–84. Wiley-Liss, New York, 1994.

[12] R. I. Dhamodharan, M. A. Jordan, D. Thrower, L. Wilson, and P. Wadsworth, *Mol. Biol. Cell,* **6,** 1215 (1995).

the opposite end.[13] The other dynamic behavior, called *dynamic instability,* is a stochastic switching between shortening and growing phases at both ends of individual microtubules.[14,15]

The two ends of individual microtubules differ both structurally and kinetically, and the rates and durations of growing and shortening at one end, called the *plus end,* are much more extensive than the rates and durations of growing and shortening at the opposite or *minus end.*[16,17] Several important parameters characterize the dynamic behavior of microtubules, including the rates of growing and shortening, the time that a microtubule spends growing, shortening, or in *pause* or *attenuation* (a phase of no detectable length change), the frequency of transition between the phases of growth, shortening, and attenuation, and the *dynamicity* (the overall rate of microscopically detectable tubulin exchange at the microtubule end). The transition from growth or attenuation to shortening is called *catastrophe* and the transition from shortening to growth or attenuation is called *rescue.* Typical values for these parameters for microtubules in interphase cells and for bovine brain microtubules polymerized to steady state *in vitro* are shown in Table I. Growing and shortening behavior may be due to a stochastic gain and loss of a stabilizing "cap" at both microtubule ends, which is thought to consist of a short region of GTP- or GDP-P_i-liganded tubulin at the ends of the microtubules.[18–23]

Both treadmilling (flux) and dynamic instability occur in living cells[24–28] and both behaviors appear to be fundamental to the ability of microtubules to function in microtubule-dependent organization and movement. With the recent development of sophisticated methods for detecting the dynamics of individual microtubules in cells, it has become clear that their dynamics

[13] R. L. Margolis and L. Wilson, *Cell* **13,** 1 (1978).

[14] T. J. Mitchison and M. Kirschner, *Nature* **312,** 237 (1984).

[15] T. J. Mitchison and M. Kirschner, *Nature* **312,** 232 (1984).

[16] T. Horio and H. Hotani, *Nature* **321,** 605 (1986).

[17] R. A. Walker, E. T. O'Brien, N. K. Pryer, M. F. Sobeiro, W. A. Voter, H. Erickson, and E. D. Salmon, *J. Cell Biol.* **107,** 1437 (1988).

[18] E. T. O'Brien, W. A. Voter, and H. P. Erickson, *Biochemistry* **26,** 4148 (1987).

[19] R. J. Stewart, K. W. Farrell, and L. Wilson, *Biochemistry* **29,** 6489 (1990).

[20] M.-F. Carlier, *Int. Rev. Cytol.* **115,** 139 (1989).

[21] R. Melki, M.-F. Carlier, and D. Pantaloni, *Biochemistry* **28,** 8921 (1990).

[22] M. F. Carlier and D. Pantaloni, *Biochemistry* **20,** 1918 (1981).

[23] E. T. O'Brien, W. A. Voter, and H. P. Erickson, *Biochemistry* **26,** 4148 (1987).

[24] V. I. Rodionov and G. G. Borisy, *Science* **275,** 215 (1997).

[25] L. Cassimeris, N. K. Pryer, and E. D. Salmon, *J. Cell Biol.* **107,** 2223 (1988).

[26] E. Schulze and M. Kirschner, *Nature* **334,** 356 (1988).

[27] T. J. Mitchison, *J. Cell Biol.* **109,** 637 (1989).

[28] Y. Hamaguchi, M. Toriyama, H. Sakai, and Y. Hiramoto, *Cell Struct. Funct.* **12,** 43 (1987).

TABLE I

EFFECTS OF VINBLASTINE ON THE PARAMETERS OF DYNAMIC INSTABILITY OF MICROTUBULES IN
BS-C-1 CELLS AND OF MICROTUBULES ASSEMBLED FROM PURIFIED BOVINE BRAIN TUBULIN *IN VITRO*

Parameter	Control microtubules in cells	Microtubules in cells in 32 nM vinblastine (% change)	Control microtubules *in vitro*	Microtubules *in vitro* in 0.2 μM vinblastine (% change)
Rate of growing	6.9 ± 3.9 μm/min	−20	1.0 ± 0.1 μm/min	−52
Rate of shortening	15.5 ± 10.6 μm/min	−67	14.4 ± 1.2 μm/min	−56
Growing length[a]	1.8 ± 1.6 μm	−50	2.3 ± 0.2 μm	−61
Shortening length[a]	2.4 ± 2.4 μm	−71	5.8 ± 0.7 μm	−75
Time in pause, %	36	+64	7.7	+40
Catastrophe frequency	0.03 ± 0.01/sec	−67	0.005 ± 0.001/sec	−45
Rescue frequency	0.07 ± 0.01/sec	No change	0.013 ± 0.005/sec	+107
Dynamicity[a]	7.2 ± 1.2 μm/min	−75	2.28 μm/min	−82

[a] *Growing length* is the mean microtubule length added during a growing event. *Shortening length* is the mean microtubule length lost during a shortening event. *Dynamicity* is the total microscopically measurable gain and loss of tubulin subunits at microtubule ends over time; it is a measure of overall dynamic instability.
Data in columns 2 and 3 are from R. Dhamodharan, M. A. Jordan, D. Thrower, L. Wilson, and P. Wadsworth, *Mol. Biol. Cell* **6,** 1215 (1995). Data from columns 4 and 5 are from D. Panda, M. A. Jordan, K. C. Chin, and L. Wilson, *J. Biol. Chem.* **271,** 29807 (1996). Values and estimates of error for parameters of dynamic instability in the presence of vinblastine are given in the original papers.

are finely regulated.[25,27–33] In interphase cells, microtubules turn over with half-times of ~3 min to several hours.[34–36] However, with the onset of mitosis, the interphase microtubule network disappears[37–39] and is replaced by a new population of extremely dynamic microtubules that form the mitotic spindle. Mitotic spindle microtubules are 10–100 times more dynamic than microtubules in interphase cells, and exchange their tubulin with tubulin in the soluble pool with half-times of approximately 15 sec.[34,40]

[29] T. J. Mitchison, *Annu. Rev. Cell Biol.* **4,** 527 (1988).
[30] G. J. Gorbsky, P. J. Sammak, and G. G. Borisy, *J. Cell Biol.* **106,** 1185 (1988).
[31] G. J. Gorbsky and G. G. Borisy, *J. Cell Biol.* **109,** 653 (1989).
[32] P. Wadsworth, E. Sheldon, G. Rupp, and C. L. Rieder, *J. Cell Biol.* **109,** 2257 (1989).
[33] P. J. Sammak, G. J. Gorbsky, and G. G. Borisy, *J. Cell Biol.* **104,** 395 (1987).
[34] W. M. Saxton, D. L. Stemple, R. J. Leslie, E. D. Salmon, M. Zavortnik, and J. R. McIntosh, *J. Cell Biol.* **99,** 2175 (1984).
[35] E. Schulze and M. Kirschner, *J. Cell Biol.* **102,** 1020 (1986).
[36] R. Pepperkok, M. H. Bre, J. Davoust, and T. E. Kreis, *J. Cell Biol.* **111,** 3003 (1990).
[37] E. Karsenti and B. Maro, *Trends Biochem. Sci.* **11,** 460 (1986).
[38] J. R. McIntosh and M. P. Koonce, *Science* **246,** 622 (1989).
[39] Y. Zhai, P. J. Kronebusch, P. M. Simon, and G. G. Borisy, *J. Cell Biol.* **135,** 201 (1996).
[40] L. Belmont, A. A. Hyman, K. E. Sawin, and T. J. Mitchison, *Cell* **62,** 579 (1990).

It seems highly likely that the rapid dynamics of spindle microtubules is critical to spindle function.

II. Properties of the Major Microtubule-Targeted Drugs

A. Vinblastine

Vinblastine Binding to Tubulin and Microtubules. Vinblastine (MW 811) (Fig. 1), a widely used antimitotic antitumor drug obtained from the periwinkle or vinca plant *Catharanthus roseus,* produces various effects on tubulin and microtubules that are dependent on the drug concentration.

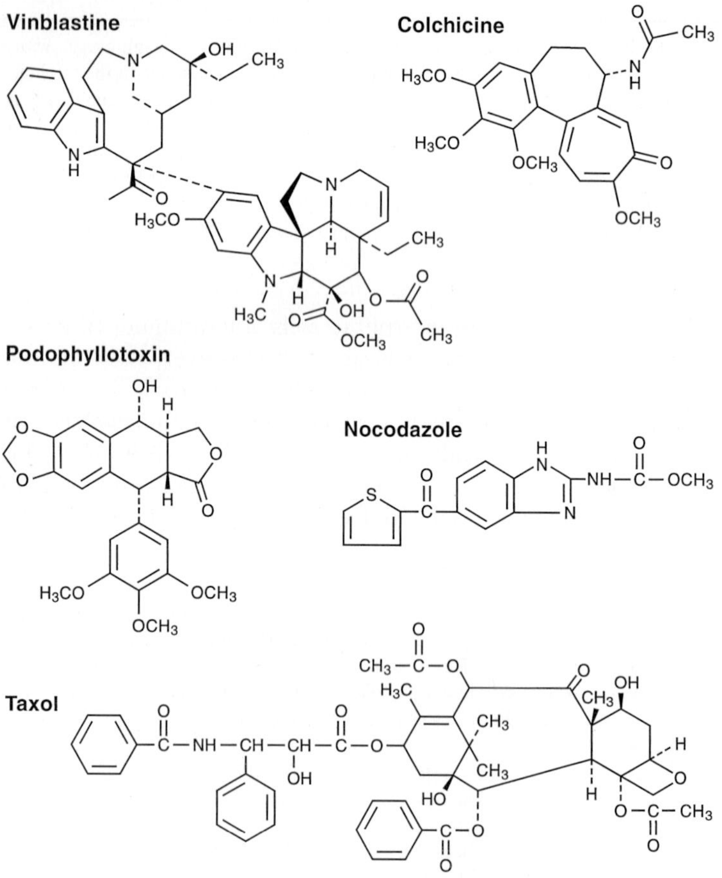

FIG. 1. Structures of antimitotic drugs.

Very high concentrations of vinblastine (≥ 10 μM) induce aggregation of tubulin into paracrystalline arrays, whereas moderately high concentrations (1–2 μM) depolymerize microtubules[41,42] and inhibit microtubule assembly.[43,44] In contrast to both of these effects, low concentrations of vinblastine (100–200 nM *in vitro,* 0.5–30 nM added to HeLa cells) suppress dynamic instability and treadmilling without changing the overall microtubule polymer mass[12,45,46] (discussed later). Vinblastine (vinblastine sulfate, MW 909, Sigma Chemical Co., St. Louis, MO; ICN Biomedicals, Irvine, CA; and other sources) is soluble in aqueous solution (1 mM) in the cold but may precipitate on warming solutions at high concentration to room temperature.

Vinblastine binds to tubulin rapidly, and the binding is reversible and independent of temperature between 0 and 37°. Solution variables including the buffer composition, magnesium concentration, and ionic strength affect the binding of vinblastine to tubulin (reviewed in Refs. 11 and 42). The binding of vinblastine to tubulin induces tubulin to self-associate in a process called *isodesmic indefinite self-association.*[47,48] The vinca alkaloid-induced self-association of tubulin appears to be responsible for the formation of vinblastine-tubulin paracrystals when cells are incubated with very high concentrations of vinblastine (≥ 10 μM)[49] and to play an important role in the stabilization of microtubule dynamics by the vinca alkaloids (discussed later).

Free vinblastine can bind directly to microtubules without first forming a complex with soluble tubulin.[41] Vinblastine does not copolymerize into the tubulin lattice of the microtubule, but binds only at the microtubule end or along the microtubule surface.[50] Vinblastine binds to tubulin in microtubules with different affinities depending on whether the tubulin is located at the ends or along the surface. Approximately 16–17 high-affinity vinblastine-binding sites per microtubule (K_d, 1–2 μM) are located at the

[41] M. A. Jordan, R. L. Margolis, R. H. Himes, and L. Wilson, *J. Mol. Biol.* **187,** 61 (1986).

[42] R. H. Himes, *Pharmacol. Ther.* **51,** 257 (1991).

[43] L. Wilson, K. Anderson, and D. Chin, *in* "Cold Spring Harbor Conferences on Cell Proliferation: Cell Motility" (R. Goldman, T. Pollard, and J. L. Rosenbaum, eds.), pp. 1051–1064. Cold Spring Harbor Laboratory Press, New York, 1976.

[44] R. J. Owellen, C. A. Hartke, R. M. Dickerson, and F. O. Hains, *Cancer Res.* **36,** 1499 (1976).

[45] R. J. Toso, M. A. Jordan, K. W. Farrell, B. Matsumoto, and L. Wilson, *Biochemistry* **32,** 1285 (1993).

[46] D. Panda, M. A. Jordan, K. Chin, and L. Wilson, *J. Biol. Chem.* **271,** 29807 (1996).

[47] G. C. Na and S. N. Timasheff, *Biochemistry* **19,** 1347 (1980).

[48] G. C. Na and S. N. Timasheff, *Biochem. Soc. Trans.* **8,** 1347 (1980).

[49] K. Fujiwara and L. G. Tilney, *Ann. N.Y. Acad. Sci.* **253,** 27 (1975).

[50] M. A. Jordan and L. Wilson, *Biochemistry* **29,** 2730 (1990).

ends of bovine brain microtubules.[51] Vinblastine also binds, with low affinity (K_d, 0.25–0.3 mM), to tubulin sites located along the microtubule surface.[41,52] The binding of the vinca alkaloids to the low-affinity sites along the microtubule surface is probably responsible for the ability of relatively high vinca alkaloid concentrations to depolymerize microtubules *in vitro* and in cells and may also play a role in tubulin paracrystal formation in cells. The potent kinetic suppression of tubulin exchange that occurs at low vinblastine concentrations (<1 μM; described later) is almost certainly due to vinblastine binding to the high-affinity binding sites at the microtubule ends.

Suppression of Treadmilling and Dynamic Instability by Low Concentrations of Vinblastine in the Absence of Significant Microtubule Depolymerization. Treadmilling studies with bovine brain microtubules reassembled to steady state *in vitro* first indicated that low concentrations of vinblastine stabilized microtubule dynamics without depolymerizing microtubules. It was found that when only 1.2 ± 0.3 molecules of vinblastine are bound per microtubule (at 0.14 μM vinblastine), treadmilling is inhibited by 50%.[51] Under the conditions of these experiments, the polymer mass was reduced by less than 5%.[50] Thus, low concentrations of vinblastine kinetically stabilize treadmilling dynamics without appreciably affecting the polymer mass.

In vitro video microscopic studies of individual bovine brain microtubules at polymer mass steady state have shown that low concentrations of vinblastine strongly inhibit dynamic instability at plus ends whereas they destabilize minus ends.[46] As shown in columns 4 and 5 of Table I, vinblastine stabilizes *plus* ends by suppressing the rate and extent of growth and shortening, decreasing the catastrophe frequency, increasing the rescue frequency, and suppressing dynamicity. It destabilizes *minus* ends by increasing the catastrophe frequency and decreasing the rescue frequency, strongly depolymerizing minus ends, without affecting the rate or extent of growth or shortening (data not shown). Thus low concentrations of vinblastine reverse the dynamics at opposite microtubule ends.[45,46] Both the kinetic *destabilization* of microtubules at minus ends and the *stabilization* at plus ends may contribute to the altered function of mitotic spindle microtubules of cells blocked in mitosis by low concentrations of vinblastine.[1,2,5]

Suppression of Microtubule Dynamics by Vinblastine in Living Cells. Vinblastine's effects on dynamics at microtubule plus ends in living cells have been studied by visualizing individual fluorescent microtubules following microinjection of rhodamine-tubulin into *Xenopus* embryonic fibroblast

[51] L. Wilson, M. A. Jordan, A. Morse, and R. L. Margolis, *J. Mol. Biol.* **159**, 129 (1982).
[52] W. D. Singer, M. A. Jordan, L. Wilson, and R. H. Himes, *Mol. Pharmacol.* **36**, 366 (1989).

cells and into African green monkey kidney cells (BS-C-1).[7,12] Both cell types are favorable for measurements of microtubule dynamics because microtubules are well separated (not bundled) and are easily distinguished from one another in the peripheral flat, thin lamella of the cells. In addition, microtubules do not undergo extensive translocations in the lamellae of these cells.

As shown in Table I, the suppressive effects of low concentrations of vinblastine on dynamics at microtubule plus ends in BS-C-1 cells are very similar to those observed *in vitro*. In both cell types (BS-C-1 kidney cells and *Xenopus* fibroblasts) it was found that vinblastine induced long periods of pause, reduced rates of growing and shortening, and reduced the overall dynamics of the plus ends of microtubules.[7,46] In BS-C-1 kidney cells, the transition frequencies were quantitated and it was found that the catastrophe frequency was reduced but the frequency of rescue was unchanged (Table I).[12]

Cellular Uptake and Loss of Vinblastine. The rate and absolute level of drug uptake into cells is an important consideration in using drugs to study microtubule assembly dynamics in cells. In uptake studies, vinblastine, when added to HeLa cells at concentrations of 2–100 nM, accumulated intracellularly many-fold, reaching concentrations of 300 nM to 3 μM.[1] The mechanism responsible for intracellular accumulation of vinblastine is not known but may involve binding to cellular tubulin; the maximum drug levels are approximately equal to the intracellular levels of tubulin (2 μM).[1] As shown in Fig. 2, the level of uptake was maximal at approximately 1–2 hr of incubation. The initial rate of uptake was not determined, but it is clear from Fig. 2 that the internal concentration must change rapidly during the initial minutes or hour of drug incubation. In HeLa cells, following the initial period of rapid uptake, the intracellular vinblastine concentration was relatively constant between 2 and 8 hr of incubation; from 8 to 48 hr the concentration diminished gradually to 25% of maximum. (The reasons for this decrease are not understood but may involve increased expression of a multidrug resistance membrane pump.) Washing of cells with vinblastine-free medium resulted in a rapid loss of drug; approximately 50% was lost during the first hour and the concentration was reduced by ≥90% 10–24 hr after washing.[53–55]

Thus, determinations of microtubule dynamics under conditions of constant vinblastine concentration in HeLa cells would require observations during the time period between 2 and 8 hr after addition of drug. The rate

[53] A. M. Lengsfeld, J. Dietrich, and B. Schultze-Maurer, *Cancer Res.* **42,** 3798 (1982).
[54] P. W. Gout, R. L. Noble, N. Bruchovsky, and C. T. Beer, *Br. J. Cancer* **34,** 245 (1984).
[55] M. A. Jordan, E. Tsuchiya, and L. Wilson (1996).

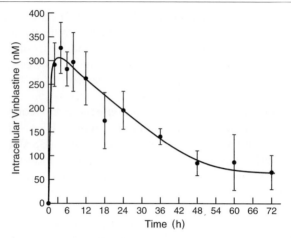

FIG. 2. The time course of uptake of 2 nM vinblastine into HeLa cells. [³H]vinblastine (2 nM, 11000 Ci/mol) was added to HeLa cells growing exponentially in suspension culture. At timed intervals from 2 to 72 hr, aliquots were sampled for determination of intracellular radioactivity as described in Section III. Data are means and standard errors of four to six individual measurements from three separate experiments.

and extent of drug uptake varies with the drug and the cell type, thus the time course of intracellular drug concentration must be determined (methods described in Section III) to ensure unchanging intracellular drug concentration.[1,54,56] For example, Gout *et al.*[54] found that vincristine, a close analog of vinblastine, was taken up slowly, and the maximal internal concentration was reached in Nb 2 node lymphoma cells only after 10 hr of incubation with vincristine. In contrast to vinblastine, vincristine was retained in these cells at high levels following washing.[54]

Examples of Studies Employing Vinblastine to Determine the Effects of Suppressing Microtubule Assembly Dynamics. In BS-C-1 African green monkey kidney cells,[12] we compared suppression of microtubule dynamics by vinblastine (3–64 nM) directly with mitotic accumulation, cell proliferation, and spindle organization. Intracellular vinblastine concentration was measured (see Section III) and uptake was found to have attained maximal and constant levels by 4 hr, thus microtubule dynamics were measured after 4 hr of vinblastine incubation. By determining cellular microtubule polymer mass (see Section IV),[57,58] we found that vinblastine (32 nM)

[56] W. D. Singer and R. H. Himes, *Biochem. Pharmacol.* **43,** 545 (1992).

[57] D. Thrower, M. A. Jordan, and L. Wilson, *J. Immunol. Methods* **136,** 45 (1991).

[58] D. Thrower, M. A. Jordan, and L. Wilson, *in* "Methods in Cell Biology" (D. J. Asai, ed.), pp. 129–145. Academic Press, San Diego, 1993.

significantly suppressed microtubule dynamics (Table I) in the absence of microtubule depolymerization.[12] The same concentration of vinblastine inhibited cell proliferation, slowed mitosis, and induced spindle abnormalities, thus strongly suggesting that these effects resulted from suppression of microtubule dynamics. As shown in Table I, 32 nM added vinblastine, yielding 9 μM intracellular concentration, induced a similar degree of suppression of dynamics as did 0.2 μM vinblastine *in vitro*. The differences in potency in cells and *in vitro* may result from isotypic or posttranslational differences in monkey kidney tubulin as compared with bovine brain tubulin, resulting in different interactions between tubulin and vinblastine. Alternatively, they may result from a higher number concentration of microtubules in cells as compared with *in vitro* or from intracellular sequestration of vinblastine in nonmicrotubule compartments, thus reducing the concentration available to bind to microtubules in cells.

Tanaka *et al.*[7] used vinblastine to test the role of microtubule dynamics in the motile behavior of neuronal growth cones of *Xenopus*. Direct measurements of the dynamics of neuronal microtubules were not possible, because microtubules undergo frequent translocations and bundling in neuronal cells. However, the authors quantitated the suppressive effects of low concentrations of vinblastine on dynamics of rhodamine-labeled microtubules in fibroblast cells of the same species and used these results to infer similar suppression of microtubule dynamics by vinblastine in neuronal growth cones. In the fibroblast cells, the effects of vinblastine were measured at a single drug concentration (10 nM) during the period of rapid uptake from the time of vinblastine addition to 8 min later. The microtubule growing rate was inhibited approximately 50% and the shortening rate by 70%. Absence of microtubule depolymerization by vinblastine was examined qualitatively by immunofluorescence microscopy. The authors concluded that dynamic microtubules were essential for the forward and persistent movement of the growth cone.

B. Colchicine

Colchicine Binding to Tubulin and Microtubules. Colchicine (Fig. 1) (MW 399, available from ICN Biomedicals, Sigma Chemical Co., and other suppliers), was isolated from the autumn crocus, *Colchicum autumnale,* and has played a fundamental role in elucidation of the properties and functions of microtubules since it was first found 30 years ago to bind to tubulin in cell extracts.[59–61] Tubulin was initially purified from brain tissue using bound

[59] G. G. Borisy and E. W. Taylor, *J. Cell Biol.* **34,** 525 (1967).
[60] G. G. Borisy and E. W. Taylor, *J. Cell Biol.* **34,** 535 (1967).
[61] L. Wilson and M. Friedkin, *Biochemistry* **6,** 3126 (1967).

[³H]colchicine as a biochemical marker.[62] Colchicine is readily soluble in aqueous solution (100 mM).

Substoichiometric Inhibition of Microtubule Polymerization by Colchicine: Copolymer Formation. Like vinblastine, colchicine inhibits microtubule polymerization *in vitro* at concentrations well below the concentration of tubulin free in solution,[63–65] indicating that colchicine inhibits microtubule polymerization substoichiometrically by interacting with the ends of the microtubules rather than by reducing the concentration of free soluble tubulin. However, in contrast with vinblastine, which binds directly to the ends of the microtubules, colchicine either cannot bind directly to microtubule ends or it does so with very low affinity. Instead, it first binds to soluble tubulin and forms a tubulin–colchicine (TC) complex, which then incorporates at the microtubule ends.[64,66] The binding reaction between colchicine and tubulin is slow, requiring a relatively long time (as much as 1 hr) to induce conformational changes in tubulin that result in the formation of the poorly reversible TC complex.[67–73] The colchicine-induced conformational changes in tubulin are likely to be responsible for the powerful effects of incorporated TC complex on tubulin exchange at microtubule ends.

When tubulin is polymerized into microtubules *in vitro* in the presence of TC complex, the TC complex becomes incorporated into the microtubules along with unliganded tubulin resulting in formation of TC–complex–tubulin copolymers.[74–76] Such data indicated that the binding of TC complex to microtubule ends does not completely prevent further tubulin addition to the ends, but rather, the ends remain competent to grow.[75] Thus, at low TC complex concentrations, incorporation of TC complexes at the microtubule ends must disrupt the tubulin lattice at or near the end in a way that impairs the efficiency of new tubulin addition but does not

[62] R. C. Weisenberg, G. G. Borisy, and W. E. Taylor, *Biochemistry* **7**, 4466 (1968).
[63] J. B. Olmsted and G. G. Borisy, *Biochemistry* **12**, 4282 (1973).
[64] R. L. Margolis and L. Wilson, *J. Biol. Chem.* **252**, 7006 (1977).
[65] R. L. Margolis, C. T. Rauch, and L. Wilson, *Biochemistry* **19**, 5550 (1980).
[66] D. Skoufias and L. Wilson, *Biochemistry* **31**, 738 (1992).
[67] D. L. Garland, *Biochemistry* **17**, 4266 (1978).
[68] A. Lambeir and Y. Engelborghs, *J. Biol. Chem.* **256**, 3279 (1981).
[69] J. M. Andreu and S. N. Timasheff, *Arch. Biochem. Biophys.* **211**, 151 (1981).
[70] J. Andreu and S. Timasheff, *Biochemistry* **21**, 6465 (1982a).
[71] J. M. Andreu and S. N. Timasheff, *J. Biol. Chem.* **258**, 1689 (1983).
[72] T. David-Pfeuty, C. Simon, and D. Pantaloni, *J. Biol. Chem.* **254**, 11696 (1979).
[73] H. W. Detrich, III, R. C. J. Williams, and L. Wilson, *Biochemistry* **21**, 2392 (1982).
[74] H. Sternlicht and I. Ringel, *J. Biol. Chem.* **254**, 10540 (1979).
[75] H. Sternlicht, I. Ringel, and J. Szasz, *Biophys. J.* **42**, 255 (1983).
[76] K. W. Farrell and L. Wilson, *Biochemistry* **23**, 3741 (1984).

completely destroy the ability of tubulin to be incorporated. In contrast, at high TC complex concentrations, tubulin addition at the microtubule ends is blocked completely.[77]

Low concentrations of colchicine or TC complex inhibit tubulin exchange at microtubule ends in the absence of appreciable loss of polymer. For example, addition of 0.2 μM colchicine or 0.1 μM TC complex to steady-state bovine brain microtubules (with associated proteins) inhibited tubulin exchange at plus ends by 50% without depolymerizing microtubules.[66,78]

Suppression of Dynamic Instability in Vitro by Low Concentrations of TC Complex. In vitro, the effects of colchicine on microtubule dynamics have been studied using preformed TC complex rather than free colchicine to eliminate the time required for formation of the active complex. TC complex kinetically stabilizes plus ends of microtubules.[79,80] TC complex (0.05–0.4 μM) strongly reduces both the growing and shortening rates to similar extents and increases the percentage of time the microtubules remain in an attenuated state, neither growing nor shortening detectably. TC complex decreases the catastrophe frequency and it increases the rescue frequency.[79] The effects of colchicine on dynamics at minus ends of individual microtubules have not been determined; however, studies of dilution-induced disassembly of microtubules in the presence of colchicine suggest that minus ends are less strongly stabilized than plus ends.[76]

Cellular Uptake and Loss of Colchicine. Like vinblastine, colchicine is accumulated intracellularly many-fold over the concentration added to the culture medium. Table II presents the uptake of radiolabeled colchicine by HeLa cells after incubation for 20 hr.

Colchicine is not Widely Used to Determine the Cellular Effects of Altering Microtubule Assembly Dynamics. Colchicine and Colcemid (demecolcine or *N*-deacetyl-*N*-methylcolchicine, an analog of colchicine often used in cellular studies[81–85]) have not been used extensively in studies of microtubule dynamics. As noted earlier, the binding of colchicine to tubulin is slow and the stoichiometry of copolymerization of colchicine with tubulin in

[77] K. W. Farrell and L. Wilson, *Biochemistry* **19,** 3048 (1980).
[78] L. Wilson and K. W. Farrell, *Ann. N.Y. Acad. Sci.* **466,** 690 (1986).
[79] D. Panda, J. E. Daijo, M. A. Jordan, and L. Wilson, *Biochemistry* **34,** 9921 (1995).
[80] A. Vandecandelaere, S. R. Martin, M. J. Schilstra, and P. M. Bayley, *Biochemistry* **33,** 2792 (1994).
[81] G. Sluder, *J. Cell Biol.* **80,** 674 (1979).
[82] J. R. Bamburg, D. Bray, and K. Chapman, *Nature* **321,** 788 (1986).
[83] W. Yu and P. W. Baas, *J. Neurosci.* **15,** 6827 (1995).
[84] C. L. Rieder and R. E. Palazzo, *J. Cell Sci.* **102,** 387 (1992).
[85] B. F. McEwen, J. T. Arena, J. Frank, and C. L. Rieder, *J. Cell Biol.* **120,** 301 (1993).

TABLE II
UPTAKE OF COLCHICINE INTO HeLa CELLS[a]

Extracellular concentration (nM)	Intracellular concentration (nM)	Increase (fold)
0.5	13.5 ± 1.1	27
7.0	178 ± 11	25
20	521 ± 50	26
100	2290 ± 157	23

[a] Intracellular concentration was measured after incubating exponentially growing HeLa cells in monolayer culture with [^{3}H]colchicine for 20 hr. Methods are described in Section III. Three independent assays were performed for each concentration. Values are means $\pm$ SE.

microtubules is complex, potentially leading to variable results on dynamics. Another disadvantage of colchicine and Colcemid is that their cellular effects are poorly reversible. Zieve et al.[86] found that after colchicine removal, HeLa cells did not reform functional spindles, and after Colcemid removal only a few cells reformed spindles.

C. Podophyllotoxin

Podophyllotoxin (Fig. 1) (MW 414, Sigma Chemical, ICN Biomedicals, and other suppliers), was isolated from the May apple *Podophyllum peltatum,* and has found clinical application in treating venereal warts. It shares structural homology with the trimethoxyphenyl ring of colchicine, binds to the "colchicine" site in tubulin, inhibits microtubule polymerization substoichiometrically, and inhibits mitosis.[43,87–89] It is readily soluble in water (0.28 mM).[90] Podophyllotoxin has often been used in cells to bring about the rapid depolymerization of cellular microtubules.[91,92]

Suppression of Microtubule Dynamics by Podophyllotoxin is Concomi-

---

[86] G. W. Zieve, D. Turnbull, J. M. Mullins, and J. R. McIntosh, *Exp. Cell Res.* **126,** 397 (1980).
[87] I. Cornman and M. E. Cornman, *Ann. N.Y. Acad. Sci.* **51,** 1443 (1951).
[88] L. Wilson, *Biochemistry* **9,** 4999 (1970).
[89] F. Cortese, B. Bhattachacharyya, and J. Wolff, *J. Biol. Chem.* **252,** 1134 (1977).
[90] S. Budaveri, ed., "The Merck Index," 12th Ed., Merck Research Laboratories, Whitehouse Station, New Jersey, 1996.
[91] A. Aszalos, G. C. Yang, and M. M. Gottesman, *J. Cell Biol.* **100,** 1357 (1985).
[92] D. K. Vaughan, S. K. Fisher, S. A. Bernstein, I. L. Hale, K. A. Linberg, and B. Matsumoto, *J. Cell Biol.* **109,** 3053 (1989).

tant with Microtubule Depolymerization, thus Limiting Use of the Drug for Studies of Alteration of Microtubule Assembly Dynamics. Podophyllotoxin binds to tubulin rapidly, reversibly, and in a relatively temperature-independent manner with a K_d of 0.28–0.83 μM.[93,94] Podophyllotoxin inhibits microtubule treadmilling and dynamic instability.[95–97] However, use of the drug to study inhibition of assembly dynamics in cells appears to be limited. In studies with HeLa cells, we found that mitotic block induced in HeLa cells by podophyllotoxin occurred in the same drug concentration range that induced significant microtubule depolymerization. For example, 75% of the cells were in metaphase at 30 nM podophyllotoxin, a drug concentration that depolymerized 75% of the mass of microtubules.[2] Thus, podophyllotoxin does not appear to be a useful drug for distinguishing the effects of stabilization or inhibition of microtubule dynamics from the effects of depolymerizing microtubules, because the two effects appear to occur concomitantly.

D. Nocodazole

Nocodazole Binds to Tubulin and, at High Concentrations, it Rapidly Depolymerizes Microtubules in Cells. Nocodazole (Fig. 1) [methyl(5,2-thienylcarboxyl-1*H*-benzimidazole-2-YL)-carbamate, MW 301, Janssen Pharmaceutical, Piscataway, NJ, also available from Sigma Chemical and ICN Biomedicals] was initially developed as a potential anticancer drug.[98–100] Although it never achieved therapeutic application, it is still widely used to study microtubule-dependent processes because of its ability to rapidly depolymerize microtubules in cells. Nocodazole is soluble in dimethyl sulfoxide at 15–30 mM, but precipitates in aqueous media at concentrations greater than 30 μM (information sheet from Janssen Pharmaceutical). Nocodazole is a benzimidazole derivative, with structural similarity to the A ring of colchicine, and it competitively inhibits the binding of colchicine to tubulin. It substoichiometrically inhibits the polymerization of tubulin into microtubules *in vitro* with a K_i of 1 μM.[98–100] Nocodazole shares with colchicine the property of increasing the GTPase activity of tubulin in the

[93] W. O. McClure and J. C. Paulson, *Mol. Pharmacol.* **13**, 560 (1977).

[94] J. K. Kelleher, *Mol. Pharmacol.* **13**, 232 (1977).

[95] M. Caplow and B. Zeeburg, *J. Biol. Chem.* **256**, 5608 (1981).

[96] M. A. Jordan and K. W. Farrell, *Anal. Biochem.* **130**, 41 (1983).

[97] M. J. Schilstra, S. R. Martin, and P. M. Bayley, *J. Biol. Chem.* **264**, 8827 (1989).

[98] J. Hoebeke, G. Van Nijen, and M. De Brabander, *Biochem. Biophys. Res. Comm.* **69**, 319 (1976).

[99] J. C. Lee, D. J. Field, and L. L. Y. Lee, *Biochemistry* **19**, 6209 (1980).

[100] M. J. DeBrabander, R. M. L. Van de Veire, F. E. M. Aerts, M. Borgers, and P. A. J. Janssen, *Cancer Res.* **36**, 905 (1976).

absence of polymerization.[101–106] This property is distinct from that of drugs that bind to the vinblastine domain and in general suppress GTPase activity in parallel with (and probably causally related to) inhibition of microtubule polymerization.[72]

The half-times for microtubule depolymerization with 20 μM nocodazole in cells range from 4 to 72 min, depending on cell type. Complete depolymerization occurs in 4 hr in MDCK cells with 4 μM nocodazole.[107] Although rates of uptake and release of nocodazole in cells have not been reported, there are indications that the rates are relatively rapid. Within 90 min of washing HeLa cells that were blocked in mitosis using 130 nM nocodazole, 80% of the cells had reformed spindles and completed mitosis.[86] Nocodazole (165–330 nM) appeared to inhibit microtubule dynamic instability in neuronal growth cones within 30 min after application, and the inhibition was reversed within 30 min of washout.[108] Nocodazole is frequently used to examine reformation of microtubules after drug removal; for example, 84% recovery of polymer was reported 2 hr following removal from MDCK cells.[107] In a study of the effects of nocodazole on cell migration (discussed further below), it was found that cell migration stopped 10 min after adding 300 nM nocodazole and resumed migration at the control rate following 10 min of washing the cells with nocodazole-free medium.[8] Thus nocodazole uptake and release is quite rapid, probably of the order of 10 min to 2 hr.

Low Concentrations of Nocodazole Inhibit Microtubule Dynamic Instability in Vitro and in Cells. Jordan *et al.*[2] found that 100 nM nocodazole induced maximal mitotic blockage of HeLa cells and abnormal spindles that retained most of their microtubules, as measured by quantitative enzyme-linked immunosorbent assay (ELISA) of stabilized cytoskeletal microtubules (see Section IV). The altered spindle organization resembled that induced by other drugs that suppress microtubule dynamics including vinblastine, taxol, colchicine, and podophyllotoxin, suggesting that the mechanism of mitotic block by nocodazole was suppression of spindle microtubule dynamics. Subsequently it was confirmed that nocodazole inhibits dynamic instability of microtubules *in vitro* (4 nM–12 μM) and in

[101] M.-F. Carlier and D. Pantaloni, *Biochemistry* **22,** 4814 (1983).
[102] C. M. Lin and E. Hamel, *J. Biol. Chem.* **256,** 9242 (1981).
[103] E. Hamel, H. H. Ho, G.-J. Kang, and C. M. Lin, *Biochem. Pharmacol.* **37,** 2445 (1988).
[104] C. M. Lin, H. H. Ho, G. R. Pettit, and E. Hamel, *Biochemistry* **28,** 6984 (1989).
[105] J. M. Andreu, M. J. Gorbunoff, F. J. Medrano, M. Rossi, and S. N. Timasheff, *Biochemistry* **30,** 3777 (1991).
[106] M. R. Mejillano, B. D. Shivanna, and R. H. Himes, *Arch. Biochem. Biophys.* **336,** 130 (1996).
[107] P. Wadsworth and M. McGrail, *J. Cell Sci.* **95,** 1990 (1990).
[108] M. W. Rochlin, K. M. Wickline, and P. C. Bridgman, *J. Neurosci.* **16,** 3236 (1996).

cells (4–400 nM), suppressing the rates of growing and shortening events and greatly increasing the percentage of time in pause.[4,109] Vasquez *et al.*[109] found that dynamics at both plus and minus ends are suppressed by nocodazole. The effects on frequencies of catastrophe and rescue appear to be complex and depend highly on the drug concentrations used.

Example: the Use of Nocodazole to Examine the Role of Microtubule Dynamics in Cell Migration. Liao *et al.*[8] used low concentrations of nocodazole to examine the role of microtubules in fibroblast locomotion. They found that 100 nM nocodazole decreased the rate of locomotion of fibroblasts by 60%. There was no change in microtubule distribution by immunofluorescence microscopy and no change in the mass of microtubules as determined by two independent methods of polymer mass quantitation (by quantitative immunofluorescence and by cell extraction/Western blotting; see Section IV). Higher concentrations of nocodazole ($\geq$300 nM) significantly depolymerized microtubules and completely stopped cell locomotion. Thus Liao *et al.* concluded that (1) microtubule dynamics are critical for the maximal speed of cell locomotion and (2) that a portion of the speed of cell locomotion is proportional to microtubule levels.

Importantly, Liao *et al.*[8] also tested the effects of low concentrations of vinblastine and of taxol on cell locomotion. These drugs similarly inhibited locomotion, thus confirming that the drug probes were likely perturbing cell function by disrupting microtubule dynamics rather than by interaction with some other cellular target. Employing additional drugs that bind to different sites on tubulin and microtubules is a valuable strategy in the use of drugs to probe cellular microtubule dynamics.

E. Taxol

Taxol (Fig. 1) (paclitaxel, MW 854, available from Calbiochem, Sigma Chemical, ICN Biomedicals, and other suppliers) is a complex diterpene isolated from the bark of the Western yew *Taxus brevifolia,* and is an effective new antitumor drug (for a review, see Ref. 110). It is a potent inhibitor of cell proliferation and appears to arrest cells in mitosis by a stabilizing action on microtubules.[3,111–113] In addition to stabilizing microtubules and greatly increasing the microtubule polymer mass in cells, Taxol

[109] R. J. Vasquez, B. Howell, A.-M. C. Yvon, P. Wadsworth, and L. Cassimeris, *Mol. Biol. Cell* **8**, 973 (1997).
[110] E. K. Rowinsky and R. C. Donehower, *N. Engl. J. Med.* **332**, 1004 (1995).
[111] D. A. Fuchs and R. K. Johnson, *Cancer Treat. Rep.* **62**, 1219 (1978).
[112] P. B. Schiff and S. B. Horwitz, *Proc. Natl. Acad. Sci. U.S.A.* **77**, 1561 (1980).
[113] W. Derry, B. L. Wilson, and M. A. Jordan, *Biochemistry* **34**, 2203 (1995).

also induces microtubule "bundling,"[112,114–116] a phenomenon that is not well understood. Taxol is readily soluble in dimethyl sulfoxide (DMSO) or methanol (>10 mM), but has limited solubility in water. Dilutions of Taxol into aqueous solutions from a stock in DMSO are initially clear at concentrations as high as 35 μM,[117] but on standing overnight the drug precipitates and the true solubility in water appears to be no higher than 0.77 μM.[118]

Taxol Binding to Microtubules. Taxol binds to microtubules in cells,[119] and it binds reversibly to microtubules reassembled *in vitro* with high affinity (K_d = 10 nM).[120] At high Taxol concentrations, 1 mol of Taxol binds per mole of tubulin in microtubules.[120–122] Taxol binds to soluble tubulin subunits with significantly reduced affinity.[121,123–125] Taxol does not inhibit the binding of colchicine, podophyllotoxin, or vinblastine to tubulin.[126,127] Taxotere (Rhone-Poulenc Rorer, Vitry-sur-Seine, France), a more water-soluble analog of Taxol, binds to the same site on microtubules as Taxol and is slightly more potent than Taxol in its effects on cells and on tubulin polymerization.[122,128]

In contrast to vinblastine, colchicine, podophyllotoxin, and nocodazole, Taxol enhances microtubule polymerization *in vitro,* promoting both the nucleation and elongation phases of the polymerization reaction, and it reduces the critical tubulin subunit concentration (i.e., soluble tubulin concentration at steady state).[126,127,129] Microtubules polymerized in the presence of taxol are extremely stable. They resist depolymerization by cold,

[114] P. F. Turner and R. L. Margolis, *J. Cell Biol.* **99,** 940 (1984).

[115] E. K. Rowinsky, R. C. Donehower, R. J. Jones, and R. W. Tucker, *Cancer Res.* **49,** 4093 (1988).

[116] J. R. Roberts, E. K. Rowinsky, R. C. Donehower, J. Robertson, and D. C. Allison, *J. Histochem. Cytochem.* **37,** 1659 (1989).

[117] C. S. Swindell, N. E. Krauss, S. B. Horwitz, and I. Ringel, *J. Med. Chem.* **34,** 1176 (1991).

[118] A. E. Mathew, M. R. Mejillano, J. P. Nath, R. H. Himes, and V. J. Stella, *J. Med. Chem.* **35,** 145 (1992).

[119] J. J. Manfredi, J. Parness, and S. B. Horwitz, *J. Cell Biol.* **94,** 688 (1982).

[120] M. Caplow, J. Shanks, and R. Ruhlen, *Biochemistry* **269,** 23399 (1994).

[121] J. Parness and S. B. Horwitz, *J. Cell Biol.* **91,** 479 (1981).

[122] J. F. Diaz and J. M. Andreu, *Biochemistry* **32,** 2747 (1993).

[123] J. F. Diaz, M. Menendez, and J. M. Andreu, *Biochemistry* **32,** 10067 (1993).

[124] M. Takoudju, M. Wright, J. Chenu, F. Gueritte-Voegelein, and D. Guenard, *FEBS Lett.* **227,** 96 (1988).

[125] S. Sengupta, T. C. Boge, G. I. Georg, and R. H. Himes, *Biochemistry* **34,** 11889 (1995).

[126] N. Kumar, *J. Biol. Chem.* **256,** 10435 (1981).

[127] P. B. Schiff and S. B. Horwitz, *Biochemistry* **20,** 3247 (1981).

[128] I. Ringel and S. B. Horwitz, *J. Natl. Cancer Inst.* **83,** 288 (1991).

[129] W. D. Howard and S. N. Timasheff, *J. Biol. Chem.* **263,** 1342 (1988).

calcium ions, dilution, and other antimitotic drugs.[126,129,130] Taxol induces the self-assembly of tubulin into microtubules at 0°,[131] in the absence of exogenous GTP,[127,132] without microtubule-associated proteins (MPA),[126] and at alkaline pH.[133]

Taxol can induce the formation of morphologically altered tubulin polymers, a phenomenon of as yet undetermined importance for cell function. For example, Taxol induces the formation of microtubules with predominantly 12 protofilaments rather than 13, both *in vitro* with pure tubulin[134] and *in vivo*.[135] Taxol also can induce the formation of hoops, ribbons, and other protofilamentous structures *in vitro*.[121,131,136]

Suppression of Treadmilling and Dynamic Instability by Low Concentrations of Taxol with no Increase in Microtubule Mass in Vitro and in Cells. The specific effects of Taxol on microtubule polymerization dynamics are complex and vary with the stoichiometry of Taxol binding to the microtubule. At low concentrations (10–50 nM), the binding of small numbers of Taxol molecules to microtubules reduces the rate and extent of shortening at microtubule plus ends.[113] For example, Derry *et al.*[113] found that at 25 nM Taxol, when only one Taxol molecule is bound for every 270 tubulin dimers in the microtubules, the rate and extent of shortening is reduced by 32%. Under these conditions, Taxol may bind at widely spaced sites along the microtubule surface, and the microtubules may shorten until a bound molecule of Taxol is reached. Then either the Taxol molecule or the tubulin complexed to Taxol may be required to dissociate before the microtubule shortens further, thus slowing the shortening rate. Low concentrations of Taxol (10–100 nM) preferentially suppress dynamics and induce a modest increase in microtubule length at microtubule plus ends, while having little or no effect on the dynamics and length change at microtubule minus ends.[137] Taxol has little effect on the rate of GTP hydrolysis.[101] Unlike vinblastine, colchicine, and nocodazole, Taxol at low concentrations does not affect the frequencies of catastrophe and rescue. Thus Taxol does not appear to directly affect gain and loss of the stabilizing GTP (or GDP-P$_i$) cap at microtubule ends.

[130] P. B. Schiff, J. Fant, and S. B. Horwitz, *Nature* **277**, 665 (1979).

[131] W. C. Thompson, L. Wilson, and D. L. Purich, *Cell Motil.* **1**, 445 (1981).

[132] E. Hamel, A. A. del Campos, M. C. Lowe, and C. M. Lin, *J. Biol. Chem.* **256**, 11887 (1981).

[133] I. Ringel and S. B. Horwitz, *J. Pharmacol. Exp. Ther.* **259**, 855 (1991a).

[134] J. M. Andreu, J. Bordas, J. F. Diaz, J. Garcia de Ancos, R. Gil, F. J. Medrano, E. Nogales, E. Pantos, and E. Towns-Andrews, *J. Mol. Biol.* **226**, 169 (1992).

[135] M. M. Mogenson and J. B. Tucker, *J. Cell Sci.* **97**, 101 (1990).

[136] C. L. Bokros, J. D. Hugdahl, V. R. Hanesworth, J. V. Murthy, and L. C. Morejohn, *Biochemistry* **32**, 3437 (1993).

[137] W. B. Derry, L. Wilson, and M. A. Jordan, *Cancer Res.* **58**, 1177 (1998).

At intermediate Taxol concentrations (100 nM–1 μM), the Taxol-binding stoichiometry to tubulin in microtubules increases, the growing rates are suppressed to the same degree as the shortening rates, and microtubules remain in a state of attenuation or pause most of the time. At very high Taxol concentrations (1–20 μM), Taxol binding to microtubules approaches saturation (1 mol Taxol/mol tubulin in microtubules), and the mass of microtubule polymer increases sharply as all of the tubulin is recruited into the microtubules. Tubulin dissociation is inhibited at both microtubule ends while the ends remain free for normal tubulin addition.[113,120,126,138,139]

Suppression of Dynamic Instability in Cells by Taxol. Jordan et al.[3] found that low concentrations of Taxol induced ~90% mitotic block at the metaphase/anaphase transition in HeLa cells. Spindle abnormalities and the absence of significant microtubule bundling or any increase in microtubule polymer mass as measured by quantitative ELISA of isolated cytoskeletons suggested strongly that the mitotic block resulted from suppression of microtubule dynamics by Taxol.

Recently, Yvon et al.[140] have found that in cells, Taxol suppresses microtubule dynamics in qualitatively the same way as it does with bovine brain microtubules *in vitro*. Human kidney carcinoma and ovarian carcinoma cells were microinjected with rhodamine-labeled tubulin; the labeled tubulin was allowed to incorporate into cellular microtubules for 90 min prior to addition of Taxol to the cells. Cells were incubated with Taxol for 4 hr prior to video microscopy to allow the intracellular Taxol concentration to reach equilibrium. Taxol accumulated in the cells 200- to 700-fold over the media concentration. Suppression of microtubule dynamics by Taxol resulted in abnormal spindle organization, some retraction of interphase microtubules from the periphery of the cell, and a slight degree of bundling in interphase.

In an examination of the role of tension on kinetochores in signaling anaphase onset, low concentrations of Taxol (0.3–3 nM) were used by Rieder et al.[6] to suppress spindle microtubule dynamics in PtK1 cells. By video microscopy of cells after addition of Taxol, they found that anaphase onset could be delayed in cells containing fully congressed chromosomes by suppressing microtubule dynamics.

Cellular Uptake and Loss of Taxol. Following a 20-hr incubation of HeLa cells with Taxol, we found that Taxol had accumulated in the cells to an extraordinary extent, 240-fold to 830-fold over the concentration

[138] M. Caplow and B. Zeeberg, *Eur. J. Biochem.* **127**, 319 (1982).

[139] L. Wilson, H. P. Miller, K. W. Farrell, K. B. Snyder, W. C. Thompson, and D. L. Purich, *Biochemistry* **24**, 5254 (1985).

[140] A.-M. Yvon, P. Wadsworth, and M. A. Jordan (1998).

TABLE III
INTRACELLULAR LEVELS OF TAXOL DURING TAXOL INCUBATION OF HeLa CELLS
AND FOLLOWING WASHING WITH TAXOL-FREE MEDIUM[a]

Taxol added to medium at time 0 (μM)	Taxol in cells at 20 hr (μM)	Uptake (fold)	Taxol in cells 24 hr after washing (μM)	Retention (%)
0.003	1.7 ± 0.2	570	0.7 ± 0.04	42
0.01	8.3 ± 2.1	830	4.0 ± 0.7	48
0.1	71.5 ± 6.2	720	27.0 ± 6.7	38
1.0	241.0 ± 27.0	240	38.6 ± 9.9	16

[a] Taxol concentration in HeLa cells after incubation for 20 hr in Taxol-containing medium and retention 24 hr after removal of Taxol from medium and washing of cells with three changes of Taxol-free medium (2 hr total washing). Data were obtained as described in Section III. Taxol (3 nM–1 μM, 33–11,000 Ci/mol) was a kind gift from the National Cancer Institute (Research Triangle Institute, NC). Values are mean ± SE from three experiments (two in monolayer, one in suspension), $n = 4$ or 6 total measurements. Data are from M. A. Jordan, K. L. Wendell, S. Gardiner, W. B. Derry, H. Copp, and L. Wilson, *Cancer Res.* **56,** 816 (1996).

added in the medium (Table III), significantly greater than the accumulation of colchicine or vinblastine in HeLa cells. A large percentage of Taxol was retained in the cells following washing, thus the uptake of Taxol is not easily reversible.[3,141]

The Intracellular Concentration of Taxol Changes Dramatically during the Initial Hours of Taxol Incubation. Taxol is taken up rapidly in cells, but attainment of equilibrium requires several hours. The time course of uptake of [^{3}H]Taxol into HeLa cells is shown in Fig. 3. Within the first 10 min after addition of 10 nM Taxol to the medium, the intracellular concentration rose to 1.4 ± 0.3 μM, a 140-fold increase, and it reached half-maximal concentration within 100 min. By 6 hr, the intracellular concentration (12.6 ± 0.1 μM) was approaching the maximum (14.6 ± 1.4 μM at 24 hr). After addition of 100 nM Taxol to the medium, the intracellular concentration rose to 8.4 ± 1.0 μM within the first 10 min, an 84-fold increase, and it reached half-maximal values by 25 min. By 2 hr, the intracellular concentration approached equilibrium at 41 μM. Thus, experiments performed during the first hours of Taxol incubation will reflect the effects of rapidly changing Taxol concentrations.

[141] M. A. Jordan, K. L. Wendell, S. Gardiner, W. B. Derry, H. Copp, and L. Wilson, *Cancer Res.* **56,** 816 (1996).

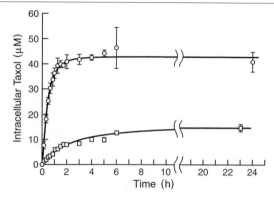

FIG. 3. The time course of uptake of 10 nM (squares) and 100 nM (circles) Taxol into HeLa cells. [³H]Taxol (350–3500 Ci/mol), a gift of the National Cancer Institute, NSC 125973) was added to HeLa cells growing exponentially in scintillation vials. At timed intervals from 10 min to 24 hr, samples were prepared for determination of intracellular radioactivity as described in Section III. Values are means and standard errors of four separate experiments.

III. Determination of Intracellular Drug Levels

We have used two methods employing radiolabeled drugs for measuring intracellular drug concentrations; they produce equivalent results. A third method using high-performance liquid chromatography (HPLC) to determine amounts of nonradiolabeled drug in cells is also described.

1. *Monolayer or suspension culture in flasks.*[1] Seed cells for suspension culture or into monolayer culture flasks (225 cm²) at a density such that they will not become superconfluent during the course of the experiment (for example, for a 20-hr drug incubation with HeLa cells, use 4×10^5 cells/ml). After 1–2 days, replace medium with medium containing ³H-labeled drug (final specific activity 30–11,000 Ci/mol). Radiolabeled vinblastine, colchicine, and podophyllotoxin can be obtained from one or more of the following: Amersham (Arlington Heights, IL), New England Nuclear (Wilmington, DE), and Moravek Biochemicals (Brea, CA); radiolabeled Taxol was obtained as a gift from the National Cancer Institute (Research Triangle Institute, NC) and is also available from Moravek Biochemicals; radiolabeled nocodazole is not currently available commercially. Reserve $2 \times$ 10-μl aliquots of the final dilution of drug in medium for determination of specific activity. After the desired duration of drug incubation (minutes to days), collect cells by centrifugation (after release from monolayer by scraping with a rubber policeman). Centrifuge in 15-ml tubes, ICN clinical centrifuge, 5 min, setting 2–3. Add [¹⁴C]hydroxymethylinulin (Amersham) (0.0156 Ci/ml final concentration) to suspensions immediately

prior to centrifugation for determination of extracellular drug trapped in the pellet and for determination of total cell volume in the pellets. Hydroxymethylinulin is not significantly internalized by living cells. Thus, total cell volume can be determined by subtracting extracellular volume (as indicated by trapped hydroxymethylinulin) from the total volume of the pellets of centrifuged cells.

Remove 100-μl aliquots from supernatant for determination of excluded drug and add to scintillant for counting. Aspirate remaining supernatant as completely as possible and discard. Remove 100-μl aliquots of the semifluid pellet (by pipetting) and add to scintillant (for example, Ready-Protein by Beckman, Fullerton, CA) for determination of intracellular drug concentration. An alternative method for measuring cell and pellet volumes (using radiolabeled H_2O and radiolabeled hydroxymethylinulin) has been described.[56]

2. *Monolayer culture directly in sterilized scintillation vials.* Seed cells directly in sterile (autoclaved) polylysine-coated (50 μg/ml, 2 hr, 37°, followed by a rinse with sterile water and a rinse with medium) glass scintillation vials at 3×10^5 cells/ml. Twenty-four hours later, aspirate medium and replace with 2.5 ml of medium containing [3H]drug at the desired concentration. Remove $2\times$ 10-μl aliquots of each suspension for determination of specific activity. Following incubation with [3H]drug, aspirate medium and wash the cells quickly three times with 2.5 ml buffer [0.1 M piperazine-N,N'-bis(2-ethanesulfonic) acid, 1 mM ethylene glycol-bis(β-aminoethyl ether)-N,N,N',N'-tetraacetic acid (EGTA), 1 mM MgSO$_4$, pH 6.9, 37°]. Lyse cells by adding 1 ml distilled water and add 10 ml scintillation fluid (Beckman Ready-Protein) for determination of radioactivity. Incubate duplicate vials under the same conditions using nonradiolabeled drug to determine the cell number at each time point and each drug concentration. Determine total cell volume by multiplying the number of cells by the volume of an average cell (calculated from measurements of cell radius by microscopy after trypsinizing the cells so they become spherical). For example, we found that the average volume of a HeLa cell was 2.4 pl. Similarly, using [3H]H$_2$O, Singer and Himes[56] found the water volume of a B16 melanoma cell to be 1.09 pl. Adherence of [3H]drug to scintillation vials should be determined; with Taxol and vinblastine, we found that it was <0.5% of total radioactivity.[141]

3. Singer and Himes[56] measured the uptake of nonradiolabeled vinca alkaloids into B16 melanoma cells by HPLC analysis after extracting the drugs from the cellular protein. Cells were grown in 100-mm culture dishes to a density of ~1×10^7 cells/dish. Medium was then replaced with 4 ml of fresh medium containing 0.1 or 1 μM drug. After the desired time of incubation (37°) the medium was removed and the plates were washed

by centrifugation, and 2.7 ml of supernatant was evaporated (60°) under a stream of N_2. The residue was dissolved in 50% methanol in 10 mM KH_2PO_4 (pH 4.5) and the amount of vinca alkaloid present was determined by HPLC.

IV. Determination of Cellular Microtubule Polymer Mass

To investigate whether an observed cellular drug effect is related to (1) microtubule depolymerization, (2) increase in microtubule polymer, or (3) alteration of microtubule dynamics, quantitative determination of the cellular microtubule polymer mass following drug incubation is critical. Several methods have been devised for this purpose. We have isolated microtubules from cultured mammalian cells and quantitated the tubulin content of both the microtubules and the soluble tubulin pool with a competitive ELISA. We detergent-lysed the cells into a microtubule stabilizing buffer containing glycerol, DMSO, protease inhibitors, and GTP; collected cytoskeletons by high-speed centrifugation; and then solubilized tubulin from the microtubule-containing pellets for ELISA by homogenization in depolymerization buffer and incubation for 1 hr on ice. Step-by-step details are described by Thrower et al.[57,58]

Liao et al.[8] developed a quantitative microscopic assay based on the fluorescence intensity of microtubules in migrating cells fixed and stained with antibodies to tubulin. Cells were incubated with drug for 1 hr, extracted for 1 min in a casein-containing buffer to remove soluble tubulin,[8] and fixed in −20° methanol prior to immunofluorescence staining. Images of stained cells obtained with a silicon intensified target (SIT) camera were digitized, and the microtubule level was quantified using the "area brightness measurement" function of the Image-1 software program. The brightness was assessed at a constant percentage (60%) of the distance from the nucleus to the leading edge of migrating cells. The authors found that at this distance from the nucleus, fluorescence intensity was linear with respect to microtubule density. Cellular microtubule levels were also measured independently, by sodium dodecyl sulfate–polyacrylamide gel electrophoresis (SDS–PAGE) and Western blot analysis of polymer fractions isolated from drug-treated cells. The density of the bands in immunoblots was standardized using a range of purified brain tubulin amounts loaded on the same blot. The microtubule levels determined by the two methods were equivalent. Note that the applicability of the brightness intensity method to nonmigrating cells of heterogeneous morphology has not been reported.

A third method, developed by Minotti et al.,[142] involves metabolically

---

[142] A. M. Minotti, S. B. Barlow, and F. Cabral, J. Biol. Chem. **266,** 3987 (1991).

labeling cells with radiolabeled amino acids followed by lysis in a microtubule-stabilizing buffer, centrifugation to separate soluble from polymerized tubulin, resolution of the proteins in each fraction by two-dimensional gel electrophoresis, and quantitation of the tubulin by liquid scintillation counting of spots excised from the gels.

Microtubule mass has been determined by electron microscopy in axons by counting the number of microtubules per cross-section of axon and multiplying by the average length of the axons.[108]

V. General Considerations

Cell-Type Specific Differences in the Effects of Drugs on Microtubule Assembly Dynamics

It is well known in cancer therapy that cells of different types, different tissues, and at different stages of the cell cycle vary widely in their responses to antimitotic drugs. Thus Taxol is effective primarily against ovarian and mammary carcinomas, whereas vincristine and vinblastine are effective primarily against leukemias and lymphomas. There are at least two sources of cell-type differences in drug response: (1) differential drug uptake and (2) differential expression of tubulin isotypes or microtubule-associated proteins; both of these result in altered drug effects on cellular microtubule dynamics. We have found, for example, that rates of microtubule growth and shortening are inhibited to a similar degree following incubation of human ovary cells with 30 nM Taxol and human kidney cells with 100 nM Taxol.[140] Similarly, kidney cell proliferation is less sensitive to a given concentration of Taxol than are the ovary cells. The difference can be partially explained by differences in Taxol uptake by kidney and ovarian cells; the kidney cells accumulate several-fold lower intracellular concentrations of Taxol.[140] The contribution of differences in tubulin isotype composition or microtubule-associated proteins has not yet been determined.

Tubulin isotypes show marked tissue-, cell-, and tumor-specific patterns of expression, and evidence is accumulating that the tubulin isotype composition of a cell plays an important role in determining the cellular effects of drugs that alter microtubule dynamics.[143–146] Tubulin isotypes have different affinities for binding to antimitotic drugs including colchicine and estramus-

[143] J.-P. Jaffrezou, C. Dumontet, W. B. Derry, G. Duran, E. Tsuchiya, L. Wilson, M. A. Jordon, and B. Sikic, *Oncology Res.* **7,** 517 (1995).
[144] M. Kavalleris, D. Y.-S. Kuo, C. A. Burkhardt, D. L. Regl, M. D. Norris, M. Haber, and S. B. Horwitz, *J. Clin. Invest.* **100,** 1 (1997).
[145] S. Ranganathan, K. Tew, and G. Hudes, *Cancer Res.* **56,** 2584 (1996).
[146] W.-P. Lee, *Arch. Biochem. Biophys.* **319,** 498 (1995).

tine,[147,148] and the dynamics of microtubules formed of purified $\alpha\beta$ tubulin isotypes are differentially suppressed by Taxol.[149]

Differences in Drug Response of Neighboring Cells in Culture

An important consideration in cellular studies of microtubule dynamics is that the degree of drug-induced morphologic alteration in microtubules (such as depolymerization induced by high concentrations of vinblastine or nocodazole) can vary considerably from cell to cell in a single population[7,12] perhaps as a result of variable drug uptake or variable drug susceptibility throughout the cell cycle. Thus, it is important to obtain measurements of assembly dynamics from a sufficient number of cells following drug incubation.

Use of Drug Combinations that Act Additively or Synergistically

Combinations of drugs that suppress microtubule dynamics appear to act synergistically to inhibit tumor cell proliferation and thus show significant clinical promise.[150–154] Similarly, combinations of drugs that suppress microtubule dynamics, such as vinblastine and Taxol together or nocodazole and Taxol together, may prove advantageous for cell biological studies to distinguish the effects of suppression of microtubule dynamics from the effects of changes in microtubule polymer mass.

Acknowledgment

We thank Ms. Kim Wendell, Ms. Etsuko Tsuchiya, and Dr. Douglas Thrower for excellent assistance in measuring drug uptake. Dr. Richard H. Himes, Dr. Vivian Ngan, Dr. Dulal Panda, and Mr. Keith DeLuca kindly read the manuscript and made valuable suggestions. The work was supported by grants to MAJ and LW from NIH, CA 57291 and NS13560.

[147] A. Banerjee and R. F. Luduena, *J. Biol. Chem.* **266,** 1689 (1991).

[148] N. Laing, B. Dahllof, B. Hartley-Asp, S. Ranganathan, and K. D. Tew, *Biochemistry* **36,** 871 (1997).

[149] W. B. Derry, L. Wilson, I. A. Khan, R. F. Luduena, and M. A. Jordan, *Biochemistry* **36,** 3554 (1997).

[150] G. R. Hudes, F. E. Nathan, C. Khater, R. Greenberg, L. Gomella, and C. Stern, *Sem. Oncol.* **22,** 41 (1995).

[151] G. R. Hudes, R. Greenberg, R. L. Krigel, S. Fox, R. Scher, S. Litwin, P. Watts, L. Speicher, K. Tew, and R. Comis, *J. Clin. Oncol.* **10,** 1754 (1992).

[152] V. C. Knick, D. Eberwein, and C. Miller, *J. Natl. Cancer Inst.* **87,** 1072 (1995).

[153] A. Photiou, P. Shah, L. Leong, J. Moss, and S. Retsas, *Eur. J. Cancer* **33,** 463 (1997).

[154] P. Garcia, D. Braguer, G. Carles, and C. Briand, *Anti-Cancer Drugs* **6,** 533 (1995).

Section III

Other Cytoskeletal Systems

[23] Purification and Assay of a Septin Complex from *Drosophila* Embryos

By KAREN OEGEMA, ARSHAD DESAI, MEI LIE WONG,
TIMOTHY J. MITCHISON, and CHRISTINE M. FIELD

Introduction

The septins are a family of homologous proteins first identified in *Saccharomyces cerevisiae* as the products of the CDC3, CDC10, CDC11 and CDC12 genes. All four gene products localize to the bud neck during cell division, and mutations in each of these genes cause defects in cytokinesis, in bud morphogenesis, and in the localization of chitin deposition. In *S. cerevisiae,* the presence of the septins correlates with the ability to form a series of 10-nm-wide electron dense striations spaced at 28-nm intervals (called the *neck filaments*) at the cytoplasmic face of the plasma membrane during cytokinesis (reviewed in Ref. 1).

Septins have subsequently been identified in other fungi, *Drosophila, Xenopus,* humans, and mice. A mutation in one of the *Drosophila* septins, *pnut,* also results in a cytokinesis defect.[2] This is especially intriguing given the apparently very different mechanisms used by yeast and metazoans to accomplish cytokinesis.[3] Septin family members display $\geqq 26\%$ identity over their length. The sequences of all known septins contain a P-loop and additional motifs that define the GTPase superfamily.[4] In addition, most septin polypeptides contain 36–90 amino acids of predicted coiled-coil at their COOH termini. Although there is a high level of conservation between individual septins from different metazoan species, there is no one-to-one correspondence between the septins found in metazoa and those found in yeast.[1,5] This situation is in contrast to that found for actin and tubulin and suggests that either the specific functions of the individual septins were not conserved during evolution, or that the expansion of the septin families occurred after the divergence of the major phylogenetic lines.[1]

---

[1] M. S. Longtine, D. J. DeMarini, M. L. Valencik, O. S. Al-Awar, H. Fares, C. De Virigilio, and J. R. Pringle, *Curr. Opin. Cell Biol.* **8,** 106 (1996).
[2] T. P. Neufeld and G. M. Rubin, *Cell* **77,** 371 (1994).
[3] S. L. Sanders and C. M. Field, *Curr. Biol.* **4,** 907 (1994).
[4] C. M. Field, O. S. Al-Awar, J. Rosenblatt, M. L. Wong, B. Alberts, and T. J. Mitchison, *J. Cell Biol.* **133,** 605 (1996).
[5] J. A. Cooper and D. P. Kiehart, *J. Cell Biol.* **134,** 1345 (1996).

We have purified a complex from *Drosophila* embryo extracts that contains three septins, Pnut, Sep1, and Sep2. Hydrodynamic and sequence data suggest that the complex has a molecular mass of 340 kDa and therefore may be composed of a heterotrimer of homodimers. The purified complex forms filaments and filament assemblies *in vitro,* copurifies with one molecule of bound guanine nucleotide per septin polypeptide, and can bind and hydrolyze exogenously added guanosine triphosphate (GTP).[4] Similar septin complexes have recently been purified in our laboratory from human cells and from the yeast *S. cerevisiae* (Sally Cudmore and Jennifer Frazier, unpublished observations). In this article, we describe two methods we have used to purify a septin complex from *Drosophila* embryo extracts, and the procedures we have used to assay the ability of the septin complex to form filaments *in vitro* and to bind and hydrolyze GTP.

Purification of Septin Complex

The two purification procedures we have used immunoisolation and conventional chromatography, are outlined in Fig. 1. The starting material for both isolation methods is a high-speed supernatant (HSS) made from homogenized 0–4 hr *Drosophila* embryos. For the immunoisolation procedure we have typically used about 40 ml of a 1:5 (prepared by adding 5

A. Immunoaffinity chromatography   **B. Conventional Purification**

1. Pre-bind antibody (raised against the C-terminal 14 amino acids of pnut) to protein A beads.

2. Incubate the beads with *Drosophila* extract (high speed supernatant) for 1 hour to bind septin complex.

3. Wash beads extensively to remove unbound extract proteins. Pour the beads into a small column

4. Elute septin complex by o/n incubation in buffer containing 200 μM C-terminal peptide.

Eluted septin complex

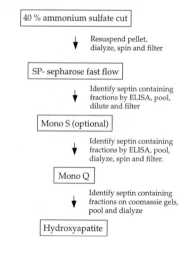

FIG. 1. Two purification strategies for the *Drosophila* septin complex: (A) immunoaffinity isolation and (B) conventional purification.

ml of extract buffer for every gram of embryos) HSS; the yield for this procedure is about 100 μg of septin complex. For the conventional purification, we usually begin with about 400 ml of a 1:3 HSS and obtain about 1 mg of purified septin complex.

Preparation of Drosophila High-Speed Supernatant

For historical reasons, the buffers we have used to make extracts for the immunoisolation and conventional purification are different. However, we find little effect of varying the salt concentration or including detergent on either the amount of septin complex extracted or its solubility. This could be due to the fact that the septin complex we are purifying is part of the maternal load and therefore might not be associated with the cortex. In other systems, extract conditions may need to be optimized. By dialyzing septin containing ammonium sulfate cuts into different buffers, we have found that the septins are more soluble in buffers between pH 7 and 9 than in low pH buffers (pH 5 or 6).

Solutions

Protease inhibitor stock: 1 mM benzamidine-HCl, 1 mg/ml each phenanthroline, aprotinin, leupeptin, and pepstatin A (This stock is used at dilutions between 1:100 and 1:500 as noted.)

Embryo wash: 0.04% Triton X-100, 0.04% NaCl

50% bleach

0.5 M Na$_3$EDTA stock: 0.5 M Ethylenediaminetetraacetic acid adjusted to pH 8.0 with sodium hydroxide

0.5 M K$_3$EGTA stock: 0.5 M [Ethylenebis(oxyethylenenitrilo)]tetraacetic acid adjusted to pH 8.0 with potassium hydroxide

0.5 M K-HEPES stock: N-2-Hydroxyethyl piperazine-N′-2-ethanesulfonic acid adjusted to pH 7.6 with potassium hydroxide

For immunoisolation procedure:

Extract buffer: 5 mM Tris-HCl, pH 7.9, 0.5 mM Na$_3$EDTA, 0.5 mM Na$_3$EGTA, 0.1% Nonidet P-40 (NP-40), 1 mM phenylmethylsulfonyl fluoride (PMSF), and 1: 100 protease inhibitor stock

Stocks of 1 M Tris-HCl, pH 7.9, and 1 M dithiothreitol (DTT)

For conventional purification:

Extract buffer: 50 mM K-HEPES, pH 7.6, 75 mM KCl, 1 mM K$_3$EGTA, 1 mM MgCl$_2$, 1 mM PMSF, 1: 100 protease inhibitor stock

Special Equipment. Collection sieves and motorized Dounce homogenizer.

Procedure

1. Wash embryos off the collection plates with embryo wash onto the top sieve in a stack of two metal testing sieves (top sieve No. 18, bottom

sieve No. 140, available from Fisher Scientific, Springfield, NJ). The first sieve removes whole flies and large particulates and the second sieve collects the embryos. Once you begin the procedure, work quickly to prevent the embryos from becoming anoxic.

2. Dechorionate the embryos in the sieve by placing the sieve in a tray containing 50% bleach. Incubate for 2 min.

3. Rinse the embryos extensively with distilled water and then remove as much of the excess water as possible by blotting with paper towels held underneath the metal filter and by patting the top of the embryo cake with paper towels.

4. Scoop the embryos out into a tared weigh boat to determine the yield. Add either 5 (immunoisolation) or 3 (conventional purification) ml/g of embryos of the appropriate ice-cold extract buffer. From this point on all steps are carried out at 4°. Resuspend the embryos somewhat by stirring and then homogenize immediately at 4° by five or six passes with a motorized Teflon Dounce homogenizer.

5. Spin the extract for 15 min at 27,000g in a SS34 rotor (Sorvall, Newtown, CT) at 4°. Collect the supernatant. (If you are preparing extract for immunoisolation, at this point add 1 M Tris-HCl, pH 7.9, and 1 M DTT to bring the supernatant to 50 mM Tris-HCl, pH 7.9, 1 mM DTT, final). In both cases, respin the supernatant a second time for 1 hr at 110,000g in a Ti 50.2 rotor (Beckman, Fullerton, CA).

6. Collect the supernatant, carefully avoiding the loose material on top of the pellet. Use the HSS immediately or freeze it in liquid N_2 in 12-ml aliquots in 15-ml conical tubes and store at −80°.

Immunoaffinity Isolation

The first method we have used to purify the septin complex is immuno-affinity chromatography with an antibody made to the C-terminal 14 amino acids (NVDGKKEKKKKGLF) of the *Drosophila* septin peanut (Fig. 1A).[4] We refer to this antibody as KEKK. More details on the immunoaffinity chromatography can be found in Chapter [41] on purification of cytoskeletal proteins using peptide antibodies in this volume.[6]

Solutions

IP buffer: 20 mM Tris-HCl, pH 7.9, 75 mM KCl, 0.5 mM Na$_3$EDTA, 0.5 mM Na$_3$EGTA, 8% sucrose
Wash buffer: 20 mM Tris-Cl, pH 7.9, 400 mM KCl, 1 mM Na$_3$EDTA, 1 mM Na$_3$EGTA, 1 mM DTT, 8% sucrose

[6] C. M. Field, K. Oegema, Y. Zheng, T. J. Mitchison, and C. E. Walczak, *Methods Enzymol.* **298**, [41], (1998) (this volume).

Elution buffer: 75 mM KCl, 20 mM K-HEPES, pH 7.6, 0.5 mM Na$_3$EDTA, 1 mM Na$_3$EGTA, 1 mM DTT

1. Equilibrate 0.8 ml of protein A-Affiprep beads (Bio-Rad, Richmond, CA) or protein A-agarose (GIBCO-BRL, Gaithersburg, MD) by washing three times with 10 resin volumes of IP buffer.
2. Add 100 μg of KEKK IgG to the beads in 0.8 ml of IP buffer. Incubate this slurry for 30 min at room temperature with gentle agitation.
3. Wash the beads three times with 10 volumes of IP buffer.

The remaining steps are performed at 4°C.

4. Incubate the beads for 1 hr with 40 ml of *Drosophila* HSS, prepared as described earlier, rotating gently.
5. Wash the beads three times batchwise with 20 resin volumes of wash buffer.
6. Pour the bead suspension into a small disposable column and drain by gravity.
7. Wash the resin five times by pipeting on 1 column volume of elution buffer and allowing the column to drain by gravity.
8. Add one column volume of elution buffer containing 200 μM KEKK peptide, allow to drain by gravity. Stopper the column and incubate overnight (10–16 hr) at 4°.
9. Elute the complex with several column volumes of elution buffer containing 200 μM KEKK peptide. The majority of the complex elutes in the first two column volumes.

The septin complex obtained is about 90% pure (Fig. 2A); the yield is approximately 1 μg septin complex per 1 μg of KEKK antibody.

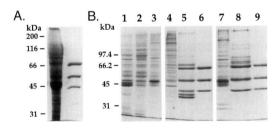

Fig. 2. Coomassie-stained gels of fractions from the two purification schemes. (A) Immuno-isolation. *Drosophila* embryo extract (left lane), immunoisolated septin complex (right lane). (B) Fractions from the conventional purification (see text for details). Lane 1, HSS; lane 2, SP-Sepharose load; Lane 3, Mono S load; lanes 4 and 7, two pools after Mono S fractionation; lanes 5 and 8, the pools in lanes 4 and 7, respectivley, after fractionation on Mono Q; lanes 6 and 9, the pools in lanes 5 and 8, respectively, after fractionation on hydroxyapatite. [From C. M. Field, O. S. Al-Awar, J. Rosenblatt, M. L. Wong, B. Alberts, and T. J. Mitchison, *J. Cell Biol.* **133**, 605 (1996).]

Conventional Purification

We have also purified the septin complex by conventional chromatography. This approach can be useful when large quantities of septin complex are required and may also translate to other systems where immunoisolation is less effective. The conventional purification is outlined in Fig. 1B.

Solutions

S buffer: 20 mM potassium phosphate, pH 7.0, 1 mM K$_3$EGTA, 1 mM MgCl$_2$, 0.1 mM GTP, 1 mM DTT, 1:500 protease inhibitor stock

Q buffer: 20 mM Tris-HCl, pH 7.4 at room temperature (this corresponds to a pH of 8.0 at 5°, 1 mM K$_3$EGTA, 1 mM MgCl$_2$, 0.1 mM GTP, 1 mM DTT, 1: 500 protease inhibitor stock

HA dialysis buffer: 10 mM potassium phosphate, pH 7.2, 100 mM KCl, 0.1 mM GTP, 1 mM DTT, 1: 500 protease inhibitor stock

HA10 buffer: 10 mM potassium phosphate, pH 6.8, 100 mM KCl, 0.1 mM GTP, 1 mM DTT, 1: 500 protease inhibitor stock

HA500 buffer: 500 mM potassium phosphate, pH 6.8, 100 mM KCl, 0.1 mM GTP, 1 mM DTT, 1: 500 protease inhibitor stock

Special Equipment. FPLC (Pharmacia Biotech Inc., Piscataway, NJ) or equivalent chromotography apparatus.

Columns. For cation-exchange chromatography, you will need SP-Sepharose Fast Flow resin (Pharmacia Biotech) and an appropriately sized column housing. Alternatively, 5-ml HiTrapS columns can be connected in series (Pharmacia Biotech). The use of a prepacked 8-ml Mono S column (Pharmacia) as a second step is optional.

Anion exchange is performed on a 1-ml Mono Q column (Pharmacia Biotech). This step is critical, resulting in about 100-fold purification of the septin complex.

The final chromatographic step is fractionation on ceramic hydroxyapatite. For this we have used pre-packed 1-ml Econo-Pac CHT-II cartridges from Bio-Rad, which are convenient and inexpensive.

Procedure
40% AMMONIUM SULFATE CUT

1. Thaw the *Drosophila* HSS in a water bath at 37°. Gently invert the tubes during thawing; place on ice just as the last ice crystals melt. We typically begin with about 400 ml of extract, prepared and frozen as described earlier. The protein composition of the HSS is shown in Fig. 2B, lane 1.

All subsequent steps are performed at 4°.

2. Pool the extract and add ammonium sulfate to 40% saturation (add 22.6 grams per 100 ml of extract). Allow the salt to dissolve while stirring.

3. After the salt has dissolved, wait 20 min. Spin the extract at 10K (8000g) for 10 min in a GSA rotor (Sorvall, Newtown, CT) to collect the precipitate. Discard the supernatant.

4. Resuspend the pellets in one-quarter of the starting extract volume of S buffer containing 50 mM KCl. Dialyze for 4 hr against 20 volumes of S buffer containing 50 mM KCl. Measure the conductivity of the dialysate and add a volume of S buffer containing 0 mM KCl sufficient to bring the conductivity down to that of S buffer containing 50 mM KCl.

5. Clarify the extract by spinning 35K (110,000g) for 1 hr in a 50.2Ti rotor, then filter through a 0.45-μm filter. The protein composition of this material, the SP-Sepharose load, is shown in Fig. 2B, lane 2.

SP Sepharose Chromatography

6. Load the clarified, resuspended ammonium sulfate pellet onto a column of SP-Sepharose Fast Flow resin preequilibrated into S buffer containing 50 mM KCl. To estimate the size of the column you will need (in milliliters of resin), divide the estimated total number of milligrams of protein you will load by 50. Because usually less than 50% of the protein will bind to the column, this will result in $\leq$25 mg of bound protein per milliliter of resin. Our resuspended pellet is usually about 15 mg/ml protein. When we start with 400 ml of extract, we have about 100 ml of about 15 mg/ml protein at this stage and we load this onto a 30-ml column.

7. Wash with 3 column volumes of S buffer containing 50 mM KCl and then elute with a 10 column volume linear gradient between 50 and 500 mM KCl. If the Mono S step (see below) is omitted, elute with a shallower, 20 column volume gradient to improve resolution.

8. Perform enzyme-linked immunosorbent assays (ELISAs) using the KEKK antibody to locate the septin peak. (For details on ELISAs, see Ref. 7 and the protocol section of the Mitchison Lab Web Page http://util.ucsf.edu:80/mitchi/.) Put 50 μl of TBS (150 mM NaCl, 20 mM Tris-Cl, pH 7.4) in each well of a 96-well microtiter plate. Do eight quadrupling dilutions of each fraction by adding 16.7 μl of each fraction to the well at the top of each column and then doing a series of four-fold dilutions by passing 16.7 μl sequentially to each of the subsequent rows. Sequential dilutions are necessary because the protein-binding capacity of the plastic

[7] E. Harlow and D. Lane, *in* "Antibodies: A Laboratory Manual," pp. 1–726. Cold Spring Harbor Laboratory Press, New York, 1988.

is low and the septin complex must compete with other proteins for binding. This means that the amount of septin that binds to the plate depends on the total protein concentration of the fraction (which can vary considerably in the region where the septins elute). At higher dilutions (lower total protein concentrations) the binding surface is less limiting and the amount of septin that binds is more representative of the amount in the fraction. For this reason, we follow the peaks in the highest dilutions where we can still see the signal. Let the plates incubate at room temperature for 30 min to allow the protein to bind to the plastic. Process the plates using standard procedures,[7] but shortening each of the incubation steps to about 30 min. The septin complex elutes at about 145 mM KCl.

MONO S CHROMATOGRAPHY (OPTIONAL). We have obtained slightly better purification by including a Mono S step at this point in the protocol, but reasonable purificaiton can also be obtained without this step.

9. Pool the septin-containing fractions from the SP-Sepharose Fast Flow column. Dilute with S buffer containing 0 mM KCl back to a conductivity corresponding to S buffer containing 50 mM KCl, filter through a 0.45-μm filter, and load onto an appropriate size Mono S column equilibrated in S buffer containing 50 mM KCl. The protein composition of the Mono S load is shown in Fig. 2B, lane 3. Wash the column with 3 column volumes of S buffer containing 50 mM KCl and then elute with a 30 column volume gradient from 50 to 500 mM KCl in S buffer.

10. Locate the septin-containing fractions as before (step 8). Pool the septin-containing fractions. We find that most of the septin complex is split between two separate pools that elute from the Mono S at slightly different salt concentrations between 130 and 200 mM KCl.

11. Dialyze the septin-containing pool(s) overnight into Q buffer containing 40 mM KCl. If there is precipitate, spin 50K (100,000g) for 10 min in the TLA 100.3 rotor. Filter the supernatant through a 0.45-μm filter. The protein compositions of the two Mono S pools after filtration are shown in Fig. 2B, lanes 4 and 7. The pool in lane 7 eluted first and the pool in lane 4 eluted a few fractions later. We keep these two pools separate for the remainder of the purification.

MONO Q CHROMOATOGRAPHY

12. Load each of the pools from the Mono S separately onto a Mono Q column equilibrated in Q buffer containing 40 mM KCl. Wash with 3 column volumes of this buffer and then elute with a 30 column volume gradient between 40 and 500 mM KCl. The septin complex elutes from Mono Q somewhat variably between about 100 and 175 mM KCl. If the Mono S step is omitted, we see two separate peaks of septin complex

eluting between these salt concentrations. We are not yet sure what the origin of this heterogeneity on the ion-exchange columns is due to. We are currently in the process of comparing the purified septins from the two pools in our assays to see if there is any functional difference.

13. Locate the septin complex peak by analyzing 30 μl of each fraction on a 10% polyacrylamide gel stained with Coomassie.

14. Pool the septin-containing fractions and dialyze into HA dialysis buffer. The two septin pools from the Mono S shown in lanes 4 and 7 (Fig. 2B) are shown after fractionation on Mono Q in lanes 5 and 8, respectively.

HYDROXYAPATITE CHROMATOGRAPHY

15. Load each Mono Q pool onto a CHT-II cartridge that has been equilibrated in HA10 buffer. Wash with 3 column volumes of this buffer. Elute with a 20 column volume gradient between HA10 buffer and HA500 buffer. The septin complex elutes between 250 and 270 mM phosphate. The protein composition of the final septin pools corresponding to the two initial Mono S pools shown in lanes 4 and 7 (Fig. 2B) are shown in lanes 6 and 9, respectively.

We typically perform this purificaiton over 3 days. Late on the first day we do the ammonium sulfate cut and set up the dialysis. The next day we do the SP-Sepharose and Mono S chromatography and set up dialysis overnight into Q buffer. On the final day of the purificaiton we do the Mono Q and hydroxyapatite chromatography.

Assay of Septin Complex

Filament Formation

Filament formation by the septin complex is assayed by negative stain electron microscopy. Fractions are applied to glow discharged, carbon-coated grids. After 2–3 min, the grids are rinsed in water and stained with 1% uranyl acetate in 50% methanol for 30 sec. The excess uranyl acetate is then wicked away.

We have observed filament formation by septins purified both by immunoisolation (in preparations still containing elution peptide) and by conventional chromatography. A montage of images is shown in Fig. 3. Septins purified by both procedures are similar and frequently appear as a dense mat of filaments (Fig. 3A; estimated septin concentrations between about 50–200 μg/ml). Figure 3B shows an area of a grid where the filaments have adsorbed to the grid at lower density. Filaments measuring 7–9 nm in diameter and of variable length (up to 350 nm) are observed (see later

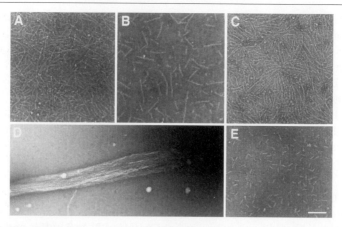

FIG. 3. Visualization of septin filaments and filament assemblies by negative stain electron microscopy (see text for details). (A) A mat of filaments characteristic of septin complex purified by either procedure. This micrograph was taken of conventionally purified septins after the Mono Q step. (B) An area of a grid where the septin filaments have adsorbed at lower density (immunoisolated septins). (C) Alignment of pairs of septin filaments is seen in this fraction off a Mono Q column. (D) Septin assembly formed by concentration in a microconcentrator (E) Monomers and dimers resulting from dilution before adsorption to the grid. Bar = 100 nm. [From C. M. Field, O. S. Al-Awar, J. Rosenblatt, M. L. Wong, B. Alberts, and T. J. Mitchison, *J. Cell Biol.* **133**, 605 (1996).]

section describing filament measurement). If the septin-containing fractions are diluted 5- to 10-fold before adsorption to the grid, short filaments 26–52 nm in length are the primary species observed (Fig. 3E). The septins show a tendency to associate laterally, forming large assemblies in a concentration-dependent fashion. Figure 3D shows an assembly formed when the septin complex is concentrated in a microconcentrator. A more paracrystal-like assembly can also be formed by less concentrated septin solutions if polyethylene glycol is added to a final concentration of 1.5% (not shown). In some cases, in fractions off the Mono Q column (Fig. 2B, lanes 5 and 8), we have observed alignment of filament pairs (Fig. 3C). We do not know if this pairing is due to the presence of additional proteins in the Mono Q fraction or is simply due to the greater concentration of septins in this pool.

Measuring the Lengths of Septin Filaments. When the septin filaments adsorb to the grid at lower density, as in Fig. 3B, their lengths can be measured. Negatives are digitized and filament lengths measured using image analysis software (such as NIH image). Alternatively, negatives can be printed at a magnification of about three and the filament lengths measured with a ruler. Only straight filaments, 250–300 per grid, are measured. Because the longer filaments were more apt to be curved, their frequency is underestimated in this analysis. Plotting number of filaments versus filament length reveals a distribution with a periodicity of approximately 26 nm (Fig. 4). We therefore conclude that the filaments result from the polymerization of a 26-nm subunit. The dilution data fit this hypothesis, with the short filaments (26 and 52 nm) shown in Fig. 3D probably being monomers and dimers of the septin complex.

Measuring the Nucleotide Content of the Septin Complex

Because each of the septins contains a P-loop and other motifs conserved among GTPases, we wanted to determine the nucleotide content of the isolated septin complex. To do this, immunoisolated septin complex is concentrated and denatured, and the released nucleotide is analyzed by Mono Q chromatography. The total septin polypeptide per experiment ranged from 60 to 400 pmol.

Solutions

33% Polyethylene glycol (PEG), MW 8000
Denaturing buffer 1: 100 mM Tris-HCl, pH 7.2, 8 M urea
100 mM NH$_4$HCO$_3$
500 mM NH$_4$HCO$_3$
Standard mixtures containing 100–1000 pmol of ADP, ATP, GDP, and CTP

Special equipment. HPLC (Gilson, Worthington, OH) 1-ml Mono Q column.

Procedure

1. Concentrate the septin complex by pelleting it in a tabletop TL-100 ultracentrifuge (Beckman) for 20 min at 300,000g (in a TLA 100.2 or 100.3 rotor). We typically started with 0.9–1.7 ml of septin complex isolated by immunoaffinity. The protein concentration varied from 20 to 100 μg/ml.

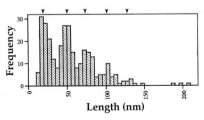

FIG. 4. Histogram of septin filament length measurements from the sample shown in Fig. 3B. Arrowheads indicate histogram peaks.

The sedimentation can be done with or without the addition of PEG to 3%. Without the PEG addition, a variable amount (25–50%) of the complex sediments. Presumably this is long filaments and larger, laterally aligned structures. The addition of PEG results in sedimentation of >90% of the septin complex. If contaminating proteins are present after immunoisolation, they do not cosediment. We have seen similar nucleotide content with both methods.

2. Resuspend the pellet from step 1 in 30 μl of denaturing buffer 1. Heat to 100° for 1 min. This releases the nucleotide from the protein. Dilute the sample by adding 30 μl of distilled H_2O.

3. Remove one-tenth of the sample for protein quantitation using SDS–PAGE. (See step 10.)

4. Deproteinize the remaining sample by filtration through a 10,000-kDa cut-off spin filter unit (Millipore Ultrafree-MC 10,000 NMWL, Bedford, MA).

5. Wash the filter two times with 30 μl of distilled H_2O, saving the filtrate.

6. Pool the filtrate and load onto a 1-ml Mono Q column preequilibrated in 100 mM NH_4HCO_3.

7. Wash column with 10 ml of 100 mM NH_4HCO_3.

8. Elute the column with a 30-ml gradient from 100 to 500 mM NH_4HCO_3.

9. Peaks are integrated using Gilson HPLC software. Standard mixtures of nucleotides are run separately and used to calibrate the integrated OD profile.

10. Protein concentration is determined by densitometry of SDS–PAGE gels that had been stained with Coomassie Blue, destained, and scanned into Adobe Photoshop. For each band, the average intensity is measured and the local background subtracted. The normalized density is then multiplied by the area of the band in pixels to give the integrated amount of protein. A standard curve is measured on the same gel using tubulin and bovine serum albumin as standards.

On average, 1.1 mol of guanine nucleotide was released per mole of septin polypeptide. Figure 5 shows a typical elution profile; the average GDP/GTP ratio was 2.6. These data suggest that each septin polypeptide in the isolated complex binds one molecule of guanine nucleotide tightly.

Assay of Nucleotide Exchange and Hydrolysis by the Septin Complex

To test whether the guanine nucleotide bound to the septins exchanges with free nucleotide, the septin complex was incubated with [α-^{32}P]GTP. At various time points samples were taken, bound nucleotide released, and

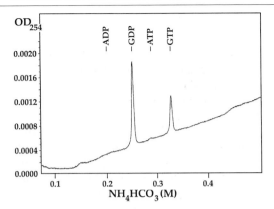

FIG. 5. Analysis of nucleotides released from the purified septin complex. Nucleotides released by urea treatment were fractionated on a Mono Q column with a gradient of NH_4HCO_3.

analyzed for exchange by thin layer chromatography (TLC) and PhosphorImaging.

Solutions

Exchange buffer A: 20 mM K-HEPES, pH 7.6, 50 mM KCl, 1 mM Na$_3$EGTA, 3 mM MgCl$_2$, 1 mM DTT

Exchange Buffer B: 20 mM K-HEPES, pH 7.6, 50 mM KCl, 1 mM Na$_3$EGTA, 2 mM Na$_3$EDTA, 1 mM DTT

Denaturing buffer 2: 8 M urea, 20 mM Tris-HCl, 5 mM Na$_3$EDTA, pH 7.6

A stock solution of 2 mM each GTP and GDP

1.4 M LiCl

Special Reagents and Equipment. [α-^{32}P] GTP. Polyethyleneimine-impregnated (PEI) cellulose plates (Machery-Nagel, Duren, Germany) and an appropriate size chamber for TLC. A setup for UV irradiation at 6000 W/cm^2 can be constructed by mounting 6 model G15T8 15-W germicidal bulbs (General Electric, Winchester, VA) in an appropriate holder. This is basically a UV transilluminator with the glass shield removed. It should be mounted in an enclosed space.

Procedure. This procedure was performed using two different starting materials, the septin complex still attached to protein A beads (method A) and the complex eluted from protein A beads with peptide (method B). Both methods gave similar results. We have not tested the complex isolated by conventional chromatography in this assay.

Method A

1. Equilibrate the septin complex still attached to antibody beads by washing three times with 10 volumes of exchange buffer A. (All steps are performed at room temperature unless indicated.)

2. Divide the complex into aliquots containing 20–30 μl of beads (15–40 pmol of Pnut polypeptide) in 100 μl of buffer.

3. Add $[\alpha\text{-}^{32}P]$GTP (500,000 cpm) to each aliquot along with cold GTP to a final concentration of 2 μM. Cold competitor nucleotide, ATP, or GTP can be added to control aliquots at a concentration of 200 μM.

4. Incubate at 23° for varying amounts of time (between 30 min and 20 hr) with occasional agitation. We have also performed this experiment at 4°, but observed very little exchange.

5. Wash the beads four times with 500 μl of ice-cold exchange buffer A, collecting the beads by centrifugation in a bench top PicoFuge (Stratagene, La Jolla, CA). These washes should be rapid, 2 min total for the four washes. After the last spin, remove as much of the residual buffer as possible.

6. Denature the protein, releasing bound nucleotide, by adding 30 μl of denaturing buffer 2 to each aliquot of beads.

7. Remove one-tenth of the sample for nucleotide analysis by TLC. Quantitate the amount of septin complex in the rest of the sample as described in step 10 of the previous protocol.

8. Analyze the nucleotide content by TLC.

    a. Cut a 20 × 20 PEI cellulose plate in half; 1 cm up from the long edge mark a series of pencil spots 1–1.5 cm apart.

    b. Combine each sample with a mixture of cold GDP and GTP (to a final concentration of 2 μM) or the cold nucleotides can be spotted on to the TLC plate separately (1 μl of 2 mM stock) and allowed to dry before addition of sample. The cold nucleotide acts as a carrier and can be visualized by UV absorption.

    c. Spot the nucleotide mixture across the bottom of a PEI plate (one sample per pencil spot) and allow to dry. The maximum volume per spot is 10 μl; 5 μl is ideal. Run the plate for about 5 cm in water. This does not separate the nucleotides, but moves the urea away from the nucleotide spots, improving the uniformity in R_F values.

    d. Chromatograph the samples using 1.4 M LiCl. PEI is a positively charged ion-exchange matrix, therefore, the species with the greater negative charge migrates more slowly. Generally 12-cm movement of the LiCl front is sufficient to separate GDP and DTP; R_F values (approximate): GDP, 0.25; GTP, 0.15. Migration of nucleotides can be monitored with a handheld UV light.

e. Labeled GDP and GTP spots can be detected and quantitated by PhosphorImaging or conventional autoradiography followed by densitometry. Nucleotide exchange is detected by accumulation of total labeled nucleotide (GDP and GTP). Hydrolysis is detected by the accumulation of label in the GDP form.

As shown in Fig. 6A, approximately 0.3 mol of exogenously added guanine nucleotide is bound per mole of Pnut polypeptide (0.1 mol/mol septin polypeptide if all are exchanging equally) after 18 hr at 23°. Exchange is much slower at 4° (0.008 mol/mol after 18 hr, not shown). Binding of exogenous nucleotide is specific for guanine nucleotide since it is efficiently competed by addition of cold GTP but not cold ATP (Fig. 6B).

We have also tested the use of EDTA-containing buffers (exchange buffer B), which have been shown to promote exchange for other GTPases.[8] In the presence of 2 mM EDTA, exchange is approximately 5-fold more rapid and goes fruther toward completion than in Mg^{2+}-containing buffers. Under these conditions, the exchanged GTP does not hydrolyze. When we add Mg^{2+} back to the protein after promoting exchange in EDTA, no hydrolysis is triggered. This suggests that the active site is damaged by prolonged incubation with EDTA.

The rate of exchange we observe is low when compared to other GTPase superfamily members. For example, the half-life for exchange of GDP bound to Ras with exogenous GTP is approximately 60 min at 30°.[8] One interesting possibility is that GTP exchange is suppressed in some physiologically relevant way in this 340-kDa heterotrimeric septin complex.

Hydrolysis of the bound nucleotide can also be examined by analysis of the fraction of radioactivity in GDP and GTP by TLC. This is shown in Figs. 6A and 6B, where the black symbols or bars represent diphosphates, and the white symbols or bars represent triphosphates. The majority (>75%) of bound guanine nucleotide (added as GTP) is recovered as GDP, and only GDP accumulates over time.

METHOD B

1. Incubate 100 μl of freshly eluted septin complex (5–20 pmol of Pnut polypeptide) with [α-^{32}P]GTP (500,000 cpm) for varying amounts of time at 23°.
2. Chill sample on ice and add PEG to 3%. Let sample incubate on ice for 20 min.
3. Sediment at 200,000g for 15 min at 4° in a TLA 100.4 rotor.
4. Rinse the pellet twice with 50 μl of exchange buffer A. The pellet is quite tight and not easily resuspended without denaturant.

[8] A. Hall and A. J. Self, *Methods Enzymol.* **256,** 67 (1995).

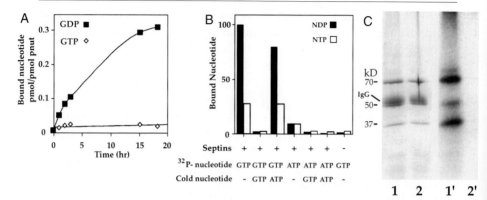

FIG. 6. Nucleotide exchange and hydrolysis. (A) Time course of nucleotide binding and hydrolysis by the septin complex. The septin complex bound to protein A beads was incubated with [α-³²P]GTP at 23°. At various times the beads were washed with cold buffer, bound nucleotide was released with urea, and analyzed by TLC. The majority of bound nucleotide was recovered as GDP (black squares). Bound nucleotide was normalized to the amount of Pnut polypeptide on the beads, measured by gel densitometry. (B) Specificity of nucleotide binding. The septin complex bound to antibody beads was incubated with [α-³²P]GTP or [α-³²P]ATP for 2 hr at 23° and processes as in part (A). Random IgG beads that had been incubated in embryo extract in parallel with KEKK antibody were used as a control (− septins). Bound nucleotide was recovered as a mixture of NDP (black bars) and NTP (white bars). (C) Photocross-linking of bound nucleotide. Lanes 1 and 1′, the Coomassie-stained SDS–PAGE gel and the corresponding autoradiograph, respectively; lanes 2 and 2′, a parallel experiment in which 200 μM cold GTP competitor was added in the initial incubation. [From C. M. Field, O. S. Al-Awar, J. Rosenblatt, M. L. Wong, B. Alberts, and T. J. Mitchison, *J. Cell Biol.* **133**, 605 (1996).]

5. Resuspend pellet in 30 μl of denaturing buffer 2.
6. Analyze sample as described in steps 7 and 8 of method A.

UV Photocross-Linking

To determine which septin polypeptides are exchanging nucleotide, UV photocross-linking (short-wavelength) is performed. This procedure is carried out with septins bound to protein A beads after washing away free [α-³²P]GTP, or with septins eluted from protein A beads in the presence of free [α-³²P]GTP; similar results are obtained.

1. Incubate aliquots of septin beads (20–30 μl of beads, 15–40 pmol of Pnut polypeptide) (method A, step 1) in carrier free [α-³²P]GTP (2 μCi/aliquot) for 2–3 hr at 23°.
2. Wash the beads four times by resuspending in 1 ml of ice-cold exchange buffer A, recovering the beads by centrifugation in a PicoFuge. Remove buffer so as to achieve a 50% slurry.

3. Add 2-mercaptoethanol to 0.3% to reduce nonspecific photocross-linking. Transfer aliquots as single drops onto Parafilm in a plastic dish floating on an ice bath.
4. Irradiate the drops for 5 min with UV light at a distance of 6 cm (setup described earlier).
5. Analyze labeled septins by SDS–PAGE and autoradiography.
6. Photocross-linking can also be performed on septin complex in solution. To do this, septin complex and carrier-free $[\alpha\text{-}^{32}P]GTP$ are preincubated for 1 hr at room temperature. The mixture is then subjected to photocross-linking in the presence of 0.3% 2-mercaptoethanol. The protein is TCA precipitated to remove noncovalently bound nucleotide and analyzed as described earlier.

As shown in Fig. 6C, radioactive GTP became covalently cross-linked to Pnut and Sep1 with Sep2 incorporating much less label. This cross-linking was efficiently competed by excess (200 μM) unlabeled GTP (Fig. 6C, lane 2′). The experiment in Fig. 6C was performed with septin complex on protein A beads. IgG provides a random polypeptide control.

Conclusions

The septins are emerging as an exciting new class of conserved polymerizing proteins that are clearly important for cytokinesis and which may also have other roles in cortical organization.[1,5] In this article, we describe two methods for the purification of a septin-containing complex from *Drosophila* embryos and the techniques we have used to assay filament formation, assembly into larger structures, nucleotide content, and the ability of the complex to exchange and hydrolyze GTP. Further work will be needed to characterize the polymerization dynamics of the septins *in vitro* and *in vivo* and to understand the role of the septin GTPase.

[24] Purification, Assembly, and Localization of FtsZ

By AMIT MUKHERJEE and JOE LUTKENHAUS

Introduction

FtsZ forms a cytokinetic ring that carries out cell division in prokaryotes.[1] This ring, designated the FtsZ or Z ring, is functionally analogous to the contractile ring that carries out cytokinesis in eukaryotes. Immuno-electron microscopy of *Escherichia coli* first demonstrated that FtsZ forms a ring at the leading edge of the constricting septum. The dynamics of the Z ring led to the proposal that FtsZ functions in bacteria as a cytoskeletal protein, similar to actin or tubulin in eukaryotes.[2,3] The Z ring was also demonstrated in *Bacillus subtilis* using immunoelectron microscopy.[4] More recently, the Z ring has been observed in *E. coli* by using fluorescence microscopy. Both indirect immunofluorescence and FtsZ tagged with green fluorescent protein (GFP) have been used.[5–7] The Z ring has also been demonstrated in *B. subtilis* and in an archaebacterium, *Haloferax volcanii,* using indirect immunofluorescence.[8,9] The presence of the Z ring in such evolutionarily divergent bacteria suggests that it is utilized by all prokaryotes to carry out cytokinesis.

FtsZ has limited sequence similarity to eukaryotic tubulins over a region of about 200 amino acids.[10] Included in this region is a highly conserved 7-amino-acid motif,GGGTGTG, in FtsZ that is homologous to the sequence, GGGTGSG, found in tubulin and thought to be involved in guanosine triphosphate (GTP) hydrolysis.[10–12] Indeed, studies with purified FtsZ from *E. coli, B. subtilis, H. volcanii,* and *Mycoplasma pulmonis* showed

---

[1] J. Lutkenhaus and S. G. Addinall, *Annu. Rev. Biochem.* **66,** 93 (1997).

[2] E. Bi and J. Lutkenhaus, *Nature* **354,** 161 (1991).

[3] J. Lutkenhaus, *Mol. Microbiol.* **9,** 403 (1993).

[4] X. Wang and J. Lutkenhaus, *Mol. Microbiol.* **9,** 435 (1993).

[5] S. G. Addinall, E. Bi, and J. Lutkenhaus, *J. Bacteriol.* **178,** 3877 (1996).

[6] X. Ma, D. W. Ehrhardt, and W. Margolin, *Proc. Natl. Acad. Sci. U.S.A.* **93,** 12998 (1996).

[7] J. Pogliano, K. Pogliano, D. S. Weiss, R. Losick, and J. Beckwith, *Proc. Natl. Acad. Sci. U.S.A.* **94,** 559 (1997).

[8] P. A. Levin and R. Losick, *Genes Dev.* **10,** 478 (1996).

[9] X. Wang and J. Lutkenhaus, *Mol. Microbiol.* **21,** 313 (1996).

[10] A. Mukherjee and J. Lutkenhaus, *J. Bacteriol.* **176,** 2754 (1994).

[11] P. de Boer, R. Crossley, and L. Rothfield, *Nature* **359,** 254 (1992).

[12] A. Mukherjee, K. Dai, and J. Lutkenhaus, *Proc. Natl. Acad. Sci. U.S.A.* **90,** 1053 (1993).

that it binds and/or hydrolyzes GTP.[9–14] The *E. coli* FtsZ, like tubulin, polymerizes in the presence of GTP/GDP (guanosine diphosphate) further supporting the view that FtsZ is a prokaryotic homolog of tubulin.[10,15,16] Importantly, FtsZ polymers are dynamic and regulated by GTP hydrolysis.[17]

In this article we describe the protocol that we routinely follow to purify FtsZ from *E. coli*. We also describe the assay for its GTPase activity, the methods used for FtsZ assembly, and a procedure for localization of FtsZ within bacterial cells. The methods described here are modifications of published procedures.[5,10,12]

Procedures

Purification of FtsZ

FtsZ is purified as described in the following subsections. The sequence of steps in the purification scheme is depicted in Fig. 1.

Overproduction of FtsZ. To overproduce FtsZ, *E. coli* W3110 is transformed with the plasmid pKD126, which contains *ftsZ* downstream of the *tac* promoter.[18] Freshly transformed colonies are used as the basal level of *ftsZ* expression from this plasmid is toxic and plasmid variants are readily selected that no longer express *ftsZ*. A point to be noted is that FtsZs from other bacteria have also been overproduced in *E. coli*. A problem often encountered is that the basal expression of the foreign *ftsZ* is quite toxic. This problem has been circumvented by cotransforming plasmid pBS58, which is a low copy number plasmid carrying the *E. coli ftsQ, ftsA,* and *ftsZ* genes.[9] This plasmid is compatible with pBR322 derivatives and suppresses toxicity by slightly raising the levels of the *E. coli* FtsZ and FtsA proteins.

Luria–Bertani (LB, 100 ml) containing 100 μg/ml of ampicillin is inoculated with a colony of W3110 (pKD126) and grown overnight with shaking at 37°. The overnight culture is diluted 100× into 3 liters of fresh LB medium with ampicillin (100 μg/ml) and grown at 37°. When the OD$_{600}$ reaches 0.3, IPTG is added to a final concentration of 0.5 mM and the culture grown for another 3 hr. Cells are then chilled, harvested at 4°, washed once with cold 10 mM Tris-HCl (pH 7.9) and the cell pellets are frozen in an ethanol–dry ice bath and stored at −70°.

[13] D. RayChaudhuri and J. T. Park, *Nature* **359,** 251 (1992).
[14] X. Wang and J. Lutkenhaus, *J. Bacteriol.* **178,** 2314 (1996).
[15] H. Erickson, *Cell* **80,** 367 (1995).
[16] H. P. Erickson, D. W. Taylor, K. A. Taylor, and D. Bramhill, *Proc. Natl. Acad. Sci. U.S.A.* **93,** 519 (1996).
[17] A. Mukherjee and J. Lutkenhaus, *EMBO J.* **17,** 462 (1998).
[18] K. Dai, A. Mukherjee, and J. Luktenhaus, *J. Bacteriol.* **176,** 130 (1994).

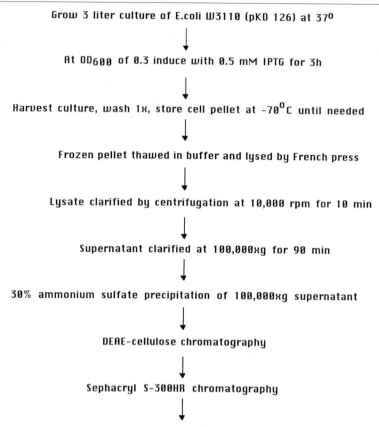

Grow 3 liter culture of E.coli W3110 (pKD 126) at 37⁰

At OD_{600} of 0.3 induce with 0.5 mM IPTG for 3h

Harvest culture, wash 1x, store cell pellet at -70^{0}C until needed

Frozen pellet thawed in buffer and lysed by French press

Lysate clarified by centrifugation at 10,000 rpm for 10 min

Supernatant clarified at 100,000xg for 90 min

30% ammonium sulfate precipitation of 100,000xg supernatant

DEAE-cellulose chromatography

Sephacryl S-300HR chromatography

FtsZ dialyzed against storage buffer, concentrated and stored at -70^{0}

FIG. 1. Purification scheme for FtsZ.

Lysis of Bacterial Cells and Fractionation of Cell Extracts. From this step onward all operations are done at 4° unless otherwise mentioned. The frozen cell pellet is thawed in 30 ml of buffer A [50 mM Tris-HCl, pH 7.9, 50 mM KCl, 1 mM ethylenediaminetetraacetic acid (EDTA), 10% glycerol] and lysed by passage two times through a French press. The resulting extract is centrifuged at 10,000g for 10 min to remove unbroken cells and debris. The supernatant is then centrifuged at 100,000g for 90 min to remove the membrane fraction. The supernatant is collected (taking care not to withdraw any loosely floating material, if any, near the membrane pellet) and 30% ammonium sulfate (16.6 g/100 ml of supernatant) is slowly added with stirring. After stirring for a further 20 min the suspension is centrifuged at 12,000g for 20 min. The pellet is dissolved in buffer A and dialyzed overnight against buffer A with two changes of 1 liter each.

DEAE-Cellulose Chromatography. The dialyzed ammonium sulfate fraction is loaded onto a DEAE-cellulose (Sigma, St. Louis, MO) column (bed volume of 25 ml) equilibrated with buffer A, washed with five bed volumes of buffer A, and eluted with 200 ml of a 50 mM–1 M gradient of KCl at a flow rate of 0.5 ml/min. Fractions of 2 ml are collected and the peak fractions of FtsZ, eluting at around 200–250 mM KCl, are pooled and then concentrated with Centriprep-10 (Amicon, Beverly, MA).

Sephacryl S-300HR Chromatography. The FtsZ recovered from the DEAE-cellulose run is loaded onto a Sephacryl S-300HR (Pharmacia Biotech, Piscataway, NJ) column (1.6 × 70 cm) and eluted with buffer A at a flow rate of 0.3 ml/min. A maximum of 50 mg of protein is loaded onto the column per run. The peak fractions are pooled and dialyzed against buffer containing 50 mM HEPES-NaOH (pH 7.2), 0.1 mM EDTA, and 10% glycerol; concentrated using Centriprep-10 to about 8–10 mg/ml; aliquoted; and stored at −70°. FtsZ is greater than 95% pure as observed by Coomassie Blue-stained gels.

About 80–100 mg of FtsZ is obtained by this method of purification. Purified FtsZ stored at −70° for 2 years shows no loss of GTPase or polymerization activities (described later). Furthermore, freezing and thawing FtsZ up to five times did not affect either of these activities.

Monitoring FtsZ during Purification. FtsZ is monitored during purification by sodium dodecyl sulfate–polyacrylamide gel electrophoresis (SDS–PAGE) and Coomassie Blue staining.

Protein Assay. Protein is assayed with the Bio-Rad Protein Assay Reagent (Bio-Rad Labs, Hercules, CA) using bovine serum albumin (BSA) as a standard.

GTPase Assay

The GTPase activity of FtsZ is measured in either 50 mM HEPES-NaOH (pH 7.2) or 50 mM MES-NaOH (pH 6.5) with 10 mM MgCl$_2$, 200 mM KCl. A standard reaction contains FtsZ at 200 μg/ml (5 μM) and the reaction is initiated by adding 1 mM [γ-^{32}P]GTP (100–200 cpm/pmol) (Amersham, Arlington Heights, IL) and incubating at 30°. The GTPase activity requires Mg^{2+} and is stimulated by K$^+$.[11–13] For optimal activity, 200 mM KCl is used as the rate decreases at lower KCl concentrations. To measure the amounts of released ^{32}P, a 5-μl aliquot of the reaction mixture is withdrawn at 5-min intervals and added to 95 μl of 0.55 M HClO$_4$ containing 0.55 mM KH$_2$PO$_4$ followed by the addition of 150 μl of 20 mM Na$_2$MoO$_4$. The mixture is vortexed and after 2 min at room temperature,

300 μl of isopropyl acetate (Acros Organics, Geel, Belgium) is added, vortexed for 15 sec, and spun for 2 min in an Eppendorf centrifuge to separate the two phases. The organic phase (upper phase) (100 μl) is withdrawn and added to 5 ml of Econo-Safe (Research Products Intl. Corp., Mount Prospect, IL) and counted in a liquid scintillation counter. Alternatively, the formation of $[\alpha$-^{32}P]GDP from $[\alpha$-^{32}P]GTP can be determined by separating them on polyethyleneamine cellulose plates as done by other workers.[11,13]

A considerable lag in the GTPase activity has been reported for FtsZ purified by other protocols.[11,12] With the above purification scheme, which retains bound GDP, there is no discernible lag as reported for the *B. subtilis* and *E. coli* FtsZ.[9,13] Also, as reported for the *B. subtilis* FtsZ the specific activity is dramatically dependent on the FtsZ concentration. Little activity is observed below 40 μg/ml (1 μM) but the specific activity increases with increasing FtsZ concentration until 120 μg/ml (3 μM), above which it is constant. At 5 μM FtsZ the number of moles of GTP hydrolyzed per mole of FtsZ per minute at pH 6.5 and 30° with 0, 50, and 200 mM KCl is 1.0, 3.5 and 8.5, respectively.[17]

FtsZ Polymerization

The polymerization of FtsZ obtained by an earlier purification procedure occurs readily in the presence of DEAE-dextran as demonstrated by electron microscopy and negative staining.[10] The polymerization occurs with GDP or GTP but only if DEAE-dextran is added. With FtsZ obtained utilizing the preceding purification procedure, polymerization occurs readily without DEAE-dextran and is strictly GTP dependent.[17] Polymerization occurs between pH 6.0 and 7.2 but appears more efficient at slightly acidic pH. Importantly, the polymers formed at pH 6.5 can be quantitatively assayed by centrifugation. Bramhill and Thompson[19] described a sedimentation and light scattering assay for FtsZ polymerization at pH 7.5. However, it was later reported that high concentrations of unbuffered GTP (5–10 mM) added to the polymerization reaction resulted in a drop of the pH from 7.6 to near the isoelectric point of FtsZ (pI is 4.59) causing FtsZ precipitation.[16]

Polymerization at pH 6.5. FtsZ is diluted in 50 mM MES-NaOH (pH 6.5), 200 mM KCl, 10 mM MgCl$_2$ to a final concentration of 200 μg/ml (5 μM). The reaction is initiated by the addition of 1 mM GTP and incubation at 30°. Polymers are rapidly formed and are present until the GTP is consumed (approximately 15 min at this KCl concentration). Without KCl, or at lower concentrations of KCl, the polymers last longer as the

---

[19] D. Bramhill and C. M. Thompson, *Proc. Natl. Acad. Sci. U.S.A.* **91,** 5813 (1994).

rate of GTP hydrolysis is reduced. The polymerization at pH 6.5 occurs readily at 30° and 37° and with or without 200 mM KCl.

Polymerization of FtsZ with DEAE-Dextran. A reaction mixture containing FtsZ at 200 μg/ml in 50 mM HEPES-NaOH (pH 7.2), 10 mM MgCl$_2$, 20 μg/ml of DEAE-dextran, and 1 mM GTP is incubated at 37° for 1 hr. The staining was done exactly as described earlier.[12,17] The time of incubation is not critical because polymers are formed with GDP or GTP in the presence of DEAE-dextran.

Quantitative Assay of FtsZ Polymers by Centrifugation. FtsZ is diluted to 200 μg/ml in 50 mM MES-NaOH (pH 6.5), 50 mM KCl, 10 mM MgCl$_2$ at room temperature (usually 100–200 μl final volume). Polymerization is initiated by adding GTP to 1 mM. The sample is then immediately centrifuged at 80,000 rpm for 10 min at 25° in a Beckman (Fullerton, CA) TLA 100.2 rotor. The supernatant is withdrawn and the amount of FtsZ in the pellet is determined. About 50% of FtsZ is recovered in the pellet, whereas with GDP or ATP less than 5% is in the pellet.[17] The KCl concentration is 50 mM to reduce the rate of GTP hydrolysis. This prevents the GTP from being consumed during the course of the centrifugation because it is known that FtsZ polymers disassemble on GTP exhaustion.[17] Alternatively, we have used GTP concentrations as high as 2.5 mM to extend the time during which polymers are observed.

EM Notes

Preparation of Carbon-Coated Copper Grids for Electron Microscopy

1. Prior to coating with carbon, the surface of 300 mesh size copper grids (EMS, Fort Washington, PA) is pretreated using a grid coating pen (EMS).
2. Carbon rods (3-mm point carbon) are evaporated at 2×10^{-6} torr in a coater (Ernest Fullum, Latham, NY) onto a freshly cleaved mica surface.
3. The carbon-coated mica surface is then put into distilled water at an angle allowing the carbon film to float off into the water.
4. A copper grid held with forceps is immersed under the floating carbon film and lifted. Excess water on the grid is then blotted off with a filter paper and allowed to air dry.

Reagents

A 1% uranyl acetate (EMS) solution is prepared in water and stored at 4° for up to 4 weeks. A 1 mg/ml stock solution of DEAE-dextran (Sigma) is prepared in sterile water and kept at room temperature for 3 months.

Buffers and water used for electron microscopy is filtered through 0.22-μm filters.

Staining and Visualization of Polymers by Electron Microscopy

A 10-μl aliquot of the polymerization reactions described previously is placed on a carbon-coated grid for 2 min and blotted dry by placing a filter paper on the side of the grid. The grid is then immediately stained by adding a few drops of 1% uranyl acetate solution and immediately blotted dry. The grids are then viewed in a JEOL (Tokyo, Japan) transmission electron microscope (model 100CXII). The staining pattern of the FtsZ polymers with uranyl acetate varies from grid to grid and even within the same grid, resulting in both negative and positive staining. Typically FtsZ forms protofilaments that are about 7 nm wide, which tend to align along their long axis to form structures about 20 nm wide that are composed of several protofilaments. A typical example of polymers formed at pH 6.5 is shown in Fig. 2.

Localization of FtsZ by Fluorescence Microscopy

Two approaches utilizing fluorescence microscopy have been used to monitor FtsZ localization within bacterial cells. The first approach uses indirect immunofluorescence of fixed cells and the second utilizes FtsZ tagged with GFP. The first approach requires fixing and permeabilizing cells and FtsZ antibodies and is based on a procedure developed in R. Losick's laboratory.[5] The second approach uses FtsZ tagged with GFP as described by Ma et al.[6] The advantage of this second approach is that live cells can be viewed; however, the tagged FtsZ cannot support cell growth, indicating that it lacks some essential function. Also, the level must be

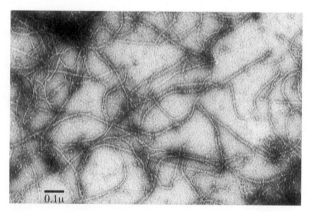

FIG. 2. Visualization of FtsZ polymers by negative staining.

carefully controlled as excess expression of the fusion protein results in a block to division and the mislocalization of FtsZ in spiral and other abnormal structures.[6] Below we describe the procedure we have used for analyzing the localization of FtsZ in *E. coli*. Similar procedures have been used for FtsZ localization in other bacteria.[8] In addition, the same procedure has been used to localize less abundant proteins in *E. coli*.[20–22]

Procedure

Cells are fixed in the growth medium by the addition of fixative prepared in phosphate buffer (Fig. 3). The fixative is prepared by adding 100 μl of 16% paraformaldehyde (which is dissolved by heating and addition of a small amount of NaOH) and 1 μl of 25% glutaraldehyde to 20 μl of 1 M Na$_3$PO$_4$ (pH 7.4). Fixative (100 μl) is added to 500 μl of culture, mixed gently, and left at room temperature for 10 min followed by 50 min on ice. The cells are washed three times with PBS (10 mM Na$_3$PO$_4$, pH 7.4, 150 mM NaCl, 15 mM KCl) at room temperature by centrifugation and resuspended in 400 μl (or less, depending on the cell density desired) of GTE (50 mM glucose, 10 mM EDTA, 20 mM Tris-HCl, pH 7.5). While cells are fixing, a multiwell slide (ICN Biochemicals, Costa Mesa, CA) is coated with polylysine by pipetting 10 μl of 0.1% poly-L-lysine to each well and letting stand 2–3 min. The polylysine is removed and the slide washed two times with distilled water, all done with a pipettor. The slide is then air dried and 15 μl of lysozyme (100 μg/ml) freshly prepared in GTE is added to the fixed cells, mixed, and 10-μl aliquots are immediately placed in the wells on the slide. After 1–10 min, the lysozyme is removed and each well is washed two times with PBS and air dried completely. The permeabilized cells are rehydrated for 2–4 min with 10 μl of PBS (all subsequent steps are done by the addition and removal of 10 μl of reagent to the wells with a pipettor). The PBS is replaced with blocking solution: PBS containing 100 μg/ml of BSA. After 10 min at room temperature the blocking solution is replaced with the primary antibody in blocking solution (1 : 1000 dilution of the antibody depending on the titer). The slide is placed in a plastic petri dish along with a few pieces of Whatman filter paper that have been wetted with distilled water and incubated overnight at 4°. The wells are washed 10 times with PBS and secondary antibody—10 μl of a 1 : 100 dilution of a Cy3-conjugated anti-rabbit immunoglobulin G antibody (Jackson Immunoresearch, West Grove, PA) in blocking solu-

[20] S. G. Addinall and J. Lutkenhaus, *J. Bacteriol.* **178,** 7167 (1996).
[21] S. G. Addinall, C. Cao, and J. Lutkenhaus, *Mol. Microbiol.* **25,** 303 (1997).
[22] D. S. Weiss, K. Pogliano, M. Carson, L. M. Guzman, C. Fraipont, M. Nguyen-Disteche, R. Losick, and J. Beckwith, *Mol. Microbiol.* **25,** 671 (1997).

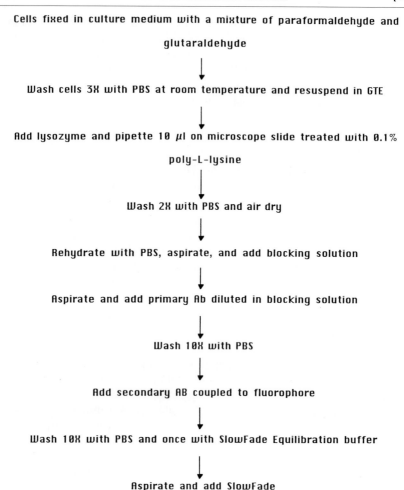

Fig. 3. Steps for immunolocalization of FtsZ.

tion—is added and incubation continued at room temperature for 1 hr. The wells are washed 10 times with PBS and once with SlowFade equilibration buffer (Molecular Probes, Eugene, OR) and finally 8 μl of SlowFade is added. The slide is stored in the petri dish at $-20°$ until viewed. To visualize nucleoids, an extra wash with SlowFade equilibration buffer containing 2.0 μg/ml of 4′,6-diamidino-2-phenylindole (DAPI) is included. Samples are viewed with a Nikon (Melville, NY) Optiphot fluorescence microscope equipped with a $100\times$ objective and a

100-W mercury lamp source. For the Cy3 fluorophore a 590 barrier filter (G-2A, Nikon) is used.

Comments

The preceding protocol describes our standard procedure and various steps can be altered to try and improve the results. The results for FtsZ immunostaining are not too sensitive to the lysozyme treatment, i.e., it works with 2-fold changes in the concentration or increased time of treatment. For other antigens we observe that the results are more sensitive to changes and optimal conditions have to be explored.

[25] Purification and Assembly of FtsZ

By CHUNLIN LU and HAROLD P. ERICKSON

Introduction

FtsZ is a key molecule involved in bacterial cell division in all species of eubacteria and archaebacteria that have been examined.[1-3] FtsZ forms a cytoskeletal ring at the site of septation, and powers the constriction of the septum. A major interest in FtsZ is its homology to tubulin.[4] Not only does FtsZ share significant sequence similarity with tubulin,[5] it assembles into protofilament sheets and minirings that are very similar to tubulin polymers.[6] Study of FtsZ is important, both to understand the mechanism of bacterial cell division and for what it can teach us about tubulin.

FtsZ has been cloned from more than two dozen bacteria, and several of these FtsZ proteins have been produced as expression proteins in *Escherichia coli*. The genomic *E. coli* FtsZ (EcFtsZ) is abundantly produced at 15,000 molecules per cell (making it 0.4 mg/ml in the cytoplasm),[6a] but for purification purposes it is generally overexpressed from an expression plasmid. The pET expression plasmid described next is one of the highest

[1] H. P. Erickson, *Trends Cell Biol.* **17,** 362 (1997).
[2] L. I. Rothfield and S. S. Justice, *Cell* **88,** 581 (1997).
[3] W. D. Donachie, *Annu. Rev. Microbiol.* **47,** 199 (1993).
[4] H. P. Erickson, *Cell* **80,** 367 (1995).
[5] A. Mukherjee and J. Lutkenhaus, *J. Bacteriol.* **176,** 2754 (1994).
[6] H. P. Erickson, D. W. Taylor, K. A. Taylor, and D. Bramhill, *Proc. Natl. Acad. Sci. U.S.A.* **93,** 519 (1996).
[6a] C. Lu, J. Stricker, and H. P. Erickson, *Cell Motil. Cytoskeleton* **40,** 71 (1998).

producing expression systems we have seen, giving 70 mg of purified protein per liter of bacterial culture.

Expression and Purification of EcFtsZ

pET Vector for Expression of EcFtsZ

This pET vector was first described by Bramhill and Thompson,[7] and was easily duplicated in our laboratory. The FtsZ coding sequence was amplified from genomic E. coli DNA using Pfu, a proofreading polymerase (Strategene, LaJolla, CA), and adding NdeI and BamHI restriction sites. This construct was ligated into pET11 plasmids (Novagen, Madison, WI), amplified in E. coli strain DH5α, purified, and used to transform strain BL21(DE3). The transformed BL21(DE3) cells have been kept as a glycerol freezer stock at −80° for 2 years, but should be regenerated by transformation from stored plasmid every year or two.

For bacterial growth, 50 ml of L broth containing 200 μg/ml ampicillin is inoculated from the freezer stock and grown at 37° overnight. The cells are pelleted by centrifugation and resuspended in 1 liter (500 ml in each of two 2-liter baffled flasks) of L broth with 200 μg/ml ampicillin. Shaking is continued at 250 rpm at 37°. Expression is induced by adding 0.5 mM isopropyl thio-galactopyranoside (IPTG) when the cultures reached $A_{600} \sim 1.2$, approximately 90 min after inoculation. (This absorbance is higher than the $A_{600} = 0.6$ that is usually recommended, but it had no effect on FtsZ expression.)

Bacteria are typically collected after 2–3 hr by centrifuging at 6000 rpm at 4° for 10 min (GSA rotor, Sorvall) and suspended in cold lysis buffer [50 mM Tris, pH 8.0, 100 mM NaCl, 1 mM ethylenediaminetetraacetic acid (EDTA), 1 mM phenylmethylsulfonyl fluoride (PMSF); 20 ml per liter of bacterial culture]. Lysozyme is added to 0.4 mg/ml and the mix is incubated on ice for 120 min. With other expression proteins we have used Triton X-100 to facilitate lysis of bacteria, but we found that Triton changed the properties of FtsZ and have avoided its use. Magnesium is added to 10 mM and the bacteria are frozen overnight at −20°. Bacteria are thawed, kept on ice, and sonicated twice for 30 sec with a 30-sec interval at a power level of 50% (4710 series. Ultrasonic Homogenizer, Cole-Parmer). DNase I is added to 10 μg/ml and incubated for 60 min at 4°. Cell walls and insoluble debris are removed by centrifugation at 35,000 rpm at 4° for 30 min (Type 45 rotor, Beckman, Fullerton, CA). FtsZ protein is soluble in the E. coli cytoplasm and represents more than half of the total soluble protein.

[7] D. Bramhill and C. M. Thompson, Proc. Natl. Acad. Sci. U.S.A. 91, 5813 (1994).

Purification by Ammonium Sulfate Precipitation

The volume of supernatant from the last centrifugation is measured (V_0). A volume of saturated ammonium sulfate (dissolved in water at room temperature) equal to one-quarter V_0 is added and mixed well, bringing the supernatant to 20% saturation. This is incubated on ice for 20 min, and the protein is pelleted at 15,000 rpm at 4° for 10 min (Type 45 rotor, Beckman). This pellet, designated EcFtsZ$_{20}$, is resuspended in 10 ml of TNEM buffer (50 mM Tris-HCl, pH 8.0, 100 mM NaCl, 1 mM EDTA, 5 mM MgCl$_2$) and agitated on ice for 1 hr until it is completely dissolved. An additional volume of saturated ammonium sulfate equal to one-twelfth V_0 is added to the supernatant, raising the level of ammonium sulfate to 25% saturation. This is incubated on ice and centrifuged as above. This second pellet (EcFtsZ$_{20/25}$) is dissolved in 10 ml of TNEM buffer. EcFtsZ$_{20}$ and EcFtsZ$_{20/25}$ are reprecipitated with 25% saturated ammonium sulfate. The final pellets are resuspended in 5 ml of TNEM buffer, centrifuged at 30,000 rpm at 4° for 10 min (Type 45 rotor, Beckman) to remove insoluble materials, and then aliquoted and stored at −80°. EcFtsZ is stable and can be repeatedly thawed and frozen without changing its GTPase activity and ability to assemble.

In several experiments we noted that the efficiency of the second 25% ammonium sulfate precipitation depended on the concentration of FtsZ. If the FtsZ pellet were suspended in a larger volume, giving lower protein concentrations, then a higher ammonium sulfate concentration was required for efficient precipitation.

An average of 70 mg of EcFtsZ was obtained from 1 liter of bacterial culture. EcFtsZ$_{20}$ represented about 30% and EcFtsZ$_{20/25}$ about 70% of the total expressed FtsZ. EcFtsZ$_{20}$ had significantly lower activity than EcFtsZ$_{20/25}$ for binding and hydrolysis of GTP. We believe that EcFtsZ$_{20}$ contains mostly inactive FtsZ, and that EcFtsZ$_{20/25}$ is the best preparation of active FtsZ.[6a]

Further Purification

At this stage EcFtsZ appeared relatively pure, with only a few weak contaminating bands visible on sodium dodecyl sulfate–polyacrylamide gel electrophoresis (SDS–PAGE) (Fig. 1). To further purify FtsZ, other groups have used gel filtration or ion-exchanged columns.[8–10] However, we found that ion-exchange chromatography on DEAE-Sephacel or Mono Q (Phar-

[8] D. RayChaudhuri and J. T. Park, *Nature* **359**, 251 (1992).
[9] P. de Boer, R. Crossley, and L. Rothfield, *Nature* **359**, 254 (1992).
[10] A. Mukherjee, K. Dai, and J. Lutkenhaus, *Proc. Natl. Acad. Sci. U.S.A.* **909**, 1053 (1993).

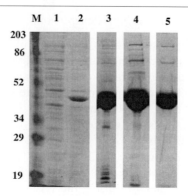

Fig. 1. SDS–PAGE analysis of expression and purification of EcFtsZ. Lane M, prestained protein markers (Bio-Rad, Richmond, CA). Their size is indicated on the left. Lane 1, total cell lysate before IPTG induction; lane 2, total cell lysate after induction; lane 3, FtsZ$_{20}$ (40 μg); lane 4, EcFtsZ$_{20/25}$ (60 μg); lane 5, EcFtsZ$_{20/25}$ after passage over a phosphocellulose column (30 μg). Samples were analyzed on 12% SDS gels and stained with Coomassie Blue. The gels were overloaded to show minor contaminants.

macia, Piscataway, NJ) did not result in any apparent increase in purity. We also tested phosphocellulose columns (P11, Whatman, Clifton, NJ), which are widely used for purification of tubulin free of MAPs. Like tubulin, EcFtsZ$_{20/25}$ did not bind to a phosphocellulose column (in 25 mM MES, pH 6.3, 0.5 mM MgCl$_2$, 0.1 mM EDTA), but the flow-through fraction showed no obvious improvement in purity (lane 5 in Fig. 1). The GTPase activity and assembly of EcFtsZ purified with DEAE-Sephacel or phosphocellulose was the same as that prepared by ammonium sulfate precipitation with no additional steps.

Protein Concentration

Because the FtsZ proteins have no tryptophan residues and few tyrosine residues, absorbance at 280 nm is not an accurate measure of protein concentration. Indirect protein assays such as the Lowry, Bradford, and BCA assay give variable color with different proteins, so we calibrated these assays by quantitative amino acid analysis.[6a] In the BCA assay the ratio of color was 0.75 for EcFtsZ/BSA. We also calibrated the Bradford assay, in which the ratio of color was 0.82 for EcFtsZ/BSA. In our experience, the BCA assay gives the most consistent results.

GTP Hydrolysis

A 4× reaction buffer (50 mM Tris-HCl, pH 7.5, 200 mM KCl, 20 mM Mg acetate, 2.0 mM GTP) is prepared. FtsZ is transferred to 50 mM Tris-

HCl, pH 7.5, using a 10-ml prepacked Econo-Pac 10DG column (Bio-Rad). All buffers, additives, and the EcFtsZ sample are kept on ice while setting up the reactions. Any desired additives are added and the mixture incubated on ice for 10 min. The reaction is initiated by mixing 13 μl of EcFtsZ sample, 5 μl of 4× reaction buffer, and 2 μl of 0.25 μCi/μl [α-^{32}P]GTP. [We found that Amersham GTP was of uniformly high quality when analyzed by thin-layer chromatography (TLC) whereas several lots obtained from ICN contained only GDP.] The reaction mixture is transferred to a water bath at 37°. At 2- or 5-min intervals, 1 μl is withdrawn using a 1-μl disposable micropipette (Drummond Scientific Co.) and applied to a 10- × 10-cm polyethyleneimine cellulose TCL plate (POLYGRAM® CEL 300 PEI/ UV$_{254}$, Machery-Nagel, Germany). The spots are air dried.

Plates are developed vertically in 0.75 M KH$_2$PO$_4$ (pH 3.4) in a closed chamber at room temperature. The development is stopped when the leading edge of solvent reaches the top of the plate. Plates are air dried, wrapped with Saran Wrap, and exposed to a PhosphorImager plate (IP, FUJIX) for 10–30 min at room temperature. The imager plate is scanned and data are analyzed using MacBAS1000 v1.01 software (Fuji Photo Film Co.). To avoid errors due to evaporation of sample during the reaction or imprecision in the amount of sample loaded, the data are always read as the ratio of GDP/GTP; this ratio, multiplied by the known amount of GTP in the starting solution, gives the amount of GDP produced, which is eventually expressed as moles GDP produced per mole of FtsZ.

One EcFtsZ$_{20/25}$ molecule can hydrolyze 2.5 molecules of GTP per minute under the conditions described above. EcFtsZ$_{20}$ was eight times less active, hydrolyzing 0.3 molecules of GTP per minute. The concentrations of protein, GTP, and salts, as well as temperature, affect the turnover values significantly.[6a,8–10]

Assembly of EcFtsZ$_{20/25}$

Three papers have described assembly of FtsZ *in vitro*.[5–7] Two of these studies used the polycation DEAE-dextran to stimulate assembly and stabilize the polymers.[5,6] One reported assembly of purified EcFtsZ[7]; it was later discovered that lowered pH, between 5.5 and 6.0, was an essential factor in this assembly.[6] We describe below conditions for producing three different polymer forms of EcFtsZ$_{20/25}$. These conditions have been useful for defining the types of polymers, but should not be considered definitive. Rather they should be taken as starting points for investigating new assembly conditions.

We use two buffers for assembly: MEMK6.5 (100 mM MES adjusted to pH 6.5 with KOH, 1 mM ethylene glycol-bis(β-aminoethyl ether)-

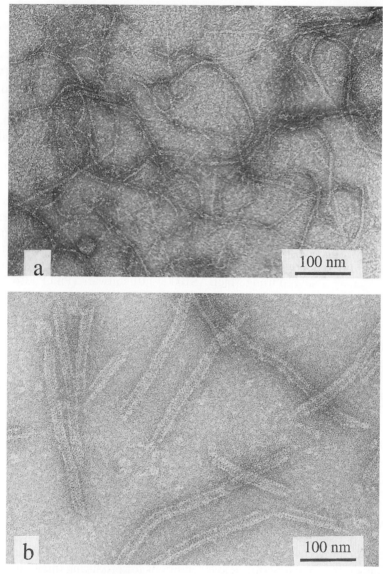

FIG. 2. Polymers assembled by EcFtsZ$_{20/25}$. (a) Protofilaments; 0.3 mg/ml EcFtsZ$_{20/25}$ was assembled in MEMK6.5 with 1 mM GTP at 10° for 10 min. (b) Tubes; 1.7 mg/ml DEAE-dextran and 2.8 mg/ml EcFtsZ$_{20/25}$ were assembled in MEMK6.5 with 1 mM GTP at 37° for 5 min. (c) Protofilament sheets were assembled by adding 10 mM EDTA to the reaction mixture in Part (b), and incubating for another 3 min at 37°. (d) Tubes assembled in PB7.7 containing 1 mM GDP; 1.2 mg/ml DEAE-dextran and 2.0 mg/ml EcFtsZ$_{20/25}$ were assembled at 37° for 5 min. The arrows indicate the circular profiles of tubes that have broken and present an end-on view.

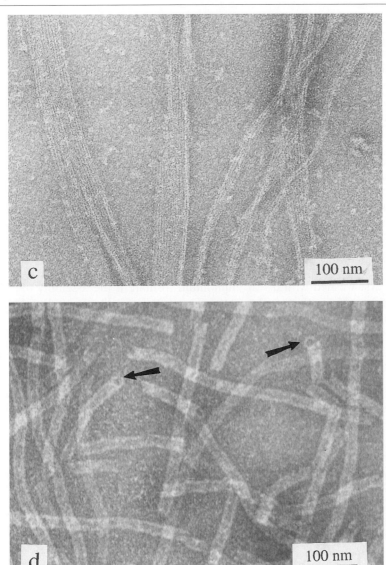

FIG. 2. (*continued*)

N,N,N′,N′-tetra-acetic acid [EGTA], 2.5 mM magnesium acetate) and PB7.7 (physiologic buffer: 35 mM MOPS adjusted to pH 7.7 with KOH, 80 mM glutamate, 350 mM potassium acetate, 2.5 mM magnesium acetate; initially the buffer contained 50 mM trehalose and 2 mM putrescine, but these were found not to be important for assembly). EcFtsZ is transferred to assembly buffer using 10-ml prepacked Econo-Pac 10DG columns (Bio-Rad), and GTP, GDP, or GTPγS is added to 1 mM.

Electron Microscopy of FtsZ Polymers

Assembly of FtsZ *in vitro* has been analyzed primarily by negative staining electron microscopy. We use grids coated with a thin carbon film that are subjected to glow discharge to render the surface hydrophilic. A small drop of the assembled FtsZ is applied to the grid, removed, and drained with filter paper. Three drops of uranyl acetate (a 2% aqueous solution, unadjusted pH, filtered through a 0.2-μm pore filter) are washed over the grid while it was held at a 30-deg angle from the vertical. Grids are first washed with 0.1 M ammonium formate if the assembly buffer is incompatible with uranyl acetate. The grid is finally drained with filter paper and stored. Grids are scanned at a magnification of 20,000×, and selected areas are photographed at 50,000×.

Protofilaments

The structure of the FtsZ protofilament was deduced by image reconstruction of protofilament sheets.[6] It is a linear array of FtsZ subunits, 4.3 nm apart and 5 nm in diameter. Single protofilaments have now been obtained from EcFtsZ$_{20/25}$ in MEMK6.5 with 1 mM GTP at an EcFtsZ$_{20/25}$ concentration of 50–300 μg/ml. Specimens prepared from this FtsZ solution on ice, or after 10 min at 10°, showed single protofilaments and short narrow sheets consisting of two to four protofilaments (Fig. 2a). DEAE-dextran was not necessary for this assembly. At 37° the protofilaments disappeared and were replaced by irregular aggregates with no well-defined internal structure.

Protofilament Sheets

The protofilament sheet is a two-dimensional array of straight protofilaments forming a flat, orthogonal lattice with dimensions of 4.3 × 5.3 nm.[6] We have discovered at least two conditions under which two-dimensional protofilament sheets can assemble. The original assembly was obtained in a buffer essentially the same as MEMK6.5, with 0.1 mg/ml DEAE-dextran and 0.5 mg/ml EcFtsZ.[6] Negatively stained specimens were prepared after

5–60 min at 37°. In that study we noted that pH was crucial in determining the conformation of the polymer; assembly in similar conditions but at pH 7.2 produced tubes (see below). The ratio of DEAE-dextran to FtsZ is also crucial. At a ratio of 0.2 we obtained sheets, but at a ratio of 0.6 (1.7 mg/ml DEAE-dextran to 2.8 mg/ml $EcFtsZ_{20/25}$) the assembly only produced tubes (Fig. 2b). However, this assembly mixture reverted to protofilament sheets if 10 mM EDTA was added, chelating Mg^{2+} ions. This defines our second assembly condition, which produced the largest and most uniform sheets. The reaction was initially incubated for 5 min at 37° without EDTA, forming tubes. EDTA was then added and incubation continued for 3 min. The resultant sheets are shown in Fig. 2c. The transition from tubes to sheets is apparently very rapid, and has not yet been visualized.

Tubes

Tubes or rods about 20–23 nm in diameter were described in two studies,[2,9] but the lattice structure was not determined. We have now concluded that tubes consist of protofilaments in the curved conformation, forming a long, closely packed spiral (C. Lu and H. P. Erickson, in preparation). The tubes are therefore constructed from protofilaments in the curved conformation, and are related to minirings.[6] Tubes and minirings are the same diameter, 20–23 nm. Minirings are confined to a plane, and therefore form a flat circle, while the curved protofilament in tubes forms a shallow helix (C. Lu and H. P. Erickson, in preparation). Tubes were assembled under several conditions, including PB7.7 containing 1 mM GDP (Fig. 2d). These tubes assembled while the protein solution was on ice, and persisted at 37°. Some end-on views of broken tubes show the hollow tubular structure.

Straight and Curved Conformations are Favored by GTP and
  GDP, Respectively

Both curved and straight protofilaments can be obtained under a variety of solution conditions, but the conformation seems to be controlled most importantly by the state of the guanine nucleotide. Single straight protofilaments and sheets are favored by high GTP, by EDTA (which blocks GTP hydrolysis), and by nonhydrolyzable GTP analogs. Tubes are assembled in GDP, but do not require 100% GDP. A ratio of GDP/GTP > 0.3 favors tubes in several conditions. These points will be discussed in more detail by C. Lu and H. P. Erickson (in preparation).

Section IV

Cell-Free and Genetic Systems

[26] High-Resolution Video and Digital-Enhanced Differential Interference Contrast Light Microscopy of Cell Division in Budding Yeast

By E. D. Salmon, Elaine Yeh, Sid Shaw, Bob Skibbens, and Kerry Bloom

Introduction

The budding yeast *Saccharomyces cerevisiae* has a number of advantages for determining the molecular mechanisms of cell division and the function of the associated cytoskeletal components. Yeast genetics is powerful and the cell division cycle is of only 90-min duration for wild-type cells.[1] Many different mutants are now known for various aspects of the cell cycle, chromosome replication, the assembly of the mitotic machinery, chromosome segregation, and cytokinesis.[2]

Imaging nuclear, spindle, and astral microtubule dynamics in budding yeast during the cell division cycle is a major challenge for the light microscopist because of their small cell size. Wild-type haploid cells are 5–8 μm in diameter and have a dense cell wall. The preanaphase nucleus is about 2 μm in diameter. The mitotic spindle forms within the nucleus between two spindle pole bodies (SPBs) that are embedded in the nuclear envelope[3] and mitosis occurs without nuclear envelope breakdown.[3-6]

We have developed two methods for high-resolution live cell imaging of nuclear and spindle dynamics in budding yeast by differential interference contrast (DIC) light microscopy (Fig. 1).[7] The first uses the video-enhanced DIC (VE-DIC) microscopy methods we developed earlier for imaging

[1] A. Murray and T. Hunt, "The Cell Cycle." Oxford University Press, New York, 1993.

[2] B. Byers, *in* "Cytology of the Yeast Life Cycle" (J. N. Strathern, E. W. Jones, and J. R. Broach, eds), p. 59. Cold Spring Harbor Laboratory Press, New York, 1981.

[3] J. R. Pringle and E. W. Jones *in* "The *Saccharomyces Cerevisiae* Cell Cycle" (J. N. Strathern, E. W. Jones, and J. R. Broach, eds.), p. 97. Cold Spring Harbor Laboratory Press, New York, 1981.

[4] S. L. Shaw, E. Yeh, E. D. Salmon, and K. Bloom, *J. Cell Biol.* **139**, 985 (1997).

[5] M. Winey and B. Byers, *Trends Genet.* **9**, 300 (1993).

[6] M. Winey, C. L. Mamay, O. T. Et, D. N. Mastronarde, T. H. Giddings, Jr., K. L. McDonald, and J. R. McIntosh, *J. Cell Biol.* **129**, 1601 (1995).

[7] E. Yeh, R. V. Skibbens, J. W. Cheng, E. D. Salmon, and K. Bloom, *J. Cell Biol.* **130**, 687 (1995).

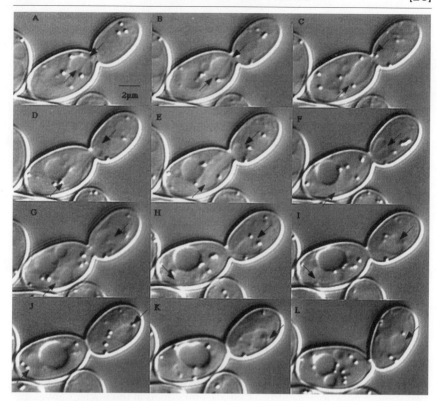

FIG. 1. Digitally enhanced DIC images of a dividing yeast *S. cerevisiae*. The spindle extends through the center of the nucleus and is formed from an overlapping bundle of microtubules extending from opposite spindle pole bodies (arrows). (A) Preanaphase. (B–K) Anaphase. (L) Cytokinesis. The preanaphase nucleus in part (A) is about 2.5 μm in diameter. [Reproduced with permission from E. Yeh *et al.*, *J. Cell Biol.* **130,** 687 (1995). Copyright © 1995 by Rockefeller University Press.]

individual native 25-nm-diameter microtubules *in vitro*[8] and *in vivo*[9] and individual 250-nm-wide kinetochores and their kinetochore microtubule bundles in vertebrate mitotic tissue cells.[10] The electronics in video cameras and digital image processors can make visible fine structural detail at the resolution of the light microscope (about 250 nm in green light) or isolated

[8] R. A. Walker, E. T. O'Brien, N. K. Pryer, M. Soboeiro, W. A. Voter, H. P. Erickson, and E. D. Salmon, *J. Cell Biol.* **107,** 1437 (1988).
[9] L. U. Cassimeris, N. K. Pryer, and E. D. Salmon, *J. Cell Biol.* **107,** 2223 (1989).
[10] R. V. Skibbens, V. P. Skeen, and E. D. Salmon, *J. Cell Biol.* **122,** 859 (1993).

complexes like microtubules that are one-tenth the resolution limit.[11–16] The second method we term digitally enhanced DIC (DE-DIC), where DIC images are acquired using a slow-scan cooled charge-coupled device (CCD) camera, and contrast enhancement of fine structural detail is achieved using digital image processing.[7,17] VE-DIC has the advantage that images can be viewed and recorded at or near video rates of 30 frame/sec. DE-DIC microscopy has the advantage that DIC imaging can be combined with multiwavelength fluorescent imaging using the same camera to combine the structural resolution of DIC with the selectivity and sensitivity of fluorescent molecular probes, like expressed fusion proteins containing green fluorescent protein (GFP)[4] or the DNA in the nucleus and mitochondria labeled with 4′,6-diamidino-2′-phenylindole (DAPI).[3]

In this article, we describe the specimen preparation methods we currently use for imaging yeast by either method. Details of the microscope design, microscope alignment, cameras, shutters, and computer image processing equipment for either the VE-DIC or multimode DE-DIC and fluorescence imaging systems we use have been recently described in detail elsewhere.[16–18] Here, we focus on the procedures we use with these instruments for imaging budding yeast and give some sample applications.

Specimen Preparation

At least five factors are important for high-resolution DIC and fluorescence imaging of yeast: the high refractive index of the thick cell wall, the importance of immobilizing the yeast, the optical thickness and quality of the specimen preparation, the need for nutrients for normal growth and cell cycle times, and minimizing background fluorescence from the medium.

We grow the yeast to logarithmic phase in rich YPD medium (2% glucose or galactose, 2% peptone, and 1% yeast extract) before experimental manipulation (e.g., switching from glucose to galactose to induce transcription and protein synthesis using GAL promoters) and specimen preparation.

[11] R. D. Allen, N. S. Allen, and J. L. Travis, *Cell Motil.* **1**, 291 (1981).
[12] R. D. Allen and N. S. Allen, *J. Microsc.* **129**, 3 (1983).
[13] S. Inoué, *J. Cell Biol.* **89**, 346 (1981).
[14] B. J. Schnapp, *Methods Enzymol.* **134**, 561 (1986).
[15] E. D. Salmon, R. A. Walker, and N. K. Pryer, *BioTechniques* **7**, 624 (1989).
[16] E. D. Salmon and P. T. Tran, *Methods Cell Biol.* **56**, 153 (1997).
[17] E. D. Salmon, S. L. Shaw, J. Waters, C. M. Waterman-Storer, P. S. Maddox, E. Yeh, and K. Bloom, *Methods Cell Biol.* **56**, 185 (1997).
[18] S. L. Shaw, E. Yeh, K. Bloom, and E. D. Salmon, *Current Biol.* **7**, 701 (1997).

To reduce the "flare" produced by the refractive index mismatch between the cell wall and the medium, the yeast cells are pressed into a thin (about 10 μm) layer of 25% gelatin (an idea suggested by Tim Stearns, Stanford University) containing minimal yeast growth medium (0.67% yeast nitrogen base, 2% glucose or galactose supplemented with 0.5% casamino acids, and 16.5 μg/ml adenine). Minimal medium supports growth without introducing significant autofluorescence into the images. This method (Fig. 2) also immobilizes the yeast to prevent Brownian movements that would hinder resolution. The thin layer of gelatin is also optically homogenous and does not significantly alter the optical path between the condenser and the objective.

Additions to the medium, such as drugs or metabolites, are either incorporated into the original gelatin solution or added to cells prior to mounting. For example, labeling of nuclei with DAPI is performed by adding 5 μl of 1 mg/ml DAPI in DMSO to a 1-ml logarithmic phase culture for 30 min. The DAPI-containing medium is exchanged for fresh medium prior to imaging.

Specific procedures for specimen preparation are as follows:

1. Grow cells to early to mid-logarithmic phase, i.e., $A_{600} = 0.3$–0.6. The phase of cells is important. Older cells are usually more vacuolated. Spin cells briefly to concentrate if necessary.
2. Clean glass slides and #1 22- × 22-mm coverslips with EtOH.
3. Melt minimal medium + 25% gelatin (SIGMA-G2500) in microwave.
4. To make slide preparations, we pipette 5 μl of heated gelatin solution onto a clean slide. A second slide is placed on top, forming a sandwich.
5. Let solidify by cooling for 15 min at −20°.
6. Gently pry the slides apart with a razor blade, yielding one slide with a thin gelatin layer.
7. Cells from logarithmic phase culture (1–5 μl) are pipetted onto the gelatin and a coverslip is firmly applied.
8. Immediately cover with clean coverslip.
9. Seal coverslip to slide along edges with Valap. (1:1:1 Vaseline: lanolin: paraffin).

FIG. 2. Yeast cells are immobilized in a thin layer of gelatin for high-resolution imaging. See text for details.

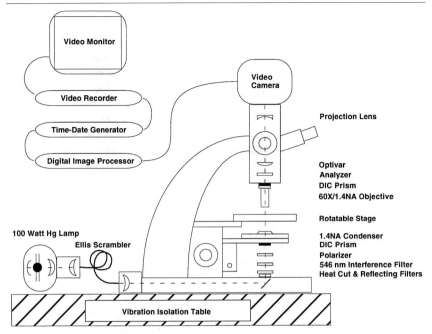

FIG. 3. Schematic diagram of a high-resolution VE-DIC microscope system. A research polarized light microscope on a vibration isolation table (Newport Corp., Irvine, CA, model VW-3660 with 4-in.-high tabletop) is equipped with an illuminator having a bright 100-W Hg light source and Ellis fiber optic light scrambler (Technical Instruments, Inc., Woods Hole, MA), heat cut, heat reflection and 546-nm interference filters, efficient polars, rotatable stage, and DIC optics (we use either Zeiss or Nikon optical components). The Optivar contains a telescope lens for focusing on the objective back focal plane during microscope alignment. The specimen image is projected at high magnification (380×) onto the faceplate of a video camera (model C2400 Newvicon, Hamamatsu Photonics, Bridgewater, NJ), enhanced using analog electronics in the camera, then passed through a digital image processor (Argus 10 or Argus 20, Hamamatsu Photonics) for further enhancement (background subtraction, exponential averaging, and contrast improvement). This image is passed through a time-date generator (Panasonic model WJ-810) and then recorded on a S-VHS videocassette recorder (Panasonic model AG1980 for real time, or Panasonic model AG6750A-P) or memory optical disc recorder (Panasonic model TQ2028F), and the final image is displayed on a monochrome video monitor (Panasonic WV-5410). [Reproduced with permission from E. D. Salmon and P. T. Tran, *Methods Cell Biol.* **56,** 153 (1997). Copyright © 1997 by Academic Press.]

Microscope Alignment for Either VE-DIC or DE-DIC

For high-resolution imaging by either VE-DIC (Fig. 3) or by DE-DIC (Fig. 4), it is important to use matched oil immersion high-numerical-aperture (NA = 1.25–1.4) objective and condenser lenses and the DIC prisms designed for these lenses.[16]

MULTIMODE DIGITAL MICROSCOPE SYSTEM
FOR
MITOSIS STUDIES IN VIVO AND IN VITRO

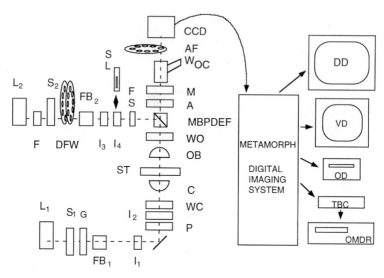

NIKON FXA LIGHT MICROSCOPE

FIG. 4. Schematic diagram of a high-resolution multimode DE-DIC and fluorescence digital imaging system. Component parts are L_1, 100-W quartz halogen lamp; S_1, Uniblitz shutter (Vincent Associates, Rochester, NY); G. ground glass diffuser; FB_1, manual filter changers including XG4 heat cut and green interference filters; I_1, field iris diaphragm; I_2, condenser iris diaphragm; P, AP, high-transmission Nikon polaroid polarizer and removable analyzer; WC and WO, DIC Wollaston prisms for condenser and objective; C, Nikon NA = 1.4 CON A, Achr-Apl condenser; ST, rotatable stage with focus position controlled by z-axis stepper motor (Mac2000, Ludl Electronic Products, Ltd., Hawthorne, NY); OB, 20×/NA = 0.75 or 60×/NA = 1.4 Nikon objectives; MBPDEF, epi-filter block with multiple bandpass dichromatic mirror and emission filter (Chroma Technology Corp., Brattleboro, VT); L_2, 100-W HBO Hg lamp; F, KG4 heat cut filter; S_2, DFW, shutter and dual eight-position filter wheel (Metaltek, Raleigh, NC), one wheel containing neutral density filters (Melles Griot, Irvine, CA), and the other a series of narrow bandpass excitation filters (Chroma Technology Corp.); FB_2 manual filter changer; I_3, epi-condenser iris diaphragm; I_4, epi-field diaphragm slider; SL, slit (25 μm width, Melles Griot), cemented to Nikon pinhole slider for photoactivation; FS, filter slider; M, optivar magnification changer, 1×–2×; OC, oculars; AFW, four-position filter wheel (Biopoint filter wheel, Ludl Electronic Products, Ltd.) with one position containing a Nikon high-transmission polarizer for the analyzer and another position an optical glass flat of the same thickness as the analyzer); CCD, cooled CCD camera (Hamamatsu Photonics); DD, 1024- × 768-pixel, 20-in. digital graphics display monitor (ViewSonic); VD, RGB video display monitor (Sony); MetaMorph digital imaging system (Universal Imaging Corp., West Chester, PA) with a 166-MHz Pentium computer having both EISA and PCI bus, 128-MByte RAM memory, Imaging Technology AFG digital and video image processing card, Hamamatsu C4880 CCD controller card, Matrox MGA Melenium 4-Meg RAM graphics display card, SVGA graphics display to S-VHS converter card (Hyperconverter, PC video conversion Corp.,

High magnifications are needed for yeast imaging because video and digital cameras have a much lower resolution capability than the eye. The magnification required depends on the resolution of the camera detector and the resolution of the microscope (0.2–0.25 μm for green light and NA = 1.4 objective and condenser). We typically use a total magnification to the camera of 380× (magnification of the objective × bodytube × projection lens to video adaptor) for the 1-in. Newvicon or Chalnicon video cameras and 150× for the 0.5-in. CCD video cameras and our Hamamatsu C4880 slow scan cooled CCD camera.[16]

The microscope must be aligned for Köhler illumination to achieve proper alignment for the DIC prisms and specimen illumination:

1. The field diaphragm should be in focus and centered on the specimen. It is best to close the field diaphragm down around the specimen as far as possible to reduce scattered light and loss of contrast. However, the field diaphragm is usually adjusted to be just outside the field of view of the video camera.
2. The condenser diaphragm is opened to match the size of the objective aperture. This is adjusted by using a telescope or a combination of an eyepiece plus Bertrand lens to view the objective back focal plane or aperture where the image of the condenser diaphragm is in focus.
3. The image of the light source is focused and centered at the condenser diaphragm. Again, this is adjusted by viewing the objective aperture.

We adjust the objective DIC prism to produce the DIC "shadow" centrast images in our microscopes by sliding the objective DIC prism with its translation screw.[16] First the prism is adjusted to zero bias retardation by finding the setting that makes the background light extinguished. Then the screw is turned, moving the prism, and brightening the background, till one edge of the nucleus becomes maximally dark; the other edge will be brighter than the background. This setting produces a bias retardation of about 1/10–1/20 of the wavelength of green light (see Salmon and Tran[16] for more details).

San Jose, CA), 1.4-MByte floppy drive, 2-GByte hard drive, 1-Gbyte Iomega Jaz drive, Hewlett Packard SureStore CD Writer 4020I, Ethernet card, parallel port cards for controlling shutter S_1 and driving laser printer; eight serial port card for controlling Metaltek filter wheel, Ludl z-axis stepper, CCD camera, and OMDR; OD, Pinnacle Micro1.3 GByte optical drive; TBC, time-base corrector; OMDR, Panasonic 2028 optical memory disk recorder. Video is also recorded on a Panasonic AG-1980P S-VHS recorder and a Panasonic AG-6750A time-lapse S-VHS recorder. Microscope is mounted on a vibration isolation table (Newport Corp., model VW-3660 with 4-in.-high tabletop). [Reprinted with permission from E. D. Salmon *et al.*, *Methods Cell Biol.* **56,** 153 (1997). Copyright © 1997 by Academic Press.]

We select cells or rotate the stage so that the DIC "shadow" contrast is perpendicular to the mother-bud axis. Contrast in DIC is maximum in one direction and near zero in the perpendicular direction.

Heat cut and heat reflection filters in the illumination light path are used to block infrared from the specimen. In addition, we use a 540-nm ± 20-mn bandpass green interference filter. Contrast in DIC is improved by using monochromatic light and this filter also blocks blue light which is phototoxic to yeast.

It is critical for good image contrast that sufficient immersion oil, free of air bubbles, be used between the objective and coverglass and the condenser and slide.

For high-resolution imaging, the objective aperture should also be fully and evenly illuminated with light.[19] This requires a bright light source because of the high magnifications, loss of light through the polarizer and analyzer optics, and the green bandpass filter.

For VE-DIC we use an HBO 100-W Hg burner, lamphouse, and Ellis fiber optic scrambler (Fig. 3).[16] Video cameras have short exposure times, about 33 ms, and require very bright light at high magnifications for high-quality images not deteriorated by either camera or photon noise.[16] The Hg source is about 10× or more brighter than the conventional 100-W quartz-halogen tungsten illuminator in part because of the bright line at 546 nm. The arc is small and very inhomogeneous. This problem is corrected by scrambling the light through a 1-mm-diameter fiber optic and using the end of the fiber as the light source.[16,19–21]

For DE-DIC, we typically use exposures of 600 ms on the cooled CCD camera.[7,17] This is a gain of 20-fold in exposure over the video rate cameras and corresponding lower light levels are required. The 100-W quartz-halogen illuminator is sufficient and a glass light diffuser is used to make the illumination of the objective aperture uniform (Fig. 4).

VE-DIC Image Recording

We routinely obtain VE-DIC images using a very-high-quality research-grade Newvicon video camera, a real-time digital image processor, and a time-data generator (Fig. 3). Unlike single microtubules,[8,11,14] the nucleus and spindle have sufficient contrast using only the camera electronics.[7] The digital image processor is mainly used to perform background subtraction to remove from the image dirt on the camera detector and out-of-focus images

[19] S. Inoué, *Methods Cell Biol.* **30,** 112 (1989).
[20] G. W. Ellis, *J. Cell Biol.* **101,** 83a (1985).
[21] S. Inoué and K. Spring "Video Microscopy," 2nd Ed. Plenum Press, New York, 1997.

of dirt, and the effects of uneven illumination in the microscope.[12,15,16] In other words, if the camera and microscope are clean, and care is used to produce even and uniform illumination of the specimen, the digital image processor is not needed (which is sometimes difficult to achieve on a routine basis).

Monochrome images can be recorded in real-time into the luminance (intensity information) channel of a S-VHS video cassette recorder (VCR) (Fig. 3). S-VHS has about 400 TV lines of horizontal resolution, about twice the resolution of standard VHS.

Most often, we perform time-lapse image acquisition and recording to an optical memory disk recorder (OMDR) at 2-, 4-, or 8-sec intervals depending on the experiment (Fig. 3) because the OMDR is much easier to use for frame-by-frame motion analysis than the VCR. This is a high-resolution monochrome machine with more than 450 TV lines of horizontal resolution. Custom-written software in a PC computer controls, through a parallel port, a shutter in the illumination path and image acquisition into the OMDR. We found that for the Newvicon camera it is important to open the shutter for several frame times before acquiring an image. The shutter is opened, there is a delay of about 100 ms, an image is captured to the OMDR, then the shutter is closed. Shuttering is important for experiments where cells are followed over several hours since accumulated exposures with intense green light eventually are phototoxic. Focus is easy to maintain by manually opening the shutter and readjusting the fine focus on the microscope while viewing the image on the monitor before the OMDR (Fig. 3).

To obtain digital images from the video records for publication (e.g., Fig. 1), we acquire 8-bit (256-graylevel) images into either the MetaMorph PC digital imaging system (Fig. 4) or into a Macintosh computer using NIH image software and a Scion image capture board as described by Shaw et al.[22] They are then transferred to Adobe Photoshop and then Canvas (Deneba Systems, Inc.) for formatting, contrast adjustment, labeling, and then printing on a Tektronics Phaser IIX dye sublimation digital printer.[22]

In principle, either digital imaging system could be used to capture VE-DIC images directly from the camera (and real-time image processor) and stored on digital medium. One advantage of our OMDR is that it can store 16,000 video images (about 0.25 MB/image) per disk, or a total of 4 GB and recall any image nearly instantaneously for viewing or analysis. Newer OMDR models from Panasonic and Sony have larger disks; however, the OMDR may no longer be needed in the near future with the recent avail-

[22] S. L. Shaw, E. D. Salmon, and R. S. Quatrano, *BioTechniques* **19**, 946 (1995).

NO MID-ANAPHASE PAUSE

A. KINETICS

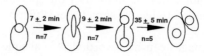

B. SPINDLE DYNAMICS

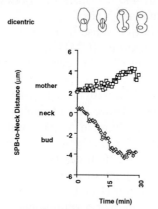

MID-ANAPHASE PAUSE

C. KINETICS

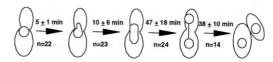

D. SPINDLE DYNAMICS

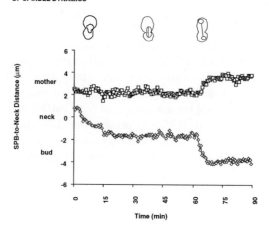

ability of very fast PC or Macintosh computers and fast removable digital storage medium like the 1-GB JAZ drives (Iomega).

DE-DIC Image Acquisition and Processing

In our multimode digital imaging system (Fig. 4), digital images are acquired with a Hamamatsu C4880 slow-scan, cooled ($-30°$) CCD (TC-215 detector). The properties of this camera and practical considerations of imaging with full-frame transfer CCD cameras are described elsewhere.[17] For our yeast studies, the important aspects of this camera are the high quantum efficiency (35–60% between 500 and 600 nm) and low camera noise in the image (10 photoelectron equivalents per image). This camera produces 12-bit (4096-graylevel, 2 byte/pixel) images. Because digital images require large amounts of computer memory and disk space, we use a 300- × 300-pixel central subregion of the 1024- × 1024-pixel CCD chip for image acquisition. Thus, each DIC image requires about 200 kB of storage space. The readout of pixel values from this detector is slow, 350 kB/s. Sony has just introduced a new type of interline cooled CCD chip (HAD ICX061) that has high quantum efficiency (about 50% at the 510-nm emission peak of GFP), and can be read out at 14 MB/sec with extraordinary low noise (about 10 electrons).[17] These specifications indicate that cameras with this chip will be very useful for yeast DIC and fluorescence imaging.

The MetaMorph digital imaging system controls the microscope and camera shutters, filter wheels, z-axis stepping motor, and image acquisition by the camera.[17] MetaMorph acquires each series of one type of image, like DIC, into an image stack in which each image has the time of acquisition marked. When needed, background subtraction is performed by subtracting an averaged background reference image from each image in the stack. The averaged background reference image is obtained by averaging 25 slightly out-of-focus images of a region lacking cells.[18] Averaging removes random noise contained in an individual image.

As described earlier, we routinely acquire time-lapse DIC images of yeast cells with 600-ms exposures. Focusing is much more difficult with the slow-scan cooled CCD compared to video cameras and we focus using our eyes through the eyepieces. The cooled CCD camera is mounted exactly

---

FIG. 5. Kinetic analysis of changes in nuclear morphology and spindle dynamics in dicentric cells displaying (A, B) wild-type anaphase progression and (C, D) a mid-anaphase pause. [Modified and reprinted with permission from Yang *et al.*, *J. Cell Biol.* **136**, 1 (1997). Copyright © 1997 by Rockefeller University Press.]

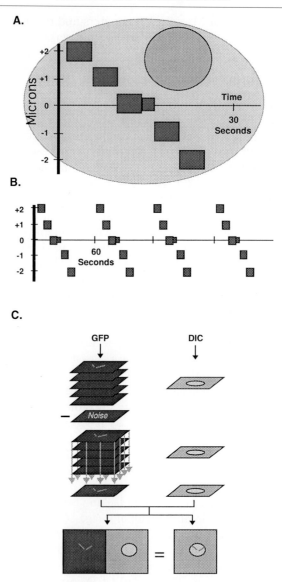

Fig. 6. Schematic of GFP, DAPI, and DIC image acquisition and processing. (A) Five fluorescence images are taken at 1-μm intervals. A single DIC image (and DAPI image) is taken in the same focal plane as the middle fluorescence image. The stage is returned to the original position with a slight adjustment to account for any hysteresis in the microscope focusing apparatus. (B) The acquisition regime is iterated every 60 sec for time-lapsed imaging. (C) Image processing schematic: Image stacks of raw fluorescence data are background subtracted and projected as sets of five to a two-dimensional representation of the three-dimen-

at the intermediate image plane of the microscope; a plane conjugate with the eyepiece reticle in the microscope binoculars. We usually obtain images at 30- or 60-sec intervals so that over a 90-min cell cycle, there will be 90–180 DIC images in the stack, or 18–36 MB of digital storage space.[7]

The highest quality DE-DIC images (e.g., Fig. 1) are achieved without a dichromatic epifluorescence filter and emission filter in the light path. The polarization properties of these filters degrade image quality slightly, but they also have been used as analyzers at the correct wavelengths for combining DIC with fluorescence imaging.[21]

Analysis of Spindle and Nuclear Cell Cycle Kinetics from VE-DIC Images

An important contribution of DIC recording, particularly the VE-DIC, is in the analysis from images like those shown in Fig. 1 of how mutations in proteins change the normal progression of events in the cell cycle.[7,23] This analysis has included the dynamics of spindle elongation, nuclear migration into the bud, nuclear morphogenesis, and cytokinesis. The positions of the spindle poles relative to the bud neck are tracked using custom-built computer systems. Single frame measurement (SFM) uses manual frame-by-frame tracking by eye with a video cursor,[24] whereas the other system uses a semiautomatic tracking method where the computer overlays a cursor on selected objects in the image and does the frame-by-frame tracking.[7,10] As an example, Fig. 5 shows the kinetic analysis of a mid-anaphase pause in spindle elongation and cell cycle progression that is produced by some cells containing a dicentric (two-centromere) chromosome.[23]

Combining DE-DIC with Multiwavelength Fluorescence Imaging

The methods we use for multimode digital imaging of yeast cells are described in detail by Shaw et al.[18] The analyzer used for DIC imaging absorbs 50–85% of unpolarized light depending on the transmission effi-

[23] S. S. Yang, E. Yeh, E. D. Salmon, and K. Bloom, J. Cell Biol. **136,** 1 (1997).
[24] N. R. Gliksman, S. F. Parsons, and E. D. Salmon, J. Cell Biol. **5,** 1271 (1992).

sional data set. DIC and GFP images are overlaid after adjusting for a spatial offset in horizontal and vertical directions. [Reproduced with permission from Shaw et al., Current Biol. **7,** 101 (1977). Copyright © 1997 by Current Biology Ltd.]

ciency of the analyzer.[16,21] We use a filter wheel mounted just before the camera (Fig. 4) to insert the analyzer in the light path for DIC images and remove it from the light path and insert a clear filter of equivalent optical thickness for fluorescence imaging.[17,18] The filter wheel is controlled by journals written in the MetaMorph digital imaging system.[17]

These MetaMorph journals also control the appropriate shutters and the selection of excitation wavelength for fluorescence imaging using a filter wheel in the epi-illumination path in combination with a multiple-bandpass dichromatic mirror and emission filter (Figs. 4 and 6).[17,18] The images in Fig. 7, for example, are part of a time-lapse series where, at 1-min intervals, a DIC and a DAPI image are recorded for focus on the spindle and nucleus in center of the cell. In the same interval, a z-series of five dynein-GFP images at 1-μm steps through the cell are also recorded to obtain images of most or all the astral microtubules (Fig. 6).[18] The DIC, DAPI, and GFP images are each acquired into their own image stacks and stored. After recording, the appropriate averaged background image is subtracted from each image. For each time interval, the GFP microtubule images are combined into one image to reveal the distribution of astral microtubules in the cell (Figs. 6 and 7).[18]

Nuclear movements relative to the mother, bud, and neck can be correlated with astral microtubule dynamics and spindle elongation in several

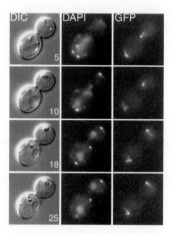

FIG. 7. Multimode time-lapse imaging of spindle elongation and nuclear division in budding yeast. Four sets of images were selected from a time-lapse series showing DIC, DAPI, and GFP images of the same specimen (time = minutes from anaphase onset). See Shaw *et al.*[4] for details on microtubule dynamic instability and Shaw *et al.*[18] for details on multimode imaging. Bar = 5 μm. [Modified and reproduced with permission from Shaw *et al.*, *Current Biol.* **7,** 101 (1997). Copyright © 1997 by Current Biology Ltd.]

ways. The DIC, DAPI, and GFP image stacks can be combined into a single monochrome stack of three frames (Fig. 7) or converted into 8-bit image stacks and combined into a single 24-bit RGB stack where the DAPI image is blue, the GFP image is green, and the DIC image is red. Selected regions of image frames are also easily montaged in MetaMorph. For printing or slides, images or montages of images (e.g., Fig. 7) are transferred to Adobe Photoshop and then to Corel Draw7 (Corel, Jericho, NY) for final contrast adjustment, labeling, and printing. Microsoft PowerPoint has also been very useful in making figures and slides. Image stacks can also be montaged in MetaMorph and movies made to display dynamic information.[17]

[27] Production of M-Phase and I-Phase Extracts from Mammalian Cells

By DUANE A. COMPTON

Introduction

The structural organization of cells is radically altered during mitosis and meiosis. These profound structural changes are necessary for the efficient segregation of the genetic material packaged as chromosomes as well as the equal distribution of all other cellular components. These structural changes are completely reversible such that the disassembly of each cellular organelle at the onset of mitosis or meiosis provides the necessary building blocks for the assembly of that organelle in each daughter cell at the completion of cytokinesis.[1,2]

Our understanding of the structural events of cell division has been greatly facilitated by the development of systems that reproduce the structural events of mitosis and meiosis *in vitro*. These cell-free systems have been used to examine the assembly/disassembly properties of a diverse array of cellular structures including the Golgi apparatus,[3] nuclear envelope and nuclear pores,[4] chromosomes,[5] spindle apparatus,[6,7]

[1] T. J. Mitchison, *Curr. Opin. Cell Biol.* **1**, 67 (1989).
[2] C. L. Rieder, *Curr. Opin. Cell Biol.* **3**, 59 (1991).
[3] T. Misteli and G. Warren, *J. Cell Biol.* **130**, 1027 (1995).
[4] J. Newport, *Cell* **48**, 205 (1987).
[5] T. Hirano and T. J. Mitchison, *Cell* **79**, 449 (1994).
[6] K. E. Sawin and T. J. Mitchison, *J. Cell Biol.* **112**, 925 (1991).
[7] R. Heald, R. Tournebize, T. Blank, R. Sandaltzopoulos, P. Becker, A. Hyman, and E. Karsenti, *Nature* **382**, 420 (1996).

centrosomes,[8] and nuclear lamina.[9–12] The most widely used cell-free systems for analysis of mitosis- and meiosis-specific structural changes are based on extracts prepared from eggs derived from marine organisms (i.e., *Xenopus laevis*).[13,14] Eggs derived from these organisms are abundant, contain large pools of many structural proteins, and synchronously proceed through the cell cycle. Despite these advantages, extracts prepared from these cells perform meiosis-specific or at the very least early embryo-specific events, and in some cases may not reflect events that occur during mitosis in somatic cells. Given that there are several distinct structural differences between mitosis in somatic cells and meiosis in gametic cells,[15] it is important for investigators in the field to have access to cell-free systems that recapitulate the structural events in mitosis as well as meiosis. In this article, the protocol used in the author's laboratory to prepare functional extracts from synchronized cultured mammalian cells is described. These extracts offer several advantages to extracts prepared from the eggs of marine organisms in that they are easy to prepare, exhibit highly reproducible behavior from day to day, are extremely easy to manipulate, and represent a bonafide mitotic system. Thus, this cell-free system based on synchronized cultured cells offers a valuable alternative to extracts prepared from gametic cells for the analysis of the structural events of mitosis.

Procedure

The preparation of mitotic extracts from cultured cells is very easy and should be feasible in any laboratory capable of cell culture and equipped for basic cell and molecular biology. Two different protocols are described. The first protocol maintains the extract in a mitotic state and has been tailored to monitor the assembly of microtubule asters[16,17] as well as the process of nuclear disassembly.[10,11] The second protocol permits the extract to exit mitosis and has been tailored to monitor both the spontaneous

---

[8] R. E. Palazzo, E. Vaisberg, R. W. Cole, and C. L. Rieder, *Science* **256,** 219 (1992).
[9] B. Burke and L. Gerace, *Cell* **44,** 639 (1986).
[10] F. A. Suprynowicz and L. Gerace, *J. Cell Biol.* **103,** 2073 (1986).
[11] J. Nakagawa, G. T. Kitten, and E. A. Nigg, *J. Cell Sci.* **94,** 449 (1989).
[12] J. W. Newport, K. L. Wilson, and W. G. Dunphy, *J. Cell Biol.* **111,** 2247 (1990).
[13] M. J. Lohka and J. L. Maller, *J. Cell Biol.* **101,** 518 (1985).
[14] A. W. Murray, *Methods Cell Biol.* **36,** 581 (1991).
[15] C. L. Rieder, J. G. Ault, U. Eichenlaub-Ritter, and G. Suder, *in* "Chromosome Segregation and Aneuploidy" (B. K. Vig and A. Kappas, eds.), p. 183. Springer-Verlag, New York, 1993.
[16] T. Gaglio, A. Saredi, and D. A. Compton, *J. Cell Biol.* **131,** 693 (1995).
[17] T. Gaglio, A. Saredi, J. B. Bingham, M. J. Hasbani, S. R. Gill, T. A. Schroer, and D. A. Compton, *J. Cell Biol.* **135,** 399 (1996).

assembly of the NuMA protein[18] and the reassembly of the cell nucleus[9,11] following the completion of mitosis. These two protocols follow a similar basic theme, and in principle could be modified to monitor the assembly/ disassembly of a variety of cellular structures during mitosis.

Reagents

    200 mM Thymidine (sterile filtered)
    0.1 μg/ml Nocodazole in dimethyl sulfoxide (DMSO)
    20 mM Cytochalasin B in 95% ethanol
    Phosphate-buffered saline (PBS)
    KHM: 78 mM KCl, 50 mM HEPES, pH 7.0, 4 mM MgCl$_2$, 2 mM ethylene glycol-bis(β-aminoethyl ether)tetraacetic acid (EGTA), 1 mM dithiothreitol (DTT)
    KPM($-$): 50 mM KCl, 50 mM PIPES, pH 7.0, 10 mM EGTA, 2 mM MgCl$_2$
    KPM($+$): 50 mM KCl, 50 mM PIPES, pH 7.0, 10 mM EGTA, 2 mM MgCl$_2$, 1 mM DTT
    3 mM ATP in either KHM or KPM($+$)
    Protease inhibitors: 1 mg/ml pepstatin in methanol, 10 mg/ml chymostatin in DMSO, 10 mg/ml antipain, 10 mg/ml leupeptin, and 10 mg/ml phenylmethylsulfonyl fluoride (PMSF)
    5 mM Taxol in DMSO

Culture and Harvest of HeLa Cells

One of the single most important determinants for the production of a high-quality mitotic extract from cultured cells is the mitotic index of the harvested cells. We have found that a mitotic index of >80% following shake-off is essential for the assembly of mitotic asters or the spontaneous assembly of NuMA. The protocol outlined in this article typically yields mitotic indices of ~95%.

Obtaining sufficient quantities of cells with the necessary mitotic index is not difficult if one is careful with several key variables during cell culture. First, it is essential that the cells be maintained in log-phase growth at all times including prior to seeding the large flasks for the experiment. Cells that are not in log-phase growth have variable growth rates, and do not synchronize in the cell cycle efficiently. Second, it is critical that the cells be at ~90% confluence at the time of mitotic shake-off. Undergrowth of the cells results in low cell yields; overgrowth of the cells leads to poor attachment of nonmitotic cells, which become easily dislodged during the

---

[18] A. Saredi, L. Howard, and D. A. Compton, *J. Cell Sci.* **110,** 1287 (1997).

shake-off step reducing the mitotic index of the harvested cells. In practice, we obtain ~90% confluence at the time of shake-off if the thymidine treatments are initiated when the cells are ~40% confluent in the culture flasks. Initiating the thymidine treatments at this density provides ample surface area to prevent overcrowding following two cell division events that will occur during the cell synchrony.

HeLa cells are cultured in Dulbecco's modified Eagle's medium (DMEM) containing 10% fetal bovine serum at 37° in a humidified incubator with a 5% CO_2 atmosphere. In a typical experiment, two 100-mm culture dishes at ~80–90% confluence are split equally into 12 T-150 culture flasks. Two to three days later thymidine is added to a final concentration of 2 mM and the cells incubated for 24 hr. The medium containing the thymidine is then aspirated off, the cells washed once with complete medium, fresh medium is applied, and the cells incubated for 8 hr. Thymidine is again added to a final concentration of 2 mM and the cells incubated for 12–16 hr. The medium containing the thymidine is then aspirated off, the cells washed once with complete medium, and fresh medium is applied. These cells are incubated for 5 hr and nocodazole is added to a final concentration of 40 ng/ml. Mitotic cells accumulate 2.5–5 hr after the addition of nocodazole and are dislodged from the culture flask by vigorously rapping the culture flask against a laboratory bench two or three times (shake-off). The efficiency of the shake-off is easily monitored by phase contrast microscopy and the mitotic cells are harvested by collection of the culture medium. Under these conditions, one can expect to harvest between 10^7 and 10^8 cells.

Preparation of Cell Extracts

Method 1: M-Phase Extract. The culture medium containing the mitotic cells harvested by mitotic shake-off is supplemented with cytochalasin B at a final concentration of 20 μg/ml and incubated for 30 min at 37°. This step is important to ensure efficient cell rupture during the subsequent homomgenization step. Following incubation, the cells are collected by centrifigation at 1500 rpm at room temperature in a tabletop clinical centrifuge, the medium is discarded, the cell pellets placed on ice, and all subsequent steps of the procedure performed at 4°. The cell pellets are gently resuspended in 20 ml of cold PBS containing 20 μg/ml cytochalasin B. The cells are fragile at this point due to the disruption of both microtubules and actin filaments so it is important to resuspend these cell pellets gently to avoid prematurely rupturing the cells. After carefully resuspending the cells, they are collected by centrifugation for 5 min at 1500 rpm using an SS34 rotor in a Sorvall (Newtown, CN) high-speed centrifuge at 4°. This wash step with PBS containing cytochalasin B is repreated again, and a final wash step is performed with cold KHM buffer containing 20 μg/ml cytochalasin B, 20 μg/ml PMSF, and 1 μg/ml each of chymostatin, leupeptin, antipain, and pepstatin. The final cell pellet is resuspended at a concentration of 3×10^7 cell/ml in the same buffer and subjected to homogenization using a Dounce homogenizer (tight pestle). The efficiency of cell rupture is monitored by phase contrast microscopy and typically requires between three and six sets of 20 pestle strokes each. Separating the homogenization strokes into sets is important to prevent frictional heating of the extract that can occur if the homogenization strokes are all performed in one series. The crude cell extract is then subjected to centrifugation at 100,000g for 20–30 min at 4°. We currently use a Beckman (Fullerton, CA) TLX tabletop ultracentrifuge for this purpose although we have produced indistinguishable results using a Beckman airfuge assembled in a cold room (run at 25 psi). Following centrifugation, the supernatant is carefully collected and supplemented with adenosine triphosphate (ATP) to a final concentration of 2.5 mM (prepared as Mg^{2+} salts in KHM buffer). For ease of use we add bulk ATP prepared in KHM buffer, but addition of an ATP regenerating system (e.g., phosphoenol pyruvate and pyruvate kinase) is equally effective. The extract prepared in this manner has a total protein concentration of ~5 mg/ml, and can be considered a bonafide "mitotic extract" based on two criteria. First, the mpm-2 antibody, which recognizes mitosis-specific phosphoepitopes, is approximately 100× more reactive with proteins in this extract compared with proteins prepared from nonmitotic cells. Second, the activity of the mitosis-specific protein kinase p34cdc2/cyclin B is 7–10× higher (using histone H1 as a substrate) in this extract compared to an extract prepared from nonmitotic cells.

We have used these mitotic extracts to analyze the requirements for the formation of mitotic asters, which represent a subassembly of the mitotic spindle.[16,17] In this case, the mitotic extract is supplemented with Taxol at a final concentration of 10 μM and incubated at 33° for 30–60 min. Following incubation, the assembly products are analyzed in two ways. First, the morphology of the mitotic asters is determined by indirect immunofluorescence microscopy using antibodies directed against tubulin and other mitotic spindle components (i.e., NuMA). For this analysis 5 μl of the extract is diluted into 25 μl of KHM buffer at room temperature and spotted onto a poly-L-lysine coated glass coverslip. The structures formed in the extract are permitted to adhere to the coverslip for 1–2 min, and then fixed by immersion of the coverslip in −20° methanol and processed for indirect immunofluorescence microscopy. Second, the efficiency with which various proteins associate with the mitotic asters can be determined by separating the extract into soluble and insoluble (mitotic aster-associated) fractions.

This is easily done by subjecting the extract to centrifugation at 10,000g for 10 min at 4° in a microcentrifuge. Proteins in the supernatant and pellet fractions are then solubilized in equal volumes of SDS–PAGE sample buffer and the fate of any protein determined by immunoblot analysis.

Method 2: I-Phase Extract. The mitotic extract prepared using method 1 will exit mitosis and enter a nonmitotic or interphase-like state when protein dephosphorylating conditions are permitted to prevail. Technically, this is easily accomplished in the extract by either not adding exogenous sources of ATP or by adding protein kinase inhibitors such as Staurosporine to the extract.

For the preparation of I-phase extracts the mitotic cells harvested by mitotic shak-off are collected by contrifugation at 1500 rpm at room temperature using a tabletop clinical centrifuge. The medium is then discarded, the cell pellets placed on ice, and all the subsequent steps performed at 4°. The cell pellet is gently resuspended in 20 ml of cold KPM(−) buffer, and subsequently collected by centrifugation for 5 min at 1500 rpm using an SS34 rotor in a Sorvall high-speed centrifuge at 4°. The wash with KPM(−) is repeated and followed by a final wash step in KPM(+) containing 20 μg/ml cytochalasin B, 50 μg/ml PMSF, and 5 μg/ml each of chymostatin, leupeptin, antipain, and pepstatin. The higher concentration of the protease inhibitors used in this method is essential to prohibit proteolysis because this method produces an extract with a higher protein concentration. The final cell pellet is resuspended at a concentration of 5×10^7 cell/ml in KPM(+) containing cytochalasin B and protease inhibitors and the cells homogenized using a Dounce homogenizer (tight pestle). As with method 1, the efficiency of homogenization is monitored by phase contrast microscopy, and the homogenization strokes are separated into sets of 20 to prevent frictional heating of the extract. The crude cell extract is then subjected to centrifugation at 100,000g for 20–30 minutes at 4° using either the Beckman TLX tabletop ultracentrifuge or the Beckman airfuge. Following centrifugation, the supernatant is carefully collected and supplemented with 1 μg/ml nocodazole to prevent microtubule polymerization. The extract prepared in this manner is ~8–10 mg/ml, and on incubation at 33° exits mitosis as judged by the approximately 100× reduction in the reactivity of the mpm-2 antibody for mitosis-specific phosphoepitopes and the 7–10× reduction in the activity of the mitosis-specific protein kinase p34cdc2/cyclin B toward histone H1.

Similar to method 1, we have used the extract prepared under these conditions to monitor the spontaneous assembly of the NuMA protein under nonmitotic conditions.[18] In this case, the assembly products are analyzed in three ways. Both the morphology of the structures that form and the efficiency with which proteins are converted from the soluble fraction to the insoluble fraction are analyzed exactly as described in method 1 by indirect immunofluorescence microscopy and immunoblotting, respectively. The structures that form under these conditions have also been analyzed by transmission electron microscopy. For this analysis, the insoluble pellet fraction was fixed with 2% glutaraldehyde in 0.1 M sodium cacodylate buffer, pH 7.4, for 12 hr at 4° either with or without prior extraction with 0.5% Triton X-100. Following fixation, the pellet was rinsed in 0.1 M sodium cacodylate buffer, postfixed with 1% OsO_4 in 0.1 M sodium cacodylate buffer for 30 min at room temperature, and *en bloc* stained in 2% aqueous uranyl acetate. This sample was then dehydrated through a graded series of ethanols and propylene oxide, and embedded in epon (LX112). Sections (60–70 nm) were prepared, stained with 2% uranyl acetate in methanol for 20 min followed by 5 min in Reynold's lead citrate, and examined at 80 or 100 kV on a JEOL 100CX.

Extract Manipulation

One distinct advantage to working with mitotic extracts prepared using these methods is that they are relatively stable and are easy to manipulate. The freshly prepared extracts retain their full biological activity if maintained on ice for up to 4 hr (the longest time point examined in my laboratory). This feature makes the extract particularly amenable to immunodepletion and/or performing a number of sequential experiments on the same day. At this point in time, we have not found conditions under which the extracts retain biological activity following freezing, which prevents one from storing the extract for extended periods. Thus, the extracts must be prepared fresh for each experiment which, while inconvenient at times, has not proved to be a substantial obstacle because the extracts are simple to prepare and highly reproducible from day to day.

Depletion of specific proteins from these extracts prior to incubation is relatively easy provided one has a suitable antibody that is both specific for the target protein and capable of binding to the protein under native conditions. Immunodepletions from the extract are performed by adsorbing 5–100 μg of antibody onto ~25 μl of either protein A-conjugated agarose or protein G-conjugated agarose (Boehringer Mannheim, Indianapolis, IN). Both monoclonal (IgG and IgM) and polyclonal antibodies have been used successfully in the author's laboratory for the depletion of proteins from the extract, but the ability of a monoclonal antibody to bind to a protein in its native conformation in a crude cell extract will be dependent on the specific epitope recognized by that antibody. IgMs may be coupled to protein A-conjugated agarose by using goat anti-murine IgM specific antibody (Vector Labs, Burlingame, CA). The antibody is mixed with the agarose matrix and incubated at 4° on a slowly rotating wheel for 2–6 hr. Following adsorption, the antibody-coupled agarose is washed in either KHM or KPM(+) buffer depending on the extract being used and then packed by centrifugation to remove the excess fluid. Efficient depletion of the target protein is routinely achieved using sequential depletion reactions, although the total quantity of packed agarose should not exceed 40 μl per 100 μl of extract. First, half of the antibody-coupled agarose is resuspended with the mitotic extract and incubated with gentle agitation on a rotating wheel for 1 hr at 4°. Following this incubation the agarose is removed from the extract by sedimentation at 15,000g for 10 sec and saved. The extract is recovered and used to resuspend the other half of the antibody-coupled agarose and another incubation peformed with gentle agitation on a rotating wheel for 1 hr at 4°. Following this incubation the agarose is removed by sedimentation at 15,000g for 10 sec and pooled with the agarose pellet from the initial depletion reaction. The extract is recovered and the assembly process of interest is induced as described previously. This protocol has been successfully used to deplete six different proteins from the extract both alone and in various combinations and in each case the depleted extract has retained full biological activity.

Discussion

In summary, the protocols contained in this chapter describe how to use synchronized cultured cells as a source of material for the production of extracts to examine mitosis-specific structural changes in cells. These extracts are both easy to prepare and easy to manipulate, which makes them an excellent system for the molecular analysis of the structural events of mitosis. The structural assemblies that we have examined are highly complex, indicating that the extracts prepared using the protocols described here are capable of performing complex multistep assembly reactions. Indeed, during our studies we have found that these extracts maintain the fine balances of both protein kinase/phosphatase reactions as well as complex microtubule-dependent motor functions. In using these cell extracts we have focused our attention on the mitosis-specific alterations of the soluble proteins (100,000g soluble).[16–18] It is important to point out, however, that other investigators have used similar extracts to examine the structural changes associated with more complex substrates (i.e., nuclei) in the context of these extracts.[9–11] Thus, the extracts described in this chapter offer an alternative to amphibian eggs for the preparation of cell-free extracts for the analysis of structural events during mitosis. These extracts represent a bonafide mitotic system that complements the meiosis-specific systems derived from amphibian eggs and should be amenable to the analysis of many different cell structures depending on the investigators interests.

Acknowledgments

I wish to thank Tirso Gaglio, Alejandro Saredi, Mary A. Dionne, Vicki Mountain, and Louisa Howard who have been indispensable in the development of the cell-free systems described in this article. The work performed in the authors laboratory has been funded by grants from the American Cancer Society (RPG-95-010-03-CB, JFRA-635) and National Institutes of Health (GM-51542).

[28] Organelle Motility and Membrane Network Formation in Metaphase and Interphase Cell-Free Extracts

By Viki J. Allan

Introduction

One of the main roles for microtubule motors in animal cells during interphase is to move membranes. Not only do microtubule motors transport vesicles between different membranous organelles in the exo- and endocytic pathways, but they also move whole organelles, such as the endoplasmic reticulum (ER) and Golgi apparatus.[1,2] Microtubule motors are therefore intimately involved in the whole process of membrane traffic during interphase. As the cell enters mitosis, many organelles change their structure and/or position, and membrane traffic in both the exo- and endocytic pathways is inhibited. Concomitantly, the amount of microtubule-based organelle motility changes, leading to a substantial inhibition during metaphase in many cell types and organisms.[3]

While major steps have been made toward understanding the control of the cell cycle itself, many of the downstream regulatory mechanisms, including the regulation of organelle movement, are not fully understood. The ability to recapitulate cell cycle-controlled events *in vitro* has proved invaluable in the study of such complex processes.

[1] V. Allan, *FEBS Lett.* **369,** 101 (1995).
[2] V. Allan, *Sem. Cell Dev. Biol.* **7,** 335 (1996).
[3] A. Robertson and V. Allan, *in* "Progress in Cell Cycle Research" (L. Meijer, S. Guidet, and M. Philippe, eds.), Vol. 3, p. 59. Plenum Press, New York, 1997.

Xenopus laevis egg extracts have been the cell-free extract of choice in many studies for a number of reasons. Of particular importance is the fact that laid eggs are naturally arrested in metaphase II of meiosis, and many eggs are laid over a short time. It is fairly simple, therefore, to make several milliliters of concentrated metaphase extract from one batch of eggs. Because organelle movement can be reconstituted from frozen extracts, this means that the same extract can be used for multiple experiments. The other major advantage is that the cell cycle status can be manipulated *in vitro*, allowing a single batch of extract to be used to study both interphase and metaphase phenomena. This system has proved particularly useful for investigating the regulation of organelle movement and ER network formation *in vitro*, since the interphase and metaphase samples differ only in their posttranslational state.[4,5] Another considerable advantage of these extracts is that they allow the correlation of motility with features such as motor phosphorylation state under virtually identical assay conditions.[4,5]

In this article, a method is described for preparing extracts that can be converted from metaphase to interphase *in vitro*. This method is simple in that it does not require any specialized equipment for electrically activating eggs, nor reagents such as nondegradable cyclins. Alternative methods are given elsewhere.[6–8] Other necessary techniques are also included, such as the preparation of organelle and cytosol fractions, and their use in motility assays; the purification of microtubule motors from both interphase and metaphase extracts; and a method for monitoring the cell cycle status of extracts. In addition, because it is important to be able to identify the membranes and membrane networks that are being observed in the motility assays, a protocol for immunofluorescence is included.

Preparation of Metaphase- and Interphase-Arrested *Xenopus* Egg Extracts

Principle

The natural metaphase arrest of laid *Xenopus* eggs is generated by the presence of cytostatic factor (CSF), which maintains high levels of $p34^{cdc2}$ mitotic kinase activity.[9] When the egg is fertilized, an influx of calcium causes the inactivation of both CSF and $p34^{cdc2}$, releasing the egg from

[4] V. J. Allan and R. D. Vale, *J. Cell Biol.* **113**, 347 (1991).
[5] J. Niclas, V. J. Allan, and R. D. Vale, *J. Cell Biol.* **133**, 585 (1996).
[6] A. Murray, *Methods Cell Biol.* **36**, 581 (1991).
[7] V. Allan, *Methods Cell Biol.* **39**, 203 (1993).
[8] D. D. Newmeyer and K. L. Wilson, *Methods Cell Biol.* **36**, 607 (1991).
[9] N. Sagata, N. Watanabe, G. F. Van de Woude, and Y. Ikawa, *Nature* **342**, 512 (1989).

metaphase arrest and allowing progression into interphase. Care must therefore be taken to prevent any rise in calcium concentration when making an extract from unfertilized eggs (a CSF extract), or else the metaphase arrest may be lost. On the other hand, the addition of calcium to a CSF extract provides a convenient means for preparing an interphase extract, as described later. The methods for extract preparation are based on those developed by Murray et al.[10] from an original procedure described by Lohka and Masui.[11]

Procedure

Materials for Extract Preparation

PMSG (Pregnant mare serum gonadotropin): 200 U/ml in sterile water

HCG (Human chorionic gonadotropin): 1500 U/ml in sterile water

MMR: 100 mM NaCl, 2 mM KCl, 1 mM MgSO$_4$, 2 mM CaCl$_2$, 5 mM HEPES, 0.1 mM ethylenediaminetetraacetic acid (EDTA), pH 7.8, with NaOH

Cysteine: 2% (w/v) cysteine free base in H$_2$O; made up *just before use* and adjusted to pH 7.8 with NaOH

XB: 100 mM KCl, 0.1 mM CaCl$_2$, 1 mM MgCl$_2$, 10 mM HEPES, 50 mM sucrose, pH to 7.7 with KOH

XB/EGTA: 100 mM KCl, 0.1 mM CaCl$_2$, 2 mM MgCl$_2$, 5 mM ethylene glycol-bis(β-aminoethyl ether)tetraacetic acid (EGTA), 10 mM HEPES, 50 mM sucrose, pH to 7.7 with KOH (prepared by adding an extra 1 mM MgCl$_2$ and 5 mM EGTA to XB)

Energy mix (20×): 150 mM creatine phosphate, 20 mM adenosine triphosphate (ATP), 2 mM EGTA, 20 mM MgCl$_2$, stored at $-20°$

Protease inhibitor stock: 10 mg/ml leupeptin, peptstatin, and chymostatin in dimethyl sulfoxide (DMSO), stored at $-20°$

Cytochalasin D stock: 1 mg/ml in DMSO, stored at $-20°$

Versilube F-50 silicone oil (from Andpak-EMA, San Jose, CA, or Atochem Chimie UK Ltd., Newbury, UK)

Egg Production. For detailed information on frog care and handling, plus a list of American suppliers, see *Methods in Cell Biology*, Volume 36. We purchase *Xenopus* from Blades Biological Co. (Edenbridge, UK). To induce mature female frogs to lay eggs, the frogs are first injected with 100 U PMSG into the dorsal lymph sac 3–10 days before you require the eggs. The frogs are then kept without feeding until ovulation is stimulated by injecting 750 U of HCG into the dorsal lymph sac, about 16 hr before

---

[10] A. Murray, M. Solomon, and M. W. Kirschner, *Nature* **339**, 280 (1989).

[11] M. J. Lohka and Y. Masui, *J. Cell. Biol.* **98**, 1222 (1984).

the eggs are needed. The frogs are then placed in 100 mM NaCl in separate containers and kept at 20° in a quiet place. The frogs will start laying eggs after approximately 12 hr.

Preparation of CSF Extracts. If more than one frog has been used, it is best to keep the batches of eggs separate throughout the whole procedure to avoid one bad batch spoiling a large quantity of pooled extract. All procedures are performed at room temperature unless noted. Collect the laid eggs and wash gently in MMR several times in a 10-cm glass petri dish. Use a cut-off, flamed Pasteur pipette (with an opening of approximately 3 mm) to remove any strings of eggs, frog skin, excrement, and bad eggs. The jelly coat surrounding the eggs must be removed before the extract can be made. This is achieved by swirling the eggs gently in several changes of 2% cysteine solution until the eggs pack tightly at the bottom of the dish (this takes about 5 min). The eggs are very fragile without their jelly coat, and it is important to work quickly but gently from now on. Rough treatment will result in eggs bursting or in the eggs entering interphase due to calcium entry (activation).

Wash the eggs several times in MMR and remove necrotic eggs with the Pasteur pipette. Transfer into XB in a clean glass petri dish and swirl the eggs gently. Wash three more times with XB, and twice with XB/ EGTA containing 1 μg/ml protease inhibitors. The EGTA prevents a rise in calcium concentration when the eggs are crushed, and so maintains the extract in a metaphase state with high CSF and p34^{cdc2} mitotic kinase activity.

The eggs are now ready to be transferred into Beckman (Fullerton, CA) SW50 Ultraclear tubes containing 1 ml XB/EGTA + 10 μg/ml protease inhibitors + 10 μg/ml cytochalasin D. Allow the eggs to sink to the bottom of the Pasteur pipette before releasing them under the surface of the buffer in the centrifuge tube to minimize damage to the eggs and to avoid diluting the cytochalasin D. Allow the eggs to settle in the centrifuge tube and then remove as much buffer as possible. Add 1 ml of Versilube F-50 oil on top of the eggs. The eggs are now ready for a "packing" spin, during which the eggs compact (without rupturing) at the bottom of the tube while excess buffer is displaced by the Versilube F-50. Spin in a clinical centrifuge at 150g for 60 sec and then increase to 600g for a further 30 sec. The eggs should still be intact, but rather squashed. Remove the excess buffer and as much of the oil as practicable. As a result of this packing spin, the final extracts will be as concentrated as possible (~75–90 mg/ml protein).

A higher speed spin is then used to crush the eggs and at the same time stratify the extract into a yolk/pigment granule pellet, a cytoplasmic extract fraction, and a lipid layer. Place the SW-50 tubes inside Falcon 10-ml white plastic culture tubes (which act as adaptors), and then spin at 10,000 rpm in a Sorval (DuPont, Stevenage, UK) HB-4 or HB-6 rotor for 10 min at 16° with Sorval rubber adaptors. Collect the cytoplasm from the tube by puncturing just above the pigment/yolk layer with an 18-gauge needle and syringe. Add an additional 10 μg/ml protease inhibitors, 1 μg/ml cytochalasin D, and 1/20 volume of energy mix to the cytoplasmic extract and place on ice. If the extract is to be frozen, add sucrose to 150 mM (from a 2.0 M stock) and snap freeze in appropriate aliquots. We routinely freeze 50-μl aliquots and store them in liquid nitrogen.

To convert a CSF extract to an interphase extract, add 1/50 volume of activation mix (10 mM CaCl$_2$, 100 mM KCl, 1 mM MgCl$_2$) and cycloheximide to 100 μg/ml final (to prevent cyclin synthesis), and incubate at room temperature for 45–60 min. Then, if desired, an extra 0.4 mM EGTA can be added. As a control, CSF extract is incubated for the same amount of time after the addition of 1/50 volume of mock activation mix (100 mM KCl, 1 mM MgCl$_2$). Extracts are then placed on ice.

It is important to note that CSF extracts vary in their ability to maintain a metaphase arrest when thawed and incubated further at room temperature. Each extract should therefore be tested for p34^{cdc2} mitotic kinase activity as described later. In addition, such testing checks that enough calcium has been added to release the CSF arrest (Fig. 1). If the loss of p34^{cdc2} mitotic kinase activity is slow or incomplete, a higher concentration of calcium should be used for activation. Any extract that loses its metaphase arrest can still be used for preparing metaphase microtubule motors (see below), because this procedure uses phosphatase inhibitors to ensure that phosphorylation states are maintained during incubations.

Preparation of Cytosol and Organelle Fractions

Cytosol and organelle fractions are prepared in the same way from both interphase and CSF extracts. Dilute the extract with two volumes of acetate buffer (100 mM potassium acetate, 3 mM magnesium acetate, 5 mM EGTA, 10 mM HEPES, 150 mM sucrose, adjusted to pH 7.4 with KOH; both the sucrose and HEPES should be of the purest grade possible) containing energy mix. We typically combine 50 μl extract, 95 μl acetate buffer, and 5 μl 20× energy mix in a polycarbonate 200-μl centrifuge tube (Beckman) and then centrifuge at 55,000 rpm (120,000g$_{av}$) for 30 min at 4° in a TLA 100 rotor in a Beckman TL100 tabletop ultracentrifuge. Collect 120 μl cytosol and then discard any remaining cytosol without disturbing the fawn-colored organelle layer resting on top of the amber pellet (which consists of glycogen and ribosomes). The cytosol fraction usually contains 15–25 mg/ml protein. Resuspend the organelle layer carefully in 17 μl acetate buffer (giving a final volume of ~20 μl), avoiding the amber pellet. The

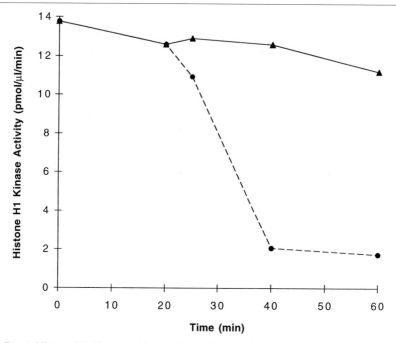

FIG. 1. Histone H1 kinase activity can be used to test the cell cycle status of egg extracts. This graph shows the H1 kinase activity of a CSF extract incubated for 20 min at room temperature, then divided into equal aliquots, one of which is treated with mock activation mix (▲—▲), the other with activation mix containing calcium (●- -●). The histone kinase activity drops rapidly following the addition of 0.2 mM calcium, indicating that the extract has become interphasic.

membranes may also be pelleted through a cushion containing 300 mM sucrose in acetate buffer.[4]

To reduce the cytosolic contamination of the organelle fraction, flotation gradients have been used successfully.[5,12] The crude organelle layer (prepared as above, but to a 15-μl final volume) is mixed thoroughly with 135 μl of acetate buffer containing a total of 2.0 M sucrose plus 1 mM Mg-ATP and 1 μg/ml protease inhibitors. Phosphatase inhibitors can also be included (such as 1 μM microcystin), although nonspecific inhibitors such as β-glycerophosphate may well strip some peripheral membrane proteins and motors from the organelles. The resuspended organelle fraction is transferred to ultracelear centrifuge tubes (5× 41 mm) and overlaid first with 200 μl acetate buffer containing 1.5 M sucrose (plus ATP, protease inhibitors, and phosphatase inhibitors, as required) and then with 300 μl

[12] V. Allan, J. Cell Biol. 128, 879 (1995).

acetate buffer (which contains 150 mM sucrose). The gradient is then centrifuged at 39,000 rpm (150,000g_{av}) for 3 hr at 4° in an SW50 rotor (with high-speed adaptors), after which the membranes are collected in minimal volume from the 1.5 M/150 mM sucrose interface.

Both cytosol and organelle fractions can be kept on ice for several hours, and can be frozen successfully for less demanding uses. For comparing metaphase and interphase characteristics, however, we always use freshly prepared fractions.

Isolation of Soluble Microtubule Motors from Metaphase and Interphase Cytosols

Principle

One of the biggest advantages of using *Xenopus* extracts to study the regulation of motor function is that the phosphorylation state and the relative activity of the motors can be determined under the same conditions. Reasonable quantites of motors such as cytoplasmic dynein can be prepared from interphase and metaphase extracts.[5] In addition, the protocol can be scaled so that motors can be isolated from as little as 100 μl of metabolically labeled extract.[5]

Procedure

Materials

Buffer A: 120 mM sodium β-glycerophosphate, 15 mM sodium pyrophosphate, 15 mM HEPES, 15 mM EGTA, 7.5 mM MgCl$_2$, pH 7.4 (with KOH); add 1 mM dithiothreitol (DTT), 10 μg/ml protease inhibitors, 10 μg/ml cytochalasin D, and 1.5 μM microcystin just before use

Taxol-stabilized microtubules prepared from purified brain tubulin

Hexokinase: 500 U/ml

AMP.PNP: 100 mM

Glucose: 1 M

Taxol: 2 mM stock in DMSO

Buffer B: 80 mM sodium β-glycerophosphate, 10 mM sodium pyrophosphate, 10 mM HEPES, 10 mM EGTA, 5 mM MgCl$_2$, pH 7.4 (with KOH); add 1 mM DTT, 20 μM Taxol, 1 μg/ml protease inhibitors, 10 μg/ml cytochalasin D, and 1 μM microcystin just before use

Buffer C: As for buffer B with only 4 μM Taxol and 1 μg/ml cytochalasin D

Sucrose solutions: 20% (w/v) sucrose in buffer C. For the final gradient step you will need 10 and 25% sucrose (w/v) in 25 mM PIPES, 1 mM MgCl$_2$, 1 mM EGTA, pH 6.8 (with KOH); add 1 mM DTT, 1 μg/ml protease inhibitors, and 100 nM microcystin just before use

Thaw CSF extracts and convert to interphase if required. Dilute extract with two volumes of buffer A. Stand on ice for 10 min (to depolymerize any microtubules formed during the activation step), then prepare cytosol fractions by centrifugation in an appropriate rotor (depending on volume of extract) at 120,000g_{av} for 30 min at 4°.

Collect cytosol and add hexokinase (10 U/ml final), AMP.PNP (0.4 mM final), and glucose (20 mM final). Warm quickly to room temperature, then add Taxol (20 μM final) and Taxol-stabilized microtubules (150 μg/ml final) and mix by pipetting through a cut-off tip. Incubate for 20 min at room temperature to allow microtubule polymerization, ATP depletion, and binding of motors to microtubules. Recover the microtubules with bound motors by centrifugation through a cushion of 20% sucrose in buffer C in a Beckman TLS55 rotor at 32,000 rpm (70,000g_{av}) for 10 min at 22°. Remove supernatant (keep for analysis by SDS–PAGE if desired), discard cushion, then resuspend the pellet in buffer B at room temperature to wash the microtubules. The volume used to resuspend the pellet should equal the starting volume of crude extract. Centrifuge as above, and resuspend the pellet in 0.5 starting extract volume of buffer B containing 100 mM KCl and 5 mM MgATP. Incubate for 25 min at room temperature to release the motors, then centrifuge as above, but without the sucrose cushion. For small volumes, adaptors are available that allow 200 μl thick-wall polycarbonate tubes to be used in the TLS55 rotor. Collect the motor fraction and load onto 10–25% sucrose gradients and centrifuge at 39,000 rpm (150,000g_{av}) in the SW50 for 5 hr at 4°. Alternatively, the TLS55 may be used.[5] For small-scale preparations we use 4- $\times$ 51-mm ultraclear tubes in the SW50 with adaptors. In that case, the gradients are prepared by diffusion of a step gradient consisting of 25, 22.5, 20, 17.5, 15, 12.5, and 10% sucrose (60 min at room temperature). Collect fractions from the gradients: cytoplasmic dynein should be in the bottom third, with kinesin and related proteins in the upper half.

Monitoring Cell Cycle Status of Extracts

Principle

The cell cycle status of extracts can be conveniently assessed by monitoring the activity of histone H1 kinase, since it has been shown that p34[cdc2]

is the kinase responsible for phosphorylating histone H1.[13] To test the stability of the CSF arrest, samples are taken every 10–15 min over a 1-hr incubation at room temperature. The success of the calcium-induced entry into interphase is monitored similarly (see Fig. 1). This method is also suitable for testing p34^{cdc2} kinase activity in cytosol and organelle fractions and combinations thereof. It should also be used to check whether any treatment (the addition of a kinase or phosphatase inhibitor, for instance) that affects membrane movement is actually affecting the cell cycle status of the extract. The H1 kinase assay is adapted from that described by Félix and co-workers.[14]

Procedure

Materials

EB: 80 mM sodium β-glycerophosphate, 15 mM MgCl$_2$, 20 mM EGTA, pH 7.3 (with NaOH); just before use add 1 mM DTT and 1 μg/ml protease inhibitors

Histone H1: 20 mg/ml (Sigma) in water

ATP: vanadate-free, sodium salt (Sigma, St. Louis, MO); stock of 100 mM is made to 100 mM MgCl$_2$ and titrated to approximately pH 7.0 with NaOH

[γ-^{32}P]ATP: 4,500 Ci/ml

Histone kinase cocktail: for 250 μl (enough for about 20 test samples, with controls) combine 25 μi histone H1, 4.2 μl Mg-ATP, 2.5 μl [γ-^{32}P]ATP, and 218.3 μl EB

Phosphoric acid: 150 mM

Industrial methylated spirits

Method. Because this protocol uses ^{32}P-labeled material, appropriate care should be taken at all stages, and local radiation safety regulations should be followed. Two microliters of concentrated extract or 6 μl of cytosol are mixed well with EB to give a total volume of 40 μl and then immediately frozen in liquid nitrogen and stored at $-80°$. After defrosting, duplicate 5-μl aliquots are placed in 500-μl microfuge tubes, with 5 μl EB as a blank. Using a positive displacement repeating pipette, add 5 μl of histone kinase cocktail to each sample and incubate for 15 min at room temperature. Place samples on ice, then spot 6 μl from each reaction onto small squares of phosphocellulose P81 cation exchange paper at room temperature (Whatman, Maidstone, UK). Allow to air dry, then wash the

[13] T. Langan, J. Gautier, M. Lohka, R. Hollingsworth, P. Moreno, P. Nurse, J. Maller, and R. Sclafini, *Mol. Cell Biol.* **9,** 3860 (1989).

[14] M.-A. Félix, P. Clarke, J. Coleman, F. Verde, and E. Karsenti, *in* "The Cell Cycle: a Practical Approach" (P. Fantes and R. Brooks, eds.), p. 253. IRL Press, Oxford, 1993.

filter papers three times in 500 ml of 150 mM phosphoric acid, with agitation, then twice in large volumes of industrial methylated spirits. Filters are then air dried and placed in plastic scintillation vials containing 3 ml scintillation cocktail and counted on the ^{32}P channel of a scintillation counter,

Calculate the average cpm for each extract, subtracting background (obtained from the EB blanks). Determine the specific activity of the [^{32}P]ATP by counting 5 μl of the histone kinase cocktail spotted directly onto a P81 square (without washing). The histone kinase activity is then calculated as pmol phosphate incorporated/μl extract/min. Remember that the cytosol fractions have been diluted with two volumes of acetate buffer, and so 3 μl cytosol corresponds to 1 μl of extract for these calculations.

As an alternative to using phosphocellulose paper, the histone kinase activity can be assessed following SDS–PAGE using a slightly different method described in detail by Murray.[6]

Motility Assays

Microscopy

Simple, disposable microscope flow cells are made from 18-mm^2 No. 1 or No. 1.5 coverslips and microscope slides.[15] Two thin strips of vacuum grease (Apiezon M or silicone vacuum grease), 1 cm apart, are extruded onto the slide through a cut-off 18-gauge needle to form the sides of the chamber. Sellotape (Scotch tape) or fragments of coverslip are used as spacers. Often, the biggest problem with getting reproducible motility *in vitro* is finding a good source of coverslips. Coverslips that look clean by eye are often covered with pits, lumps, strange whorl patterns, and glass fragments when observed by video-enhanced differential interference contrast (VE-DIC), while apparently dirty coverslips are often fine for VE-DIC. Added to this, quality can vary within a batch, and the coverslips get worse with time. We try to avoid using a box of coverslips that has been open for more than 1 month. Storage in ethanol may help. An additional problem is that the amount of microtubule gliding across the glass surface may vary with the source of coverslip, so it is worth testing coverslips from a few manufacturers to determine which one is optimal for your intended use.

Differential interference contrast is the method of choice for observing membrane motility since organelles and microtubules can be observed simultaneously in real time. Fluorescence could also be used, but in practice the fact that both the microtubules and organelles of interest must be labeled

[15] R. D. Vale and Y. Y. Toyoshima, *Cell* **52,** 459 (1988).

in some way with different fluorescent dyes, and that photobleaching and concomitant free-radical-induced damage can drastically reduce microtubule motor activity, makes real-time fluorescence less applicable for the study of organelle movement.

Many different microscope systems may be used successfully for VE-DIC (e.g., Refs.7, 16, and 17 and elsewhere in this volume). The major requirements are a bright, even illumination system (usually provided by a 100-W mercury arc, sometimes with a G. W. Ellis fiber optic scrambler[16,17]) coupled with high numerical aperture (NA), strain-free optics. Currently, we are using an Olympus BX60 microscope equipped with DIC optics (Wollaston prisms specially designed for VE-DIC: U-DICV and U-ODPA60V), a 1.4-NA achromatic aplanatic oil immersion condenser and a 60×, 1.4-NA planapochromat objective. Illumination is provided by a 100-W mercury arc lamp (using a U-UCLHG/XEB collector lens) and passed through glass heat filters and a 546-nm narrowband interference filter. To obtain the high magnification necessary, DIC images are passed through a 5× photo projection lens onto a Hamamatsu (Enfield, UK) Newvicon camera (C2400-07).

Image Processing and Acquisition. The use of video cameras such as the Newvicon (or a CCD camera of similar sensitivity) generates enough contrast to visualize microtubules. The image is usually degraded, however, by dirt in the optics and on the faceplate of the camera. Background subtraction, image contrast enhancement, and frame averaging all improve the image (see Salmon and co-workers,[16,17] for reviews). We are using a Hamamatsu Argus 10 image processor to perform these tasks, and the images are then recorded onto S-VHS tape and displayed on a high-resolution RGB monitor.

Numerous methods exist for analyzing rates of movement and for grabbing individual frames from live images or videotapes for the production of figures (e.g., Ref. 7). The Retrac program (available from Dr. Nick Carter, e-mail: N.Carter@mcri.ac.uk) provides a convenient means of performing both these tasks, using a Data Translation DT-55 Quick Capture board in a Pentium-based PC. Retrac also allows computer control of the Panasonic AG-5700 S-VHS recorder.

Observation and Analysis of Organelle Movement

Xenopus egg cytosol is an excellent source of tubulin for microtubule polymerization (at room temperature!) and also provides active microtubule motors. Interphase cytosol will spontaneously nucleate microtubules,

[16] E. Salmon, *Trends Cell Biol.* **5,** 154 (1995).
[17] T. Salmon, R. Walker, and N. Pryer, *BioTechniques* **7,** 624 (1989).

which grow very long. Metaphase cytosols rarely nucleate microtubules unless nucleation sites are provided in the form of centrosomes or Taxol-stabilized microtubule seeds, for instance.[4] Procedures for preparing centrosomes are given elsewhere in this volume. Taxol-stabilized seeds are prepared by shearing Taxol-stabilized purified tubulin microtubules through a 27-gauge needle. It is clear that if microtubule numbers differ greatly between interphase and metaphase samples, any quantitative comparison of organelle movement will be invalid.

Standard Assay. For both interphase and metaphase samples, 9 μl of cytosol is combined with 0.5-μl of organelle fraction and 0.5-μl microtubule seeds (25 μg/ml stock) and mixed by pipetting a few times without introducing air bubbles. The mixture is then flowed into the microscope chamber and incubated for 2 min or more, coverslip down, in a humid chamber. During this time, microtubules start to polymerize and attach to the coverslip surface via the activity of microtubule motors and MAPs adsorbed onto the glass. This provides a convenient two-dimensional microtubule network for analyzing organelle movement.

Two types of moving organelles are seen in interphase assays: vesicles and membrane tubules. The latter undergo fusion to give extensive membrane networks that have been identified as endoplasmic reticulum[12] using immunofluorescence (see below) and electron microscopy. The identity of the moving vesicles is not known. In metaphase samples, however, both types of membrane movement are reduced by more than 90%.[4,5]

Membrane tubules can also be extended if membrane attaches to the tip of a growing microtubule (via a "tip attachment complex" or TAC[18]). The proportion of membrane tubules that extend via TACs is small in the preparations described here, but do occur occasionally. It is easy, with practice, to determine whether a membrane tubule is truly moving along a microtubule, however. Fractionating extracts according to different protocols can greatly reduce the number of motor-driven movements while correspondingly increasing the apparent activity of TACs,[18] although it is not understood why this happens.

Amount of Organelle Movement. Vesicle movement can be quantitated by counting motile events in two ways. First, the total number of movements per field can be averaged over a number of randomly selected fields. This method clearly relies on a similar number of microtubules being present under all experimental conditions. This can normally be achieved, at least up to a 30-min incubation, by the use of microtubule seeds. Second, the

[18] C. M. Waterman-Storer, J. Gregory, S. F. Parsons, and E. D. Salmon, *J. Cell Biol.* **130**, 1161 (1995).

number of movements can be counted over time on individual microtubules whose length is subsequently measured. The resulting data are expressed as movements/μm microtubule/min (e.g., Ref. 4).

The formation of endoplasmic reticulum networks depends on the combination of membrane movement and fusion. The extent of the tubular ER network across the coverslip surface is therefore an historical record of the amount of ER movement along microtubules. ER network extent can be assessed after a set incubation time by determining the length of ER tubules per field, or by counting the number of three-way junctions in the ER network per field. In both cases, data from 20 randomly selected fields are averaged. When these methods are used to compare activity under different conditions, the relative values are very similar to those obtained if ER tubule movements are counted directly.[12]

Direction of Organelle Movement. There are many ways of determining the direction of membrane movement, most of which involve using easily recognizable nucleating structures such as centrosomes, demembranated sperm nuclei (which provide basal bodies), or salt-washed axonemes. Protocols for using such nucleation sites in *Xenopus* egg extracts are given elsewhere.[4,5,7,12] One major problem with such assays, particularly in interphase extracts, is that spontaneously nucleated microtubules swamp the microtubules of interest. We are now using a different approach,[19] as described below, which uses kinesin-coated beads to reveal microtubule polarity. The kinesin beads are prepared according to Howard *et al.*[20]

Organelle movement is observed and recorded in a field with a suitable density of microtubules. Acetate buffer (15 μl) containing 50 μg/ml casein, 50 μg/ml cytochrome *c*, 20 μM Taxol, and 0.1% Triton X-100 is flowed through to stabilize the microtubules and solubilize the membranes. Kinesin-coated beads are prepared by incubating 1 μl of a suspension of 0.1-μm carboxylated beads (Polyscience, Warrington, PA; diluted 1/100 in acetate buffer containing 50 μg/ml casein, 50 μg/ml cytochrome *c*) with 1 μl of purified pig brain kinesin. The beads are then diluted with 8 μl of acetate buffer/casein/cytochrome *c* containing 1 mM ATP and flowed into the microscope chamber. It is important to record the field during all washes, so that you can be sure which microtubules remain (and that you have not changed field by accident!). The movement of the beads reveals the polarity of the microtubules, and this information is then used to determine the direction of organelle movement in the previously recorded sequence.

[19] J. Lane and V. Allan, unpublished results (1997).
[20] J. Howard, A. Hunt, and S. Baek, *Methods Cell Biol.* **39,** 137 (1993).

Immunofluorescence

This section describes a method that has been used successfully to identify endoplasmic reticulum networks formed in interphase extracts[12] and to characterize motile domains in rat liver Golgi fractions that are enriched in secretory products.[21] Obviously, this procedure will only be useful if appropriate antibodies are available.

Cytosol and organelle fractions are combined as above, but are then incubated for appropriate times in flow cells made using 12- × 12-mm coverslips with 0.075-mm-thick plastic strips as spacers. Surface tension is sufficient to hold the coverslip in place, even when the flow cell is incubated inverted. Acetate buffer (100 μl) containing 20 μM Taxol is then gently flowed through the chamber using filter paper to draw the liquid through. It is important not to flow too fast. The microtubules and membrane networks are then fixed by gently flowing through 100 μl of 0.2% glutaraldehyde in acetate buffer, followed by incubation for 20 min at room temperature. After rinsing by flowing through 100 μl acetate buffer, the coverslips are carefully lifted off the spacers using fine forceps. Excess buffer is removed by blotting the edge of the coverslip with filter paper, and then the samples are postfixed in methanol at −20° for 5 min. Hold the coverslip sample side down at an angle of 45 deg to the methanol surface, and immerse the coverslip gently. The angle is important, because you will see the aqueous buffer sinking at this point: if the coverslip is vertical, the membranes are washed off by this flow. This step serves both to attach the membranes firmly to the glass surface and to permeabilize the membranes. The coverslips are then processed for immunofluorescence according to standard methods (e.g., Refs. 12 and 21). This protocol has also been modified for electron microscopy.[12]

Conclusions

Xenopus laevis egg extracts have proved their value in the study of organelle movement and its regulation. One important aspect is that such regulation appears to cross species, since the movement of rat liver membrane fractions is active in interphase *Xenopus* egg cytosols, but inhibited in metaphase cytosols, as is the case for endogenous *Xenopus* organelles.[4] The system does have its disadvantages, however, since it is hard to study motile processes in the intact egg, and it is also common for antibodies to mammalian proteins not to recognize their *Xenopus* counterparts. In many cases, though, the benefits greatly outweigh the problems, and such extracts

[21] V. Allan and R. Vale, *J. Cell Sci.* **107**, 1885 (1994).

are likely to play an important part in future studies on many aspects of motor protein function.

Acknowledgments

Part of the work described in this article was supported by The Lister Institute for Preventive Medicine, The Wellcome Trust (043846 and 048894), and The Royal Society. VJA is a Lister Institute Research Fellow.

[29] Organelle Motility and Membrane Network Formation from Cultured Mammalian Cells

By CAROL MCGOLDRICK and MICHAEL SHEETZ

Introduction

Most membrane-bound organelles in eukaryotic cells rely on the microtubule-motor machinery to maintain their distribution and organization within the cell.[1,2] The endoplasmic reticulum (ER) utilizes the microtubule machinery to extend membrane tubules that fuse with each other to form polygonal membrane networks.[2,3] The Golgi apparatus also extends tubules and forms networks[1,4,5] and is disassembled on loss of microtubules.[6] ER tubulovesicular extension and network formation have been reconstituted *in vitro*.[7,8] Such *in vivo* networks are suitable for studying calcium loading of ER, ribosome–ER interactions, and a host of other questions about ER function that may depend on the native organization of the membrane.

The *in vitro* reconstitution of organelle motility on microtubules has identified a minimal complex required for motility. This organelle–motor complex requires the kinesin or cytoplasmic dynein motor molecules, an accessory protein such as dynactin, and a vesicle receptor protein such as kinectin (reviewed in Ref. 9). In addition to the motor complex that trans-

[1] J. Lippincott-Schwartz, J. G. Donaldson, R. Schweizer, E. G. Berger, H. P. Hauri, L. C. Yuan, and R. D. Klausner, *Cell* **60,** 821 (1990).

[2] M. Terasaki, L. B. Chen, and K. Fujiwara, *J. Cell Biol.* **103,** 1557 (1986).

[3] C. Lee and L. B. Chen, *Cell* **54,** 37 (1988).

[4] J. Lippincott-Schwartz, L. Yuan, C. Tipper, M. Amherdt, L. Orci, and R. D. Klausner, *Cell* **67,** 601 (1991).

[5] V. J. Allan and R. D. Vale, *J. Cell Biol.* **113,** 347 (1991).

[6] W. C. Ho, V. J. Allan, G. Van Meer, E. G. Berger, and T. E. Kreis, *Eur. J. Cell Biol.* **48,** 250 (1989).

[7] S. L. Dabora and M. P. Sheetz, *Cell* **54,** 27 (1988b).

[8] R. D. Vale and H. Hotani, *J. Cell Biol.* **107,** 2233 (1988).

[9] R. B. Vallee and M. P. Sheetz, *Science* **271,** 1539 (1996).

ports components between the organelle compartments, there are proteins that stabilize specific organelle compartments on the microtubules. Moreover, both the velocity and frequency of organelle movement are regulated by unknown factors that can cause motility to vary by orders of magnitude.[10] The analysis of organelle movement *in vitro* will aid in our understanding of the interaction between the proteins that anchor organelle membranes and the motor proteins that transport the organelles.

This article describes how the activities of intracellular organelle membranes can be studied by reconstitution of purified intracellular components *in vitro*. The purification protocol described is suitable for the analysis of organelle motility and for membrane network formation. The major improvements made from previous procedures for network formation and organelle motility[11] are the stimulation of intracellular motility with cAMP before cell lysis, the blocking of the glass surface with casein to increase vesicle binding to microtubules, and the use of Nycodenz (Accudenz) gradients instead of sucrose gradients for organelle separation.

Stock Solutions

Buffers

PMEE: 35 mM piperazine-N, N'-bis(2-ethanosulfonic acid) (PIPES), 5 mM MgSO$_4$ · 7H$_2$O, 1 mM ethylene glycol-bis(β-aminoethyl ether)-N, N, N, N'-tetraacetic acid (EGTA), 0.5 mM ethylenediaminetetraacetic acid (EDTA), adjusted to pH 7.4 with KOH and filter sterilized

PMEE': PMEE with protease inhibitors (1×) and dithiothrethiol (DTT) (1 mM) added just prior to use

Protease Inhibitors

Protease inhibitor cocktail 400×: Pepstatin 0.2 mg/ml, N^α-p-tosyl-L-arginine methylester (TAME) 2.0 mg/ml, L-1-chloro-3-(4-tosylamido)-4-phenyl-2-butanone (TPCK) 2.0 mg/ml all in 100% ethanol; leupeptin 0.2 mg/ml and soybean trypsin inhibitor 2.0 mg/ml in deionized water; store at −20°

Phenylmethylsulfonyl fluoride (PMSF) 100× stock made to 100 mM in 2-propanol; store at room temperature

Other Stocks

Taxol: 4 mM stock in 100% dimethyl sulfoxide (DMSO); store at −20° or −80°

---

[10] S. F. Hamm-Alvarez, P. Y. Kim, and M. P. Sheetz, *J. Cell Sci.* **106,** 955 (1993).

[11] J. J. McIlvain and M. P. Sheetz, *Methods Enzymol.* **219,** 72 (1992).

Mg-ATP: 100 mM ATP stock at pH 7.0 diluted to 10 mM in the presence of 10 mM Mg^{2+} prior to use; keep on ice

GTP 200×: 200 mM stock adjusted to pH 7.0 with NaOH; store at $-20°$

Tubulin: Purified tubulin is commercially available (Cytoskeleton) or porcine brain tubulin can be phosphocellulose-purified[12]

Casein: 10 mg/ml stock in water; dilute in PMEE′ to 5 mg/ml prior to use

Dibutyryl cAMP: added to 40 ml of prewarmed minimal essential medium (MEM) culture media prior to use at a final concentration of 1 mM

Methods

Cell Lines

Our laboratory routinely uses chicken embryo fibroblast cells to assay both vesicle motility and network formation *in vitro*.[11,13,14] *Xenopus* egg extracts,[5] African green monkey kidney cells (CV-1), and Madin Darby canine kidney (MDCK) cell lines also reliably produce networks *in vitro*.[13]

Culture and Harvest of CEF Cells

Chicken embryo fibroblast cells (CEFs) are prepared from embryonic day 12 chickens as previously described.[7,13,15] The procedure for culturing the CEF cells requires the growth of primary or secondary CEFs in MEM containing Earle's salts (GIBCO, Grand Island, NY), with 5% fetal bovine serum, and 40 units/ml penicillin and 40 μg/ml streptomycin.[7,11,13] The cultures are grown in 850-cm^2 rollerbottles for 2–3 days at 37°. This will yield approximately 2×10^8 cells per preparation.

The spent medium is discarded and 10 ml of fresh medium containing dibutyryl-cAMP (1 mg/ml) is added to each of the rollerbottles. This first step is optional, but it has been shown to increase *in vitro* motility.[16]

After 1 hr at 37° the CEFs are incubated with 10–12 ml of 0.05% (v/v) trypsin and 0.53 mM EDTA (GIBCO) for 15 min. The cells are collected in 50-ml conical tubes and the bottles are rinsed with 12 ml of media, which is also added to the conical tubes. The cells are pelleted at low speed (1000g) for 5 min. The pellet is then washed in 30 ml of PMEE′ and spun

[12] R. C. Williams and J. C. Lee, *Methods Enzymol.* **85**, 376 (1982).
[13] J. J. McIlvain, C. Lamb, S. Dabora, and M. P. Sheetz, *Methods Cell Biol.* **39**, 227 (1993).
[14] S. L. Dabora and M. P. Sheetz, *Cell Motil. Cytoskelet.* **10**, 482 (1988a).
[15] B. Kelly and M. J. Schlessinger, *Cell* **15**, 1277 (1978).
[16] J. Kumar, H. Yu, and M. P. Sheetz, *Science* **267**, 1834 (1995).

at low speed again for 5 min. The cell pellet from this spin is resuspended in an equal volume of PMEE' (usually about 700 μl).

Homogenization

The cells are homogenized by four passes through a prechilled Balch ball-bearing homogenizer using an 8.004-mm-diameter ball.[17] The number of passes through the homogenizer is important (one up/down cycle of the syringes is equivalent to one pass) because the microsome size is critical for network formation.

Cell Fractionation

After homogenization, the nuclei and unbroken cells are removed by centrifugation at 1000g for 10 min. The supernatant (S1) is centrifuged at 100,000g for 30 min to yield S2 and P2. The P2 contains microsomal membranes and is resuspended in 40 μl of PMEE'. The S2 is then incubated at 37° for 20 min with 1 mM guanosine triphosphate (GTP) and 20 μM Taxol to polymerize the endogenous tubulin. The polymerized microtubules are pelleted in an airfuge at 15 psi for 5 min, or alternatively by centrifugation for 10 min at 100,000g in a TLS 55 swinging bucket rotor (Beckman, Fullerton, CA). The microtubule pellet is discarded. The number of microtubules present affects the quality of the membrane-network formation, and it is easier to remove the endogenous tubulin and add back a fixed concentration of polymerized microtubules. Figure 1 provides an overview of the isolation protocol for network components.

Organelle Motility

Flow chambers can be formed by attaching two pieces of double-sided tape approximately 0.4–0.8 mm apart on a microscope slide. The chamber is formed when a coverslip is placed over the tape leaving the ends of the chamber open. The chambers can hold 6- to 8-μl volume and are filled by capillary action. For optimal imaging conditions, it is important that the coverslips be very clean. Coverslips are washed in 20% nitric acid for 20 min and rinsed for another 20 min in deionized water. The cleaned coverslips are then stored in 70% ethanol.

Network Formation

Microtubules are polymerized from purified tubulin (0.5 mg/ml) in the presence of GTP (1 mM) and Taxol (20 μM) at 37° for 20 min. Two

---

[17] W. Balch and J. E. Rothman, *Arch. Biochem. Biophys.* **240**, 413 (1985).

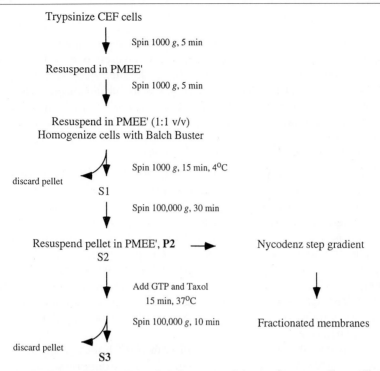

Trypsinize CEF cells

Spin 1000 g, 5 min

Resuspend in PMEE'

Spin 1000 g, 5 min

Resuspend in PMEE' (1:1 v/v)
Homogenize cells with Balch Buster

discard pellet

Spin 1000 g, 15 min, 4°C

S1

Spin 100,000 g, 30 min

Resuspend pellet in PMEE', **P2** ⟶ Nycodenz step gradient
S2

Add GTP and Taxol
15 min, 37°C

Spin 100,000 g, 10 min   Fractionated membranes

discard pellet

S3

Fɪɢ. 1. Flowchart of the cell fractionation protocol for *in vitro* organelle motility and network formation.

microliters of the polymerized microtubules, 3 μl of S3, 5 μl of P2, and 2 μl of Mg-ATP are added to each flow chamber. The slides are incubated for 1–2 hr at 37° in a humidified chamber. Depending on the preparation, the P2 fraction may need to be diluted in order to have a reasonable membrane density. Undiluted membranes will form dense polygonal networks and it may be difficult to distinguish individual membrane tubules. Network formation is monitored by video-enhanced DIC microscopy since microtubules are below the limit of resolution of the light microscope. The DIC optics and imaging components for motility assays have been previously described.[18,19] An example of network formation can be seen in the article by McIlvain and Sheetz in this series.[11]

[18] B. J. Schnapp, *Methods Enzymol.* **134,** 561 (1986).
[19] R. D. Allen, N. S. Allen, and J. L. Travis, *Cell Motil.* **1,** 291 (1981).

Vesicle Fractionation

To study the motility of individual organelle membranes, further fractionation of the P2 is necessary. The P2 is layered over a 0–30% step Nycodenz gradient consisting of 10, 15, 20, 25, and 30% Nycodenz (w/v) and is centrifuged at 100,000g for 90 min. ER membrane is enriched at the 25–30% interface whereas Golgi membrane is enriched 20–25% interface (Aupadahyaya, Glazier and Sheetz, unpublished observations, 1997).

Identification of Specific Membrane Populations

Fluorescence microscopy provides an alternative method for visualizing networks and enables the identification of specific organelle networks. For example, $DiOC_6(3)$ is used to stain ER networks both *in vitro*[11] and *in vivo*.[3,20] Similarly, the Golgi membranes can be detected after network formation and analysis by using the fixation and staining protocols described by McIlvain and Sheetz.[11] Alternatively, Golgi membranes can be specifically labeled by incubating trypsinized CEF cells in culture medium containing C6-*N*-NBD-ceramide prior to homogenization and fractionation of the cell. This dye is a fluorescent label specific for Golgi membranes[21,22] and enables network formation to be imaged by video DIC and fluorescence microscopy.

The advent of green fluorescent protein (GFP) allows one to express chimeric fluorescent membrane proteins.[23,24] For example, the expression of chimeric GFP–Golgi membrane proteins in mammalian cells has been used to monitor the diffusional mobility of membrane proteins.[25] Presumably, fluorescent membrane proteins could be used to differentiate the different membrane populations during fractionation and to detect network formation of specific organelle membrane fractions.

Critical Parameters

Microtubules

Microtubules have a tendency to glide making it hard to form networks. This problem can be solved by using anti-tubulin antibodies or polylysine

[20] C. Lee, M. Ferguson, and L. B. Chen, *J. Cell Biol.* **109,** 2045 (1989).
[21] N. G. Lipsky and R. E. Pagano, *Science* **228,** 745 (1985).
[22] R. E. Pagano, M. A. Sepanski, and O. C. Martin, *J. Cell Biol.* **109,** 2067 (1989).
[23] D. C. Prasher, V. K. Eckenrode, W. W. Ward, F. G. Prendergast, and M. J. Cormier, *Gene* **111,** 229 (1992).
[24] M. Chalfie, Y. Tu, G. Euskirchen, W. W. Ward, and D. C. Prasher, *Science* **263,** 802 (1994).
[25] N. B. Cole, C. L. Smith, N. Sciaky, M. Terasaki, M. Edidin, and J. Lippincott-Schwartz, *Science* **273,** 797 (1996).

to attach the microtubules to the coverslip. Tubulin polyclonal antibody (Sigma, St. Louis, MO) at a 1:40 dilution is added to the flow chamber for 5 min. Excess antibody is then washed away with PMEE' containing 1 mM GTP and 20 μM Taxol. This is followed by flowing in 10 μl of polymerized microtubules and incubating up to 15 min at room temperature. The unbound microtubules are washed away in the PMEE' with GTP and Taxol. The motor fraction (S3), membrane fraction (P2), and Mg-ATP are then added and incubated for 1–2 hr at 37°.

Casein Coating of Slides

The addition of casein prevents nonspecific binding of the vesicular fraction to the glass, thereby increasing the number of vesicles available for network formation.[26] After flowing in microtubules, casein at 5 mg/ml is added and let sit for 2 min. [Excess casein is washed away by rinsing with 2× PMEÉ containing GTP and Taxol. Then 3 μl of S3, 5 μl of P2, and 2 μl of Mg-ATP are added as above.]

Other critical parameters in forming good networks include the extent of cell lysis, the freshness of the protease inhibitor cocktail, and the concentrations of the microtubule and membrane stocks and are discussed by McIlvain et al.[13]

Discussion

Quantitative Analysis of Motility

To make comparisons of different treatments on a given sample, it is important to be able to quantify the frequency of organelle movements from S-VHS tapes of the experiments (2–4 min of video per field and 4–6 fields per sample). This is accomplished typically by counting the number of organelles that move more than 1–2 μm along microtubules in the field (organelles that stop on the microtubule and start moving again are not counted twice). The run length and the velocity can be measured as well as the number of vesicles moving. Logically, the frequency of moving organelles in a given video microscope field depends on many parameters including (1) vesicle concentration, (2) motor concentration, (3) cofactor concentration, (4) the relative activity of each of the first three components, (5) microtubule concentration, (6) adenosine triphosphate (ATP) concentration, (7) temperature, and (8) ionic strength. By keeping factors 1–3 and 5–8 constant for a given preparation, it is possible to study the factors that influence the relative activity of specific components. When treating vesicles

[26] B. J. Schnapp, T. S. Reese, and R. B. Bechtold, *J. Cell Biol.* **115**, 40a (1991).

with blocking antibodies, it is important to control for possible aggregation of the vesicles. This can be done by counting the number of diffusing vesicle profiles that are seen per unit time in the sample medium and then normalizing to the vesicle concentration. Different video fields have different microtubule concentrations; therefore, it is important to normalize the frequency of movements to the total length of microtubules in the field.

Sources of Variability

In considering the major difficulties encountered in these assays, the biggest problems are usually traced to the cell preparation. There is some normal "biological variability" between preparations, but in many experiments little or no motility is observed due to errors in the early stages of cell preparation. For example, not using fresh protease inhibitors will result in poor motility. Although the baseline level of motility can be increased by pretreating cells with cAMP, it should be noted that the actual number of vesicles moving at any given time is low particularly if one considers that those vesicles are only moving for a short distance. The challenge of observing a statistically significant number of events is often great for control samples (variously estimated to be between 30 and 100 movements), much less for inhibited samples. Still, variability between preparations precludes pooling data, except to look at relative activation or inactivation.

Conclusion

It is important to recognize that these assays are designed for understanding the important factors involved in controlling *in vivo* organelle motility. Carboxylated latex spheres will pick up motors nonspecifically and move with them, and it is likely that vesicles could be manipulated to take up motors nonspecifically. Thus, it is important in these studies to be able to verify conclusions about function in crude supernatant fractions or better yet in microinjected cell systems. As many of the potential cofactors are identified and probes (antibodies, peptides, or modified proteins) are produced, it will become easier to utilize the *in vitro* motility assay to determine the role of specific molecules in the organelle transport process.

[30] *In Vitro* Motility Assay for Melanophore Pigment Organelles

By STEPHEN L. ROGERS, IRINA S. TINT, and VLADIMIR I. GELFAND

Introduction

The cytoplasm of eukaryotic cells is a highly ordered, yet dynamic environment. This organization is maintained primarily by the microtubule cytoskeleton and its associated motor proteins. Microtubule-based motors have been shown to mediate the directional transport and steady-state distributions of numerous organelles.[1] The study of these processes *in vivo* suggests that cells precisely regulate microtubule-dependent organelle transport,[2] but molecular mechanisms of this control remain largely unknown. One of the most valuable approaches for investigating various aspects of motor protein function has been the reconstitution of motility *in vitro*. Microscope-based motility assays allow observation of motor-mediated transport in real time under experimentally defined conditions. Ideally, cell-free motility systems should be biochemically defined, consisting of only those cellular components necessary to reconstitute transport. They require (1) microtubules; (2) a defined cargo, such as purified organelles; (3) motor proteins, supplied as either purified protein or in a cell extract or stably attached to the cargo; and (4) a nucleotide, usually adenosine triphosphate (ATP), to fuel motor activity. Investigations of regulated organelle transport additionally require the ability to selectively activate or inhibit motor activity.

Here we describe a method for reconstitution of a defined cell-free system of regulated organelle transport using pigment organelles from cultured *Xenopus laevis* melanophores. Melanophores are large cells present in the dermal layers of fish and amphibians, and are named for the black melanin-containing pigment granules that occupy their cytoplasm (for review, see Ref. 3). The physiologic role of melanophores involves the simultaneous transport of these membrane-bound pigment organelles, termed *melanosomes,* either to aggregate at the center of the cell or disperse throughout the cytoplasm. The net effect of this transport is to give the animal the ability to change color, appearing darker when the cells have dispersed pigment, or lighter when the cells have aggregated pigment.

[1] H. Goodson, C. Valetti, and T. Kreis, *Curr. Opin. Cell Biol.* **9,** 18 (1997).
[2] C. Thaler and L. Haimo, *Int. Rev. Cytol.* **164,** 269 (1996).
[3] L. Haimo and C. Thaler, *Bioessays* **16,** 727 (1994).

Melanosomes are transported along a radially organized microtubule cytoskeleton, which, as in most cell types, is oriented with the minus-ends associated with a perinuclear microtubule-organizing center and the plus-ends extending out to the cell periphery.[4] A kinesin-related protein was implicated as the motor for pigment dispersion by microinjection of function blocking antibodies raised against the motor domain of kinesin heavy chain.[5] Pharmacologic evidence has suggested a role for cytoplasmic dynein as the motor driving pigment aggregation.[6] Investigations using detergent-permeabilized cells have shown that the direction of pigment transport is modulated by the intracellular second messenger, cAMP.[7,8] Stimulation of these cells with the appropriate hormonal stimulus increases intracellular cAMP levels, triggering pigment dispersion via the activation of cAMP-dependent protein kinase (PKA).[9] Conversely, a decrease of cAMP levels within the cell causes pigment aggregation through the action of protein phosphatases.[10] Melanophores, therefore, offer a potentially ideal system for examination of the mechanisms and regulation of organelle motility.

In this article, we describe protocols for the purification of pigment granules from an immortalized *Xenopus* melanophore cell line, as well as for an *in vitro* motility assay itself. We hope that this system will prove useful in understanding the regulatory pathways that govern intracellular transport in melanophores as well as in other motile systems.

Materials

Melanophore Cell Culture

We use an immortalized clonal *Xenopus* melanophore cell line as a source of pigment organelles. This represents a significant advantage over the use of primary cultures for two reasons: (1) the cell line is uncontaminated by other cell types, and (2) the quantity of material is only limited by the costs of tissue culture and time spent in the hood. The cell line was initially generated by Daniolos and co-workers[11] and has been adapted in our laboratory for growth in a simplified medium that consists of 67%

[4] U. Euteneuer and J. McIntosh, *Proc. Natl. Acad. Sci. U.S.A.* **78**, 372 (1981).
[5] V. Rodionov, F. Gyoeva, and V. Gelfand, *Proc. Natl. Acad. Sci. U.S.A.* **88**, 4956 (1991).
[6] T. Clark and J. Rosenbaum, *Proc. Natl. Acad. Sci. U.S.A.* **79**, 4655 (1982).
[7] M. Rozdzial and L. Haimo, *J. Cell Biol.* **103**, 2755 (1986).
[8] M. Rozdzial and L. Haimo, *Cell* **47**, 1061 (1986).
[9] P. Sammak, S. Adams, A. Harootunian, M. Schliwa, and R. Tsien, *J. Cell Biol.* **117**, 57 (1992).
[10] B. Cozzi and M. Rollag, *Pigment Cell Res.* **5**, 148 (1992).
[11] A. Daniolos, A. Lerner, and M. Lerner, *Pigment Cell Res.* **3**, 38 (1990).

Leibowitz L-15 medium, 29% water, 4% fetal calf serum, 5 μg/ml insulin, and 0.1 mg/ml each of penicillin and streptomycin. Stock cultures of melanophores are maintained in 75-cm^2 plastic tissue culture flasks at 27° and are split 1:6 every 5 days using a solution of 0.05% trypsin and 1 mM ethylenediaminetetraacetic acid (EDTA) in 0.7× Ca-Mg-free phosphate-buffered saline (PBS; 140 mM NaCl, 2.7 mM KCl, 10 mM Na$_2$HPO$_4$, 1.7 mM KH$_2$PO$_4$ diluted with water to 70%). For melanosome preparation, cells are plated onto 10-cm plastic tissue-culture plates 5–7 days prior to the experiment. Pigment aggregation or dispersion may be induced in these cells with 10 nM melatonin or 100 nM α-melanocyte-stimulating hormone (MSH), respectively. The hormones are applied to the cells after first rinsing them with 70% PBS followed by incubation in serum-free 70% L-15 medium containing the appropriate treatment. Aggregation must be carried out under low light because *Xenopus* melanophores are light sensitive and will disperse their pigment as a result of bright illumination.[11,12]

Melanosome Purification

Cell Harvesting and Homogenization

Pigment granules are isolated in IMB50 buffer [50 mM imidazole, pH 7.4, 1 mM (EGTA); 0.5 mM EDTA, 5 mM magnesium acetate, 175 mM sucrose, 1 mM dithiothreitol (DTT)]. If melanosomes are to be used in the motility assays, IMB50 is supplemented with 150 μg/ml of bovine milk casein throughout the entire purification to reduce the binding of pigment granules to glass and plastic surfaces during isolation and motility assays. Buffers used for cell lysis are also supplemented with protease inhibitors [10 μg/ml of each leupeptin, pepstatin, and chymostatin and 1 mM phenylmethylsulfonyl fluoride (PMSF)]. Protease inhibitors are omitted from all other steps of purification.

One 10-cm plate of confluent melanophores is carefully washed several times with 70% PBS to remove traces of growth medium and briefly rinsed with IMB50. Cells are scraped from the plate with a rubber policeman into 3 ml of IMB50 supplemented with protease inhibitors. All subsequent steps are performed at 0–4° in a bucket of ice. Cells are lysed by 5–10 passes through a 27-gauge hypodermic needle attached to a 1-ml syringe. We have found that the shearing force generated by trituration of the suspension through a needle is sufficient to release the cell contents without breaking the organelle membranes (see below). Other methods of cell lysis, including

[12] T. Moriya, Y. Miyashita, J. Arai, S. Kusunoki, M. Abe, and K. Asami, *J. Exp. Zool.* **276,** 11 (1996).

sonication and homogenization in glass-Teflon homogenizers, are unsatisfactory because membranes are stripped from melanosomes, liberating their core particles of free melanin. The extent of cell lysis may be monitored microscopically by comparing the ratio of released nuclei to intact cells; a typical cell-shearing procedure will lyse 75–90% of the cells in suspension. Unbroken cells and large cell fragments are removed from the lysate by centrifugation at $600g$ (2000 rpm) in an HB-6 rotor (Sorvall, Newtown, CT) for 5 min.

Melanosome Purification

Melanosomes have dense, insoluble cores of melanin that make them the densest organelles in melanophores. Centrifugation through an isotonic density gradient is, therefore, the method of choice for separation of melanosomes from other cellular components.

We have used 80% Percoll in IMB50 to separate melanosomes from soluble cytosolic proteins and unpigmented organelles. A 10-ml cushion is prepared by combining 8 ml Percoll solution with 2 ml of 5× IMB50 supplemented with 150 μg/ml casein, as described earlier. Cell lysates are overlaid on top of this cushion in a 15-ml round-bottomed polypropylene tube. The gradients are centrifuged at $4000g$ (5000 rpm) in an HB-6 rotor (Sorvall) for 15 min. Following centrifugation, melanosomes form a loose pellet at the bottom of the tube, while other organelles are concentrated at the Percoll–buffer interface and soluble proteins are retained in the top phase. The interface between Percoll and buffer often contains variable amounts of melanosomes that fail to sediment. We believe that these melanosomes are bound to other, less dense, cellular components and the density of the aggregates is, therefore, insufficient to enter the Percoll cushion. Melanosomes are recovered by aspirating the top phase and the Percoll–buffer interface with a Pasteur pipette and washing the walls of the tube several times by filling the top half of the tube with distilled water and aspirating it repeatedly. The remainder of the Percoll cushion is then aspirated and melanosomes are resuspended in 30 μl IMB50 per 10-cm culture plate by gently tapping the side of the tube.

We found that it is essential for successful motility assays not to pellet melanosomes too tightly, because the vigorous pipetting necessary to resuspend a tight pellet decreases motility, possibly by damaging the membranes. If, however, melanosomes are to be used for biochemical analysis, the rate of centrifugation can be increased to $16,000g$. Tight packing of the pellet allows complete removal of the Percoll cushion, and as a result a more concentrated melanosome sample can be obtained.

To estimate the yield of melanosomes during purification, we compared the absorbance of melanin (measured at 550 nm) in original cell lysate versus purified melanosomes. We found that melanosome recovery in a typical purification procedure is between 40 and 50%. Most melanosomes are lost at the Percoll–buffer during the density centrifugation step, and the remainder are lost as a result of incomplete cell lysis.

Characterization of Purified Melanosomes

Electron Microscopy

Transmission electron microscopy was used to analyze the purity of melanosomes pelleted through Percoll gradients. Melanosome pellets were fixed with glutaraldehyde, postfixed with OsO_4, embedded in Medcast resin, and thin sectioned. Microscopy of these sections demonstrated numerous spheric melanosomes with a diameter of about 0.5 μm. No other cellular organelles were present (Fig. 1). In many cases a typical membrane bilayer was seen to encapsulate the melanin core of the organelles (Fig. 1, inset). These results indicate that the organelles isolated by density gradient centrifugation are pure and morphologically intact.

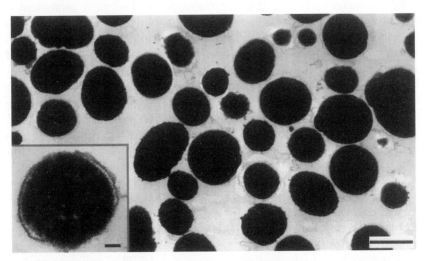

FIG. 1. Electron micrograph of a thin-sectioned pellet of purified melanosomes. Bar = 0.5 μm. No other organelles are present in the purified melanosome fractions. Inset, a melanosome at higher magnification showing a bilayer membrane encapsulating the electron-dense melanin core. Bar = 0.1 μm. [Reproduced from S. Rogers, I. Tint, P. Fanapour, and V. Gelfand, *Proc. Natl. Acad. Sci. U.S.A.* **94**, 3720 (1997). Copyright © 1997 National Academy of Sciences.]

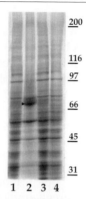

FIG. 2. Comparison of the protein composition of purified melanosomes and melanin incubated in a cell extract. A Coomassie Blue-stained 6–10% gradient SDS gel loaded with whole-cell extract (lanes 1 and 3), purified melanosomes (lane 2), and cytosolic proteins bound to naked melanin (lane 4). The 70-kDa band marked with an arrowhead is tyrosinase. [Reproduced from S. Rogers, I. Tint, P. Fanapour, and V. Gelfand, *Proc. Natl. Acad. Sci. U.S.A.* **94,** 3720 (1997). Copyright © 1997 National Academy of Sciences.]

Protein Composition

SDS–PAGE shows that melanosome fractions contain a characteristic protein pattern with several major components. The most prominent of them has a molecular weight of 70,000 and is dramatically enriched in melanosome fractions compared to the crude cell extract (Fig. 2, compare lanes 1 and 2). In addition to this component, major cellular proteins with molecular weights of about 45,000 and 55,000 copurify with melanosomes. The 70-kDa melanosomal protein was found to cross-react on Western blots with an antibody, PEP-7, directed against murine tyrosinase, a key transmembrane enzyme responsible for melanin synthesis.[13] This result provides biochemical support for retention of the melanosome membrane and seems to be an accurate indicator of membrane integrity during isolation. If, for example, cells are homogenized by sonication instead of shearing through a syringe, the tyrosinase band is essentially absent from melanosome preparations. In addition, we observed that many cytosolic proteins, otherwise absent from the melanosome pellet, cosediment with melanin in this case. To control for the possibility that proteins that copurify with melanosomes may in fact bind to the membrane-stripped melanin, we prepared pigment by boiling melanophores in 1% SDS and then washing melanin from cellular proteins by repeated centrifugation and resuspension in 1% SDS. After removal of the detergent, naked melanin was tested for its affinity for cytoplasmic proteins by incubation in cell extract in IMB50,

[13] M. Jimenez, K. Tsukamoto, and V. Hearing, *J. Biol. Chem.* **266,** 1147 (1991).

and reisolation using Percoll gradient centrifugation, as with intact melano-somes. The bound proteins were analyzed by SDS–PAGE. We found that almost all proteins from the cell extract bind tightly and cosediment with melanin (Fig. 2, lanes 3 and 4), whereas the tyrosinase enrichment seen with intact organelles is absent (Fig. 2, compare lanes 2 and 4). Nonspecific binding of protein to melanin may be explained by the fact that melanin is a cationic polymer of tyrosine; its charge and the hydrophobic nature of its aromatic side chains may be responsible for binding numerous proteins from the cell extract.[14] This fact, in addition to the observation that melano-somes are enriched for specific organelle marker enzymes, proves that they have intact membranes to protect them from nonspecific binding of cytosolic proteins to melanin.

Motor Proteins

The presence of microtubule motors was analyzed using a peptide anti-body HIPYR, which recognizes a peptide sequence conserved in most kinesin-related proteins.[15] In purified melanosome fractions, HIPYR recog-nizes a doublet of proteins with the molecular weight of 95,000 and 85,000 (Fig. 3, lane 4). The molecular weight of this doublet suggested that these proteins are the motor subunits of a heterotrimeric kinesin-related protein, kinesin-II.[16,17] This hypothesis was directly confirmed by blotting with a monoclonal antibody K2.4[16] against the 85-kDa subunit of kinesin-II (Fig. 3, lane 8). Because HIPYR cross-reacts with conventional *Xenopus* kinesin only weakly,[15] we specifically probed melanophores with a poly-clonal antibody, HD, raised against the motor domain of conventional kinesin.[5] Western blotting with this antibody, known to cross-react with *Xenopus* kinesin, showed that conventional kinesin heavy chain is absent from melanosome preparations (Fig. 3, lane 2). Kinesin-II is, therefore, the only motor protein of the kinesin superfamily that can be detected on melanosomes by Western blotting. Note here that kinesin, as with many other proteins, readily binds to membrane-stripped melanin (see above), and if melanosomes are not handled carefully or are pipetted too vigorously, nonspecific association of kinesin with pigment can be observed.

Melanosomes were also examined for the presence of cytoplasmic dynein by Western blotting for dynein intermediate chain using an antibody, 74.1.[18] The 83-kDa *Xenopus* protein was detected in crude cell extract as

---

[14] T. Sarna, J. Hyde, and H. Swartz, *Science* **192**, 132 (1976).

[15] K. Sawin, T. Mitchison, and L. Wordeman, *J. Cell Sci.* **101**, 303 (1992).

[16] D. Cole, S. Chinn, K. Wedaman, K. Hall, T. Vuong, and J. Scholey, *Nature* **366**, 268 (1993).

[17] J. Scholey, *J. Cell Biol.* **133**, 1 (1996).

[18] J. Dillman and K. Pfister, *J. Cell Biol.* **127**, 1671 (1994).

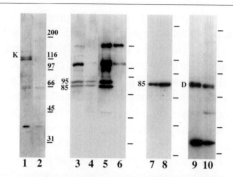

FIG. 3. Western blots of purified melanosomes probed with antibodies against motor proteins. Kinesin was probed with polyclonal HD-kin5 in cell extract (lane 1) and melanosomes (lane 2). Kinesin heavy chain is marked with "K." Kinesin-related proteins were detected with the pan-kinesin antibody HIPYR in cell extract (lane 3), melanosomes (lane 4), and cytosolic proteins that cosediment with microtubules in the presence of AMP-PNP (lane 5) or ATP (lane 6). The 85- and 95-kDa subunits of kinesin-II are indicated. Kinesin-II was probed with the monoclonal K2.4 in cell extract (lane 7) and melanosomes (lane 8). The position of the 85-kDa chain is indicated. Cytoplasmic dynein was detected with the monoclonal 74.1 against dynein intermediate chain in cell extract (lane 9) and melanosomes (lane 10). The position of the 83-kDa intermediate chain is indicated with "D." The positions of the molecular weight standards are marked to the right of each blot. [Reproduced from S. Rogers, I. Tint, P. Fanapour, and V. Gelfand, *Proc. Natl. Acad. Sci. U.S.A.* **94,** 3720 (1997). Copyright © 1997 National Academy of Sciences.]

well as on melanosomes (Fig. 3, lane 10). Cytoplasmic dynein is, therefore, a component of melanosome membranes.

Motility Assays

Polarity-Marked Microtubules

We purify bovine brain tubulin by two cycles of temperature-induced assembly and disassembly followed by phosphocellulose chromatography using the methods of Weingarten *et al.*[19] and store aliquots under liquid nitrogen.

Melanosome transport is reconstituted along axoneme-nucleated microtubules in order to determine polarity of movement. Axonemes are purified from sea urchin sperm using the procedure of Gibbons *et al.*[20] and stored

[19] R. Weingarten, A. Lockwood, S. Huo, and M. Kirschner, *Proc. Natl. Acad. Sci. U.S.A.* **73,** 1858 (1975).

[20] I. Gibbons, A. Lee-Eiford, G. Mocz, C. Phillipson, W. Tang, and B. Gibbons, *J. Biol. Chem.* **262,** 2780 (1987).

in liquid nitrogen in 10-μl aliquots in BRB80 (80 mM PIPES, pH 6.9, 1 mM EGTA, 1 mM magnesium chloride) at about 2 mg/ml. Microtubules are nucleated from the axonemal plus ends using N-ethyl maleimide (NEM)-treated tubulin as described by Hyman *et al.*[21] Ten microliters of BRB80, 5 μl of axonemes, 10 μl of NEM-modified tubulin, and 20 μl of unmodified tubulin (both from 5 mg/ml stocks) are combined in a microfuge tube, brought to 1 mM guanosine triphosphate (GTP), and gently mixed. Microtubules are induced to grow from the plus-end of the axonemes by incubation in a water bath at 37° for 45 min. Then 300 μl of BRB80 supplemented with 10 μM Taxol is added to stabilize microtubule polymer. This concentration of axonemes is suitable for perfusion in the motility assay.

Substrate Preparation

Unlike other organelle fractions, the high buoyant density of melanosomes causes them to sink to the bottom of microscope perfusion chambers. The microtubules in the chamber must, therefore, be fastened to the bottom surface in order to maximize the probability of interactions with melanosomes. Attachment of the microtubules is not a trivial matter for this assay; melanosomes are large (0.5–1 μm) and sticky, and will not exhibit motility on microtubules that are fastened to glass by previously published protocols. To circumvent this problem, we have developed a novel method to anchor axonemes for observation using aminosilanized glass slides as substrates.

Standard microscope slides (Corning, Corning, NY) are first cleaned overnight in chromic acid. On the day of the experiment, they are thoroughly washed with deionized water and dried in an oven. A solution of 10% 3-amino-propyltriethoxysilane (Sigma, St. Louis, MO) in acetone is prepared in a plastic Coplin jar. Slides are transferred to the silane solution and incubated at room temperature for 15 min. The solution is decanted and unbound silane is removed by washing the slides with acetone three times. The slides are then transferred to a clean baking dish, covered with tinfoil, and cured at 110° for 1 hr to cross-link the silane to the glass. Slides prepared in this manner may be stored in an airtight box for a few days. The cationic charge on the glass following this treatment is not enough to fasten individual microtubules well, but grips axonemes tightly.

Microscopy

Melanosome interactions with microtubules are conveniently viewed using video-enhanced differential interference contrast (DIC) microscopy.

[21] A. Hyman, D. Drechsel, D. Kellogg, S. Salser, K. Sawin, P. Stefen, L. Wordeman, and T. Mitchison, *Methods Enzymol.* **196,** 478 (1991).

Our microscope system consists of a Nikon Microphot SA (Nikon, Melville, NY) upright microscope equipped with DIC optics, a Nikon 60×, 1.4-NA PlanApo objective lens, and a 1.4-NA oil-immersion condenser. For illumination we use a 150-W Nikon metal halide illumination system (NMH-1) coupled to the microscope through a liquid light guide fiber. Images are collected with a Hamamatsu NewVicon video camera using a 4× magnifier lens. A Hamamatsu Argus-10 video processor is used for contrast enhancement and frame averaging. Images are recorded onto S-VHS tapes using a Panasonic (Secaucus, NJ) AG-6730 time lapse video recorder with 1 : 3 compression.

Motility Assay

Motility assays are performed within 5-μl flow chambers constructed with aminosilanized slides and 22-cm^2 square No. 1 coverslips (Corning) separated by two parallel strips of Apiezon M grease (M&I Materials Ltd., Manchester, England) mixed (1% by mass) with 20-μm polystyrene beads (Duke Scientific, Palo Alto, CA) to act as spacers. The chambers are filled with BRB80 supplemented with 10 μM Taxol and imaged by video-enhanced DIC microscopy. Five to 10 μl of the axoneme-nucleated microtubules are perfused into the chamber and monitored visually until they attach at a density of about one axoneme per field. Unbound axonemes and free microtubules are removed and the glass blocked by perfusion of IMB50 and 10 μM Taxol and 15 mg/ml casein.

Melanosomes are typically diluted 10-fold to an OD$_{550}$ of about 0.16 into 20 μl IMB50 containing 2 mM ATP. Five-microliter aliquots of this dilute melanosome suspension are perfused into the chambers for each assay. To keep the organelles in suspension, fresh perfusions of melanosomes are made every 5 min.

Perfused melanosomes drift in Brownian motion until encountering microtubules, at which point they attach and exhibit motility that is bidirectional with respect to microtubule polarity (Fig. 4). The movement of individual melanosomes is usually processive, continuing in a single direction over many microns. On rare occasions, some pigment granules are observed to pause and reverse direction on a single microtubule during the course of their movement. Plus-end-directed melanosomes move with an average velocity of 0.65 μm/sec ($\pm$0.21 SD), while minus-end-directed melanosomes travel at an average velocity of 1.15 μm/sec ($\pm$0.43 SD). Pigment granule motility is absolutely dependent on nucleotide hydrolysis, because omission of ATP or substitution of 5'-adenylyl-β,γ-imidodiphosphate (AMP-PNP), a nonhydrolyzable analog of ATP, causes melanosomes to bind tightly to microtubules in a rigor-like state. Melanosome motility in our system does not require soluble proteins; about 85% of Percoll-

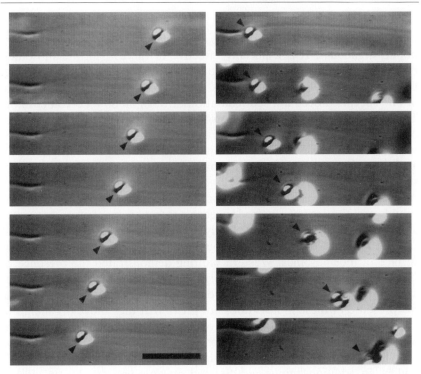

FIG. 4. Bidirectional motility of a single melanosome along an axoneme-nucleated micro-tubule observed by video-enhanced DIC microscopy. Portions of a video sequence showing (A) minus-end directed motility and (B) plus-end directed motility of the same melanosome (arrowheads). The elapsed time between frames is 2 sec. Bar = 6 μm. [Reproduced from S. Rogers, I. Tint, P. Fanapour, and V. Gelfand, *Proc. Natl. Acad. Sci. U.S.A.* **94,** 3720 (1997). Copyright © 1997 National Academy of Sciences.]

purified melanosomes exhibit motility on contact with microtubules. Addi-tion of soluble melanophore proteins does not increase the fraction of motile organelles.[22]

Melanosomes can be purified from cells induced either to disperse pigment with 100 nM MSH or aggregate pigment with 10 nM melatonin. When the polarity of movement of these two populations of organelles was directly compared, we found that melanosomes from the cells dispersing pigment move preferentially toward the plus-ends, while the movement of organelles from cells aggregating pigment was biased toward the minus-end.[22] These results demonstrate that melanosome-bound motors retain their regulated states during purification and in the motility assay.

[22] S. Rogers, I. Tint, P. Fanapour, and V. Gelfand, *Proc. Natl. Acad. Sci. U.S.A.* **94,** 3720 (1997).

Discussion

Density gradient centrifugation is an effective way to purify membrane-bound organelles. It is especially effective in the case of melanosomes, because melanin makes the density of these organelles significantly higher than the density of other organelles. Melanosomes can easily be purified in one specific procedure, sedimentation through an 80% Percoll cushion. This step effectively removes all soluble proteins and nonmelanosomal organelles. Moreover, this purified fraction is highly active—melanosomes exhibit vigorous bidirectional motility along microtubules. Western blot analysis demonstrated that the major component of the purified melanosomal fraction is tyrosinase, the key enzyme in melanin biosynthesis.

Analysis of this motility demonstrated that, unlike most other systems, soluble proteins are not required to reconstitute movement of melanosomes along microtubules. This result clearly indicates that all the proteins necessary for motility, including microtubule motors, are stably attached to the surface of melanosomes. Western blotting demonstrated the presence of two motors associated with purified organelles, a minus-end-directed motor, cytoplasmic dynein, and a plus-end-directed motor, kinesin-II. Of course, the presence of these motors does not prove that they move organelles *in vivo* or even *in vitro*, but the rates of the *in vitro* motility are consistent with known properties of these motors, and no other motors could be detected in the organelle fraction using broadly reactive pan-kinesin antibodies. In addition, minus-end-directed melanosome motility *in vitro* is blocked by micromolar vanadate, an inhibitor of cytoplasmic dynein, and is unsupported by substitution of GTP for ATP, which is consistent with the known nucleotide specificity of dynein.

Our *in vitro* motility assays with purified melanosomes have shown that motors bound to the surface of organelles can retain their regulated state. Therefore, it is possible to isolate two populations of melanosomes—one committed to aggregation and the other committed to dispersion. Direct biochemical comparison of these two populations will provide useful information about mechanisms regulating organelle transport.

Acknowledgments

We thank Dr. M. Lerner for the *Xenopus* melanophore cell line. The following colleagues have kindly provided us with antibodies: Fatima Gyoeva (kin5-HD), Jon Scholey (K2.4), Tim Mitchison and Ken Sawin (HIPYR), Kevin Pfister (74.1), and Vincent Hearing (anti-PEP7). This study was supported by grants from the National Science Foundation (MCB 95-13388) and the National Institutes of Health (GM 52111).

[31] Purification of Phagosomes and Assays for Microtubule Binding

By JANIS K. BURKHARDT

Introduction

Phagocytosis is an important means by which the body clears away infectious microorganisms, senescent cells, and harmful environmental particulates.[1] The specialized endocytic organelle formed when a phagocyte engulfs a particulate is termed a phagosome. Under normal circumstances, newly formed phagosomes fused with late endosomes and lysosomes, forming phagolysosomes, where the internalized particles are degraded by lysosomal hydrolases.[2] Because phagosomes can be readily isolated from cells by taking advantage of the properties of the particles they contain, extensive analysis of the protein composition of these organelles has been performed. This analysis shows that phagosomes and phagolysosomes closely resemble their endocytic counterparts.[3,4] With time within the cell, the protein composition of phagosomes "matures," as fusion with lysosomes results in the transfer of lysosomal proteins into the phagosome compartment, and cell surface proteins are retrieved from it.

The fusion of lysosomes with incoming phagosomes is a microtubule-dependent process, because drugs that depolymerize microtubules inhibit the transfer of lysosomal contents into the phagosome fraction.[5] Video analysis reveals that both lysosomes and phagosomes normally move bidirectionally along microtubules, and that this movement is inhibited by microtubule depolymerization.[6] The phagosome therefore represents an organelle where interaction with microtubules *in vivo* has been documented, and where the physiologic role of this interaction is relatively clear, in that

---

[1] S. Greenberg and S. C. Silverstein, *in* "Fundamental Immunology" (W. E. Paul, ed.), p. 941. Raven Press, New York, 1993.

[2] J. K. Burkhardt, A. Blocker, A. Jahraus, and G. Griffiths, *in* "Trafficking of Intracellular Membranes," NATO ASI Series (M. C. Pedroso de Lima, N. Düzgünes, and D. Hoekstra, eds.), Vol. H91, p. 212. Springer Verlag, Heidelberg, 1995.

[3] M. Desjardins, L. A. Huber, R. G. Parton, and G. Griffiths, *J. Cell Biol.* **124,** 677 (1994).

[4] M. Desjardins, J. E. Celis, G. van Meer, H. Dieplinger, A. Jahraus, G. Griffiths, and L. A. Huber, *J. Biol. Chem.* **269,** 32194 (1994).

[5] A. Blocker, F. F. Severin, A. Habermann, A. A. Hyman, G. Griffiths, and J. K. Burkhardt, *J. Biol. Chem.* **271,** 3803 (1996).

[6] A. Blocker, F. F. Severin, J. K. Burkhardt, J. B. Bingham, H. Yu, J.-C. Olivio, T. A. Schroer, A. A. Hyman, and G. Griffiths, *J. Cell Biol.* **137,** 113 (1997).

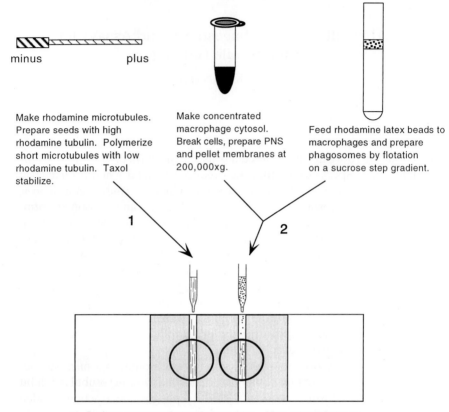

Make rhodamine microtubules. Prepare seeds with high rhodamine tubulin. Polymerize short microtubules with low rhodamine tubulin. Taxol stabilize.

Make concentrated macrophage cytosol. Break cells, prepare PNS and pellet membranes at 200,000xg.

Feed rhodamine latex beads to macrophages and prepare phagosomes by flotation on a sucrose step gradient.

1

2

Perfuse components into assay slides. Allow microtubules to adsorb to the microscope coverslip, forming a dense lawn. Then add phagosomes premixed with cytosol, etc.

Incubate 20 min, Wash.
Count bound phagosomes using fluorescence microscope

FIG 1. Assay for phagosome–microtubule binding. Macrophages grown on tissue culture dishes are allowed to take up fluorescent latex beads. The cells are then homogenized and phagosomes are prepared by the method developed originally by M. G. Wetzel and E. D. Korn, *J. Cell Biol.* **43,** 90 (1969), modified as described in Blocker *et al., J. Biol. Chem.* **271,** 3803 (1996). Taking advantage of the low buoyant density of the beads relative to that of cell membranes, the bead-containing phagosomes are floated away from impurities in a sucrose step gradient. The phagosomes are then mixed with cytosol or with cytosolic fractions whose activity is to be measured. A perfusion chamber is prepared on a microscope slide, and a lawn of rhodamine microtubules is perfused in and allowed to adhere to the glass (1). The

microtubule interactions facilitate the movement of these large organelles through the cytoplasm, increasing dramatically the likelihood that they will contact and fuse lysosomes. Because phagosomes can be purified to a high degree by a rapid and simple procedure, these organelles are well suited for studying the organelle-specific components involved in microtubule-based interactions. The preparation of phagosomes containing latex beads is described next, along with a simple assay for measuring their binding to microtubules *in vitro*. Figure 1 outlines the assay.

Preparation of Phagosomes

Materials

Cells. J774A.1 mouse macrophages are grown as monolayers in Dulbecco's modified Eagle's medium (DMEM) containing 10% newborn calf serum, 100 U/ml penicillin, and 100 μg/ml streptomycin, and passaged by scraping. One day prior to use, the cells are plated in 500 cm^2 culture dishes to give 50–70% confluence the following day. The density of cells is critical to a good phagosome yield, because cells that are too dense are lost from the plates during washing. In addition, care must be taken that dishes are level, to ensure a uniform cell lawn. For preparation of cytosol, cells are grown as spinner cultures to 5×10^5/ml in Joklik's modified Eagle's medium containing 2 g/liter NaHCO$_3$, and supplemented as described previously.

Latex Beads. To prevent nonspecific binding of latex beads to microtubules, should any uninternalized beads contaminate the phagosome preparation, the beads are blocked with fish skin gelatin (FSG; Sigma, St. Louis, MO) prior to internalization. Carboxylate-modified, orange fluorescent latex beads (Molecular Probes, Eugene, OR) are incubated for 15 min to 1 hr with 1 mg/ml FSG in 50 mM 2-[N-morpholino]ethanesulfonic acid (MES) buffer, pH 6.7, with rocking. Covalent coupling is performed by adding 0.2 volumes of 10 mg/ml 1-ethyl-3-(3-dimethylaminopropyl)carbodiimide (EDAC), prepared fresh in water, and incubating for an additional 90 min. Coupled beads are washed twice in 50 mM MES buffer, pH 6.7, by spinning for several minutes in a microfuge at full speed, and resuspended at 1% solids (w/w) in phosphate-buffered saline (PBS)/1% FSG/0.02% NaN$_3$. Aliquots stored at 4° are good for approximately 1 month. Before use, beads should be washed once in complete medium.

---

phagosome/cytosol mixture is then perfused into the chamber (2), free phagosomes are washed away, and the phagosomes remaining bound to microtubules in each field are counted using a fluorescence microscope. [Adapted from J. K. Burkhardt *et al., in* "Trafficking in Intracellular Membranes," NATO ASI Series (M. C. Pedroso de Lima, N. Düzgünes, and D. Hoekstra, eds.), Vol. H91, p. 212. Springer Verlag, Heidelberg, 1995.]

Buffers

PBS: Phosphate-buffered saline (Ca^{2+} and Mg^{2+} free)

Homogenization buffer: 0.25 M sucrose, 3 mM imidazole, pH 7.4

Buffered sucrose solutions: 62, 25, and 10% sucrose (w/w) in 3 mM imidazole, pH 7.4; adjust percentages using a refractometer and store at $-20°$

Protease inhibitor cocktails: 0.5 mg/ml leupeptin, 4 mg/ml aprotinin in water; 0.5 mg/ml N-tosyl-L-phenylalanine chloromethyl ketone (TPCK), 1 mg/ml pepstatin in MeOH; store both in aliquots at $-20°$; both stocks are 1000×

Procedure

1. Dilute beads to 0.05% solids (w/w) in complete medium containing serum, warmed to 37°. Add beads to cells, allowing 40 ml of medium per 500-cm^2 plate. Incubate with gentle rocking for the desired time at 37°. Wash several times with warm PBS, adding the buffer gently to the sides of the dish so as not to disturb the cells, and chase with complete medium for an additional time period, if desired. To obtain a relatively uniform population of phagolysosomes, a pulse time of 1 hr and a chase time of 1 hr is recommended. For very short pulse times, the number of input beads must be increased 2- to 3-fold.

2. Working in a cold room or on an ice bucket covered by an aluminum plate to conduct cold, wash the cells 3 × 5 min with ice-cold PBS to remove uninternalized beads. Cover plates with approximately 20 ml PBS, and collect cells by gentle scraping. Transfer cells to a tube, wash plates again with PBS, and pool.

3. Centrifuge cells 5 min at 1000 rpm (~200g), at 4°. Wash 1× with PBS. Resuspend in cold homogenization buffer and pellet again at 1200 rpm (~250g) × 10 min.

4. Add protease inhibitors to homogenization buffer. Resuspend cells in approximately 1 ml homogenization buffer per 500-cm^2 plate and homogenize by passage several times through a 22-gauge needle. Monitor cell lysis carefully using a phase contrast microscope, and stop homogenization when nuclear breakage is detected. Remove nuclei by centrifugation at 1400 rpm (~350g) for 10 min.

5. Collect the postnuclear supernatant. Add protease inhibitors to all sucrose solutions. Adjust postnuclear supernatant to 35–40% sucrose by adding an equal volume of 62% sucrose and transfer to an SW40 tube, adding no more than 4–5 ml per tube. Overlay with 6 ml 25% sucrose and fill tube with 10% sucrose.

6. Centrifuge in SW40 at 24,000 rpm (100,000g) for 60 min at 4°.

7. Carefully collect phagosomes from the 25%/10% sucrose interface.

To concentrate phagosomes, readjust phagosome fraction to 40% sucrose and transfer to a smaller centrifuge tube (e.g., SW60). Overlay with homogenization buffer and centrifuge at 20,000 rpm × 20 min. Collect as before, aliquot, and flash freeze in liquid nitrogen. Store at −80° or in liquid nitrogen.

Microtubule Binding Assay

Materials

Phagosomes. Prepared as described above. Use within 2 hr of thawing.

Rhodamine Microtubules. Polarity marked microtubules are prepared as described by Howard and Hyman,[7] in PMEE containing 10 μM Taxol. Although the assay does not take direct advantage of the polarity marking, we find that the brightly labeled seeds of these microtubules bind better to the coverslips in the assay than rhodamine-labeled microtubules polymerized in the absence of seeds.

Cytosol. Concentrated cytosol is prepared from J774 cells as described in Blocker *et al.*[5] Briefly, 41 liters of cells are pelleted at 1500 rpm in a Sorvall GS3 rotor, resuspended in 400 ml ice-cold PBS, transferred to 50-ml conical tubes, and pelleted at 1500 rpm for 10 min. This PBS wash step is repeated. The cells are then resuspended in a total of 20 ml PMEE (see below) containing 0.25 M sucrose and 1 mM dithiothreitol (DTT), divided between two 15-ml conical tubes, and pelleted at 2500 rpm for 15 min. The resulting pellets are then resuspended in slightly less than one pellet volume of PMEE/sucrose/DTT and the cells are homogenized by passage through a 22-gauge needle, with breakage monitored using a phase contrast microscope. When the majority of cells are broken with minimal nuclear breakage, a postnuclear supernatant is generated by centrifugation at 6000g for 15 min at 4°. This is then cleared of particulates by ultracentrifugation at 200,000g for 30 min at 4°, and the resulting supernatant is aliquotted (20- to 100-μl aliquots are convenient) and snap frozen in liquid nitrogen until needed. Each liter of spinner culture at 5×10^5 cells/ml should generate approximately 0.5–1 ml cytosol, with a total protein concentration of 30–50 mg/ml.

Stock Buffers

5× PMEE: 175 mM PIPES, 25 mM MgSO$_4$, 5 mM ethylene glycol-bis(β-aminoethyl ether)-N,N,N',N'-tetraacetic acid (EGTA), 2.5 mM ethylenediaminetetraacetic acid (EDTA), pH to 7.4 with KOH

---

[7] J. Howard and A. A. Hyman, *Methods Cell Biol.* **39,** 105 (1993).

5× PMEE/casein: 5× PMEE, 1 mg/ml casein (Sigma); freeze in
1-ml aliquots

DTT: 1 M dithiothreitol stock in water; store frozen in aliquots; do
not refreeze

Taxol: 4 mM stock in dimethyl sulfoxide (DMSO); store frozen in
aliquots; do not refreeze

Antifade stocks: 10 mg/ml glucose oxidase (Sigma) in PMEE/50%
glycerol; 10 mg/ml catalase (Sigma) in PMEE/50% glycerol; 1 M
glucose in water

Store all in aliquots at $-20°$.

Working Buffers

Microtubule dilution buffer: 1× PMEE, 10 μM Taxol; keep at room
temperature

5× Assay buffer: To 1 ml 5× PMEE/casein, add 5 μl each: aqueous
protease inhibitor cocktail, methanolic protease inhibitor cocktail,
DTT, and Taxol; keep at 4° during use

Wash buffer: Mix 200 μl 5× complete assay buffer with 800 μl water,
5 μl glucose, 2.5 μl catalase, and 2.5 μl glucose oxidase; mix by
inverting several times, and incubate at 37° for 10 min; keep tube
closed at room temperature during use; discard after a few hours

Other Supplies

Assay Slides. Small perfusion chambers with a total volume of about
3 μl each are made by applying three parallel strips of double-sided tape
across the short axis of a microscope slide, with a spacing of about 2–3 mm
between tape strips (Fig. 1). Coverslips, preferably 12 mm round, Assistent
brand (available from Carolina Biological Supply, Burlington, NC) are then
applied to the tape and sealed gently using a blunt tool, taking care not to
crush the delicate ends of the chamber. By using three strips of tape and
two coverslips per slide, as shown in Fig. 1, duplicate assays can rapidly be
performed side by side. These chambers can be prepared in advance and
stored without touching one another.

Humidified Boxes. Small boxes should be prepared with damp tissues
in the bottom, and a support above which fits several microscope slides.

Procedure

1. Dilute microtubules appropriately (see below) in microtubule dilution buffer, and perfuse into assay chambers, using 2–3 μl per chamber.
Place slides in humidified boxes and allow microtubules to adsorb to glass
for 5 min.

2. Prepare assay mixes. See the note below regarding the titration of assay components. For a typical 10-μl mix, add the following in the order given:

4–6 μl water
2 μl 5× assay buffer
1–2 μl cytosol
1–2 μl phagosomes

3. Perfuse 4 μl of wash buffer through each chamber, and then add 4 μl assay mix. Always pipette into the same side of the chamber, and withdraw from the other side using the rounded side of a filter paper wedge. Pipette and withdraw smoothly, avoiding excess liquid at side of addition, and removing only the excess at the other side. With practice this can be performed very consistently.

4. Return slides to the humidified boxes. Cover and incubate at room temperature for 20 min.

5. Wash each chamber with 6 μl wash buffer.

6. Observe in a fluorescence microscope at 40×, examining only the coverslip surface. Be sure to note the uniformity of the microtubule lawn. A poor lawn will give erratic results; in this case a fresh preparation of microtubules should be used. Count 10–12 random fields from each chamber, avoiding atypical regions that sometimes form near the edges of the chambers.

Note: Several components in the assay require initial titration before optimal conditions are reached. The microtubule dilution that gives a homogeneous lawn should be determined empirically. Similarly, the phagosomes should be diluted to give a number that is easily counted in each field. Finally, the volume of cytosol added is critical because, in our hands, binding is inhibited at high cytosol concentrations. Once these titrations have been performed, the assay is reproducible and rapid, making it suitable for biochemical analyses such as screening column fractions. This is especially true since both the phagosomes and the cytosol can be stored frozen until needed, and the microtubules can be rapidly prepared from frozen components.

Assay Variations to Test Specific Aspects of Phagosome–Microtubule Interactions

Anchorage of Microtubules to Resist Motor Activity

In testing the role of nucleotides in phagosome–microtubule interactions, it was observed that the microtubule lawn can be compromised by the action of microtubule motor proteins in the presence of adenosine triphosphate (ATP). This effect, along with other conditions that might disturb the microtubule lawn, can be overcome by anchoring microtubules to the glass using poly-L-ornithine.[8] Following binding of microtubules to poly-L-ornithine, it is best to block additional binding sites with aspartic acid prior to addition of organelles, as detailed previously.[5] Using this approach, we could show that the binding of phagosomes to microtubules is independent of nucleotide addition.[5]

Screening for Soluble Binding Factors

The involvement of specific cytosolic proteins in supporting phagosome–microtubule binding can be tested either by depleting these proteins from the cytosol or by testing protein fractions. When the protein concentration of the fraction to be tested is low, casein should be added to the assay mix to minimize nonspecific interactions. In this way, we showed that binding of phagosomes to microtubules depends on the activity of a nonmotor microtubule-associated protein with an apparent size by gel filtration of approximately 150 kDa.[5]

Analysis of Phagosome Membrane Proteins

The role of phagosome proteins in mediating binding to microtubules can be assayed by controlled proteolysis of phagosomes, or by stripping phagosomes with salt or high pH, and then reisolating the organelles for use in the assay. Using this approach, we found that specific proteins on the phagosome membrane are required for microtubule binding, and that these likely to be integral membrane proteins.[5] This approach is being used in conjunction with two-dimensional gel analysis of the phagosome proteins[9] to identify the specific proteins involved. In other studies, the role of phagosome membrane proteins was tested by altering the time during which the phagosomes were allowed to "mature" within the macrophage before they were purified.[5] Early phagosomes bound much more efficiently to microtubules than did later, more lysosome-like phagosomes, again arguing that the relevant proteins on the phagosome surface are likely to be among those which are removed from the compartment with time.

Measuring Phagosome Motility

With minor modifications to the assay, primarily the use of a more dilute microtubule mixture, more concentrated cytosol, and blue-dyed latex

---

[8] R. Urrutia, D. B. Murphy, B. Kachar, and M. A. McNiven, *Methods Cell Biol.* **39,** 253 (1993).
[9] J. Burkhardt, L. Huber, H. Dieplinger, A. Blocker, G. Griffiths, and M. Desjardins, *Electrophoresis* **16,** 2249 (1995).

beads that fluoresce only weakly in the red channel, Blocker *et al.*[6] have adapted the microtubule binding assay for the study of microtubule-based motility, and shown that cytoplasmic dynein and its accessory factor dynactin are required for minus-end directed movement, while kinesin and the receptor protein kinectin are involved in plus-end movement. Late phagosomes moved several-fold more actively along microtubules than did early phagosomes, consistent with an overall mechanism whereby early phagosomes are more often tethered to microtubules, becoming motile as they mature.

Detecting Interactions with Actin Filaments

Finally, a modification of the basic assay has also been used to measure phagosome interactions with actin filaments.[10] Though these studies are still relatively preliminary, they indicate that phagosomes can interact with actin filaments through the activity of both ATP-independent and ATP-dependent actin binding proteins. These studies hold the promise of developing a unified assay system for measuring the interaction of a purified membrane organelle with both actin and microtubules.

[10] M. A. Shonn, A. Blocker, J. K. Burkhardt, G. Griffiths, D. Weiss, and S. A. Kuznetsov, *Mol. Biol. Cell* **6:S,** 272a (1995).

[32] Magnetic Bead Assay for Characterization of Microtubule–Membrane Interactions

By Jochen Scheel and Thomas E. Kreis

Introduction

The functions of interphase microtubules in intracellular membrane traffic and the spatial arrangements of cytoplasmic organelles are mediated by microtubule binding proteins. In addition to the microtubule-based motor proteins of the dynein and kinesin superfamilies, which can move organelles and vesicular carriers,[1,2] additional proteins are required to regulate spatially and temporally interactions of these various membranous struc-

[1] R. D. Vale, *in* "Guidebook to the Cytoskeletal and Motor Proteins" (T. E. Kreis and R. D. Vale, eds.), pp. 175–212. Oxford University Press, 1993.
[2] R. B. Vallee, *Proc. Natl. Acad. Sci. U.S.A.* **90,** 8769 (1993).

METHODS IN ENZYMOLOGY, VOL. 298

Copyright © 1998 by Academic Press
All rights of reproduction in any form reserved.
0076-6879/98 $25.00

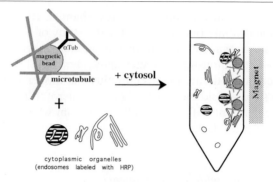

FIG. 1. Scheme of the *in vitro* microtubule binding assay. An affinity matrix consisting of microtubules bound via a monoclonal antibody against tubulin (1A2) to magnetic beads is incubated with cytosolic proteins and a fraction of membranes containing the organelles [e.g., endosomes labeled by internalized horseradish peroxidase (HRP)] to be tested. Microtubule beads with bound organelles are retrieved with a magnet and the unbound fraction is then removed. The amount of bound specific marker is finally determined to quantify the extent of organelle–microtubule binding. (Components are not drawn to scale.)

tures with microtubules.[3,4] Biochemical identification and characterization of factors involved in these processes require the reconstitution of these microtubule–organelle interactions *in vitro*. An *in vitro* assay in which specific binding of membranes to microtubules is reconstituted has to be easy to perform, and the different components (i.e., microtubules, membranous organelles, and soluble cytosolic factors) should be manipulatable independently. We have established such an assay system, using microtubules immobilized on magnetic beads, specifically labeled organelles, and cytosol, for characterizing the binding of cytoplasmic organelles to microtubules. The procedure is simple, quantitative, versatile, and it is essentially independent of the physical properties of the organelles to be tested. This *in vitro* assay has been successfully applied for the identification of a cytoplasmic linker protein (CLIP-170) involved in binding of endosomes to microtubules.[4]

Methods

The *in vitro* binding assay procedure is described schematically in Fig. 1. An affinity matrix is produced by binding Taxol-stabilized microtubules to magnetic beads using 1A2, a monoclonal antibody against α tubulin.[5]

[3] T. A. Schroer and M. P. Sheetz, *J. Cell Biol.* **115,** 1309 (1991).
[4] P. Pierre, J. Scheel, J. E. Rickard, and T. E. Kreis, *Cell* **70,** 887 (1992).
[5] T. E. Kreis, *EMBO J.* **6,** 2597 (1987).

This matrix can be incubated with the fraction of organelles and soluble cytosolic factors to be tested under defined conditions for binding to the immobilized microtubules. The bound fraction is subsequently retrieved by a magnet, the unbound fraction removed, and the distribution of the marker of interest measured in the bound and supernatant fractions to quantify binding.

Preparation of the Microtubule Affinity Matrix on Magnetic Beads

Tubulin is isolated from bovine or porcine brain by three cycles of polymerization and depolymerization followed by phosphocellulose chromatography.[6] Taxol-stabilized microtubules are prepared with this purified tubulin as previously described.[7] Monoclonal antibody against α tubulin, 1A2, is purified by affinity chromatography as described.[5] Magnetic beads are chemically activated using tosylchloride, and the linker antibody (anti-mouse-Fc) is covalently bound to the surface of the beads.[8] The affinity-purified 1A2 is then bound to the linker antibody followed by binding of Taxol-stabilized microtubules to 1A2.

Procedures. Monodisperse magnetic beads, 4–5 μm in diameter (from Dynal, Hamburg, Germany) are transferred into dry acetone by sequential washes in 50, 75, and 100% acetone and washed three times in dry acetone. Beads are handled as suspension of 20–100 mg/ml for chemical activation. Tosyl activation is achieved by incubation of the beads in dry acetone with 2.2 mM tosylchloride and 4.5 mM pyridine for 20 hr at room temperature (commercial preparations of tosyl-activated beads or beads carrying linker antibodies are also available from Dynal). The activated beads are washed 5$\times$ in acetone, transferred into 1 mM HCl and stored in 1 mM HCl at 2 mg/ml at 4° (activated beads can be stored in 1 mM HCl for at least 3 months at 4°). Further handling of the beads is in suspension of 2 mg/ml if not stated otherwise. (The concentrations of beads referred to in the text are based on the input amount of beads and the assumption that losses of beads during the procedure are negligible.)

Activated beads are washed (all washing steps before coating the beads with microtubules are followed by centrifugation: 10,000g for 5 min; cf. later in Comments section) three times with 0.1 M sodium borate, pH 9.5, and incubated overnight at room temperature with sheep antibodies directed against the Fc-domain of mouse IgG (e.g., from Dianova, Hamburg, Germany) at a final concentration of 15 μg/ml. The bead matrix is then washed three times with PBS-B [phosphate-buffered saline (PBS) con-

[6] T. Mitchison and M. Kirschner, *Nature (Lond.)* **312,** 232 (1984).
[7] J. E. Rickard and T. E. Kreis, *J. Cell Biol.* **110,** 1623 (1990).
[8] K. E. Howell, R. Schmid, J. Ugelstad, and J. Gruenberg, *Methods Cell Biol.* **31A,** 263 (1988).

taining 5 mg/ml bovine serum albumin (BSA)], before overnight incubation at room temperature with 20 μg/ml affinity purified 1A2 to generate the "1A2 beads" (beads carrying antibodies are stable in PBS-B supplemented with 0.05% sodium azide for 2–3 months at 4°). All further handling of the beads is at room temperature.

The 1A2 beads are washed twice with PBS-B and twice with PEMT (PEM: 0.1 M PIPES-KOH, 1 mM ethylene glycol-bis(β-aminoethyl ether)-N,N,N',N'-tetraacetic acid (EGTA), 1 mM MgSO$_4$, pH 6.8, supplemented with 20 μM Taxol) and incubated with Taxol-stabilized microtubules (0.5 mg/ml final concentration) for 20–30 min. The mixture is gently pipetted three to five times to break aggregates between beads that may have formed during the preceding incubations. Microtubule beads are then retrieved with a magnet and the supernatant is removed. Damage or loss of the microtubules bound to the beads by the pipetting is compensated for by incubating the microtubule beads with 0.2 mg/ml of tubulin in PEMT to allow elongation of microtubules on the beads. (This step is not always necessary but yields a microtubule matrix that shows less variation between experiments.) These final microtubule beads are retrieved with a magnet, the supernatant is removed, and the beads are finally resuspended directly in the solutions used for the binding assay. The microtubule beads cannot be stored and have to be prepared freshly for the binding assays.

We have only used the magnetic beads with attached microtubules in binding assays, but in principle these beads can be used just as well in a similar manner with any other cellular component bound (i.e., other cytoskeletal filaments, organelles, DNA).

Comments. Magnets for the assays described here are available commercially (Dynal or from CPG, New Jersey) or can be "homemade." In the latter case the strength of the magnetic field has to be tested to allow quick but gentle retrieval of beads in aqueous solutions; ideally, untreated beads suspended in a 1.5-ml plastic tube should be completely retrieved within 15–60 sec (complete retrieval of microtubule beads may take a few minutes).

Magnetic beads should be washed by centrifugation (10,000g for 5 min) whenever possible; repeated retrieval with a magnet may magnetize the beads and lead to aggregation. The magnet should only be used once the microtubules have been bound to the beads. Beads coated with antibodies can be handled using standard micropipette tips, whereas the microtubule beads should be pipetted only with tips with a wide opening (e.g., cut the tip off).

In some situations it may be preferable to add the extra tubulin during the assay to allow the elongation of the microtubules on the beads and binding in one reaction; in this case the microtubule beads are retrieved after incubation with Taxol-stabilized microtubules, the supernatant is removed, and the beads are resuspended in the solution for the binding assay supplemented with 0.1 mg/ml purified tubulin. It is necessary to have the beads prebound with microtubules for obtaining reproducible results; anti-tubulin coated beads and polymerization of tubulin present in the cytosol fraction does not, in our hands, lead to the formation of a good microtubule matrix.

Preparation and Manipulation of Organelles

Organelle fractions can be prepared by standard procedures used for subcellular fractionation; the preparations of endosomes and Golgi membranes are briefly described here. Often postnuclear supernatants can be used for testing the binding of organelles (e.g., endosomes, Golgi membranes), provided specific markers are available. Some organelles (Golgi preparations) can be stored frozen, whereas others (endocytic carrier vesicles) need to be prepared freshly.

Procedures. Endosomes are labeled by internalization of HRP into BHK cells at 37°.[9] Specific labeling of early endosomes is achieved by incubating BHK cells for 3 min in GMEM (Glasgow minimal essential medium) containing 5 mg/ml HRP. Late endosomes can be labeled by a 10-min pulse with HRP-containing medium, followed by a 30-min chase in medium lacking the marker. Endocytic carrier vesicles are labeled by internalization of HRP during 10 min into nocodazole-treated cells (10 μM nocodazole for 1 hr at 37°) followed by a chase in HRP-free GMEM for 25 min. Once the cells have been treated appropriately to obtain the specific labeling, they are chilled by three washes in ice-cold PBS and scraped into PBS (3 ml per one 10-cm petri dish). Cells are washed once more in ice-cold PBS and then homogenized on ice by trituration using a 1-ml Gilson pipette tip in 5 mM imidazole, 250 mM sucrose, pH 7.1 (0.3 ml per 10-cm dish). A postnuclear supernatant (PNS) is prepared by centrifugation at 200g for 5 min at 4°. The PNS is either used directly in the binding assay or endosomes are further purified by sucrose density centrifugation as described.[9] Golgi membranes (or membranes of other organelles) can be prepared by fractionation of the PNS, for example, by sucrose density gradient centrifugation as described.[10,11]

The enriched organelles (or the organelle fraction) can be treated in various ways for identifying and characterizing the membrane-associated factors regulating binding to microtubules. It is usually easiest to treat organelle fractions in the PNS, since at that stage, the reagents can easily be removed during the further purification of the organelles. The following

[9] J. Gruenberg, G. Griffiths, and K. E. Howell, *J. Cell Biol.* **108,** 1301 (1989).
[10] P. I. Karecla and T. E. Kreis, *Eur. J. Cell Biol.* **57,** 139 (1992).
[11] W. E. Balch, W. G. Dunphy, W. A. Braell, and J. E. Rothman, *Cell* **39,** 405 (1984).

treatments are basic for studying factors associated with the membranes of the organelles investigated:

1. Salt extraction of peripheral membrane proteins: 1 M KCl is added to the PNS, from which, after a 15-min incubation, organelles can be further isolated.
2. Alkylation of organelle proteins: NEM (N-ethylmaleimide) is added to the PNS to a 1 mM final concentration, and after an incubation on ice for 15 min, it is quenched with 2 mM DTT (dithiothreitol) for a further 15 min on ice.
3. Removal of cytoplasmic membrane proteins: The PNS is digested with 50 μg/ml N-tosyl-L-phenylalanine chloromethyl ketone (TPCK)-treated trypsin for 20 min at 30°; the protease is subsequently inactivated by 1 mM PMSF (phenymethylsulfonyl fluoride).

Preparation and Manipulation of Cytosol

Cytosol is prepared by high-speed centrifugation (1 hr, 120,000g) of homogenized tissue culture cells (e.g., HeLa cells) or tissue extracts (e.g., human placenta, rat or chicken liver) in PEM containing 1 μM cytochalasin D and a cocktail of protease inhibitors.[10–13] Cytosol (5–25 mg/ml) is snap frozen in liquid nitrogen and stored at $-80°$. The following treatments are used for examination of cytosolic proteins involved in organelle–microtubule interactions:

1. Heat inactivation: Cytosol is incubated at various temperatures (45–95°) for 5 min before chilling on ice and centrifugation (12,000g for 10 min at 4°).
2. Alkylation: Cytosol is treated with 1 mM NEM for 15 min on ice, before quenching of remaining NEM by addition of 2 mM DTT during 15 min on ice.
3. Inactivation of cytoplasmic dynein: 1 mM Mg-ATP and 0.1 mM sodium vanadate are added to the cytosol, which is then irradiated at 366 nm for 1 hr on ice. Vanadate is subsequently inactivated by incubation with 5 mM norepinephrin during 20 min on ice.
4. Immunodepletion of specific antigens: Cytosol is incubated for 2 hr at 4° with antibodies against specific proteins (e.g., kinesin, CLIP-170) immobilized on CNBr-activated Sepharose and the antigen subsequently removed with the beads by centrifugation (2 min at 12,000g at 4°).

[12] P. van der Sluijs, M. K. Bennett, C. Antony, K. Simons, and T. E. Kreis, *J. Cell Sci.* **95,** 545 (1990).
[13] J. Scheel and T. E. Kreis, *J. Biol. Chem.* **266,** 18141 (1991).

5. Fractionation of cytosol: Cytosol is fractionated by ammonium sulfate precipitation or various chromatographic methods for purifying a cytosolic activity involved in the binding of organelles to microtubules. Fractions of interest are either diluted in or dialyzed against PEM prior to examination in the binding assay.

A variety of other additional substances have been tested in the binding assay.[10,13] These include different proteins (BSA, immunoglobulins, gelatin) and enzymes (RNase, DNase) added to reduce unspecific binding, ions (KCl or NaCl up to 2 M), and low molecular weight compounds (nucleotides, nucleotide analogs, NEM). Substances that are not compatible with the system include those destabilizing microtubules (e.g., high concentrations of Ca^{2+}) or membranes (e.g., detergents).

Binding of Organelles to Microtubule Beads

Microtubule beads are incubated with the enriched organelle fraction of interest (e.g., endosomes, Golgi membranes) in the presence of soluble cytosolic factors. The fraction bound to the microtubule matrix is retrieved with a magnet after the appropriate time of incubation. The unbound fraction can then easily be removed and binding of the organelles quantified by using the specific markers associated with the organelles in the bound and the unbound fraction. A comparison of the effects of various treatments on the binding of endocytic carrier vesicles and Golgi membranes to microtubules is shown in Fig. 2.

Procedures. Fifty microliters of microtubule beads (2 mg/ml) are incubated with an organelle fraction (5–15 μl organelles with 0.1–0.5 μg of total protein) in PEM containing 1 μM cytochalasin D (to prevent polymerization of actin), 2 μM Taxol (to keep microtubules stabilized), and 1 mM DTT in a final volume of 100 μl for 50 min at room temperature. (The duration of the incubation may depend on the organelles examined. Binding of endosomes or Golgi membranes to microtubule beads reaches a plateau at 45–50 min of incubation.) The suspension is mixed very gently two to three times during the incubation. The bound fraction is retrieved at the end of the incubation with a magnet and the unbound fraction removed completely. Beads are then resuspended in 100 μl of PEMT containing 1% Triton X-100 to permeabilize the membranes and to solubilize the marker used for quantification. Triton is also added to the unbound fraction (1% final concentration). The amounts of marker associated with the bound and unbound fractions are finally quantified.

Comments. The amount and concentration of microtubule beads used for the assay depend on the particular application; we have used final

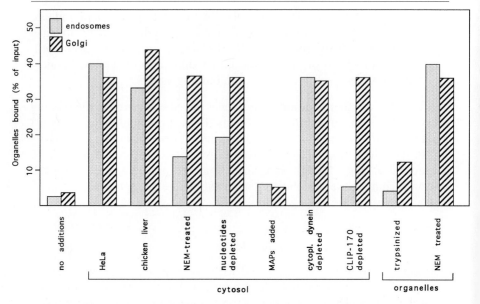

Fig. 2. Characterization of the binding of endosomes or Golgi membranes to microtubules *in vitro*. Binding of endosomes (stippled bars) and Golgi membranes (crossed bars) to microtubules immobilized on magnetic beads has been examined as described in the text. Binding was assayed in the absence of cytosol (leftmost column), in the presence of HeLa cytosol (unless mentioned otherwise) treated as indicated, or in the presence of HeLa cytosol with pretreated organelles. It is unknown to what extent free microtubules (e.g., broken, released, or cytoplasmic) in the assay interfere with binding of endosomes to the microtubules on the magnetic beads. Because some free microtubules are most likely present in the suspension during the assay, it could at least in part explain why not all of the added endosomes can be retrieved.

concentrations of 0.2–2 mg beads/ml. Increasing the concentration of beads may lead to increased aggregation during the incubation. Final assay volumes as small as 20 μl have been used successfully.

The method for mixing the suspension of microtubule beads, cytosol, and organelles during the incubation is an important variable. Settling of the beads can be reduced due to formation of "microtubule networks"; settling, on the other hand, is accelerated by aggregation of the beads, which can occur during incubation. Gentle mixing of the incubation solution helps to keep the beads in suspension. However, this mixing has to be extremely careful and is best done by lightly tapping the tube. In our experience too hard or too frequent mixing leads to aggregation of the beads or microtubule breakage; not mixing the beads at all often turns out to be the better choice.

The bound fraction is separated with the magnet from unbound material, which can be easily removed by pipetting. In most cases it is not necessary to wash the bound fraction. This requires, however, efficient retrieval to the magnet (may take a few minutes to reach completion) and complete removal of liquid. The magnet should be designed so that the beads are retrieved toward the side and not near the bottom of the tube (see Fig. 1). In addition, removal of the unbound fraction should occur slowly to prevent beads from flowing down the wall of the tube due to surface tension. Assay volumes smaller than 25 μl make it more difficult to remove the supernatant completely. If the microtubule beads with bound membranes need washing, it is important to keep in mind that they are fragile at this stage. Washing is best done by removing the tube from the magnet, adding wash buffer (PEMT containing additional factors if necessary), incubating at room temperature for a few minutes without agitation, and retrieving the beads by the magnet.

Acknowledgments

This work was supported by the Fonds National Suisse and the Canton de Genève.

[33] Reactivation of Vesicle Transport in Lysed Teleost Melanophores

By Leah Haimo

Introduction

Because the primary function of melanophores is to translocate coordinately thousands of pigment granules to or from the cell center, these cells are very amenable to an analysis of mechanisms controlling direction of organelle transport. These organelle movements result in the cell appearing alternately light and dark on aggregation of pigment to the cell center and dispersion of pigment throughout the cell, respectively. Melanophores and other types of pigmented cells are found in the dermis of lower vertebrates and are responsible for the color changes that these animals display. These cells have been the subject of numerous studies and have yielded a wealth of information concerning regulation of organelle transport. Live[1-6], trans-

[1] D. B. Murphy and L. G. Tilney, *J. Cell Biol.* **61,** 757 (1974).
[2] R. Fujii and Y. Miyashita, *Comp. Biochem. Physiol.* **51C,** 171 (1975).
[3] Y. Miyashita and R. Fujii, *Comp. Biochem. Physiol.* **51C,** 179 (1975).

formed,[7] and microinjected melanophores,[5,8] lysed cell models,[9-15] and *in vitro* reconstituted systems[16,17] have all contributed to an understanding of the molecular motors and the signal transduction pathways generating and controlling organelle transport.

In this article methods to generate useful lysed melanophores from teleost fish, which retain the ability to aggregate and disperse pigment granules on addition of adenosine triphosphate (ATP) or of both cyclic adenosine 3′,5′-monophosphate (cAMP) and ATP, respectively, will be discussed. When lysed gently with digitonin, the cells can be routinely induced to undergo a full round of transport *in vitro*, i.e., to aggregate and then disperse (or vice versa) pigment. These lysed cell models have revealed that dispersion requires protein phosphorylation, mediated by cAMP-dependent protein kinase,[13] whereas aggregation requires protein dephosphorylation, mediated by protein phosphatase 2B (calcineurin).[15] When lysed more thoroughly with Nonidet P-40 (NP-40), melanophores can transport granules in either direction, but are usually unable to undergo a cycle of aggregation and dispersion *in vitro*. Sufficient material for biochemical analysis can be obtained from this latter model and will be useful for identifying the proteins targeted by the kinase and phosphatase during bidirectional transport.

Source of Material

Melanophores are present in the dermis of fish and can be readily obtained from the thin dermal layer that grows onto the scales. We routinely

---

[4] C. D. Thaler and L. T. Haimo, *Cell Motil. Cytoskelet.* **22,** 175 (1992).

[5] P. J. Sammak, S. R. Adams, A. T. Harootunian, M. Schliwa, and R. Y. Tsien, *J. Cell Biol.* **117,** 57 (1992).

[6] D. Sugden and S. J. Rowe, *J. Cell Biol.* **119,** 1515 (1992).

[7] G. F. Graminski, C. K. Jayawickreme, M. N. Potenza, and M. R. Lerner, *J. Biol. Chem.* **268,** 5957 (1993).

[8] V. I. Rodionov, F. K. Gyoeva, and V. I. Gelfand, *Proc. Natl. Acad. Sci. U.S.A.* **88,** 4956 (1991).

[9] T. G. Clark and J. L. Rosenbaum, *Proc. Natl. Acad. Sci. U.S.A.* **79,** 4655 (1982).

[10] M. E. Stearns and R. L. Ochs, *J. Cell Biol.* **94,** 727 (1982).

[11] N. Grundstrom, J. O. G. Karlsson, and R. G. G. Andersson, *Acta Physiol. Scand.* **125,** 415 (1985).

[12] M. M. Rozdzial and L. T. Haimo, *J. Cell Biol.* **103,** 2755 (1986).

[13] M. M. Rozdzial and L. T. Haimo, *Cell* **47,** 1061 (1986).

[14] M. A. McNiven and J. B. Ward, *J. Cell Biol.* **106,** 111 (1988).

[15] C. D. Thaler and L. T. Haimo, *J. Cell Biol.* **111,** 1939 (1990).

[16] S. L. Rogers, I. S. Tint, P. C. Fanapour, and V. I. Gelfand, *Proc. Natl. Acad. Sci. U.S.A.* **94,** 3720 (1997).

[17] V. I. Gelfand, *Methods Enzymol.* **298,** 1998 (this volume).

use scales from the African cichlad, *Tilapia mossambica,* a freshwater fish that grows to 1 kg or more and is aquacultured in California. The fish are generously provided to us by Pacific Aquafarms (Niland, CA). We also use scales from the killifish, *Fundulus heteroclitus,* a small marine fish that is collected by the Marine Biological Laboratory in Woods Hole, MA. Any fish that is darkly pigmented likely contains numerous melanophores, which could be used in reactivation studies. When cells are needed only for microscopic analysis of motility, a few scales are removed from the fish. For biochemical analyses, 1 or 2 *Tilapia* or 50–100 *Fundulus* are sacrificed and the scales on the dorsal and lateral surfaces are collected. The scales are placed into either a goldfish Ringer's (125 mM NaCl, 2.5 mM KCl, 0.3 mM MgSO$_4$, 1.0 mM MgCl$_2$, 4 mM HEPES, pH 7.4) or other freshwater Ringer's (103 mM NaCl, 1.8 mM KCl, 2 mM CaCl$_2$, 0.8 mM NaHCO$_3$, 5 mM Tris, pH 7.2) or saltwater Ringer's (134 mM NaCl, 2.5 mM KCl, 2 mM CaCl$_2$, 1 mM MgCl$_2$, 0.5 mM Na$_2$HPO$_4$, 15 mM NaHCO$_3$) solution and can be kept for 1 or 2 hr at room temperature, during which time the melanophores remain viable. A layer of epidermis overlying the dermis must be removed to gain access to the dermal melanophores.

Removal of the Epidermis

One of two methods can be used to remove the epidermis from the scales of fish so that the melanophores in the underlying dermis are exposed. Proteolysis digests the extracellular matrix while ethylenediaminetetraacetic acid (EDTA) chelates divalent cations that are required for cells to retain their adhesion to the matrix; either treatment causes the epidermis to detach from the dermis.

Dispase Method

This method uses dispase (neutral protease) to digest extracellular matrix between the epidermis and the dermis. The method is highly reliable and simple (undergraduates in a cell and molecular biology teaching laboratory course are able to remove the epidermis by this method on their first try), but is only suitable when epidermis is to be removed from a few scales, as each scale must be handled individually. For larger preparations, it is necessary to use the EDTA method, discussed later.

Scales, in either goldfish Ringer's (*Tilapia*) or saltwater Ringer's (*Fundulus*), in a petri dish, are observed in a stereo dissecting microscope as they are cut longitudinally with a clean, sharp razor blade into slices to provide additional areas of access of the proteases to the extracellular matrix between the dermis and epidermis. The slices should remain joined at their proximal ends to facilitate their transfer among solutions. After slicing, the scales are transferred into 1–2 ml of freshly prepared dispase solution [32 mg/ml dispase (neutral protease, Sigma, St. Louis, MO), 10 mg/ml bovine serum albumin (BSA) in Ringer's], usually in multiwell plastic disposable dishes. The scales, in dispase solution, are gently agitated on a rotary shaker at 37° for approximately 30–40 min. The time necessary to remove the epidermis must be determined empirically, and will vary with the activity of the dispase and with the health of this tissue. Usually, the healthier and more robust the fish, the longer the time of dispase treatment required. After digestion, the scales are returned to petri dishes containing the appropriate Ringer's solution and are observed in a dissecting stereo microscope as the epidermis is pipetted off. To accomplish this, the dispo tip on a 1-ml pipettor is placed at a slight angle at the proximal edge of the tissue on the scale, and Ringer's solution is vigorously pipetted across the surface of the tissue. If dispase treatment has been sufficiently long, the stream of solution will dislodge the epidermis from the dermis. Prior to its detachment, the epidermis can be visualized, by illuminating the scales against a dark background, as a cloudy cover on the dermis. The epidermis can be seen to peel up and off each section, often as an intact, cloudy sheet of tissue and the melanophores will now appear in sharper focus. Removing the epidermis by this method requires little finesse and thus is far preferable to prior methods that used collagenase digestion, which requires skilled hand dissection, using jewelers' forceps.[12] Moreover, dispase is significantly less expensive than collagenase. After epidermis removal, scales (of either species) are placed into goldfish Ringer's. The melanophores, now accessible, are ready to be lysed and reactivated.

EDTA Method

To conduct studies on lysed melanophores that are to be followed by a biochemical analysis, it is necessary to use large numbers of scales and, accordingly, the prior method, in which each scale must be manipulated individually, would be highly impractical. Instead, incubation of scales in EDTA causes cells in the epidermis to slough off the dermis, and these epidermal cells can then be removed by vigorous shaking and washing. The drawback of this method is that EDTA is quite toxic to the melanophores, and the incubation time must be minimized to maintain cell viability.

Scales are removed from 1 or 2 large *Tilapia* or from 50–100 *Fundulus,* after first sacrificing the fish, by using a flat spatula and scraping the animal from the tail toward the head. The dislodged scales are washed several times with freshwater Ringer's (*Tilapia*) or saltwater Ringer's (*Fundulus*). The scales are then transferred from Ringer's solution into freshwater or saltwater Ringer's, as appropriate, containing 50 mM EDTA or 10 mM EDTA, respectively, and lacking $CaCl_2$. The scales are washed with several changes of the EDTA Ringer's and then vigorously agitated on a rotary shaker at high speed for 45 min (*Tilapia*) or up to 2 hr (*Fundulus*). The solution, which will become cloudy as the epidermal cells are sloughed off, should be changed several times during this period. The scales are finally washed with several changes of EDTA Ringer's and then several changes of freshwater Ringer's (both species).

The scales need to be checked to determine if the epidermis is sufficiently removed and if the melanophores are still viable. To accomplish the former, the scales should be examined by dark-field illumination in a stereo dissecting microscope. Epidermal cells will appear as a cloudy layer on the dermis. Alternatively, cells in a plane above that of the melanophores can be seen readily by DIC microscopy. To determine if cells are viable, melanophores should be challenged to aggregate and disperse pigment, using 10 μM epinephrine in freshwater Ringer's to induce aggregation or 0.25 mM isobutyl methyl xanthine (IBMX) in freshwater Ringer's to induce dispersion. If the epidermis has been removed and the melanophores are viable, they are ready to be lysed and reactivated.

Digitonin Lysis and Reactivation of Transport in Melanophores

Digitonin Lysis

This method is quite useful for lysing small numbers of cells, usually on a scale or two at a time. These can either be lysed with the scale in suspension or immobilized on a microscope slide. Before transport is to be induced, cells lysed with digitonin can be incubated with a variety of potential inhibitors (such as antibodies) for half an hour or more after which time control, lysed cells, not incubated with an inhibitor, are still able to transport pigment on perfusion with a solution containing ATP.

The following solutions are used. A 10-mM stock of epinephrine in dimethyl sulfoxide (DMSO) is stored in the dark at $-20°$. Epinephrine induces aggregation in teleost melanophores at concentrations as low as 0.1 μM, but gives optimal transport (rapid and complete) at concentrations between 10 and 100 μM. Epinephrine is light sensitive and is therefore diluted just before use into Ringer's or lysis buffer. The stock solution should be discarded when it turns moderately pink. A 250-mM stock solution of IBMX in DMSO is stored at room temperature and is diluted, when needed, into Ringer's or lysis buffer to a final concentration of 0.25–2.5 mM. Melanophores are lysed in lysis buffer [0.0016% digitonin in 30 mM HEPES, pH 7.4, 33 mM potassium acetate, 2 mM ethylene glycol-bis(β-aminoethyl ether)N,N,N′,N′-tetraacetic acid (EGTA) [note: EGTA can also be spelled out in a shorter form: ethylene glycol tetraacetic acid], 1 mM free Mg^{2+},[18] 2.5% polyethylene glycol (MW 20,000)].

To lyse melanophores in suspension, epidermis is removed using dispase (see above), and scale pieces are placed into a microfuge tube containing 1 ml of lysis buffer with either epinephrine or IBMX to induce aggregation or dispersion, respectively. All cells will either aggregate or disperse before or as they lyse and thus will exist as a uniform population when they are subsequently reactivated. The scales are then gently agitated in the lysis buffer for 2–2.5 min. The extent of agitation at this step is critical. Insufficient agitation will result in a mixed population of cells, some of which are lysed, some of which are not. Extensive agitation results in cells that cannot be reactivated. The cells, if properly lysed, are now ready to be reactivated, or to be incubated with inhibitors or other molecules prior to being reactivated.

To lyse melanophores in a perfusion chamber, epidermis is removed using dispase and scale slices are placed on a microscope slide in a drop of Ringer's solution. Vaseline, placed at the nontissue end of the slice (proximal end), helps keep the scale immobilized during perfusion. A coverslip is then elevated above the scale slice by using clay "feet." The chamber is perfused with Ringer's containing epinephrine or IBMX so that all melanophores are either aggregated or dispersed. The cells are then lysed by perfusion with lysis buffer also containing epinephrine or IBMX to ensure that the cells remain aggregated or dispersed during lysis. The melanophores are now ready to be perfused with reactivation buffer, or first to be perfused with inhibitors.

The advantage to lysing melanophores on scales in suspension is that one can usually obtain a uniform population of cells that are all well lysed. The disadvantage is that one cannot follow a given cell or cells from the live to lysed to reactivated state, as is possible to do in a perfusion chamber. The disadvantage of the perfusion chamber, however, is that the flow of perfusion solution over the scale is not uniform, and some melanophores on the scale may be lysed and reactivatable, others may be unlysed, while still others may be so extensively lysed that they cannot be reactivated. It takes some experience with the perfusion chambers to obtain reproducible results.

Reactivation of Cells Lysed with Digitonin

Melanophores that have been lysed with digitonin can be reactivated to transport pigment. Scale pieces, in a perfusion chamber, are perfused

[18] C. J. Brokaw, *Methods Cell Biol.* **27**, 41 (1986).

with several changes of lysis buffer containing 0.5–1.0 m*M* ATP and 1 m*M* cAMP to induce dispersion, or with lysis buffer containing 0.5–1.0 m*M* ATP to induce aggregation. Digitonin is retained in the buffer during reactivation so that the plasma membrane does not reseal. Transport of granules should commence within 30–60 sec of addition of nucleotide and will reach its maximal extent within 5–15 min.

After cells have been induced to aggregate or disperse *in vitro*, they are often able to reverse direction of transport. Cells that have dispersed *in vitro* are perfused with several changes of lysis buffer lacking cAMP but containing ATP. Cells that have aggregated *in vitro* are perfused with several changes of lysis buffer containing both ATP and cAMP. A complete round of transport to and from the cell center should occur routinely *in vitro* if optimal conditions have been devised for cell lysis.

Examples of cells reactivated to transport pigment *in vitro* are shown in Figs. 1 and 2, which demonstrate the usefulness of this system for identi-

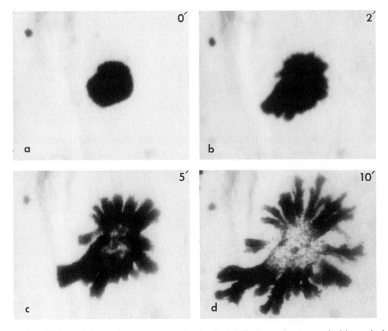

FIG. 1. Addition of the active catalytic subunit of cAMP-dependent protein kinase induces dispersion in lysed cells in the absence of cAMP. A lysed melanophore with aggregated pigment is incubated in the catalytic subunit of cAMP-dependent protein kinase and ATP for (a) 0 min, (b) 2 min, (c) 5 min, and (d) 10 min. The catalytic subunit induces pigment dispersion, thereby demonstrating that the requirement for cAMP in dispersed cells is to activate this kinase.

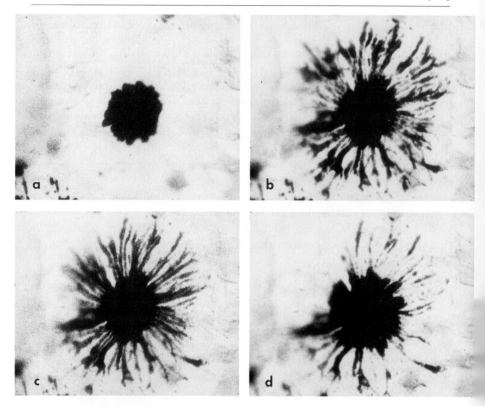

FIG. 2. The phosphatase calcineurin can rescue aggregation of pigment transport in extensively lysed cells. A lysed cell with aggregated pigment (a) is induced with cAMP and ATP to disperse pigment (b). After removal of cAMP, little aggregation occurs (c). Addition of 20 μg of calcineurin to this cell results in substantial rescue of aggregation (d).

fying the regulatory molecules responsible for controlling the direction of transport. We observed that the requirement for cAMP to induce dispersion in lysed cells can be bypassed by incubating aggregated, lysed cells in the active catalytic subunit of cAMP-dependent protein kinase (Fig. 1). Moreover, extensively lysed cells that cannot aggregate pigment can be rescued to do so by addition of the phosphatase calcineurin (Fig. 2), thereby revealing that this phosphatase controls aggregation.[15]

Lysis and Reactivation of Large Numbers of Melanophores Using NP-40

It may be desirable to follow motility assays with an analysis of biochemical changes that correlate with the direction of transport. An examination

of the proteins that become phosphorylated during dispersion and dephosphorylated during aggregation requires large amounts of melanophores.
These studies can be conducted either using live cells incubated in
[32]P-labeled inorganic phosphate and then induced to aggregate and disperse
pigment or using lysed cells incubated in radiolabeled ATP ([γ-^{32}P]ATP)
and induced to transport pigment *in vitro*. The resulting profile of
[32]P-labeled proteins is simpler when lysed rather than live cells are used.
In live cells, all phosphoproteins potentially become radioactive during the
long incubation in labeled phosphate needed to generate an endogenous
radioactive ATP pool. Against this background of abundant, labeled phosphoproteins, a minor protein that changes its phosphorylation state in
response to pigment granule transport may be difficult or impossible to
discern. In lysed cells, only a few proteins become radioactive during the
incubation in the labeled ATP, and some of these change their extent of
phosphorylation during transport. Because digitonin lysis is not effective
when using large numbers of scales, we devised methods to lyse and reactivate melanophores using NP-40. Cells lysed with this detergent can transport pigment in only one direction, as NP-40 eventually solubilizes the
pigment granule membranes after which the granules are unable to move
along microtubules.

Scales from 1 or 2 *Tilapia* or from 50–100 *Fundulus* are placed into two
or more wells of a six-well culture dish after epidermis has been removed
using the EDTA method. The melanophores are then induced to aggregate
or disperse pigment by addition of freshwater Ringer's containing either
epinephrine or IBMX, and the cells are then lysed by addition of Ringer's
containing 0.1% NP-40, protease inhibitors [10 μg/ml each of aprotinin,
chymostatin, leupeptin, and pepstatin and 1 mM phenylmethylsulfonyl
fluoride (PMSF)], and 0.5–1 mM ATP (or ^{32}P-labeled ATP). After 45 sec,
epinephrine is added to cells that had been dispersed while 1 mM cAMP
and IBMX are added to cells that had been aggregated. The scales are
incubated for an additional 10 min during which transport of pigment should
occur. The epinephrine, which binds to α-adrenergic receptors resulting in
an inhibition of adenyl cyclase, ensures that cAMP levels are reduced so
that pigment can aggregate while the IBMX ensures that cAMP is not
hydrolyzed and remains elevated so that pigment can disperse. NP-40 is
then added to a final concentration of 2% to lyse the cells completely. When
radiolabeled phosphorylated proteins are being examined, unlabeled ATP
is also included with the 2% NP-40 to prevent spurious phosphorylation
patterns from occurring after transport has ceased. Moreover, a phosphatase inhibitor (100 μM vanadate and/or 10 μM okadaic acid) is added
to the lysate to prevent dephosphorylation of proteins that had been phosphorylated during transport. The lysate is collected and is suitable for a
biochemical analysis.

Troubleshooting Reactivation

The parameters for successful reactivation of transport in lysed cells may require some effort to develop. If cells are gently lysed, they may not require exogenous nucleotide to induce transport, because their mitochondria may still be intact and function for some time. These gently lysed cells will disperse pigment when cAMP alone is added. Live cells, however, will not respond to cAMP and lysed cells depleted of an endogenous energy supply will respond to cAMP only if ATP is also added.[12] If aggregated cells do not disperse in cAMP, they should be challenged with cAMP and ATP. If they respond, they are lysed and depleted of endogenous nucleotides, as is desirable. If they do not respond, they should be challenged with IBMX. If alive, they will disperse. If lysis has been too extensive or if the reactivation buffer is inappropriate for motility, the cells will not disperse under any of the above conditions. The choice or concentrations of detergent, the length of time of lysis, or the composition of the lysis buffer may therefore need to be altered. If digitonin is ineffective in lysing the cells, Brij 58[9] or saponin[11] should be tested. If the pigment granule membrane remains intact after lysis but the cells cannot be reactivated, then altering the buffer conditions, rather than the detergent, would be recommended. Electron microscopy would be needed to assess the state of the granule membranes. A variety of other buffers support organelle

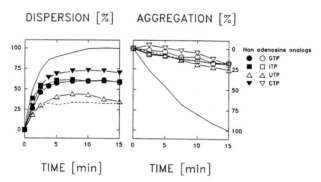

FIG. 3. Quantitation of the extent of dispersion (left panel) or aggregation (right panel) in lysed cells reactivated with various nucleotides is measured by the change in area brightness of a field of cells undergoing pigment granule transport. The extent of transport in each nucleotide (symbols) is plotted as a percent relative to transport in ATP (solid line). The change in brightness of a field of cells to which no nucleotide has been added is also shown (dashed lines). Filled symbols indicate a change in area brightness statistically different from that with no nucleotide added. Open symbols indicate a change in area brightness not statistically different from that of no nucleotide addition. Some nucleotides were able to support dispersion while none, except ATP, was able to support aggregation.

transport along microtubules in lysed cells or in reconstituted systems.[9,10,11,14,16,19] If one direction of transport, but not the other, can be reactivated *in vitro,* it is possible that a regulatory molecule, required for a change in direction, has been extracted from the cells.[15] It might be fruitful to test whether addition of a particular kinase or phosphatase to lysed melanophores can induce pigment granule movements if they do not occur otherwise.

Quantitative Measurements of Pigment Granule Transport

Pigment granules are black, absorb light, and move coordinately. Therefore, the rates and extents of this transport can be readily quantified, either in live or in lysed cells, by measuring the light absorbed by the cells as pigment granules aggregate or disperse. A field of melanophores is imaged on a microscope interfaced with an imaging system that has the capability to read the light intensity of the field. The "area brightness" function in Image1 (Universal Imaging Corp., West Chester, PA) is well suited to this task. The amount of light transmitted through the cells is measured; they are then challenged to aggregate or disperse pigment, and the change in area brightness is measured as a function of time. During aggregation, the pigment granules, which, when dispersed, are present throughout the cytoplasm, accumulate at the cell center and an increase in area brightness results because most of the cell is now free of pigment. During dispersion, the reverse is true. Accordingly, it is possible to quantify pigment granule transport to determine if particular treatments affect movement in one direction or the other. Figure 3 shows an example of lysed melanophores that have been depleted of endogenous nucleotides and then challenged to aggregate or disperse pigment in either ATP or other nucleotide triphosphates. Although some dispersion occurs with these other nucleotides, quantitation of this transport reveals that none is as effective as ATP.

Acknowledgments

This research was funded with support from the National Science Foundation (MCB-9405763) and from the Committee on Research, University of California. Dr. Catherine D. Thaler conducted the studies shown in Figs. 1 and 2, Dr. Mark Lazarro conducted the studies shown in Fig. 3.

[19] M. P. Koonce and M. Schliwa, *J. Cell Biol.* **103,** 608 (1986).

[34] Reactivatable Cell Models of the Giant Amoeba *Reticulomyxa*

By Michael P. Koonce, Ursula Euteneuer, and Manfred Schliwa

Introduction

Biological processes are often effectively studied in cell systems that are readily amenable to their observation and manipulation. *Reticulomyxa*, first described by Nauss[1] in 1949 and rediscovered in this laboratory[2] and by Hülsmann and co-workers,[3,4] is an extremely large freshwater protozoan with rapid and pervasive intracellular motility. The forms of movement seen in *Reticulomyxa* are qualitatively similar to motility observed in higher eukaryotes, but occur at rates up to 10 times faster.[5-7] *Reticulomyxa* was found by us growing on the bottom and sides of a freshwater aquarium. The organism consists of a large, naked, multinucleate cell body of thick interconnected strands 50 to 200 μm in diameter, surrounded by a dynamic, extensively branched feeding network of finer strands (Fig. 1). Typical individuals cover an area of up to 4 cm^2 but, under crowded conditions, cells can fuse to form contiguous arrays many times larger. The size and form of both the cell body and network continuously change over time.

A persistent bulk streaming of cytoplasm is prominent in the cell body, carrying organelles, vacuoles, and nuclei at rates up to 20 μm/sec.[6,7] Streaming is bidirectional, often with several lanes of countercurrent flow in a given area. Bidirectional saltations of discrete organelles (mitochondria and small, clear membrane-bound vesicles are the most prominent types) occur at rates up to 25 μm/sec.[8]

The peripheral network contains one of the most extensive and well-ordered cytoskeletal arrays yet described. Antibody localization and elec-

---

[1] R. N. Nauss, *Bull. Torrey Bot. Club.* **76,** 161 (1949).

[2] M. Schliwa, K. L. McDonald, M. P. Koonce, and U. Euteneuer, *J. Cell Biol.* **99,** 239a (1984).

[3] N. Hülsmann, *J. Protozool.* **34,** 55 (1984).

[4] M. Hauser, J. Lindenblatt, and N. Hülsmann, *Eur. J. Protist.* **25,** 145 (1989).

[5] U. Euteneuer, K. L. McDonald, M. P. Koonce, and M. Schliwa, *Ann. N.Y. Acad. Sci.* **466,** 936 (1986).

[6] M. P. Koonce, U. Euteneuer, K. L. McDonald, D. Menzel, and M. Schliwa, *Cell Motil. Cytoskelet.* **6,** 521 (1986).

[7] U. Euteneuer, K. Johnson, M. P. Koonce, K. L. McDonald, J. Tong, and M. Schliwa, *in* "Cell Movement, Vol. 2, Kinesin, Dynein, and Microtubule Dynamics" (F. D. Warner and J. R. McIntosh, eds.), pp. 155–167. Alan R. Liss, New York, 1989.

[8] M. P. Koonce and M. Schliwa, *J. Cell Biol.* **100,** 322 (1985).

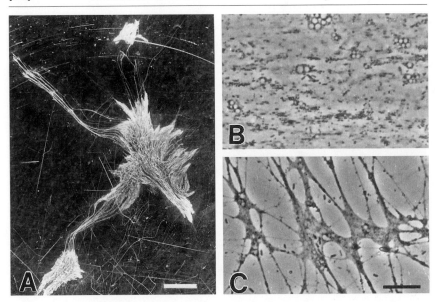

Fig. 1. Overview of *Reticulomyxa*. (A) Macrophotograph of obliquely illuminated specimen in a 10-cm petri dish, showing the large interconnected strands of the cell body. The thick strands taper and flatten to form the highly dynamic feeding network, which cannot be seen at this magnification. Panels B and C show phase contrast micrographs at higher magnification of a flattened lamellipodial region (B) and a branched network region (C). This feeding network frequently covers an area at least twice that of the cell body. Bar in (A) = 0.5 cm; (B, C) = 20 μm.

tron microscopy have demonstrated that the numerous linear elements visible in video-enhanced microscopy consist primarily of single or small bundles of microtubules,[6,7] which are primarily of the same polarity.[9] The microtubules are often associated with colinear arrays of actin filaments (Fig. 2) and, together, these give a structural support to the network strands and provide a framework on which organelles translocate.

In *Reticulomyxa*, single microtubules can be followed for distances exceeding 200 μm, and are probably much longer. Typical mammalian-like organizing centers (centrioles and centrosomal material) have not been detected by electron microscopy, nor are there any apparent microtubule focal centers in the network.[10] Since isolated network fragments can extend, branch, move, and transport, the microtubule system in *Reticulomyxa* largely appears to be self-organized. *Reticulomyxa* expresses the most un-

[9] U. Euteneuer, L. T. Haimo, and M. Schliwa, *Eur. J. Cell Biol.* **49**, 373 (1989).
[10] E. Kube-Granderath and M. Schliwa, *Eur. J. Cell Biol.* **72**, 287 (1997).

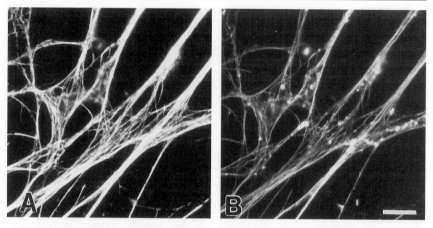

FIG. 2. Fluorescence microscopy. This image shows an area of the feeding network double-labeled with an antibody to (A) tubulin and (B) rhodamine-phalloidin. Note the near complete correspondence between the microtubules and actin filaments. Bar = 10 μm.

usual β tubulin known[11]; whether this is the reason for the unusually fast assembly kinetics of microtubules[12] and the fast rates for organelle transport is presently unknown.

Because of the dramatic motility, and the seemingly simple architecture of the peripheral network, we sought to develop a lysed cell model of this organism that could be used to examine the structural and biochemical basis of organelle transport. Lysis of the organism with a nonionic detergent results in a relatively "clean" skeletal framework with attached organelles that could be reactivated on the addition of adenosine triphosphate (ATP).[13] Three types of motility seen *in vivo* could be restored: individual saltatory particle movement, bulk streaming of aggregates, and a splaying/sliding apart of the individual microtubules.

Methods

Cell Culture

Reticulomyxa can be maintained in spring water and thrives on a diet of wheat germ or flaked fish food. Typical cultures are grown in 10-cm plastic Petri dishes, beginning with four or five fragments of cell bodies

[11] S. Linder, M. Schliwa, and E. Kube-Granderath, *Cell Motil. Cytoskelet.* **36,** 164 (1997).
[12] Y-T. Chen and M. Schliwa, *Cell Motil. Cytoskelet.* **17,** 214 (1990).
[13] M. P. Koonce and M. Schliwa, *J. Cell Biol.* **103,** 605 (1986).

and several flakes of wheat germ. Too much food will encourage rapid growth of bacteria or other protozoans and should be avoided. Healthy organisms will double in size in 2–3 days and then separate by simply streaming apart. This leaves behind an organic shell of debris, which should be removed with a pipette for long-term cultivation. Alternatively, weekly subculturing will provide continuous clean cultures. At present, we do not know of a long-term storage method that preserves cell viability.

Light Microscopy

For video light microscopy, small portions of cell bodies are placed on clean glass coverslips in 35-mm Petri dishes containing either spring water or 10 mM HEPES, 2 mM MgCl$_2$, (pH 7.0), and permitted to attach and extend networks (usually 15–30 min). Coverslips cleaned with alcohol generally work well, though pretreating with polylysine tends to make the cell body more adhesive and increases the amount of feeding network extended as a broad lamellipodium. While the network predominantly forms by extending membranous tubes which then branch, overlap, and fuse, one can generally find large flattened regions of contiguous cytoplasm (also see below). The organism's surface is hydrophobic and when brought close to an air–water interface can cause the overlying liquid to sheet off, resulting in large-scale network disruption. To minimize this effect, the central cell bodies are removed with forceps or a Pasteur pipette once a sufficient network has extended. The coverslips are then quickly brought out of the dish and inverted over coverslip spacers affixed to slides. The longitudinal sides of this sandwich are sealed with VALAP (1:1:1, Vaseline:lanolin:paraffin), leaving the ends open for perfusion. Preparations can then be viewed in an upright microscope. Unless otherwise noted, all manipulations are performed at room temperature. At this stage, perfusion of a 0.25% trypsin solution onto the living organism will result in the large-scale formation of lamellipodial sheets of cytoplasm, facilitating the observation of individual microtubule dynamics and motility.[12]

For microscopy, differential interference contrast (DIC) optics work well for visualizing organelle movement, membrane integrity, and the underlying microtubules. To maximize contrast and to view single microtubules, images from a 100× objective are projected into a high-resolution videocamera (Newvicon tube or CCD camera) and manipulated with an image processor minimally capable of background subtraction and contrast enhancement (e.g., MetaMorph, Universal Imaging Corporation). Images can be stored on VHS tape, optical disk, or captured directly onto computer drives. Hardcopy images can either be photographed from the monitor with a 35-mm camera or directly printed.

Cell Lysis

The network strands are lysed by perfusion of 5% hexylene glycol, 1 mM sodium orthovanadate, and 0.15% Brij 58 in a buffer (50% PHEM[14]) consisting of 30 mM PIPES, 12.5 mM HEPES, 4 mM EGTA, and 1 mM MgCl$_2$, pH 7.0. Initial work included a protease inhibitor cocktail of 10 μg/ml soybean trypsin inhibitor, 10 μg/ml TAME, 10 μg/ml BAME, 1 μg/ml leupeptin, and 1 μg/ml pepstatin (adapted from Ref. 15) in the lysis solution; however, this was later found to be optional. Hexylene glycol has long been used as a stabilizing agent and probably affects protein surface hydration. Vanadate, even when added to the culture medium, causes an abrupt stop of motility and likely functions to produce a rigor condition that locks organelles onto the microtubule, thus stabilizing the overall structure. The nonionic detergent, Brij 58, perforates the plasma membrane and facilitates its removal. Its action is more gentle than Triton X-100, which does not provide an effective initial lysis step. Within a few seconds of perfusion with Brij 58, motility comes to a complete stop, "holes" form in the plasma membrane, then the remaining membrane literally appears to peel off the cytoskeletal scaffold. Lysis is complete within 1 min and yields a relatively "clean" skeletal framework of microtubules, microfilaments, and attached organelles (Fig. 3). At this point, the hexylene glycol, vanadate, and Brij 58 can be removed by rinsing with 50% PHEM alone. Structurally, the network is stable in buffer for at least 1 hr after lysis (longest time tested). Both microtubules and microfilaments appear to remain intact and to retain their colinear organization.

Whole-mount electron microscopy of lysed preparations demonstrates that in most areas, the plasma membrane has been completely removed by the lysis solution (Fig. 3). Surprisingly, the organelle membranes appear to remain intact. Most of the organelles retained along the network appear attached to microtubules, often at multiple points.

Reactivated Motility

The addition of buffer containing 1 mM ATP reactivates three distinct forms of motility observed *in vivo:* organelle movements along linear tracks, splaying and bending of microtubule bundles, and bulk streaming of organelle aggregates.

Organelle Movements along Linear Tracks. Buffer containing 1 mM ATP restores the motility of approximately 50–70% of the remaining organelles (Fig. 4). Depending on the conditions, reactivation results in average

[14] M. Schliwa and J. van Blerkom, *J. Cell Biol.* **90**, 222 (1981).
[15] W. Z. Cande, P. J. Tooth, and J. Kendrick-Jones, *J. Cell Biol.* **97**, 1062 (1983).

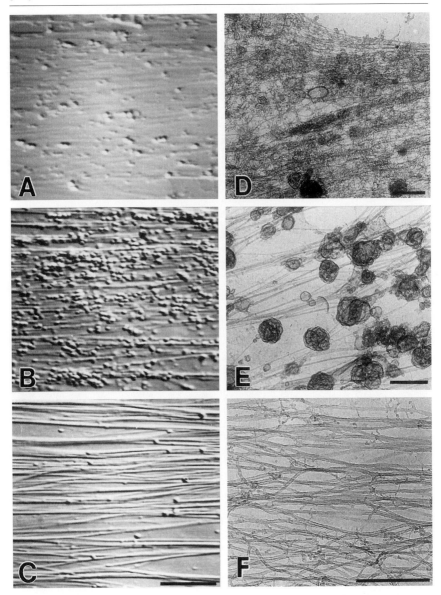

FIG. 3. Panels A–C show video-enhanced DIC images of the same region taken while the organism was alive and motile (A), immediately after lysis (B), and after a rinse with 1% Triton X-100 (C). panels D–F show whole-mount electron micrographs of similar regions: (D) an unlysed region of lamellipod, (E) fixed after lysis, and (F) fixed after Triton extraction. Bars in (A–C) = 10 μm; (D–F) = 0.5 μm.

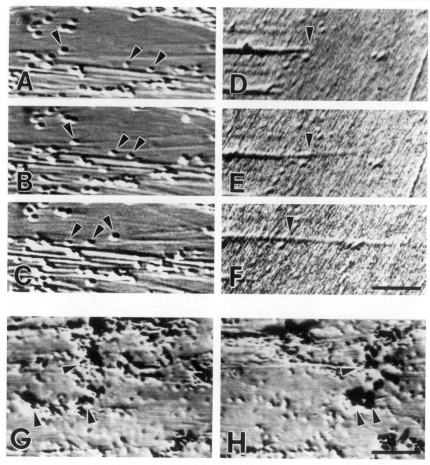

Fig. 4. Reactivated motility. Video-enhanced DIC images. *Note:* The amount of motility present in the live organism and on reactivation is not done justice with a few static images. A videotape of this organism is available that better illustrates this point.[24] A–C show a sequence of individual organelle movements. The arrowheads mark three individual particles over a 5-sec time period. D–F show microtubules protruding from the end of a mechanically severed bundle over a 10-sec period following ATP addition. The arrowhead provides a reference for the original end of the bundle. Panels G and H show two panels in which particle aggregates move from left to right. Bar in (A–C, G, H) = 10 μm; (D–F) = 5 μm.

organelle rates of up to 15 μm/sec.[13] In general, most reactivated models display equal frequencies and rates of motility in the anterograde (away from the cell body) and retrograde directions. Organelle movements only occur in association with the visible linear elements (microtubules): they are most prominent within the first few minutes of reactivation, but motility can still be found 30–40 min later.

Splaying and Bending of Microtubule Bundles. In addition to individual organelle movements, ATP causes many of the parallel microtubule bundles to detach from the coverslip and undergo active splaying and bending motions (Fig. 4). To examine splaying in more detail, lysed networks were cut into a series of 20- to 50-μm-long segments with a glass micropipette, then stripped of organelles with 1% Triton X-100. When reactivated with 1 mM ATP, microtubules actively extruded from the ends of the cut bundles and dissociated laterally from along the bundle sides.[16] This suggests that the microtubules can actively slide against one another, perhaps contributing to the dynamic form of the intact network.

Bulk Streaming of Organelle Aggregates. A third motility component of reactivated *Reticulomyxa* strands is a steady streaming motion of entire strands, parts of strands, and aggregates of organelles at rates of 1–5 μm/sec (Refs. 13 and 17 and Fig. 4). This movement is strikingly similar to the bulk cytoplasmic streaming seen *in vivo,* and to the occasional movement of microfilaments visualized with rhodamine-phalloidin in the reactivated networks.[17]

Reactivation Biochemistry

The reactivated movements occur within a pH range of 6.0–8.0 and require ATP.[13,18] No motility is induced by the nonhydrolyzable analogues ATPγS or AMPPNP (1 mM) or the nucleotides GTP, CTP, UTP, ITP, or ADP (1 mM). ATP concentrations as low as 10 μM stimulate reactivation, but organelles move slower and have more frequent pauses than with 1 mM ATP. Reactivated organelle movements are insensitive to 5 mM EHNA and sensitive only to relatively high concentrations of vanadate (100 μM).

Network Manipulations

The stability and accessibility of this lysed system greatly facilitates further network manipulation. Each of the three major components (microtubules, microfilaments, and organelles) can be selectively removed or inactivated, allowing an independent examination of component function and interactions.

Most organelles can be removed with a 1% Triton X-100 rinse (Fig. 3) or with a brief high-salt (200 mM KCl) wash. This yields a relatively "clean"

---

[16] M. P. Koonce, J. Tong, U. Euteneuer, and M. Schliwa, *Nature* **328,** 737 (1987).

[17] M. P. Koonce, U. Euteneuer, and M. Schliwa, *J. Cell Sci. Suppl.* **5,** 145 (1986).

[18] M. Schliwa, T. Shimizu, R. D. Vale, and U. Euteneuer, *J. Cell Biol.* **112,** 1199 (1991).

system of native microtubules and microfilaments, which can serve as a motility assay system for other organelles or coated beads.[17]

Actin filaments can be selectively removed by incubation with 20 μg/ ml gelsolin in the presence of 10 μM free calcium.[17] No actin fluorescence can be detected with rhodamine-phalloidin after a 10-min treatment, and fragmented staining patterns result from shorter incubations. Reactivation with ATP after gelsolin treatment yields normal organelle motility, bundle splaying, and microtubule telescoping, but organelle aggregation and bulk streaming-like movements are less apparent.

Although microtubules in this system are calcium insensitive in the absence of ATP (1 mM free calcium for 15 min had no visible effect on microtubule integrity, even in the presence of 1 mg/ml calmodulin), they will depolymerize in the presence of 0.5 M KCl. The addition of high salt causes endwise depolymerization of microtubules at an average rate of 1 μm/sec ($\pm$0.25 μm/sec). This treatment also liberates nearly all of the network-associated organelles. Rhodamine-phalloidin staining of salt-washed networks, devoid of any visible microtubules, demonstrates that significant numbers of microfilaments remain.

Beyond Reactivation

Motor Characterization

Reticulomyxa supernatants contain a high molecular weight protein that cosediments with microtubules in an ATP-sensitive fashion. While this protein is dynein-like in some respects (20S sedimentation, UV-vanadate cleavage[18–20]), the purified protein promotes a bidirectional movement of latex beads along the unipolar arrays of *Reticulomyxa* microtubules, suggesting a potentially novel mechanism for organelle transport.[19] Thus the lysed cell model can be coupled with biochemical fractionation and motility reconstitution experiments. The forces developed by these motor molecules when driving organelle movements *in vivo* have been measured using optical tweezers and were found to be on the order of a few piconewtons.[21]

Surface Transport

Latex beads can be attached to the exterior surface of a number of different cell types whereupon they will undergo rapid linear transloca-

---

[19] U. Euteneuer, M. P. Koonce, K. K. Pfister, and M. Schliwa, *Nature* **332**, 176 (1988).

[20] U. Euteneuer, K. B. Johnson, and M. Schliwa, *Eur. J. Cell Biol.* **50**, 34 (1989).

[21] A. Ashkin, K. Schutze, J. M. Dziedzic, U. Euteneuer, and M. Schliwa, *Nature* **348**, 346 (1990).

tions.[22] This work demonstrates a transmembrane connection between the cell exterior and the underlying cytoskeleton, and couples that linkage with motility machinery. Beads attached to the exterior surface of *Reticulomyxa* also undergo motility qualitatively similar to the underlying organelle transport. Orokos *et al.*[23] have recently demonstrated that this surface transport mechanism can survive the lysis procedure and will reactivate on the addition of ATP. This provides an experimental model in which to examine the mechanochemical and structural basis for this pervasive form of movement.

Acknowledgments

Work in the author's laboratories is supported in part by grants from the NIH (GM51532) to MPK and the Deutsche Forschungsgemeinschaft (SFB 266 and Schl 175/7) to MS.

[22] R. A. Bloodgood, *Biol. Cell.* **76,** 291 (1992).
[23] D. D. Orokos, S. S. Bowser, and J. L. Travis, *Cell Motil. Cytoskelet.* **37,** 139 (1997).
[24] M. Schliwa and M. P. Koonce, *in* "Video Supplement 2" (J. M. Sanger and J. W. Sanger, eds.), *Cell Motil. Cytoskelet.* **17,** 356 (1990).

Section V

Force Production Assays

[35] Design and Use of the Centrifuge Microscope to Assay Force Production

By RONALD J. BASKIN

Introduction

The centrifuge microscope was invented nearly seventy years ago as a method for studying the physical properties of "protoplasm." Cells and cellular components were observed directly while being subject to increasing centrifugal forces.[1] The separation of granules with different densities was observed in sea urchin (*Arbacia*) eggs spun with a centrifugal force of 5500*g*. Cell viscosity was estimated using Stokes law.[2] Years later, an improved version of the centrifuge microscope was developed and used for studies of cell propulsion and cytoplasmic streaming.[3,4] The propulsive force of single, swimming Paramecium cells was calculated to be approximately 7×10^{-4} dynes.

More recently, a centrifuge microscope-based motility assay using isolated myosin was developed by Oiwa and his co-workers.[5] The motor protein myosin was adsorbed onto microspheres and the centrifugal force required to halt movement of these spheres on actin cables was determined. In addition, the velocity of microsphere movement as centrifugal force was varied was measured allowing the determination of a force-velocity curve. A different approach has been developed for studying kinesin force and motility and uses a demembranated sperm assay in the centrifuge microscope.[6] We will discuss these approaches and suggest the advantages of using more recent motility assays[7] in future work with the centrifuge microscope.

The centrifuge microscope is an alternative method to a laser trap-based motor force assay. The force applied to the motor proteins can be accurately determined and regulated by varying the angular velocity of the motor assay system located on the centrifuge microscope rotor. Also, the dependence of bead (or sperm) movement velocity on motor protein force

[1] E. Harvey and J. Newton, *Appl. Phys.* **9**, 68 (1938).
[2] R. H. J. Brown, *J. Exper. Biol.* **17**, 21 (1940).
[3] K. Kuroda and N. Kamiya, *Exper. Cell Res.* **184**, 268 (1989).
[4] E. Kamitsubo, Y. Ohashi, and M. Kikuyama, *Protoplasma* **152**, 148 (1989).
[5] K. Oiwa, S. Chaen, E. Kamitsubo, T. Shimmen, and H. Sugi, *Proc. Natl. Acad. Sci. U.S.A.* **87**, 7893 (1990).
[6] K. Hall, D. G. Cole, Y. Yeh, and R. J. Baskin, *Biophys. J.* **71**, 3467 (1996).
[7] S. M. Block, L. S. B. Goldstein, and B. J. Schnapp, *Nature* **348**, 348 (1990).

is easily observed using a centrifuge microscope, and a force-velocity curve is readily calculated. In contrast, laser trap-based assays, when used to calculate a force-velocity curve, involve the combination of a laser trap with a dual-beam interferometer.[8] To derive the motor velocity from the measured bead velocity, it is necessary to characterize the elasticity of the bead–microtubule linkage. (Scatter in the data results from the Brownian motion of tethered beads and from linkage heterogeneity.) Finally, the centrifuge microscopes produced so far are considerably less expensive to construct than a laser trapping system; however, as discussed later, a high-precision centrifuge microscope has yet to be built and the cost of such a system would probably be near that of a laser-based system.

Centrifuge Microscope

Our centrifuge microscope (Fig. 1) is constructed from a Zeiss light microscope with the stage replaced by a motor-driven rotor plate constructed from aluminum. The rotor, including the assay cell mounting holes, was machined from a plate of 0.25-in. aluminum that was precision ground ($\pm$0.0001 in.) to ensure that the surfaces were parallel. The rotor has an effective radius of 10.5 cm and is driven by a 12-V d.c. motor (Fig. 2). (We used a heavy-duty d.c. motor capable of drawing several amps of current. There would be a considerable advantage in using a motor system capable of feedback control of rotation velocity.) It can be rotated at angular velocities of 200–5000 rpm (4–2000g). During rotation, kinesin-driven sperm moving on the lower surface of the sample cell top coverslip remain within the focus of a 40×, 0.75-NA objective. The depth of field is approximately 1.17 μm, which is larger than the vertical movement (wobble) of the rotor. The rotor holds four assay cells (Fig. 3) and a timing unit permits selection of any one for observation. A delay circuit permits observation of different regions of an assay cell.

A stroboscopic light source model 438 nanopulser, Xenon Corp., Woburn, MA) was triggered to flash in synchrony with the rotating stage by a photonic sensor control unit. Flash rise time was approximately 20 ns (Fig. 4) yielding a quantifiable but slightly blurred particle image. The computer was located several feet away from the nanolamp because pulse-generated radiation will cause disruption of the computer program. Particle movement is analyzed following enhancement of the video signal using an Imagen signal enhancer (Imagen Corp.), which is obtained using a DAGE-MTI SIT 68 video camera and recorded on a video recorder (Sony Corp.). Video images are captured by a frame grabber (LG-3, Scion Corp., Freder-

[8] K. Svoboda and S. M. Block, *Cell* **77**, 773 (1994).

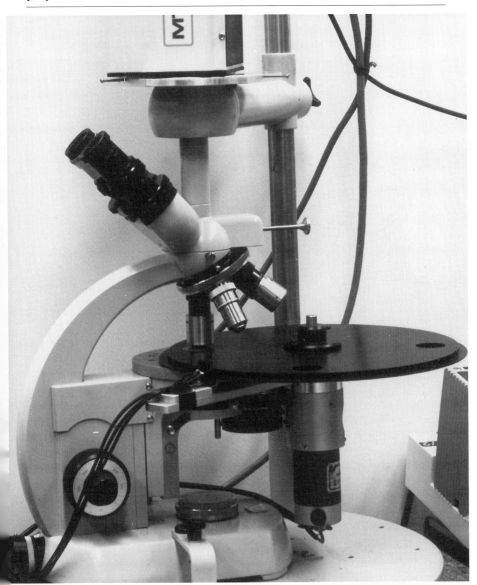

FIG. 1. Photograph of centrifuge microscope. Zeiss microscope stage has been removed and a motor-rotor assembly mounted in its place. Video camera is mounted on a trinocular head. Timing light and photocell are mounted on the edge of the rotor.

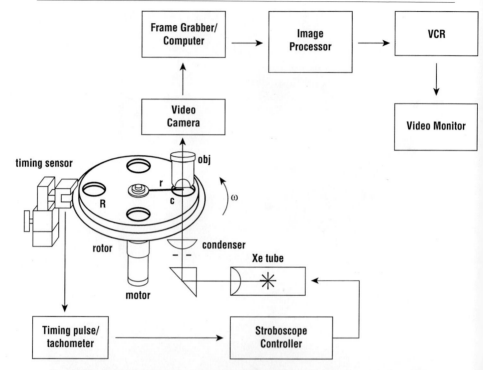

Fig. 2. A schematic diagram of the centrifuge microscope depicting the signal pathways in the experimental system. The strobe light flashes, illuminating the rotating motility assay cell (c), (triggered by the pulse and timing delay unit). The light beam passes through the microscope objective, into the camera. The camera output is directed into the image processing computer.

ick, MD) and processed using NIH IMAGE v.1.45 software. A particle (bead or sperm head) is observed as centrifugal force is increased and velocity is determined from distances measured in time-sequenced captured frames. Particle position was measured with a precision of ± 0.4 μm. Final magnification was approximately 2000× and the field of vision was 2500 μm^2. (Any individual wishing to construct their own centrifuge microscope is welcome to consult me via e-mail at rjbaskin@ucdavis.edu.)

Theoretical Considerations

To explain the nature of forces operating on a particle that is rotating at an angular frequency ω, on the rotor (microscope stage) we must consider an inertial frame of reference and a rotating frame of reference.[9] Within

[9] E. F. Taylor, "Introductory Mechanics," Chaps. 6 and 9. John Wiley & Sons, 1963.

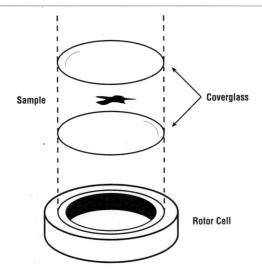

FIG. 3. Diagram of assay cell. Cell is machined to fit into rotor. Sample is placed between coverslips, which are sealed onto cell ledge.

the fixed, inertial frame of reference, a test particle (sperm or bead) at a radial distance r from the axis of rotation will experience an inward radial acceleration, $a_I = \omega \times (\omega \times r)$ (centripetal acceleration), which keeps the particle on a circular path. The equation of motion in this inertial frame of reference is thus given by $F = ma_I = m\omega \times (\omega \times r)$, where m is the mass of the particle being rotated. For this particle to remain stationary in the rotating frame of reference, we have a balance of forces: $F - m\omega \times (\omega \times r) = 0$. Such a force, directed outwardly on the particle in the rotating frame of reference is called a *fictitious* or *pseudoforce*. A particle in the rotating frame of reference experiences this force in a very real way: it is directed away from the center of the rotor and, hence, is termed centrifugal.

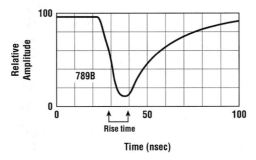

FIG. 4. Rise time of the Xenon N-789B nanolamp.

If instead of remaining fixed in the rotating frame of reference, the particle moves with a constant radial velocity, v_r, relative to the rotor coordinates, then the velocity of a particle in the inertial reference frame is given by

$$v_I = v_r + \omega \times r$$

Correspondingly, the acceleration in the stationary reference frame is

$$
\begin{aligned}
a_i &= (dv_I/dt) \\
&= (dv_r/dt) - \omega \times v_I \\
&= a_r + 2(\omega \times v_r) + \omega \times (\omega \times r)
\end{aligned}
$$

In the inertial frame of reference, the total force acting on the particle can be written as

$$F = ma_I = m[a_r + 2(\omega \times v_r) + \omega \times (\omega \times r)]$$

As before, if we transpose the latter two terms, we see that in the rotating frame of reference, there are now two fictitious or pseudoforces operating:

$$F_{rot} = ma_r = F - 2m(\omega \times v_r) - m\omega \times (\omega \times r) = 0$$

The quantity $-2m(\omega \times v_r)$ is the Coriolis force, directing a force component transverse to the motional direction, r; and $-m\omega \times (\omega \times r)$ is the centrifugal force pushing the particle away from the center of the turntable.

Thus, if the velocity of the particle movement in the rotating frame is much smaller than the angular velocity of rotation, as is the case in most experiments involving the motility of motor proteins (i.e., velocities in the range of 1 μm/sec), the Coriolis force will be negligible in comparison with the centrifugal force.

Centrifuge Microscope-Based Assays

Two different motility assays, capable of estimating motor protein forces, have been used in the centrifuge microscope. The goal of both of these assays was to measure the force developed by molecular motors and determine the relationship between force and velocity for different motors.

The first assay estimates the force developed by myosin molecules adsorbed to a latex bead and measures the velocity of movement of these beads on actin filaments. The questions asked are these: How does the myosin force determined from this motility assay compare with the force per molecule estimated from fiber studies (1–2 pN/molecule), and how does the force-velocity curve, determined using a centrifuge microscope-based motility assay, compare to that obtained from an intact fiber (rectangular hyperbolic curve)?

The second assay measures the force developed and the velocity of movement of a demembranated sperm moving on a kinesin-coated coverslip. The goal of this assay was, as mentioned earlier, to determine the maximum isometric force/kinesin molecule (earlier studies estimated it at less than 1–2 pN/molecule), and to determine a force-velocity curve for kinesin; a relationship that can only be obtained using a motility assay. In the first assay centrifugal force is acting mainly on the latex bead and in the second assay it is acting mainly on the sperm head.

Bead Assay Preparation of Myosin-Coated Beads

Myosin is prepared from rabbit skeletal muscle[10] and stored in 50% glycerol, 0.3 M KCl, 0.2 mM NaHCO$_3$, 1 mM dithiothreitol (DTT), at $-20°$. Before use,[5] myosin is washed twice with 0.5 M KCl, 20 mM PIPES, pH 7.0, 1 mM DTT to prepare a myosin sample at a concentration of 1 mg/ml in 50 mM KCl, 20 mM PIPES, pH 7.0, and 0.1 mM DTT. Tosyl-activated polystyrene beads (Dynabeads, Dynal, Oslo) are incubated with the myosin sample (10^7 beads/ml) in 50 mM sodium carbonate/bicarbonate (pH = 9.0) for 60 min at 0°. The solution is diluted 1:9 with 20 mM PIPES (pH = 7.0) and the myosin-coated beads are precipitated by centrifugation and resuspended in a Mg-ATP solution containing 5 mM ethylene glycol-bis (β-aminoethyl ether)-N,N,N′,N′,-tetraacetic acid (EGTA), 6 mM MgCl$_2$, 1 mM adenosine triphosphate (ATP), 200 mM sorbitol, 50 mM KOH, and 20 mM PIPES (pH = 7.0).

Bead Assay Preparation of the Assay Cell

The bead assay used in the centrifuge microscope is a modified version of that described by Sheetz and Spudich.[11] An internodal cell preparation from *Nitellopsis obtusa* is fastened to the bottom of an acrylite cuvette using polyester thread and mounted on the rotor. The cell is cut open at both ends and its contents replaced with a Mg-ATP solution. Polystyrene beads (diameter = 2.8 μm, specific gravity = 1.3) coated with rabbit skeletal muscle myosin in the Mg-ATP solution are introduced into the cell. Beads were observed moving on actin cables in directions with and against the centrifugal force. The centrifugal force (load) F is calculated from $F = \delta\rho V r \omega^2$, where $\delta\rho$ is the density difference between the bead and the surrounding solution, V is the bead volume, r is the radius of centrifugation, and ω is the angular velocity of the rotor.

[10] S. V. Perry, *Methods Enzymol.* **2**, 582 (1955).
[11] M. P. Sheetz and J. A. Spudich, *Nature* **303**, 31 (1983).

Sperm-Based Assay

In a centrifuge microscope-based sperm motility assay,[6] we determine the centrifugal force necessary to stop the movement of demembranated sperm on a kinesin-coated glass surface. The centrifugal force affects the sperm head strongly due to its high density (approximately 1.6 g/cm^3) and is transmitted to the kinesin attachment site via the axoneme. Since kinesin moves toward the tail or the plus end of an axoneme, bending of the axoneme by the centrifugal force generates a torque about the point of kinesin attachment. (Under certain conditions the sperm axoneme functions as a lever arm, magnifying the effects of the centrifugal force. This results in pull-off prior to stall.)

Sea urchin kinesin (SUK) is purified from sea urchin eggs by modifications of published procedures.[12] Bovine brain kinesin is purified from fresh bovine brains,[13] which were stored on ice no longer than 2 hr prior to removal of the meninges.

Sea urchin sperm (*Strongylocentrotus purpuratus* or *Lytechinus pictus*) are demembranated by treatment with Triton X-100 in a high-salt solution by modification of published procedures.[14] They are incubated in a demembranating solution [10 mM Tris(hydroxymethyl)aminomethane (Tris) base, pH = 8.1; 0.15 M potassium acetate; 2.0 mM EGTA; 1.0 mM MgSO$_4$; 1.0 mM DTT; 0.05% (w/v) Triton X-100] for 5 min and then washed in a high-salt buffer (PMEG with 0.5 M KCl, pH = 8.1) to inactivate the dynein-mediated sperm motility. The demembranated sperm are checked for intact, nondamaged axonemes using video-enhanced differential interference contrast microscopy. The region of kinesin-axoneme attachment is easily observable as the sperm is moved. The sperm is pushed head-first as the kinesin motor moves towards the plus (tail) end until it moves beyond the location of the kinesin and detaches from the coverslip. When kinesin is moving demembranated sperm against the centrifugal force, the sperm head bends down (0.5–1.5 μm) away from the kinesin-coated coverslip.

Using the centrifuge microscope-based assay, sperm movement is observed at a series of rotor velocities corresponding to different values of centrifugal force. We monitor sperm that are moving within ±10° of the centrifugal force vector. Velocity of sperm movement is measured for three to five values of centrifugal force. Stall force (P_0) is found by applying a linear regression analysis to the data and extrapolating to zero velocity. Sperm are observed to move both with and against the direction of centrifugal force.

[12] D. Buster and J. M. Scholey, *J. Cell Sci.* **14(Suppl.),** 109 (1991).
[13] G. S. Bloom, M. C. Wagner, K. K. Pfister, and S. T. Brady, *Biochem.* **27,** 3409 (1988).
[14] B. H. Gibbons and I. R. Gibbons, *J. Cell Biol.* **54,** 75 (1972).

Surprisingly, sperm moving in the same direction as the centrifugal force are also observed to slow down as centrifugal force is increased. This is likely due to the centrifugal force inducing a torque on the sperm head and the subsequent development of a force component, which tends to push the head into the glass surface. This will result in a drag force that will slow motor-driven sperm movement.

Sperm Assay Preparation of the Assay Cell

A coverslip is hydrated in a moist chamber for approximately 2 min and is then coated with 6 μl of 5 mg/ml casein. The casein solution is prepared by stirring 5 mg/ml casein in water for 30 min at room temperature, passing the solution through a 0.2-μm filter, and finally subjecting it to a high-speed spin (12,000 rpm in a Sorvall SS-34 rotor) for 5 min to separate residual lipid from the casein. Three microliters of 20 mM Mg-ATP is added to 14 μl of a kinesin solution, which is gently applied to the casein-coated coverslip, so as not to disturb the coating. The coverslip is then incubated in the moist chamber for 3–5 min to allow kinesin to bind to the surface. The casein coating is estimated to be ~8 nm thick, allowing kinesin that is bound to the coverslip to extend beyond the casein coating. We assume that the kinesin is bound directly to the glass, since kinesin does not bind to casein in solution and other proteins may be used to aid in the binding of kinesin to a coverslip besides casein.[15] The optimal concentration of casein varies for different motor proteins. For dilute SUK it is 0.33 mg/ml.

The sample cell consists of two coverslips that are separately prepared (one as described above) and then joined into a sandwich configuration. Demembranated sperm solution (3–5 μl) is gently layered onto the kinesin-coated coverslip and allowed to sit approximately 2 min to allow the dense demembranated sperm to settle down to the kinesin coating. During incubation of the coverslip with the kinesin-nucleotide solution, the second, uncoated glass coverslip is placed on the ledge in the assay cell where silicon grease had previously been applied to provide a seal. (In some experiments the glass coverslip was glued to the assay cell using super glue, but this was found to be unnecessary because the silicone grease forms a tight seal.) An ATP regenerating system consisting of 2 μl each of creatine phosphate (50 mM) and creatine phosphokinase (10 U/ml) is placed on this uncoated coverslip. Finally the kinesin-nucleotide-demembranated sperm coverslip is placed on top of the ATP regenerating system coverslip, which was on the cell holder. The coverslips are sealed with a solution of Vaseline, lanolin,

---

[15] J. Howard, A. J. Hudspeth, and R. D. Vale, *Nature* **342,** 154 (1989).

and paraffin, which is heated to 35–37°. This prevents fluid evaporation as well as leakage once the cell is spun in the centrifuge microscope. This procedure brings the total volume of the cell to about 30 μl. (The reason for having some solution on each coverslip is to limit air bubbles in the assembled cell.)

Solution Drag Force

As the sperm moves through the solution, it must overcome the viscous drag force of the liquid, defined by $F_d = f \, v_r$ where the viscous drag coefficient, f, is multiplied by the velocity of sperm movement and is oppositely directed. The viscous drag coefficient depends on the size and shape of the particle being moved and on the solution viscosity. For the purpose of calculating the viscous drag force that opposes kinesin-driven motility, we have modeled a demembranated sperm as a combination of a spherical head, with viscous drag coefficient of $f_h = 6\pi\eta r_h$; and a long thin oblong tail, with viscous drag coefficient along the axis of the axoneme equal to $f_t = 4\pi\eta a/[\ln(2a/b) - 1/2]$,[16] where b is the radius of the tail and $2a$ its length. Using this model for a demembranated sperm, the viscous drag resisting the kinesin-driven motion ranges between 0.03 and 0.15 pN for kinesin-driven velocities between 0.2 and 1.0 μm/sec, where, considering the proximity of the sperm to the glass surface, we have used $\eta = 0.04$ poise (~4× the estimated solution viscosity); the demembranated sperm tail length, $2a = 40$ μm; the demembranated sperm tail width, $2b = 0.10$ μm; and the radius of the sperm head, $r_h = 1$ μm. Similar drag force corrections must also be applied to bead movement in bead-based assays.

Because we are measuring forces applied to demembranated sperm being moved through a buffer by kinesin motors, we can modify the mass term using the definition of volume density, $m = \delta\rho V$, where $\delta\rho$ is the volume density of the particle relative to its surrounding medium, and V is the volume of the particle. We assume that the volume density of the sperm is 1.6 g/cm^3 based on the estimates of Da Silva et al.[17] for bull sperm and the volume of the sperm is 4.2 μm^3. Typical rotational speeds range from 0 to 4000 rpm, generating forces (force $= \delta\rho V \, r\omega^2$) on the sperm heads of up to 48 pN.

[16] S. J. Broersma, Chem. Phys. 32, 1632 (1963).
[17] L. B. Da Silva, J. E. Trebes, R. Balhorn, S. Mrowka, E. Anderson, D. T. Attwood, T. W. Barbee, Jr., J. Brase, M. Corzett, J. Gray, J. A. Koch, C. Lee, D. Kern, R. S. London, B. J. MacGowan, D. L. Matthews, and G. Stone, Science 258, 269 (1992).

Functional Density of Motor Proteins

When incubating a solution of kinesin on a coverslip, only a small percentage of kinesin will bind in a functional manner. Casein coating of the coverslip surface has been shown to greatly increase the number of functional motors. Assays have been developed to estimate the number of functional motors on a glass surface. These are discussed by Howard *et al.*,[15] by Block *et al.*,[7] and by Howard *et al.*[18] In our kinesin preparations, we estimate the presence of one to two functional kinesin molecules for about every two to six thousand added to the assay (depending on the particular kinesin preparation). At the lowest kinesin concentration that will support sperm movement we estimate that >80% of the sperm are being moved by only one or two motors. These calculations have implications for the assay with the fully concentrated (high density) kinesin as well, suggesting that we probably have (at a minimum) 3–20 kinesin molecules generating motility along an axoneme.

We have never observed demembranated sperm rotating about a point as is observed in microtubule assays. In these assays this result provides strong support for the argument that one motor is available per microtubule. In the absence of these data it is difficult to establish that demembranated sperm are being moved by one kinesin molecule even though the functional density assay would suggest that this is the case. It is likely that the absence of rotation is due to the inability of Brownian forces to overcome the viscous resistance of moving an entire demembranated sperm (head plus 40-μm tail). The lack of this evidence leads, therefore, to less certainty in the number of kinesin molecules active in sperm movement (especially when using low-density kinesin preparations) and is an inherent limitation of the sperm assay.

Results of Centrifuge Microscope-Based Assays

In the bead-based assay, (centrifugal) force was plotted against bead velocity.[5] A force-velocity relationship similar to that measured in intact muscle fibers was observed. Considerable variation in the results was seen, possibly due to the fact that myosin was most likely in a filamentous state (solution ionic strength was low), and filament sizes probably varied from bead to bead and between different regions on the same bead. This may be the basis for the reported variation in bead load bearing ability and makes interpretation of the results with "negative" loads difficult.

---

[18] J. Howard, A. J. Hunt, and S. Baek, *Methods Cell Biol.* **39**, 137 (1993).

The sperm-based assay, in theory, allows the estimation of a force-velocity curve and the maximum isometric force of a kinesin molecule. This assumes, however, that the axoneme behaves as a rigid rod. The results differed from those obtained using a laser trap-based assay. Maximum isometric force per molecule was estimated at 0.89 pN and a force-velocity curve with a slope of 0.83 μm/sec-pN[6] was obtained. In contrast, the laser trap-based assays gave a maximum isometric force per kinesin molecule (stall force) of about 5.0 pN[8] and a slope of the force-velocity curve of 0.16 μm/sec-pN. As we show in the next section, the axoneme probably acts as a lever and magnifies the centrifugal force. At the same time this "lever action" force tends to push the distal portion of the sperm axoneme into the glass surface as the axoneme pivots around the attachment point, resulting in a substantial induced drag force. The centrifuge microscope-based estimates of maximum isometric force per kinesin molecule as well as the resultant force-velocity curve are strongly influenced by axoneme bending and drag.

Analysis of Results from Sperm Assay

To estimate the magnitude of the forces acting on a demembranated sperm and examine the possibility that the low values of centrifugal force that resulted in sperm pull-off were due to a levering effect of the axoneme, we modeled a sperm as shown in Fig. 5. (A complete description of this

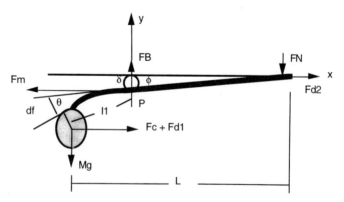

FIG. 5. Diagram of forces acting on a sperm being moved by a kinesin molecule against an applied centrifugal force. The kinesin molecule is represented by the small circle at point P. Centrifugal force (F_c) and drag forces (F_{d1} and F_{d2}) pull the sperm in the directions indicated by the arrows. Kinesin motor force (F_m) pushes the sperm in the opposite direction. Axoneme segment l_1 acts as a lever arm and a torque is developed at point P ($M = \rho - \rho_0)V$; $\delta = 2 \times 10^{-8}\ M$).

analysis is given in Ref. 6.) We assume that the demembranated sperm has an axoneme stiffness coefficient, EI, of approximately 4×10^{-22} N m^2, depending on the ATP concentration.[19] We analyze the forces developed and the torque balance in the presence of a motor protein force (F_m) and an oppositely directed centrifugal force (F_c) acting on the sperm head. There are two significant drag forces. One from the sperm head moving in solution, F_{d1}, where $F_{d1} = f_d v$, with f_d being the Stokes coefficient for the particle in the particular solvent. We let $f_d = 6\pi\eta a$, where η is the solution viscosity and a is the radius of the sperm head. The other drag force is due to the distal portion of the axoneme moving near or against the wall, F_{d2}. When the axoneme is not in contact with the coverslip, the frictional drag coefficient will be described by the Broersma[16] equation. When in direct contact with the coverslip, this drag force will be $F_{d2} = k F_N$, where F_N is the normal force exerted by the coverslip against the distal end of the axoneme, and k is the frictional coefficient between the coverslip and the axoneme. The flexibility of the axoneme will allow bending (angle θ) of the l_1 portion of the axoneme. We define the bending force as W, where

$$W = Mg \cos(\theta + \phi) + (F_c + F_{d1}) \sin(\theta + \phi) \tag{1}$$

With $C = l_1^2/3EI$, we obtain

$$\tan \theta = \frac{C Mg}{1 - C(F_c + F_{d1})} \tag{2}$$

We note that when $C(F_c + F_{d1}) = 1$, $\tan \theta =$ infinity, $\theta = \pi/2$. We calculated the force tending to pull off the sperm ($-F_B$) as a function of l_1 for different applied centrifugal forces. We used an EI value (stiffness) of 4×10^{-22} N m^2, which is approximately 10× that of the single microtubule measured by Gittes et al.[20] and in the range of direct measurements of echinoderm stiffness by Okuno and Hiramoto.[19]

We find (Fig. 6) that with $F_c = 0.5$ pN, the probability of pull-off before stall is high since the length l_1 at which binding force would exceed 5 pN is <40 μm. For higher F_c forces, the axoneme position which would allow a pull-off force > 5 pN to be developed, is shorter, assuring an even higher probability of pull-off.

The major conclusion from this analysis is that centrifugal forces in the range of 0.4–1.2 pN when applied through a 30- to 40-μm lever arm (sperm axoneme) can generate forces in excess of 5 pN, and it is this lever effect

[19] M. Okuno and Y. Hiramoto, J. Exp. Biol. **79**, 235 (1979).
[20] F. Gittes, B. Mickey, J. Nettleton, and J. Howard, J. Cell Biol. **120**, 923 (1993).

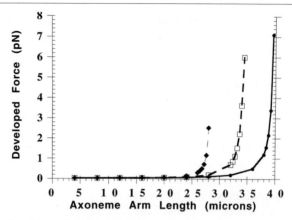

FIG. 6. Developed force $(-F_B)$ as a function of axoneme arm length (μm). EI = 4 × 10–22 N m^2. Curves are for F_c = 0.5 pN (●), 1.0 pN (□), and 1.5 pN (◆).

that causes sperm to pull off from the kinesin coated surface in the centrifuge microscope-based sperm assay.

Axoneme Surface Drag Force

Another result of the centrifuge microscope-based sperm assay was the steep slope of the force-velocity curve, that is, the rapid decrease in velocity as centrifugal force (load) is increased. The reason for this, in light of the preceding assay, is obvious. The same force tending to pull the sperm away from the kinesin molecule (or molecules) will create a torque around the region of attachment, which will push the distal portion of the axoneme against the protein-coated glass surface. As the axoneme is moved forward, this force will induce a drag on the axoneme that the motor will need to overcome in order for movement to occur. Thus at the same time the motor is resisting pull-off induced by the centrifugal force acting on the sperm head, it must also overcome drag in the distal portion of the sperm axoneme. Estimates of the magnitude of axoneme drag force depend on assumptions as to the drag coefficient, length of axoneme, and axoneme distance from the glass surface. With an axoneme to glass distance of <0.1 μm the drag coefficient is increased by about 8-fold over that in free solution and will increase further as portions of the axoneme move closer to the glass surface.

Conclusions

The centrifuge microscope is a reliable and precise method for applying a force or load to microscopic particles, but applying the load through a sperm axoneme introduces complications due to compliance and bending. This results in the generation of a torque around the point of motor attachment and a lever arm effect between the point of centrifugal force attachment and motor attachment. Increased drag as the distal portion of the axoneme is forced against the glass surface also results.

Future studies using the centrifuge microscope need to focus on an assay that will allow better quantitation of the isometric force per motor molecule and accurate determination of the single motor force-velocity relationship. One potential assay involves adsorbing microtubules to the surface of an assay cell such that they are aligned parallel to the centrifugal force vector, and then measuring the velocity of movement of a kinesin-coated bead as centrifugal force is varied. A motor protein such as myosin presents special difficulties due to the short time during an ATPase cycle that it is strongly bound to an actin filament. The centrifuge microscope is in the class of constant-tension probes as is the optical tweezers. This opens the possibility of a bead or other transported object (i.e., microtubule) being driven backward or detaching during that portion of the ATPase cycle that the motor is not strongly bound.[7] Thus assays involving more than one motor protein molecule may be required in centrifuge microscope studies.

With the availability of computer-controlled machining techniques it is possible to construct a centrifuge microscope far more stable than the current model. Feedback regulation of rotor speed would contribute to increased stability of the microscope. If one is constructed with a rotor wobble of less than 0.1 μm, a higher power objective could be used (i.e., $100\times$, NA = 1.4) and fluorescently labeled filaments and microtubules observed.

[36] Motility Assays on Molluscan Native Thick Filaments

By Yung Jin Han and James R. Sellers

Introduction

The sliding actin *in vitro* motility assay and the *in vitro* force measurement assays that have evolved from the original assay make use of the ability to image single rhodamine-phalloidin-labeled actin filaments.[1–5]

---

[1] T. Yanagida, M. Nakase, K. Nishiyama, and F. Oosawa, *Nature* **307,** 58 (1984).

These assays have revealed much about myosin's function and have been used with myosins from many sources.[6] The assays can be used with either synthetic thick filaments of myosin, with myosin bound to the surface as monomers or with enzymatically active subunits of myosin. In all of the above cases, the myosin is not imaged and is assumed to have bound randomly to the surface. To correlate directly the position of actin on the myosin thick filament with the speed and direction of movement, it is necessary to have a thick filament that is at least 2–4 μm in length, which is considerably longer than native vertebrate thick filaments. Synthetic thick filaments from vertebrates are typically heterogenous and often do not contain the precise structural characteristics as native thick filaments.[7] On the other hand, many molluscan muscles contain immense thick filaments that can be up to 50 μm in length.[8,9] These thick filaments have a core of paramyosin, a rodlike coiled-coil protein, that packs in a paracrystalline array to form a bundle.[10] Myosin is deposited around the outside of this bundle in a bipolar manner such that in the muscle, actin filaments would be drawn toward the center of the thick filament.

These molluscan thick filaments can be easily isolated using mild extraction of muscle fibers combined with differential centrifugation. They offer several advantages as a system for *in vitro* motility. They can be imaged in the light microscope using one of several techniques including video-enhanced differential interference contrast (DIC) microscopy and dark-field microscopy (Fig. 1). They offer extended linear tracks along which the movement of actin filaments can be measured. When combined with methods for holding the ends of actin filaments, such as dual optical traps this system allows for the effect of myosin head orientation on force and step size to be measured.[11] Such studies reveal that, as expected from the known polarities, actin filaments can bind to the thick filaments and move toward their center.[12,13] However, it was also observed that actin filaments

[2] S. J. Kron and J. A. Spudich, *Proc. Natl. Acad. Sci. U.S.A.* **83,** 6272 (1986).
[3] Y. Harada, A. Noguchi, A. Kishino, and T. Yanagida, *Nature* **326,** 8005 (1987).
[4] A. Kishino and T. Yanagida, *Nature* **334,** 74 (1988).
[5] J. T. Finer, R. M. Simmons, and J. A. Spudich, *Nature* **368,** 113 (1994).
[6] J. R. Sellers, H. V. Goodson, and F. Wang, *J. Musc. Res. Cell Motil.* **17,** 7 (1996).
[7] J. S. Davis, *Ann. Rev. Biophys. Biophys. Chem.* **17,** 217 (1988).
[8] A. G. Szent-Györgyi, C. Cohen, and J. Kendrick-Jones, *J. Mol. Biol.* **56,** 239 (1971).
[9] L. Castellani, P. Vibert, and C. Cohen, *J. Mol. Biol.* **167,** 853 (1983a).
[10] L. Castellani and P. Vibert, *J. Muscl. Res. Cell Motil.* **13,** 174 (1992).
[11] A. Ishijima, H. Kokima, H. Higuchi, Y. Harada, T. Funatsu, and T. Yanagida, *Biophys. J.* **70,** 383 (1996).
[12] A. Yamada, N. Ishii, and K. Takahashi, *J. Biochem.* (*Tokyo*) **108,** 341 (1990).
[13] J. R. Sellers and B. Kachar, *Science* **249,** 406 (1990).

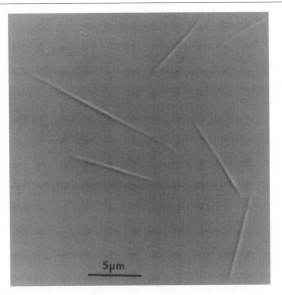

FIG. 1. Video-enhanced DIC light micrograph of isolated *Mercenaria* native thick filaments. The image was obtained with a Zeiss-Axiomat microscope equipped with a 100×, 1.4-NA objective in the critical illumination mode.[21] A NewVicon video camera (DAGE-MTI) was used. [This figure was taken from J. R. Sellers and B. Kachar, *Science* **249**, 406 (1990).]

can move away from the center of the thick filament at much reduced rates compared to the movements in the expected direction (Fig. 2).

There are several detailed methods articles for conducting the *in vitro* motility assay that describe the required equipment, the method for producing flow cells, the general procedure for setting up the assay, and the

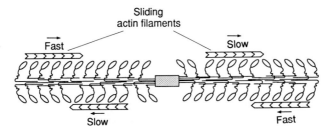

FIG. 2. Schematic diagram showing the fast movement of actin filaments toward the center of the thick filament and the slow movement of actin filaments away from the center. [This figure was taken from J. R. Sellers and B. Kachar, *Science* **249**, 406 (1990).]

methods for quantification.[14-16] In this article, the biochemical isolation of native thick filaments and the necessary modifications to the standard *in vitro* motility assay will be described. Readers are referred to the above articles for other details.

Preparation of Molluscan Thick Filaments

Buffers

Buffer A: 10 mM adenosine triphosphate (ATP), 10 mM MgCl$_2$, 1 mM ethylene glycol-bis(β-aminoethyl ether) N,N,N′,N′-tetraacetic acid (EGTA), 20 mM MOPS (pH 7.0), 3 mM NaN$_3$, 1 mM dithiothreitol (DTT), 0.1 mM phenylmethylsulfonyl fluoride (PMSF)

Isolation of Thick Filaments

This preparation was used by Kachar and Sellers[13] and was modified from a preparation by Yamada *et al.*[17] It can be completed in less than a day and can be carried out with small amounts of muscle tissue. Many molluscan species have two types of muscle that are involved in closing the shell. An example of this is the muscles involved in closing the shell of the clam. *Mercenaria mercenaria.* Two muscle masses are each attached to the top and bottom shell and each muscle mass has two visually distinct muscle types. The larger muscle has a pinkish appearance and is termed the phasic or adductor muscle. It is a fast muscle that is used to close the shell rapidly. The second, smaller muscle, which wraps around the pink muscle in a near semicircular manner, is white in color. This muscle, termed the "catch" muscle, is a specialized muscle that is used to hold the shell tightly shut for extended periods of time against predator attack with little expenditure of energy.[18] Of these two *Mercenaria* muscles, the catch muscle has by far the longer and thicker thick filaments. Myosins from the two muscles are probably derived from a single gene with alternative splicing based on data from similar molluscan systems.[19,20] The pink phasic muscle

[14] S. J. Kron, Y. Y. Toyoshima, T. Q. P. Uyeda, and J. A. Spudich, *Methods Enzymol.* **196,** 399 (1991).
[15] J. R. Sellers, G. Cuda, F. Wang, and E. Homsher, *in* "Motility Assays for Motor Proteins" J. M. Scholey, ed.), pp. 23–49. Academic Press, San Diego, California, 1993.
[16] J. R. Sellers and Jiang, *in* "Laboratory of Cell Biology Handbook" (J. Celis, ed.), 2nd Ed., Vol. 2. Academic Press, San Diego, 1998.
[17] A. Yamada, N. Ishii, T. Shimmen, and K. Takahashi, *J. Musc. Res. Cell Motil.* **10,** 124 (1989).
[18] C. Cohen, *Proc. Natl. Acad. Sci. U.S.A.* **79,** 3176 (1982).
[19] L. Nyitray, A. Jancsó, Y. Ochiai, L. Gráf, and A. G. Szent-Györgyi, *Proc. Natl. Acad. Sci. U.S.A.* **91,** 12686 (1994).
[20] C. L. Perreault-Micale, V. N. Kalabokis, L. Nyitray, and A. G. Szent-Györgyi, *J. Musc. Res. Cell Motil.* **17,** 543 (1996).

myosin has a higher rate of *in vitro* motility than does the white catch muscle myosin (Fig. 3). Therefore, care should be taken to separate the two muscle types. The preparation of native thick filaments has worked from both phasic and catch muscles from three clams, *Mercenaria mercenaria*, *Mya oranaria*, and *Spisula solidissima*, as well as from the anterior bysuss retractor muscle (ABRM) of the mussle, *Mytilus edulis*. Because live clams and mussels are commercially available, specimens can be purchased at local seafood market and even grocery stores. They can be held for short periods of time in cold room or refrigerators.

1. Dissect approximately 1 g of muscle from the mollusk taking care to choose the proper muscle type. Dice the muscle into small pieces and place them into a beaker containing buffer A. Rinse the muscle twice in buffer A. Homogenize in 5 ml of buffer A in an Omnimixer

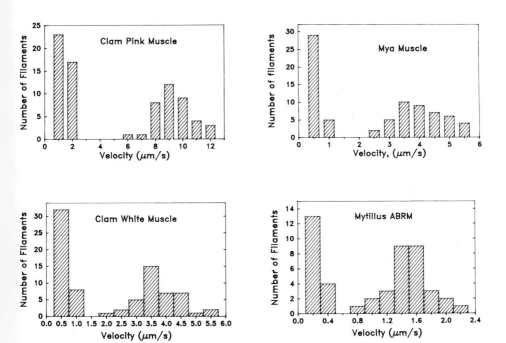

FIG. 3. Histograms of speeds of actin filament sliding on four molluscan native thick filament preparations showing a bimodal distribution. The movement toward the center of the thick filament represents the fast population and the movement away from the center of the thick filament represents the slow population. Immobile actin filaments were not scored. In each case the mean speed of the slow movement was about 10 times less than the speed of the fast population.

(Sorvall, Newtown, CT) using the small cup at a setting of 7.5 for 2× 7 sec on ice.

2. Add an equal volume of buffer A containing 0.1% Triton X-100 to the homogenate. Let it sit on ice for 5 min.
3. Sediment the homogenate at 500g for 5 min.
4. Carefully remove the supernatant using a pipette with an enlarged tip to avoid shearing, then sediment for 20–30 min at 5000g. The length of time of centrifugation is determined by the size of the thick filaments with long thick filaments pelleting more rapidly than short ones. The time should be chosen such that the thick filaments form a loose pellet, which can be easily resuspended in buffer A.
5. Repeat the two centrifugation steps (3 and 4) two or three times.
6. Gently resuspend the final pellet in a small amount (1–2 ml) of buffer A. Sediment at 500g for 5 min immediately prior to use to remove aggregates.

Thick filaments remain active in moving actin filaments for 1–2 days. The final preparation consists primarily of myosin and paramyosin in ratios characteristic of the particular muscle when examined by sodium dodecyl-sulfate–polyacrylamide gel electrophoresis (Fig. 4). A few minor bands of undetermined identity are visible as well. Staining of the preparation with rhodamine-phalloidin reveals that there is little actin present, and the band at 42 kDa in lane b has been tentatively identified as arginine kinase.[17]

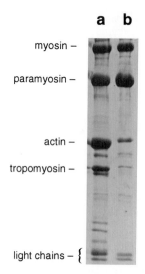

Fig. 4. Sodium dodecyl sulfate (SDS) 12.5% polyacrylamide gel of the muscle homogenate (land a) and the isolated native thick filaments (lane b) from *Mercenaria mercenaria*.

In Vitro Motility Assays of Native Thick Filaments

Buffers

> Buffer M: 20 mM KCl, 20 mM MOPS (pH 7.2), 5 mM MgCl$_2$, 1 mM ATP, 0.2 mM CaCl$_2$, 0.1 mM EGTA, 20 mM DTT, 2.5 mg/ml glucose, 0.1 mg/ml glucose 0.02 mg/ml catalase (last three items from Sigma, St. Louis, MO)

Measurement of the in Vitro Motility

The *in vitro* motility assay has been well described in several recent articles.[14-16] The method below details the modified method used for native thick filaments.

1. Native thick filaments at a concentration of 30 μg/ml in buffer A are applied as a droplet to a nitrocellulose-coated coverslip[15] that is not yet mounted into a flow cell. This allows the filaments to bind in random orientations. When native thick filaments are applied by capillary action into a mounted flow cell they tend to align with the flow and sometimes laterally associate.

2. The coverslip is mounted into a flow cell created by placing coverslip slivers (18-mm^2 No. 0 coverslips cut into about 3- to 4-mm widths) on each side of parallel tracks of Apiezon M grease (Apiezon Products, London) that are separated by about 10–20 mm.

3. The sample is washed with 0.5 mg/ml bovine serum albumin in 20 mM KCl, 20 mM MOPS (pH 7.4), 5 mM MgCl$_2$, 0.1 mM EGTA.

4. Rodamine-phalloidin-labeled actin (2 nM) is added in buffer M. Note that this buffer contains 0.1 mM CaCl$_2$ in excess of the EGTA concentration. This is necessary since molluscan myosins require bound calcium ions for activity. Lower concentrations of free calcium (10^{-6} to 10^{-5} M) can be used if desired.

5. Video-enhanced DIC images of the thick filaments can be captured using a microscope equipped with a 1.3–1.4 NA objective and condenser. The image captured in Fig. 1 used critical illumination mode[21] and were background subtracted, image averaged, and contrast enhanced using a video image processor (Argus 10 from Hamamatsu, Japan).

6. Two alternative modes of imaging native thick filaments are possible. The first uses dark-field microscopy, which in general does not give as clean images as with DIC. The second mode involves "negatively" staining of the background with fluorescent bovine serum albumin. In this latter case,

---

[21] B. Kachar, P. C. Bridgman, and T. S. Reese, *J. Cell. Biol.* **105,** 1267 (1987).

after depositing thick filaments on the coverslip, the surface is blocked with bovine serum albumin mixed with tetramethylrhodamine-labeled bovine serum albumin (Molecular Probes, Eugene, OR). This gives the background a uniform fluorescence except where thick filaments are bound. The key is to use a sufficiently small amount of labeled bovine serum albumin such that the brighter rhodamine-phalloidin-labeled actin filaments can still be observed over the background fluorescence. This has the advantage that the fluorescently labeled actin filaments and the position of the thick filament can be observed simultaneously.

7. Data can be recorded on S-VHS videotapes and can be quantified by previously described methods.[15]

Movement events will be detected as the actin filaments in the solution make contact with native thick filaments. The number of observable events can be increased by raising the concentration of native thick filaments bound to the surface or by increasing the concentration of actin filaments in the solution. Increasing the number of thick filaments may result in difficulty in obtaining movement along a single filament since an actin

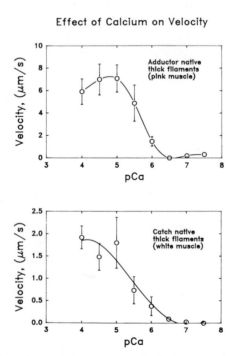

FIG. 5. Calcium dependence of the movement of actin filaments toward the center of two *Mercenaria* native thick filaments. Bars show the standard deviation of the mean.

filament can change from one filament to another if filaments cross or are aligned side by side. Increasing the concentration of rhodamine-phalloidin-labeled actin in the solution results in a higher background fluorescence, which decreases image quality.

Using the above methods, separate or simultaneous images of actin moving on native thick filaments can be obtained in the presence of calcium ions. As the assay proceeds in time, progressively more actin filaments bind to the thick filament and remain immobile. This behavior can be lessened if the thick filaments are as fresh as possible and if actin is not allowed to bind to the thick filaments under rigor conditions before introduction of MgATP. While the bipolarity of the thick filaments is not apparent from the light microscopic images, it is always obvious from observing the speeds of movements of actin filaments. Various thick filaments offer different characteristics in terms of both length and enzymatic properties. *Mercenaria* pink muscle thick filaments have a maximum length of less than 10 μm and move actin filaments at a relatively fast rate of about 9 μm/sec. Thick filaments 30 μm or longer can be isolated from the white catch muscle from this species. These thick filaments move actin filaments at a rate of about 3.5 μm/sec. Striated muscle from another clam, *Mya oranaria*, has shorter thick filaments (7–8 μm) and moves actin at about 3.5–4 μm/sec. A very long and thin thick filament (30–40 μm) can be isolated from the anterior bysuss retractor muscle of the mussel, *Mytilus edilus*. This thick filament exhibits the slowest rate of *in vitro* motility (about 1.5 μm/sec). In all the cases above the rate of actin filament sliding away from the center of the thick filament is about one-tenth of the rate toward the center. This is seen in Fig. 3 as a bimodal distribution.

Molluscan myosins are regulated by calcium binding to the essential light chain.[22,23] Therefore, it is necessary to have calcium ions present in the assay in order to observe movement. Figure 5 shows the calcium dependence of the velocity of actin filament sliding toward the center of the thick filament for the two *Mercenaria* thick filaments.

[22] S. Fromherz and A. G. Szent-Györgyi, *Proc. Natl. Acad. Sci. U.S.A.* **92,** 7652 (1995).
[23] A. G. Szent-Györgyi, *Biophys. Chem.* **59,** 357 (1996).

[37] Use of Optical Traps in Single-Molecule Study of Nonprocessive Biological Motors

By A. D. MEHTA, J. T. FINER, and J. A. SPUDICH

Introduction

Early attempts to reconstitute muscle motility using purified protein components culminated in the gliding filament assay, in which fluorescently labeled actin filaments were observed to move on a myosin-covered surface inside a microscope flow cell.[1,2] This assay has been extended to the single-molecule level by incorporation of an optical trapping microscope along with high-resolution imaging of trapped beads[3-6] or by use of glass microneedles with high-resolution imaging of attached particles.[7] In the optical trapping experiments described here, the prevailing geometry employs two beads attached to either end of a single actin filament in solution. Optical traps generated by focused laser beams are used to "grab" these beads, stretch the filament to tension, and move it into proximity of surface-attached platforms decorated sparsely with myosin molecules.

We describe the single-molecule measurements, using the gliding assay as our point of departure. We first discuss preparation of proteins, coverslips, and labeled polystyrene beads for use in optical trapping. We then provide a sketch of instrument design, described in more detail elsewhere (e.g., Refs. 8–10). Finally, we focus on experimental conditions and data analysis. We review problems in identifying single-molecule binding events and methods developed to overcome them.

[1] S. J. Kron and J. A. Spudich, *Proc. Natl. Acad. Sci. U.S.A.* **83,** 6272 (1986).
[2] S. J. Kron, Y. Y. Toyoshima, T. Q. Uyeda, and J. A. Spudich, *Methods Enzymol.* **196,** 399 (1991).
[3] J. T. Finer, R. M. Simmons, and J. A. Spudich, *Nature* **368,** 113 (1994).
[4] J. E. Molloy, J. E. Burns, J. Kendrick-Jones, R. T. Tregear, and D. C. S. White, *Nature* **378,** 209 (1995).
[5] W. H. Guilford, D. E. Dupuis, G. Kennedy, J. Wu, J. B. Patlak, and D. M. Warshaw, *Biophys. J.* **72,** 1006 (1997).
[6] A. D. Mehta, J. T. Finer, and J. A. Spudich, *Proc. Natl. Acad. Sci. U.S.A.* **94,** 7927 (1997).
[7] A. Ishijima, Y. Harada, H. Kojima, T. Funatsu, H. Higuchi, and T. Yanagida, *Biochem. Biophys. Res. Commun.* **199,** 1057 (1994).
[8] R. M. Simmons, J. T. Finer, S. Chu, and J. A. Spudich, *Biophys. J.* **70,** 1813 (1996).
[9] K. Visscher, S. P. Gross, and S. M. Block, *IEEE J. Select. Topics Quantum Electron.* **2,** 1066 (1996).
[10] A. D. Mehta, J. T. Finer, and J. A. Spudich, *Methods Cell Biol.* **55,** 47 (1998).

At the outset, we emphasize that the single-molecule measurement, unlike its predecessor, is stochastic at many levels of data collection. These include the amount of data accessible from given flow cells, the incidence of binding events from given surface-attached beads, durations of binding events, and, it turns out, the apparent amplitude of bead deflections caused by these binding events. The experimentalist samples different protein concentrations and, in a given experiment, moves a trapped filament over a field and samples different areas, an active search for stochastic binding events as opposed to passive observation of deterministic ensemble properties. Results take the form of distribution parameters with fitting errors, encompassing data that has been somehow selected for.

Motor Protein Reagents

Considerations for protein reagents are much the same as those for the gliding assay. One can examine the ~500-kDa full-length myosin or its fragments, including the 340-kDa N-terminal dimeric fragment heavy-meromyosin or the 130-kDa monomer Subfragment 1. Proteolytic fragments require more stringent purity than in earlier assays, because full-length or other contaminants could generate binding events that would be selected in an active search, even if the fragment of interest did not generate such events. The search and selection for stochastic events allows such contaminants to affect results out of proportion to their relative amount present. To date, most of our work has involved skeletal muscle HMM, prepared as described previously,[2] or full-length myosin purified to homogeneity from *Dictyostelium discoideum*. Preparation of skeletal muscle myosin and actin are as previously described.[2] Actin affinity purification to eliminate "dead" myosin protein (see below) is also as previously described,[2] and is almost always necessary, sometimes every few hours, when working with *Dictyostelium* proteins.

Coverslip Preparation

The experiment requires a flow cell, constructed by fixing a modified coverslip onto a microscope slide using adhesive spacers. The myosin is fixed to the coverslip, and the observed protein interactions occur within 1–2 μm of the coverslip surface. Unlike the gliding assay, the present one requires elevated platforms on which the myosin must be mounted to interact with the actin filament, itself attached to beads 1 μm in diameter and thus held approximately 1 μm away from the surface. To this end, 1- or 2-μm-diameter silica beads are mounted on a microscope coverslip by spreading a dilute bead solution (2–3 mg/ml) in 0.05% Triton X-100 over the coverslip. The silica bead solution must be mixed vigorously and large aggregates allowed to settle before use. After drying, we cover the beads with nitrocellulose as previously described.[2]

Microspheres

The experiments involve trapping microspheres, attaching them to ends of an actin filament, and using them as handles to stretch and move the filament. The microspheres for trapping must have high affinity for actin filaments, usually accomplished by attachment of an actin binding protein. To date, the most common method involves N-ethyl maleimide-modified (NEM) myosin, which is unable to release adenosine diphosphate (ADP) and is thus permanently bound to actin. We have also experimented with the actin-binding protein α actinin, but the links are far more likely to break when under tension than those from NEM myosin.

N-Ethyl Maleimide-Modification of Myosin

Buffers

BE: 0.1 mM NaHCO$_3$, 0.1 mM EGTA
2× LB: 20 mM imidazole hydrochloride, pH 7.4, 1 M KCl, 4 mM MgCl$_2$
1× LB: 10 mM imidazole hydrochloride, pH 7.4, 0.5 M KCl, 2 mM MgCl$_2$
SB: 0.1 mM NaHCO$_3$, 0.1 mM EGTA, 3 mM MgCl$_2$, 7 mM dithiothreitol (DTT)
HSAB: 25 mM imidazole hydrochloride, pH 7.4, 200 mM KCl, 4 mM MgCl$_2$, 1 mM EGTA, 1 mM DTT

1. Add nine volumes cold BE to approximately 20 mg of stock myosin and mix to promote filament assembly. After 10 min of storage on ice, pellet filaments at low speed (e.g., 27,000g for 10 min).
2. Dissolve pellet in 2× LB and BE to achieve a final solution of 15 mg/ml myosin in 1× LB.
3. Incubate the solution 10 min at 22°.
4. Add NEM to 1.5 mM by dilution of a stock solution (e.g., 0.1 M NEM in 1× LB).
5. Incubate 20 min at 22°.
6. Add 12 volumes SB and mix. Incubate on ice for more than 1 hr to allow filament assembly.
7. Sediment filaments at high speed (e.g., 230,000g, 10 min).
8. Resuspend to 10–30 mg/ml in HSAB. The final protein preparation can be used for many weeks.

Bead Labeling

NEM-modified myosin can be covalently cross-linked either to beads preactivated for amino group binding or to carboxylated beads activated by carbodiimide treatment. The resulting bead–actin links tend to break when the filament is stretched beyond a few pN of tension, the exact number varying from one filament to another, presumably because the cross-linking has affected the integrity of the myosin. The links remain robust below this tension, but associated compliance is sometimes in the same range as the optical traps, making bead position an imperfect probe of attached actin filament position and requiring one to correct bead movement for displacements absorbed by the linkage. An earlier study attributes this compliance to actin filament bending, indicating that any linkage along the filament axis and not at the end will suffer from this problem unless stretched to such a high tension that the filament can no longer bend.[11] Attachment to beads using end-capping proteins has been demonstrated,[12,13] but may be of limited use in this geometry because only one filament end can be attached to a bead this way. Nonspecific adsorption to beads has demonstrably produced fewer fragile links between the bead and the actin. But we find that the protein tends to exchange off such beads, into solution, and onto the surface on experimentally relevant time scales, all this despite a layer of bovine serum albumin (BSA) that should block any protein attachments to the surface.

The following is a carbodiimide-based attachment procedure.

Buffers

LS-PBS, pH 6.5: 25 mM NaCl, 2.5 mM KCl, 10 mM Na$_2$HPO$_4$, 1.8 mM KH$_2$PO$_4$, pH adjusted to 6.5

LS-PBS, pH 8.0: as above, pH adjusted to 8.0

HSAB-BSA: HSAB, as above, with 1 mg/ml essentially fatty-acid-free BSA added

1. Wash carboxylated beads three times in LS-PBS, pH 6.5. Final suspension should be approximately 2.5% solid. Beads can be sedimented using a tabletop microfuge at 10,000 rpm.
2. Dilute 1-ethyl-3-(3-dimethylaminopropyl)carbodiimide-HCl (EDC) into LS-PBS, pH 6.5, to 10 mg/ml.

---

[11] D. E. Dupuis, W. H. Guilford, J. Wu, and D. M. Warshaw, *J. Muscle Res. Cell Motil.* **18,** 17 (1997).

[12] T. Nishizaka, H. Miyata, H. Yoshikawa, S. Ishiwata, and K. Kinosita, *Nature* **377,** 251 (1995).

[13] N. Suzuki, H. Miyata, S. Ishiwata, and K. Kinosita, *Biophys. J.* **70,** 401 (1996).

3. Within a few minutes of diluting, mix equal volumes of EDC dilution and washed beads. Discard remaining EDC solution, because it cannot be used long after dilution.
4. Incubate 1 hr at 22°.
5. Pellet beads using a microfuge at 10,000 rpm, and resuspend in LS-PBS, pH 8.0. Wash three times in LS-PBS, pH 8.0.
6. Dilute NEM myosin into LS-PBS, pH 8.0, to 1–3 mg/ml. The NEM myosin should form filaments in solution. Suspend thoroughly.
7. Add equal volumes of the diluted NEM myosin and modified beads.
8. Incubate 1.5–2 hr at 22°.
9. Add an equal volume of HSAB-BSA with 0.01 mg/ml TRITC-BSA added. The TRITC-BSA allows one to image the beads in fluorescence.
10. Incubate 5–10 min at 22°.
11. Add glycine to 50 mM to block all remaining binding sites.
12. Wash several times with HSAB-BSA. Although the NEM myosin should bind to the beads in filament form, we now seek to remove any proteins which are parts of filaments but not themselves covalently bound to the bead. Hence, we shift to a high-salt wash.
13. Bath sonicate the bead solution for approximately 10 min before using. This must be repeated regularly. Beads are best if made fresh daily, but tend to bind actin filaments and hold tension well for 2 or often 3 days.

Fluorescent Actin

As in earlier assays,[2] the actin filaments must be observed in order to attach optically trapped beads to them and also to confirm that one and only one actin filament is allowed to interact with surface-bound myosin. Unlike earlier protocols,[2] we currently assemble actin filaments in the presence of tetramethylrhodamine (TRITC) phalloidin to generate brighter filaments, a requirement in these experiments because optical trapping and bright-field imaging optics will prevent the path length minimization commonly employed to optimize throughput in fluorescent imaging.

Buffer

Buffer A: 5 mM potassium acetate, 5 mM potassium phosphate, pH 7.5, 0.5 mM magnesium acetate

Filament Labeling

Add TRITC-phalloidin to 10 μM and G actin to 10 μM in buffer A. Incubate overnight to allow filament assembly.

Instrument Considerations

Beam Expansion and Steering Optics

The experimental apparatus, shown in Fig. 1, is mounted on a vibration isolated table. Detailed accounts of trap construction are described elsewhere (e.g., Refs. 8–10). We provide a sketch of our instrument here. A single Nd : YLF laser beam (wavelength 1047 nm) is split into its polarization components by a polarizing beamsplitter. A half-wave plate is positioned just after the laser output, since a rotation of laser light polarization can be used to adjust the fraction of power in each of the two split beams. The two beams must be steered separately to provide independent movement of the two optical traps.

Optical trapping requires that one almost fill the back aperture of a high-numerical-aperture (1.2–1.4) microscope objective with the roughly collimated trapping beam. Our system uses a $63 \times$ oil immersion objective of NA 1.4 (Zeiss). The large beam size and high numerical aperture translate to a sharper optical gradient at the focal point and thus a stronger trap. However, the price for high NA is typically small working distance (here 200–300 μm), preventing deep imaging. Fortunately, this is not needed in our experiment. One cannot trap particles effectively beyond 20 μm into the solution cell, since the laser focus becomes blurred due to spherical aberrations.

In the present assay, each trap requires 10–20 mW of power measured before objective incidence to create traps having stiffness 0.02–0.04 pN/nm. The strength of the optical trap lies in its weakness, as such compliant probes easily yield against the force of a single myosin molecule. Significantly less compliant probes, including more powerful optical traps or atomic force microscope cantilever tips, have been used to perturb single molecules and measure properties under strain. Examples include recent demonstrations of reversible domain unfolding within single titin molecules.[14-16] In contrast, the probes here are used merely to create an appropriate experimental geometry and are rendered as weak as possible so as to minimally perturb the interacting proteins.

To displace the trap laterally in the specimen plane, one must control the angle of incidence of the beam on the objective, which in turn maps to lateral position of the focal waist in the specimen plane. For ease of description, we now move backward from the beam focal waist to the laser.

[14] L. Tskhovrebova, J. Trinick, J. A. Sleep, and R. M. Simmons, *Nature* **387**, 308 (1997).
[15] M. Reif, M. Gautel, F. Oesterhelt, J. M. Fernandez, and H. E. Gaub, *Science* **276**, 1109 (1997).
[16] M. S. Z. Kellermayer, S. B. Smith, H. L. Granzier, and C. Bustamante, *Science* **276**, 1112 (1997).

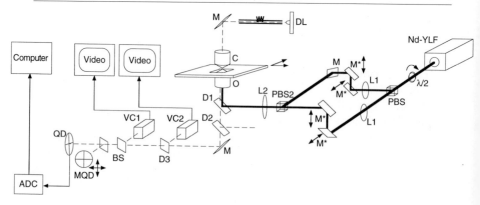

Fɪɢ. 1. A Nd:YLF laser (1047 nm) is split into two independent trapping beams by a polarizing beamsplitter (PBS). The ratio of the two beam intensities can be controlled by rotation of a half-wave plate λ/2. The beam should be parallel on contact with L1, and an additional lens is sometimes needed to eliminate beam divergence. L1 and L2 form a beam expander to enlarge equivalently both beams. In our instrument, each L1 has focal length 5 cm, L2 has focal length 75 cm, and both L1 front focal planes are coincident with the L2 back focal plane, creating a 15-fold expansion of the approximately collimated beam. The mirrors shown cause a lateral beam deflection inside of the expander, optically equivalent to an angular deflection to the left of L2. The two beams are again directed along the same axis by the polarizing beamsplitter PBS2. Dichroic beamsplitter D1 will deflect light above 1000 nm at 45-deg incidence, deflecting the trapping beam upward while allowing the downward pro-gressing imaging light to pass through. The expanded beam nearly fills the back aperture of the objective lens, which focuses the beam to an almost diffraction-limited spot in the specimen plane. The angular deflection noted earlier maps to a lateral deflection of the focal spot in the specimen plane, meaning the traps can be moved laterally without change in depth by shifting mirrors M*. The stage is moved to search for suitable microspheres in the specimen flow cell. The trapping plane is imaged using light from a diode laser DL (830 nm) running below the lasing threshold and passed through a multimode optical fiber with a static mode scrambler. This light is focused by the condenser to a spot around 100–300 μm in diameter. Fluorescence excitation via an argon laser (AL) beam (514 nm), expanded and passed through a diffusive optic, is deflected upward by dichroic beamsplitter D2, which reflects light of wavelength 500–560 nm and passes light above 570 nm.

The downward-traveling bright-field image light and fluorescent emissions pass through dichroic beamsplitters D1 and D2 before leftward deflection by a mirrow. Dichroic beamsplitter D3 will reflect light under 630 nm, primarily the red fluorescent emissions of a rhodamine dye used to label actin filaments, into VC2, a sensitive SIT camera used to image only the fluorescent emissions in the specimen plane. In our experiments, the solution polystyrene beads used for trapping and the actin filaments are labeled with TRITC, and both are visible in the fluorescent image. Around 10% of the remaining light, primarily the bright-field image, is deflected by glass coated on one side for antireflection (BS) into VC1, a CCD camera. Through this, both solution beads for trapping and silica beads fixed to the surface are visualized. The remaining light, most of the bright-field image is split between two quadrant photodetectors, QD and MQD. The image is magnified further by a 20× eyepiece (not shown), with total magnification around 600- to 700-fold at the detector point, so the image of a 1-μm-sized bead almost fills a 1-mm-diameter detector. The signals are processed, amplified, and digitized (ADC) for computer recording.

Before incidence on the objective, the beam is expanded and collimated. Moreover, its angle of incidence on the objective must be adjustable. Collimation is accomplished by focusing the light in the rear focal plane of L2. One seeks to maximize the L2 focal length without allowing the expanding beam to overfill it or the microscope objective back aperture. Adjustment of incidence angle is optically equivalent to lateral shift of the beam before L2 incidence, and this is achieved by the mirrors M*, attached to motorized translation stages controlled by computer or joystick. Lateral beam movement before L2 incidence maps to trap displacement in the specimen plane with an inverse magnification of f_2/f_{obj}, here 75 cm/3 mm = 250. Alternatively, one can use a Galilean beam expander, a diverging lens followed by a laterally movable converging one, to accomplish this in less space. L1 is used to focus the collimated laser beam in the back focal plane of L2, an approximate relation given movements of M*. However, the trapping plane remains more or less constant as the M* mirrors are moved to displace traps by tens of microns in the specimen plane. L1 and L2 together expand the beam by f_2/f_1, here 75 cm/5 cm = 15.

Imaging

The present experiment requires fluorescent imaging of both actin filaments and polystyrene beads held by traps alongside bright-field imaging of the trapped polystyrene beads and surface-bound silica spheres. The fluorescent image facilitates connection of beads to filaments, and the bright field enables appropriate positioning of filaments over the surface. Resolution limits in nanometer-scale measurements typically depend on image brightness, arguing against contrast enhancement at the expense of brightness. We use an inverted microscope, assembled from the minimum number of components required and constructed to minimize mechanical noise. This configuration allows direct mounting of bulky detectors and cameras on the vibration isolated table (Fig. 1).

The bright-field illumination source, an 830-nm diode laser, is mounted on the table and coupled there into a multimode fiber, which then passes through a static mode scrambler, also mounted on the table. The fiber is then directed near the top of the inverted microscope, where the light is expanded, collected, and focused onto the specimen through a 1.4-NA oil immersion condenser positioned just over the specimen slide. Important requirements in illumination sources are described below.

Simultaneous fluorescence and bright-field imaging requires care in spectral separation of the images, although this is less a problem when narrowband laser sources are used. Rhodamine is excited by green light and emits in the red. The brightfield light is in the near infrared, far removed from fluorescence in wavelength. The trapping light is further infrared, allowing separation of the beams using dichroic beamsplitters. Fluorescent excitation is provided by a 35-mW argon ion laser (514 nm), which is expanded and passed through a diffusive optic before incidence on the dichroic beamsplitter D2. D2 passes light above approximately 570 nm, including downward-traveling fluorescent emissions and diode laser light, but reflects light from 500 to 560 nm, including the fluorescence excitation beam. Immediately above D2, the beamsplitter D1 reflects light over 1000 nm, only the trapping beam, and passes all light involved in imaging. Below D1 and D2, a mirror directs this light into the imaging pathway.

The red fluorescent image is deflected into a silicon-intensified target (SIT) camera (VC2) by beamsplitter D3, which reflects light under 630 nm. Because fluorescent emissions are generally weak, one must minimize the number of optics separating the specimen from the camera. In addition to the objective, one lens and a 20× eyepiece are used to magnify and focus the image on the camera.

A filter reflecting light over 1000 nm at normal incidence (not shown in Fig. 1) is included to eliminate residual backscatter from the trapping light. Diode laser light will pass through D3, and approximately 10% is deflected into a CCD camera (VC1) by glass coated on one side for antireflection (BS). Several lenses are used to focus the specimen plane, the same as that imaged by fluorescence, onto the camera with approximately the same magnification as the fluorescent image. One seeks to image the same field in both ways.

Remaining diode laser light is directed at the quadrant photodetectors, specialized four-element silicon photodiode arrays with each element in one quadrant of the detector plane (S-1557, Hamamatsu, Hamamatsu City, Japan). A lens and a 20× eyepiece are used to focus an expanded image onto the detector, with magnification 600–700. This allows the image of a 1-μm bead to nearly fill a 1-mm-diameter detector. High-bandwidth I-V converters with transimpedence gains of 100 MΩ are fixed to the detectors, a requirement since the feeble currents generated are affected by stray electromagnetic interference if passed along lengthy lines before conversion to more robust voltage signals. Parasitic capacitances in the I-V converter tend to limit detector bandwidth, a potentially serious problem that can be minimized by use of printed circuits and minimal leads. Each of the four signals generated reflects the light incident on one quadrant of the detector. Hence, simple electronic arithmetic generates effective position signals in both lateral dimensions. One can observe a "silhouette" of the bead in real time using an oscilloscope, and record traces for later analysis via analog-to-digital conversion and storage on computer. To verify reliable imaging, we check for signal stability, linearity of signal with bead displacement (induced by acousto-optic modulators not discussed here), and lack of statistical correlation between diffusion in the two lateral dimensions. Regarding the last, bead diffusive motions in orthogonal directions must be independent; any correlation detected is an artifact of problematic imaging.

Position resolution is limited by sources of signal noise indistinguishable from actual bead movements. These include mechanical vibrations, fluctuation in light intensity, and electronic noise. Vibrations are typically minimized by firmly mounting the instrument on a vibration isolated table, by reducing ambient sounds, and by working at basement levels to avoid building vibration modes. Light fluctuation is minimized by use of a stable source, but one faces fundamental limits in statistical fluctuation due to the quantum nature of light energy, commonly called "shot noise." Since this noise varies with the square root of imaging light intensity and the signal varies linearly with this intensity, one can combat shot noise by making the image more bright. This also minimizes the relative contribution of electronic noise, independent of light intensity. Other sources of electronic noise include Johnson noise across the feedback resistor used in the I-V converter mounted on the detector itself. This noise scales with the square root of the resistance value and the signal scales with the resistance, allowing one to decrease its relative contribution by use of a larger resistor. However, this concern prevents one from using a small transimpedence gain and further amplifying the signal afterwards. Finally, one must exercise care in use of appropriate amplifiers to minimize noise in the input bias current or the offset voltage as appropriate. Circuit design considerations are described in more detail elsewhere.[8,10]

Alternatively, one can image the trapped bead by tracking the optical trapping laser beam in the rear focal plane of the condenser on the other side of the specimen. This technique and its variants are discussed elsewhere.[9]

Experimental Cell

Flow cells are constructed as described before[2] but with double-stick tape in place of coverslip slivers to separate the modified coverslip from the glass slide. Flow cell volumes are approximately 20–30 μl.

Buffers

AB: 25 mM imidazole hydrochloride, pH 7.4, 25 mM KCl, 4 mM MgCl$_2$, 1 mM EGTA, 1 mM DTT

AB-BSA: AB with 1 mg/ml essentially fatty-acid-free BSA

M1: 0.036 mg/ml catalase, 0.2 mg/ml glucose oxidase, 0.01% solids microspheres (above), 40 units/ml pyruvate kinase, twice final adenosine triphosphate (ATP) concentration; bring to final volume with AB-BSA.

M2: 6 mg/ml glucose, 2 mM phosphoenolpyruvate, 0.5 μM fluorescent actin; bring to final volume with AB-BSA.

Glucose oxidase, catalase, and glucose are included to slow photobleaching of the TRITC label. Pyruvate kinase and phosphoenolpyruvate are included to regenerate ATP and maintain its initial concentration. Their inclusion is redundant if [ATP] exceeds 1 mM or so. However, most binding events at such saturating [ATP] are difficult to identify as apart from thermal noise. Because ATP binding must precede myosin detachment from the actin filament, single binding events are rendered longer and thus more pronounced by limiting [ATP] to the 1–20 μM range.[3]

Enzymes are kept separate from their substrates until the moment of introduction into the experimental flow cell. Actin filaments are also kept apart from actin-adhesive microspheres until the moment of introduction, to prevent formation of a bead/filament mesh. Under optimal conditions, many beads will attach to only one filament by diffusive contact, and very few beads will attach to two filaments.

1. Infuse freshly diluted stock myosin in AB into flow cell. A good starting point is one-tenth the dilution needed to support smooth movement in the gliding filament assay. For instance, at 5–20 μM ATP, 5 μg/ml rabbit skeletal heavy meromyosin or 20 μg/ml *Dictyostelium* myosin are reasonable estimates.
2. Allow 4 min for surface attachment.
3. Wash the flow cell with AB-BSA.
4. Mix equal volumes of M1 and M2 together and the flow mixture into the cell. Excessive incubation time after mixing causes the microspheres and the actin to become entangled.

Experimental Procedure

We describe the procedure for capturing actin filaments and searching for myosin binding events. To prevent an evaporation front from inducing stray forces on trapped particles and eventually drying the flow cell, we seal open cell edges with parafilm to retard evaporation. The flow cell should not be left completely open, although one also does not want to lose much time while a liquid sealant dries. The useful life of a cell can be as short as 30 min, due to bead/filament aggregations and attachment of the polystyrene beads to the surface.

Right after sealing, we place the flow cell in the optical trapping microscope, observe the bright-field image, and search for beads to trap. Because the trap is fixed to the objective used in imaging, stage movement allows one to search the flow cell while keeping the optical probes at fixed points in the image, a more tricky maneuver with mechanical probes like micro-needles.

Ideally, one of the two beads trapped should have an actin filament already attached, as mentioned previously. One can also promote attachment by searching for free filaments and moving beads into their proximity. Labeled actin filaments are observed under fluorescence.

Once a trapped bead has an actin filament attached, we move one bead to a marked point conjugate with the center of the fixed quadrant detector (bead 1) and the other (bead 2) onto the axis connecting points conjugate with both detectors, such that the two beads are separated approximately by the length of the actin filament. After manual movement of bead 1 close to the area imaged by detector QD (Fig. 1), final centering of the bead image on the detector is achieved by computer control of the M* motors to steer the bead. We then move the stage to create a solution drag and stretch the filament along the axis connecting both beads. Once contact is made and the filament attached to both beads, the bead/filament assembly is slowly lowered into proximity of a surface-attached silica sphere so that the actin filament is draped just over the silica sphere. To stretch the filament, bead 2 is moved away from bead 1 until bead 1 is observed to move by a given distance in its trap. This algorithm is automated and performed through computer control of M* motors. The tension in the filament scales linearly with bead 1 displacement from its trap center, allowing us to stretch always to a roughly consistent tension level. Once the filament is stretched, the mobile detector MQD is moved to the image point conjugate to bead 2. MQD mobility is required, given that movement of bead 1 to the fixed detector conjugate position uses the only available degree of freedom. Once bead 1 position is so constrained, bead 2 position is fixed by the actin filament length, which is different for each bead-filament assembly. Thus, we are forced to move the detector onto the bead instead of the reverse. Once again, final centering of the bead image on detector is performed by computer, but in this case the computer controls motors mounted to MQD directly.

Observed Bead Deflections

If the position detector has a bandwidth high enough to track the full range of diffusive motion, one observes significant thermal fluctuation of the trapped bead, approximately 40–50 nm from peak to peak (Figs. 3–7) when the trap stiffness is set near 0.03 pN/nm, in the range commonly used to measure single motor activity.

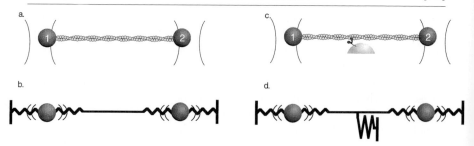

Fig. 2. (a) A single actin filament attached to two beads, labeled here as 1 and 2. (b) Schematic illustration of part (a), emphasizing compliant optical traps and bead/filament attachments. If the bead/filament linkages are significantly stronger than the optical traps, bead Brownian motion reflects filament motion. (c) Trapped actin filament attached to a single myosin molecule mounted on a coverslip-attached silica bead. (d) Schematic illustration of part (c), showing a stiff spring, which now connects the filament to the surface. The filament is clamped tightly to the surface, although the beads continue to experience significant thermal diffusion, constrained by the traps and complaint bead/filament linkages. Here, bead diffusion no longer provides a good measure of filament diffusion. The bead/filament connections, attributed in one experiment to bead rotation and actin filament bending,[11] should be as rigid as possible. Given enough tension, all compliance in the system can presumably be removed. However, the bead–actin connections typically break at tension exceeding a few pN, where the exact number varies from one filament to another. We stretch to a maximum tension most bead/filament connections can safely bear and measure connection compliance at that tension. The myosin molecule itself has an internal elasticity, which has been the subject of several measurements.[6,12,23]

The bead/filament assembly can be modeled as shown in Fig. 2. The two beads suffer constant bombardment by heated water molecules providing an effective white force noise. If the filament were stretched beyond the point where filament/bead compliance is effectively removed, meaning rendered far less than optical trap compliance, the bead/flament assembly would behave as a rigid rod and movements of the two beads would remain perfectly correlated. In this regime, displacement of either bead provides an accurate measure of filament movement. Among the connections we have used, filament/bead stiffness have ranged from 0.06 to 0.14 pN/nm, meaning the bead-to-bead stiffness is approximately half that, never negligible relative to a 0.02–0.03 pN/nm trap stiffness. Throughout this regime, thermal movements of each bead only imperfectly propagate to the other one, and the correlation coefficient between bead movements can be computed analytically.[6] Nonetheless, bead movements remain a reasonable indicator of filament diffusion.

However, once the actin filament is attached firmly to the surface through a myosin molecule, of stiffness in or above the 0.5–0.9 pN/nm range,[6,12] thermally driven bead movements become poor indicators of filament diffusion. Because the myosin molecule effectively clamps filament position, continued bead diffusion is determined by thermal forces constrained by a parallel combination of the optical trap and the filament/bead linkage. In fact, decoupling diffusion of one bead from that of the other can be used as a signature of myosin binding, and rms amplitude of this diffusion provides a measure of filament/bead linkage compliance. This compliance measurement is necessary in computing myosin-induced actin filament movement from the observed bead movement. These points are expanded below.

Working at myosin density slightly higher than that specified earlier, we often observe "staircase"-type structures when observing displacement versus time (Fig. 3), in which multiple motors pull one bead far from the center of its trap, allowing the other bead to return to baseline due to building slack in the filament. In principle, one can extract a stroke distance estimate from such data. The first motor should bind the filament and clamp tightly its position, as mentioned earlier. The associated shift in the center of bead diffusion may not reflect the stroke distance since the myosin binding event can occur when the filament was far from the center of its diffusion range,[4] a point to which we return later. Once the first myosin head is attached, probe diffusion is no longer a good indicator of filament diffusion, which should be clamped tightly. A second motor will then bind the filament and pull it to a new position. The center of bead diffusion should shift to reflect the myosin-induced movement, save for a known multiplicative factor reflecting the series elastic combination, thus allowing an estimate of the working stroke distance. However, in practice, many of these large-scale deflections do not have a clean stepwise character, many undergo sloped upward and downward transitions without clean points of transition, and the measured discrete movements tend to have a high variance. At these high myosin concentrations, multiple heads likely compete with each other to establish the equilibrium filament position, rendering contributions from single molecules ambiguous (Fig. 4).

Nonetheless, such movements demonstrate viable experimental conditions. Now one merely need reduce the surface density. Twofold changes are usually the best way to find the optimal density, and we find such a reduction to make the difference between staircases and single discrete events, such as those shown in Fig. 5.

High-resolution single-molecule experiments with processive motors can involve either a motor attached to a probe, moving on a polymer track mounted to a surface (e.g., Ref. 17) or a motor mounted on a surface, pulling a substrate polymer attached to a single probe on one end (e.g.,

[17] K. Svoboda, C. F. Schmidt, B. J. Schnapp, and S. M. Block, *Nature* **365,** 721 (1993).

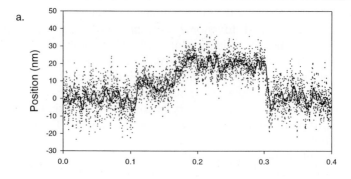

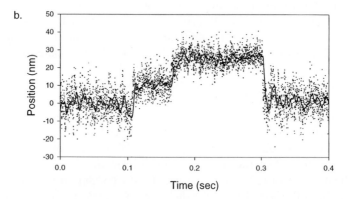

Fig. 3. Movement of beads in two optical traps, presumably driven by more than one myosin head: (a) the left trap and (b) the right one. Although the first "step" begins with a binding event that can correspond to any position in the 30- to 40-nm range of bead diffusion, the second "step" should begin with a filament tightly locked by the head already bound. In principle, one can extract the working stroke distance from such discrete steps, in this isolated event around 15 nm, occurring through the extended displacement, but this is difficult in practice. Rather, such displacements indicate an excessive surface density of myosin, which should then be reduced. More frequently, transitions are unclear, "backward" movements occur near forward ones, and one of the two beads is pushed back to its trap center as multiple heads build up slack between their attachments and the trapped bead.

Ref. 18). One observes advances in the center of probe diffusion, which reflect a protein advancing along binding sites. Observable discrete steps reflect transitions between clearly identified and biologically meaningful starting points and endpoints. The protein remains attached while it moves and thus is already attached before new substrate bindings.

[18] H. Yin, M. D. Wang, K. Svoboda, R. Landick, S. M. Block, and J. Gelles, *Science* **270,** 1653 (1995).

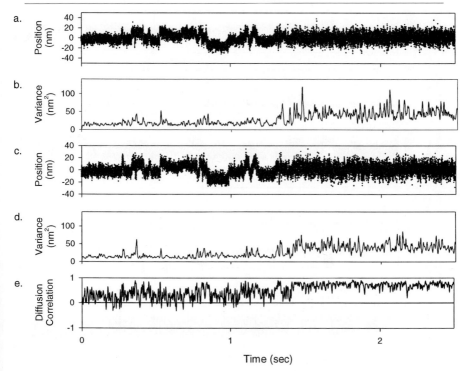

FIG. 4. Excessive protein density produces signs of competition among multiple heads to establish an equilibrium position. Shown here, presumably three or more heads are alternating probe position between a few stable points, separated by 7–10 nm. Such data are difficult if not impossible to deconvolve and may depend on uninteresting parameters such as the separation between involved myosin molecules on the surface.

In contrast, experiments with nonprocessive motors require observation of binding events between a detached motor protein and its track, which is necessary since the protein spends most of its interaction cycle detached from its polymer substrate. The "starting point" for a protein-driven probe displacement requires only that a binding site on the filament be made accessible to the surface-attached protein somewhere in the filament's range of thermal motion, not necessarily at the center.[4] Hence, *one observes meaningful endpoints without observing the corresponding starting points.* If the filament is free to rotate on time scales fast compared to observations, then myosin may bind to the filament anywhere in that range with equal probability.[4] If the filament does not rotate quickly,[5] some positions in the diffusion range will be favored for protein attachments, but these positions

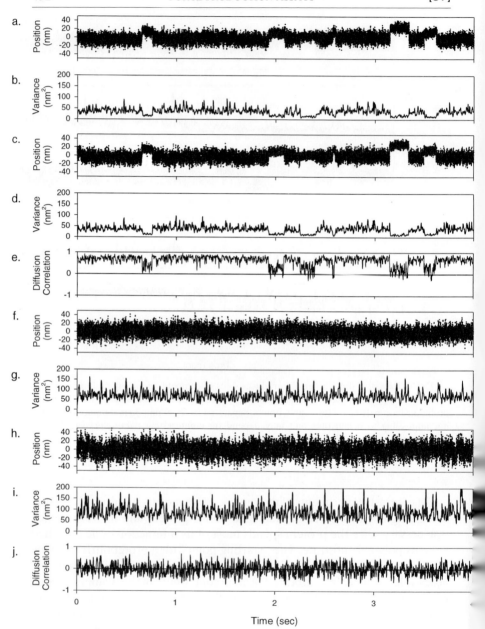

need not coincide with the center of baseline diffusion, requiring statistical analysis of many binding events to extract a meaningful starting point.

Figures 5a and 5c provide an example of probe deflections induced by myosin binding to an actin filament, while Figs. 5f and 5h show no corresponding deflection in the direction perpendicular to the actin filament, monitored simultaneously. The baseline thermal diffusion spans over 40 nm.

Aside from probe deflection, another signature of protein attachment is given by decrease in position variance,[4] which scales inversely with the stiffness of whatever constrains bead movement. At the baseline, the bead is constrained at one side by its optical trap and at the other by the second optical trap in series with compliance separating the two beads, primarily the bead–actin linkages (Fig. 2). Perpendicular to the filament axis, the bead is constrained only by its optical trap. Hence, the variance in this dimension is approximately 2-fold higher.

A myosin bound to the attached actin filament provides more stiff a constraint on bead diffusion than the optical traps. Hence, corresponding to bead deflections induced by protein binding, one can see a decrease in diffusion amplitude (Figs. 5a and 5c), clarified by a running variance computation (Figs. 5b and 5d). An automated data analysis method based on corresponding position and variance shifts has been used to estimate

---

FIG. 5. The protein density should be reduced to the point where transient binding events become visible. An example of raw position data for the two beads are shown in (a) and (c). There are five clear bead deflections, all corresponding to a visible reduction in noise amplitude. There is also one noise suppression not corresponding to an apparent deflection. This noise amplitude reduction is illustrated further in (b) and (d), with each point reflecting computed position variance for 1.5 ms (15 points) of data. The variance trace has been filtered (second-order Butterworth, f_c 70 Hz) to clarify transient drops. Such traces as displayed allow one to identify binding events, but the numbers will be underestimates due to the small window size. One should use bin sizes of around 100 points or longer to estimate quantitatively the variance under the given conditions (trap stiffnesses 0.03 pN/nm, sampling frequency (10 kHz). Another signature of myosin binding is shown in (e), displaying a correlation coefficient between bead motion, each point reflecting 5 ms (50 points) of data. Again, this display allows identification of binding events, but one must compute the coefficient over longer times (again around 100 points or more) to avoid underestimates due to subsampling. In practice, the errors due to 50-point computations are very small. Quantification of the correlation provides information regarding the stiffness of the element breaking the correlation, a series combination of the actomyosin linkage, the myosin molecule, and its surface attachment. Six binding events are visible by inspection of variance or correlation coefficient traces. (f) and (h) show diffusion of the two beads perpendicular to the filament axis. (g) and (i) show the corresponding variance, which experiences no obvious change during the binding events. (j) shows that diffusion along this axis is more or less uncorrelated, as expected since the connecting filament should not couple them in this direction.

the myosin stroke distance,[5] but the results remain controversial[19] and inconsistent with others.[4,6]

Another indicator of protein attachment exploits the correlation between thermal diffusion of the two beads attached to the actin filament, as mentioned earlier.[6] When the filament is stretched taut, motions of one bead will affect the other through the intervening filament and compliant linkages separating them. When a surface-attached myosin attaches to the filament, it reduces coupling between movement of the two beads. As shown in Fig. 5e, one observes transient drops in a running linear correlation coefficient between position of the two beads, these drops corresponding to bead deflections and decreased variance. In the direction perpendicular to the actin filament, bead diffusion remains uncorrelated (Fig. 5j).

In a sense, observation of diffusion correlation and bead position variance to detect binding events complement one another. When myosin binds the actin filament, bead diffusion is constrained by a parallel combination of the optical trap and the bead-to-surface contact (Fig. 1d), primarily the bead/filament connection. The more rigid this connection, the more pronounced the drop in variance caused by myosin binding. If the bead/filament contacts were perfectly rigid, however, both beads would move in tandem with or without myosin attached, and the diffusion correlation would provide no useful information. Loss of diffusion correlation becomes most sensitive to myosin binding when the bead/filament links are somewhat compliant. Although movement of one bead will still affect the force experienced by the other, binding of a surface-fixed myosin can impede this transmission significantly when the bead/filament links are less rigid than the filament-to-surface attachment through myosin. In fact, the diffusion correlation in the data shown in Figs. 3–7 drops to a higher value than in traces previously reported,[6] because here the actin–bead linkages are less compliant. If bead/filament connections are too compliant, even the diffusion correlation loss ceases to be a pronounced signature of myosin binding. In the extreme case, the linkages would be so weak that diffusion of the two beads would remain uncorrelated with and without myosin bound. In many of the experiments described here, both the variance and diffusion correlation drops are visible and independently reinforce the decision to score a given binding event (Fig. 5). Moreover, both signals are most pronounced when detectors are fast enough to capture the full frequency range of bead motion.

These traces illustrate the importance of high detector bandwidth, here 15 kHz, in tracking the full frequency range and thus amplitude of bead

[19] J. E. Molloy and D. C. S. White, *Biophys. J.* **72,** 984 (1997).

diffusion. As shown in Fig. 6, binding events need not involve an appreciable probe deflection from the center of its baseline diffusion, meaning that visual selection of pronounced deflections in filtered traces will generate an incomplete distribution. Our original version of this dual-beam laser trap system demonstrated the effectiveness of the approach and established a fundamentally small step size of ∼10 nm,[3] in contrast to countervailing estimates over 100 nm.[20–22] This version of the trap system, however, lacked the high detector bandwidth needed in detecting binding events that do not displace the bead notably from its baseline position (Fig. 6f). Improvements on the original design, which incorporate methods beyond visual inspection of raw or filtered bead position data, point to a stroke distance closer to ∼5 nm,[4,6] although different tactics have led to slightly higher estimates.[5,23] Moreover, simultaneous detection of both beads, attached to either end of the actin filament, has established that the filament ends move in unison and by the same amount, arguing against the role of actin filament length changes in driving actomyosin-based motility.[24]

Poor myosin preparations are characterized by "dead" heads failing to release actin in the presence of ATP and retarding gliding motion when present in relatively small amounts. In the single-molecule assay, one samples only one or a few molecules at a time and can more effectively avoid a minority of "dead" heads, which affect the gliding assay out of proportion to their numbers. Although one can still take measurements using the laser trap system, such conditions are not optimal and often become prohibitive. Such heads will bind to a suspended actin filament (Fig. 7) and sequester it, forcing the experimentalist to release the filament and search for another one. Depending on protein quality, this can happen so frequently that the process is no longer efficient. In any case, one prefers to sample heads with minimal doubts about preservation of their native function, such as those arising from the presence of many "dead heads" in a preparation.

When a large number of transients have been measured, one can tabulate statistics to extract the mechanical and kinetic properties of the myosin molecules. Measurement of bead position during binding events relative to baseline regions to either side yields a distribution of myosin-induced bead deflections. If the filament is rotating fast enough so that myosin will bind to it anywhere in its range of diffusion with more or less equal probabil-

[20] Y. Harada and T. Yanagida, *Cell Motil. Cytoskelet.* **10,** 71 (1988).

[21] Y. Harada, K. Sakurada, T. Aoki, D. D. Tomas, and T. Yanagida, *J. Molec. Biol.* **216,** 49 (1990).

[22] T. Yanagida, T. Arata, and F. Oosawa, *Nature* **316,** 366 (1985).

[23] A. Ishijima, H. Kojima, H. Higuchi, Y. Harada, T. Funatsu, and T. Yanagida, *Biophys. J.* **70,** 383 (1995).

[24] A. D. Mehta and J. A. Spudich, *NATO ASI Series H: Cell Biol.* **102,** 247 (1997).

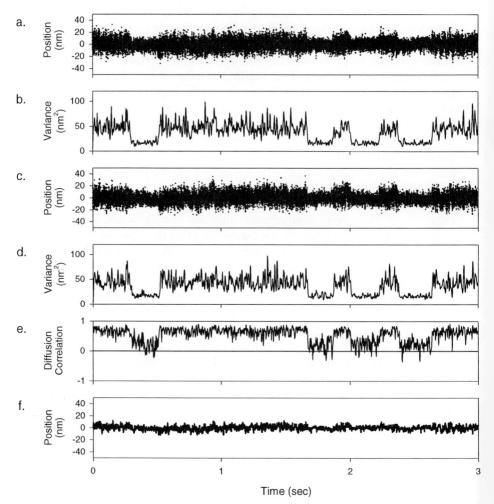

Fig. 6. Many events will not be visible from inspection of raw or filtered bead position data. (a) and (c) show the raw position data, in which one can see noise suppression during four intervals. These binding events become clear on examination of the position variance (b, d) or the diffusion correlation (e). However, almost nothing is visible in a filtered position trace (f). A bandwidth-limited detector provides a signal equivalent to a filtered one, obscuring from view events such as the four shown here. The fact that all four events seem around zero might indicate that this particular filament is not rotating quickly, consistent with a previous report.[5] Here, the accessible myosin-binding site could be located one stroke distance below the zero point. However, in most cases (e.g., Fig. 5), we and others observe high variability in apparent deflection amplitudes of binding events occurring within seconds of each other. Analysis of many events as a function of the time between them indicates that the amplitude of a given displacement is poorly correlated with that of the event preceding it, even when the binding events are separated by less than a second. This suggests indirectly that the filament is likely rotating fast on time scales compared to durations separating binding events.

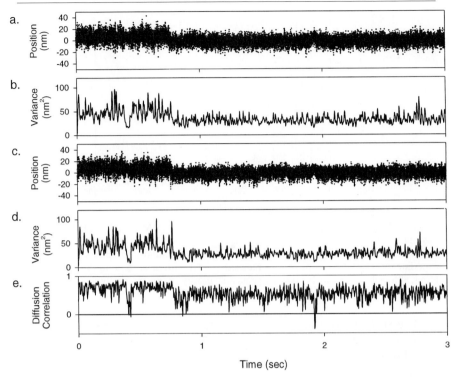

FIG. 7. Filaments gliding over a high density of myosin heads are usually arrested by small numbers of "dead" myosin heads, unable to release actin after binding in the presence of ATP (binding at time ~0.7 sec in the figure). Although the single-molecule assay is a bit more forgiving since only one or a few molecules are sampled at a time, poor protein preparations can still prevent effective measurements. Shown here, a "dead" head binds to an actin filament and sequesters it. Bead position is shown in (a) and (c), position variance in (b) and (d), and the diffusion correlation coefficient in (e). Once a filament is sequestered, one usually must abandon it and search for another one. Hence, significant numbers of damaged heads can prevent such measurements of the given preparation.

ity, one expects the distribution to resemble the shape and size of the baseline bead position distribution, except that it is shifted upward by the myosin stroke distance,[4] an idea supported by some measurements[4,6,23] and not others.[5] The baseline position distribution will be Gaussian with width determined by the stiffness of constraints on the bead, a parallel combination of its optical trap on one side and the second optical trap, in series with an actin filament and the bead/filament linkages, on the other. If the model is correct, the variance of bead deflection should match the variance of baseline bead position, consistent with many previous observations.

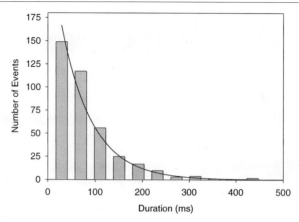

F<small>IG</small>. 8. Distribution of event durations for rabbit skeletal HMM at 5 μM ATP. The continuous line is an exponential fit with rate constant corresponding to mean time 65 ms. The expected mean time is 50 ms, based on the ATP-binding rate measured in solution kinetic experiments. The data can be fit to a reduced χ^2 of 1 with mean time in the range of 55–80 ms, although it could not be so fit to a sequential process of two kinetically comparable steps.

The stiffnesses of all springs shown in Fig. 2, including that reflecting the filament link to the surface through the myosin molecule, can be estimated using the values of position variance and diffusion correlation coefficients. As described elsewhere,[6] the diffusion correlation coefficient at the baseline and the position variance when myosin is attached provide independent estimates of the bead–actin connection stiffness, and the diffusion correlation coefficient with myosin attached provides the stiffness of the filament link to the surface, a series combination of the actomyosin connection, the myosin molecule, and its surface attachment. More direct estimates of single-molecule stiffness by measurement of stress–strain curves[12] involve larger displacements and require all system elements except the optical trap to be significantly less compliant than the molecule of interest.

Finally, the distribution of event durations will reflect the kinetics of detachment. If myosin detachment is limited by a single first-order step, one expects to see an exponential Poisson distribution.[25,26] This reflects nothing more than the probability of detachment for a given bound molecule being independent of its history, meaning that in an ensemble, a constant fraction of those remaining bound will detach at any given time. If

[25] J. T. Finer, A. D. Mehta, and J. A. Spudich, *Biophys. J.* **68,** 291s (1995).
[26] J. E. Molloy, J. E. Burns, J. C. Sparrow, R. T. Tregear, J. Kendrick-Jones, and D. C. S. White, *Biophys. J.* **68,** 298s (1995).

detachment is limited by a sequence of two similar steps, such as sequential detachment of the two heads, one expects the durations to fit a second-order Gamma distribution, a convolution of the Poisson distribution with itself. Similar analysis has been invoked in studying the kinetics and chemo-mechanical coupling of processive motors.[27–29] In examining stepwise motions, one cannot observe all of the fast transitions to capture the regions where Poisson and gamma distributions most differ. Experiments with processive motors have finessed this problem by looking instead at evolution of mean position relative to position variance during prolonged movements generated by multiple steps,[27,28] or by incorporating the likelihood and impact of missed events in the data analysis.[29] Such techniques cannot be applied straightforwardly to the isolated deflections of nonprocessive motor assays. Instead, one can test the statistically reliable subset of the data, meaning the section of a duration histogram (Fig. 8) populated significantly and including only events long enough to identify clearly, to determine which distribution better fits the data and what the range of allowed time constants are. Such analyses with rabbit skeletal heavy meromyosin at limiting ATP concentration indicate a single-step process with a decay constant matching the ATP on rate measured in solution studies[6] (Fig. 8).

Conclusion

The study of biological motors has entered a new and distinct phase with the advent of techniques to observe and manipulate single molecules. We have described here considerations regarding protein reagents, coverslips and beads, instrument design, collection of data, and identification of myosin binding events. Despite significant advances upon early experiments, the field remains nascent and developing. We expect that future work in many laboratories will continue to advance methods of protein attachment, precision of measurements, and integration of optical trapping systems with other techniques to probe single-molecule chemistry in different ways.

[27] K. Svoboda, P. P. Mitra, and S. M. Block, *PNAS* **91,** 11782 (1994).
[28] M. Schnitzer and S. M. Block, *Nature* **388,** 386 (1997).
[29] W. Hua, E. C. Young, M. L. Fleming, and J. Gelles, *Nature* **388,** 390 (1997).

[38] Versatile Optical Traps with Feedback Control

By KOEN VISSCHER and STEVEN M. BLOCK

Introduction

Structural techniques, such as X-ray crystallography and multidimensional nuclear magnetic resonance (NMR) imaging, generate static images of biomolecules, thereby supplying essential clues about how proteins may function. Ineluctably, however, such methods provide an incomplete picture of dynamic processes. A variety of novel physical approaches, including atomic force microscopy (AFM), advanced fluorescence microscopy, and optical trapping ("optical tweezers"), have recently been developed to study directly the dynamics of individual macromolecules. Already, researchers have begun to use these new tools to probe molecular details of such diverse phenomena as vesicle transport, muscle contraction, and DNA transcription, all of which have in common the fact that they are powered by the action of specialized mechanoenzymes, or motor proteins.

In this article, we describe the design and implementation of optical trapping devices for use in conjunction with *in vitro* motility assays for motor proteins, with particular reference to the microtubule-based motor, kinesin. Despite this focus, the technology described here is general purpose and has widespread applicability to problems in cellular biophysics, molecular biology, and even materials science. During the past several years, optical trapping microscopes have quickly evolved from simple grasping devices into sophisticated instruments that not only produce regulated optical forces, but also incorporate detection systems for determining the positions of objects with nanometer-scale resolution, or better. In this article, various schemes to detect submicroscopic movements of an optically trapped object are presented and compared, including arrangements that remain insensitive to the absolute location of a particle within the microscope field of view. We also describe computer-controlled schemes to synthesize multiple optical traps from a single laser beam, by rapidly time-sharing the position of one trap among different locations. When information from position sensors is combined with rapid servo control of various optical trap parameters, such as the light intensity or laser beam location, it becomes possible to implement a variety of feedback schemes with useful properties. The most widely used scheme is a position clamp, which is used to measure force with high temporal precision. Position clamps work by altering the properties of the trap dynamically in such a way as to fix the location of a trapped object while it is subjected to varying levels of force: the feedback required to maintain this *isometric* condition supplies a measure of the force. We discuss various ways of realizing such a position clamp with an optical trapping microscope, and also introduce here an alternate feedback scheme: the force clamp. Force clamps work by altering the tension exerted on an object by the optical trap dynamically in such a way as to maintain a constant tension while its position changes: the feedback required to maintain this *isotonic* condition supplies a measure of the displacement. Force clamps turn out to have special properties that make them especially well suited to studying steps taken by processive motor proteins, in particular to single motors attached to small beads via compliant linkages.

Optical Traps and Trap Stiffness

Small dielectric particles located near a tightly focused laser beam are subject to at least two distinct kinds of optical force: the *scattering force,* acting in the direction of propagation of the light on the particle, tending to push it away from the focus, and the *gradient force,* which derives from the interaction of induced dipoles in the dielectric with the electric field gradient in the laser beam, tending to pull the particle toward the focus.[1,2] For stable trapping in all three dimensions, a significant spatial gradient in the electric field is required to overcome the scattering force, which would otherwise destabilize the trap in the axial direction. Steep gradients are obtained by focusing the laser beam with a lens of high numerical aperture (NA $\gtrsim$ 1), typically an oil- or water-immersion objective. Particles become trapped along the optical axis, at a point just beyond the focal zone, where both the gradient and scattering forces are balanced: stable trapping of small objects, such as micrometer-sized polystyrene or silica beads, bacteria, single eukaryotic cells, cellular vesicles, etc., has been demonstrated.[1,3-6] If the laser spot is moved using external optics, the trapped particle will follow accordingly. To minimize photodamage to biological specimens, the wavelengths typically chosen are in the region of 800–1100 nm, where absorption is minimized for most biological specimens. For high-power use, a popular choice is the Nd:YAG laser at 1064 nm (or the Nd:YLF and Nd:YVO$_4$ lasers, which are quite similar). At reduced powers, single-mode

[1] A. Ashkin, J. M. Dziedzic, J. E. Bjorkholm, and S. Chu, *Opt. Lett.* **11,** 288 (1986).
[2] K. Svoboda and S. M. Block, *Annu. Rev. Biophys. Biomol. Struct.* **23,** 247 (1994).
[3] A. Ashkin, J. M. Dziedzic, and T. Yamane, *Nature* **330,** 769 (1987).
[4] A. Ashkin and J. M. Dziedzic, *Proc. Natl. Acad. Sci. U.S.A.* **86,** 7914 (1989).
[5] A. Ashkin, K. Schütze, J. M. Dziedzic, U. Euteneuer, and M. Schliwa, *Nature* **348,** 346 (1990).
[6] S. M. Block, D. F. Blair, and H. C. Berg, *Nature* **338,** 514 (1989).

diode lasers (or diode-pumped diode lasers, such as the MOPA) offer an alternative, at wavelengths from ~800 to 990 nm.

The forces exerted by optical traps are typically in the piconewton (pN) range for reasonable laser power levels delivered to the specimen plane (from tens to hundreds of milliwatts). At the center of the optical trap the net force is zero, and the trapping potential is harmonic (quadratic) in its immediate neighborhood: for small displacements, the restoring force increases proportional to displacement, out to ~150 nm or even more from the optical axis. To an excellent approximation, therefore, optical traps resemble Hookeian springs in this regime, characterized by a fixed stiffness. The stiffness of a trap, and the region over which it remains invariant, depend, in nontrivial ways on the optical wavelength and power, as well as the size, shape, and refractive index of the particle, and the refractive index of the surrounding medium.[1,2,7] Once stiffness has been calibrated, a trap can be used to make quantitative measurements of the forces applied to a trapped object by measuring position with nanometer-level accuracy, or better. There are now several well-established methods for measuring trap force, either directly or through measurements of position, once the trap stiffness has been determined.

Escape Force Method

Historically, the first determinations of the strength of optical traps were based on the escape force: that force required to pull an object from the trap.[3,5,6] Measurements of escape force depend critically on the shape of the potential at the very limb of the trapping zone: in this region, force varies in a highly nonlinear way with displacement[2] and stiffness is not readily ascertained. The escape force is generally applied by viscous drag,[2,5,6] produced either by pulling the trapped particle through the fluid, or by moving the fluid past the trapped particle. Several variations of the method exist. One of the simplest is to videotape a trapped particle in a fixed trap while translating the microscope stage at an ever-increasing rate, until the particle just escapes. The particle velocity immediately after escape is measured from the video record, which permits an estimate of the escape force, provided that the viscous drag coefficient of the particle is known. While somewhat crude, this technique can permit calibration of force to within ~10%. If the stage is instead moved at a fixed, known velocity, the laser trapping power can be reduced until the particle just escapes, which provides somewhat better reproducibility in measurements. Escape forces are generally different in the x-y (lateral) plane and the z (axial) direction,

[7] A. Ashkin, *Biophys. J.* **61,** 569 (1992).

so the exact escape path must be determined for precise measurements. This calibration method does not require a position detector with nanometer resolution.

Drag Force Method

When a well-calibrated position sensor is available, the stiffness near the trap center follows from $k_{trap} = F/x$, where x is the particle displacement from the trap center and F is the applied force. In practice, forces are usually applied through the viscous drag of fluid past the particle, generated by periodic movement of the microscope stage while the particle is held in a stationary trap. Either triangle waves of displacement (corresponding to square waves of force) or sine waves of displacement (corresponding to cosine waves of force) work well.[2,8–10] Once trap stiffness is determined, optical forces can be computed from knowledge of the particle position relative to the trap center, provided that measurements are made within the linear region of the trap (typically, out to~150 nm). Along with the need for a well-calibrated piezo stage and position detector, the viscous drag on the particle must be known. This can be problematic, because drag coefficients depend on the size and shape of particles, the fluid viscosity, and the possible presence of nearby walls or obstacles. Drag coefficients are generally unknown for irregularly shaped particles. The drag force method is therefore best suited to uniform spherical particles, for which explicit expressions for drag exist, including near chamber walls, which can have a profound influence.[2] For example, viscous drag increases by ~40% when a sphere approaches a wall within a distance equal to its radius, and by ~200% within a quarter radius (Faxen's law). Optical trapping work near surfaces can therefore pose serious difficulties for calibration. However, in some instances, the strong distance dependence of drag near walls may be turned to an advantage. For example, Wang and co-workers[10] used the reduction in the Brownian roll-off frequency of the power spectrum of a trapped bead near the coverglass to fix its height to within ±50 nm.

Momentum Transfer Method

Under certain circumstances, when it is possible to collect all the light scattered by a trapped particle, the transverse force acting on it may be found directly, by computing the optical momentum transfer from the angle of the beam deflected by the particle, a technique pioneered by Bustamante

[8] S. C. Kuo and M. P. Sheetz, *Science* **260,** 232 (1993).
[9] R. M. Simmons, J. T. Finer, S. Chu, and J. A. Spudich, *Biophys. J.* **70,** 1813 (1996).
[10] M. D. Wang, H. Yin, R. Landick, J. Gelles, and S. M. Block, *Biophys. J.* **72,** 1335 (1997).

and co-workers.[11] This method is based on conservation of momentum, and the fact that force is the time rate of change of momentum. Unfortunately, most single-beam optical traps use the full aperture of a high-NA lens to illuminate the specimen, so as to produce the steepest possible gradient in light. Under such circumstances, significant amounts of light may be scattered from the optical system and not recovered, precluding the use of this approach. However, in dual-beam, counter-propagating traps,[2] which typically use a lower NA, the apertures of the two objective lenses are underfilled, i.e., the $1/e^2$ diameter of the laser beam is less than the entrance pupil of the lens. In such systems, (nearly) all the scattered light is collected, so that the exit angle of this light emerging from the trapping system may be determined, and the force computed from the relationship:

$$F = \frac{dp}{dt} = \frac{I}{c}(\text{NA})\left(\frac{X}{R}\right) \tag{1}$$

where I is the light intensity, c is the speed of light, NA is the numerical aperture of the lens collecting the light from the trap, X is the displacement of the beam from the optical axis, and R is the radius of the back aperture of the lens.[11] The momentum transfer method has the singular virtue that it does not depend on knowing the viscous drag, nor, for that matter, the shape, size, or refractive index of the particle trapped.

Equipartition Method

One of the simplest and most straightforward ways of determining trap stiffness is to measure thermal fluctuations in the position of a trapped particle. The stiffness of the trap, k_{trap}, is then computed from the equipartition theorem for a particle bound in a harmonic potential:

$$\tfrac{1}{2}k_{\text{B}}T = \tfrac{1}{2}k_{\text{trap}}\langle x^2 \rangle$$

where x is the position and $k_{\text{B}}T$ is Boltzmann's constant times the absolute temperature. The primary advantage of this method is that knowledge of the viscous drag coefficient is not required. A fast, well-calibrated position detector is essential, however, precluding most video-based schemes. Although knowledge of particle size, shape, or optical properties is not necessary to compute stiffness, bear in mind that these properties do affect detector calibration in the first place. When using this method, several cautions should be borne in mind: first, the analog bandwidth of the detection system must be adequate, since low-pass filtering will tend to underesti-

[11] S. B. Smith, Y. Cui, and C. Bustamante, *Science* **271,** 795 (1996).

mate $\langle x^2 \rangle$ and thereby inflate the apparent stiffness. Second, $\langle x^2 \rangle$ is a statistically biased estimator: other independent, systematic sources of noise (electronic noise, etc.) add in quadrature and inflate $\langle x^2 \rangle$, thereby underestimating the stiffness. Third, slow drifts in the position, unless removed, may also inflate $\langle x^2 \rangle$, again leading to misestimation. Finally, thermal motion represents a weighted average over a distribution of positions near the center of the trap, in a way that depends on the trap stiffness, so care must be taken that the optical potential remains harmonic over the region explored by this motion. Although the *analog* bandwidth of the position detector should be high, for the reasons just discussed, one feature of the equipartition method is that the *digital* sampling rate at which the position of the object is updated need not be correspondingly high, i.e., possible aliasing artifacts do not represent a source of difficulty, since the position signal has random phase.

Power Spectrum Method

A third method, quite accurate when the viscous drag coefficient, β, of an object is known, is based on determining the power spectrum of its position.[2] For a particle bound in a harmonic potential at low Reynolds number (i.e., when inertia can be neglected), the power spectrum is given by Eq. (2):

$$\tilde{P}_{xx} = \frac{k_B T f_c}{k_{\text{trap}} \pi^2 (f^2 + f_c^2)} \tag{2}$$

Equation (2) is a Lorentzian with roll-off frequency $f_c = k_{\text{trap}}/2\pi\beta$. By fitting the measured power spectrum to a Lorentzian, the trap stiffness follows once the drag has been determined. A typical example of such a power spectrum is shown in Fig. 1. The use of power spectra to calibrate stiffness has been found to be of particular help in diagnosing problems with optical traps. If the trap is misaligned in some fashion, if the laser beam profile is corrupted, or if something is awry with the position detection system, the power spectrum degrades rapidly and becomes noticeably non-Lorentzian, and/or displays peaks at specific frequencies. These details are readily missed with most other methods. Because only the roll-off frequency need be determined, the power spectrum may have arbitrary amplitude scaling, so that absolute calibration of the sensor is unnecessary. However, when such a calibration is available, the amplitude of the power spectrum at zero frequency, $\langle x^2 \rangle$, provides identical information to the equipartition method. Essentially identical considerations about analog bandwidth apply here.

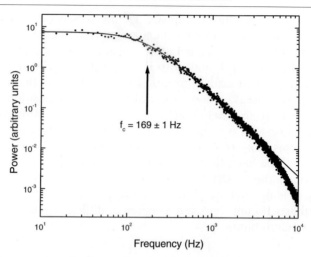

FIG. 1. Power spectrum for the position of a silica bead, 0.5 μm in diameter, held approximately 2 μm above the coverglass surface in a weak trap (solid dots). Data were sampled at 20 kHz and anti-alias filtered in hardware at 10 kHz. The fit to a Lorentzian (solid line) gives a roll-off frequency, f_c, of 169 ± 1 Hz, corresponding to a trap stiffness of ~0.006 pN/nm. Above the roll-off frequency, the power spectrum decays with a slope of 20 dB/decade up to ~6 kHz (the steeper decay beyond this point reflects contributions from the electronics).

Step Response Method

The trap stiffness may also be determined by finding the response of a particle to a rapid, stepwise displacement of the trap.[9] For small steps, x_{trap}, the response, x_{bead} is given by Eq. (3):

$$x_{\text{bead}} = x_{\text{trap}}[1 - \exp(-k_{\text{trap}}t/\beta)] \tag{3}$$

To determine the trap stiffness k_{trap}, the viscous drag, β, must be known. Here again, absolute calibration of the detector is not required. As with some of the other methods, care must be taken that the particle remains within the linear region of the trap. Although this approach provides essentially the same information as the power spectrum method, discussed earlier, it is harder to identify extraneous sources of noise or artifact. However, it is simpler computationally and more tolerant of some sources of noise, such as drift. The time constant for movement of the trap should be much faster than the characteristic damping time of the particle, β/k_{trap}, and similar considerations about analog bandwidth apply here as for the previous two methods.

With the exception of the equipartition and momentum transfer methods, all the approaches essentially measure the ratio of the stiffness to the damping, and therefore require some independent way to estimate the drag contribution. Because both the viscous drag and the optical trapping strength vary with the height over the surface, it is important to bear in mind that a calibration performed at one particular height may not be valid at another. In fact, trap stiffness has been found to vary little with height above the coverglass surface over the first few micrometers. Beyond this point, however, spherical aberrations begin to effect the distribution of light in the focal spot and stiffness changes accordingly. The power spectrum and step response methods are perhaps better suited to signal averaging than some of the others, and are therefore quite robust, statistically. The escape force and drag force methods are particularly well suited to mapping force profiles in the outermost, nonlinear regions of traps. While it is true that some methods do not require absolute detector calibrations to determine the trap stiffness, such calibrations may nevertheless be necessary for force computations, once stiffness is known.

Position Sensing

For most quantitative applications of optical traps, the position of particles must be measured with nanometer resolution or better. Position sensors have therefore become an essential feature of modern trapping microscopes used for single-molecule research. Several schemes have been developed, which we now present.

Direct Imaging Methods

Centroid Tracking. The position of an object recorded by video can be determined with surprisingly good resolution, by digitizing individual frames and running a computer-based centroid-tracking algorithm, which performs a two-dimensional cross-correlation between the image and a stored mask of the object.[12] Because all pixels representing the object of interest make weighted contributions to the cross-correlation, the position of its peak—representing the location of the centroid—can be determined with subpixel accuracy, at least in principle. In practice, the technique is limited by the timing stability of analog video signals—line to line, as well as frame to frame—which in turn depends on the actual performance achieved by specific devices in the video chain (camera, recorder, frame grabber, etc.). This can vary widely. Most tape-based VCRs produce too much jitter to record signals suitable for subpixel tracking, necessitating the use of expensive optical memory disk recorders, or direct digital re-

---

[12] J. Gelles, B. J. Schnapp, and M. P. Sheetz, *Nature* **331,** 450 (1988).

cording to hard disk. For small spheres viewed in a microscope at high magnification, a typical pixel may subtend ~20–50 nm on a side, and it is possible to achieve roughly one-tenth pixel accuracy with appropriate instrumentation, corresponding to a resolution limit in the range of 2–5 nm. Beyond this point, various instabilities during acquisition and/or playback limit the technique. For NTSC-standard video systems, the frame rate is 30 Hz, so the bandwidth of measurements is quite limited. Furthermore, it is generally necessary to do centroid processing off-line, because the computations can be quite slow. Nevertheless, centroid methods can be invaluable in helping to calibrate optical trap displacements, especially over relatively large baselines.

Quadrant Photodiode. Position can be determined by focusing the image of a particle on the face of a quadrant photodiode and recording the relative light intensity impinging on the four quadrants.[13–15] The photodiodes are used pairwise, in a differential configuration: for example, the difference in photocurrent arising from the sum of the upper two quadrants and the sum of the lower two quadrants encodes the vertical position of the particle. Normalizing (differential) amplifiers are used to remove any dependence of the output on the level. Alternatively, normalization can be performed in software. In most cases, trapped particles are imaged by a conventional light source unrelated to the laser used for trapping, e.g., by the microscope's built-in tungsten-halogen illuminator. Systems employing distinct light sources for illumination and optical trapping at different wavelengths are well suited to feedback-enhanced and/or time-shared traps, because signals are derived solely from the objects' position, with no contributions arising from the trap itself. Because an imaging photodiode detector system is fixed in space, however, the trapped object must be carefully aligned with the corresponding position of the detector at an appropriate place in the specimen plane.

Conventional microscope imaging methods are generally plagued by limiting light levels and require considerable time integration at the photodiode, so that bandwidth is typically below 1 kHz (shot noise limit). The situation can be improved somewhat by exchanging the tungsten-halogen light source for a higher luminosity xenon or mercury arc lamp,[9,13,14,16,17]

[13] J. T. Finer, R. M. Simmons, and J. A. Spudich, *Nature* **368**, 113 (1994).

[14] J. E. Molloy, J. E. Burns, J. Kendrick-Jones, R. T. Tregear, and D. C. S. White, *Nature* **378**, 209 (1995).

[15] S. Kamimura, *Appl. Opt.* **26**, 3425 (1987).

[16] J. E. Molloy, J. E. Burns, J. C. Sparrow, R. T. Tregear, J. Kendrick-Jones, and D. C. S. White, *Biophys. J.* **68**, 298s (1995).

[17] W. H. Guilford, D. E. Dupuis, G. Kennedy, J. Wu, J. B. Patlak, and D. M. Warshaw, *Biophys. J.* **72**, 1006 (1997).

although arc sources are generally less stable. Extremely high-brightness LEDs, or, equivalently, laser diodes driven by currents below the lasing threshold, represent another illumination choice, although the former are not yet sufficiently bright for most practical purposes. Alternatively, direct use of laser light—at a different wavelength from the trapping source, e.g., in the visible—can increase intensity by orders of magnitude. However, when used to supply light over an extended region, coherent laser sources produce their own problems arising from speckle and interference, necessitating some form of phase randomization. This randomization must be accomplished at frequencies higher than the upper limit of the position detector bandwidth.

Laser-Based Methods

At least several milliwatts of laser power are required in the specimen plane to produce a stable trap. Even if only a small fraction of this intense light ($\sim$1 mW) were used to measure the position, a typical sensor would not become shot noise-limited until very high frequencies (tens of kilohertz). Following this line of thinking, a number of implementations have been developed that we outline briefly.

Interferometry. The first and arguably most sensitive detector is an optical trapping interferometer.[2,18,19] In this arrangement, the Wollaston prism located behind the objective of a microscope set up for differential interference contrast (DIC) imaging splits the laser light into two orthogonally polarized beams, producing two nearly overlapping, diffraction-limited spots in the specimen plane; together, these spots function as a single optical trap. After passing through the specimen, the beams are recombined beyond the condenser in a second Wollaston prism. When no object is in the trap, or when it is exactly centered in the trap, the recombined beam has the same linear polarization as the incoming laser light. However, as the particle moves from the center of the trap, one beam is phase delayed with respect to the other, such that after recombination, an elliptical polarization is produced. The degree of ellipticity, which can be measured quite sensitively, provides a direct measure of displacement. This detection is intrinsically aligned, in the sense that the trapping and photodetection beams are one and the same. In addition to its extraordinary sensitivity, an advantage to this approach is that it is a nonimaging method; as such, the trap can be moved about within the specimen plane without a need to realign the position detection system. On the other hand, the optical trap and position detection cannot be spatially uncoupled, which precludes im-

[18] K. Svoboda, C. F. Schmidt, B. J. Schnapp, and S. M. Block, *Nature* **365,** 721 (1993).
[19] W. Denk and W. W. Webb, *Appl. Opt.* **29,** 2382 (1991).

plementing certain feedback systems described later. Moreover, the detection scheme is one-dimensional: only displacements along the Wollaston shear axis are registered.

Single Photodiode. Other methods that use the trapping light for position detection use photodiodes, and vary in the type of sensor used and its placement along the optical path. The method introduced by Ghislain and Webb[20,21] is based on a single photodiode sensor. Their scheme was specifically developed to detect axial (z) movements of particles, in contrast to most sensors, which detect primarily lateral movements. A bead or particle, trapped just below the focus, may be considered to act as a miniature lens. As the bead moves from the trap center, it deflects the laser beam. The microscope condenser (or alternatively, a second objective), magnifies this beam and relays the transmitted light to the surface of the photodiode: the magnification functions as an optical lever arm to facilitate measurement of small deflections. In contrast to the quadrant photodiode imaging system described earlier, this photodiode is not placed in a plane optically conjugate to the specimen (i.e., an image plane), where the particle would remain in focus. Instead, the location of the detector is chosen such that roughly half the optical power in the diverging cone or light illuminating it is intercepted (i.e., it is overfilled by a factor of ~ 2). At this position, it turns out that axial as well as lateral displacements are sensitively registered. Such a system can be optimized for the detection of lateral motion when the active area of the detector is physically offset from the optical axis by roughly one detector radius.[21] Lateral displacements are registered through a change in overlap of the diode and the deflected light pattern, but no information about the direction in the x-y plane can be obtained. Axial movements of a trapped particle cause the size of the light cone to change, and the resulting variation in intercepted power provides a measure of vertical displacement. Axial and lateral displacements cannot be deconvolved in this simple setup (but this may be possible, at least in principle, using a quadrant photodiode instead).

Lateral Effect Detector. Another scheme that takes advantage of the trapping light to detect position was developed by Bustamante and colleagues,[11] who positioned a lateral effect position detector (linear displacement encoder) in such a way that it captured all the transmitted light in the laser beam passing through the specimen. The transmitted power and associated beam deflection (i.e., the shift in light on the face of the detector) induced by the trapped object also permit a computation of the force, as described earlier, by measuring the change in beam momentum. For this

[20] L. P. Ghislain and W. W. Webb, *Opt. Lett.* **18,** 1678 (1993).
[21] L. P. Ghislain, N. A. Switz, and W. W. Webb, *Rev. Sci. Instrum.* **65,** 2762 (1994).

work, two counterpropagating beams were focused by matched objectives and used to form the trap, and the deflection through one objective was monitored to determine force. In a dual-beam counterpropagating trap, the laser light need not be as sharply focused as in a single-beam gradient trap, and therefore the back aperture of the objective need not be filled with light: underfilling the pupil guarantees that all of the deflected light can be captured at the condenser side, as required.

Quadrant Photodiode. To measure nanometer-scale displacements in the instrument described next, we use an alternate scheme. Laser light passing through the specimen is collected on a quadrant photodiode that is placed on the optical axis in a position optically conjugate to the back focal plane of the microscope condenser. In contrast to the scheme employed by Ghislain and Webb,[20,21] the active area of our photodiode detector is somewhat larger overall than that of the illuminating beam, which renders it insensitive to axial (z) movements. By locating the detector at this position, it turns out that its response is rendered less sensitive to the x-y position of the optical trap itself within the specimen plane: instead, the detector responds mainly to a *relative* displacement between an object and the center of the trap, wherever that trap is located. In practice, this method has been shown to work over an area $\sim$5 μm or so in diameter, after subtraction of a small, position-dependent offset, presumably due to background light. The principles of operation of this scheme are illustrated in Fig. 2. An ability to register displacements in a manner that remains nearly independent of trap location is the principal advantage of this scheme over photodiode imaging methods, which place the detector instead at a plane conjugate to the specimen. In practice, one cannot ensure the exact geometry of Fig. 2, where the beam is perfectly collimated by the trapped bead. Trap position-independent responses can nevertheless be achieved for a given particle by adjusting the axial position of the sensor while moving both the optical trap and the particle together, until the detector signal is nulled. When using multiple-beam traps or feedback-enhanced systems, uncoupling the trapping and illuminating light sources becomes necessary. For this purpose, a second laser source at a different wavelength (and lower power) may equally well be focused to a diffraction-limited spot, overlapped with the location of a trapping beam in the specimen plane, and used for position detection, taking the identical approach.

How do various detection methods compare in terms of stability? Photodiode-based, direct imaging approaches are particularly sensitive to mechanical vibrations occurring in the arm of the apparatus holding the detector, although they are comparatively less sensitive to vibrations arising in the illumination arm of the microscope. In contrast, the optical trapping interferometer is relatively insensitive to vibrations of the condenser arm

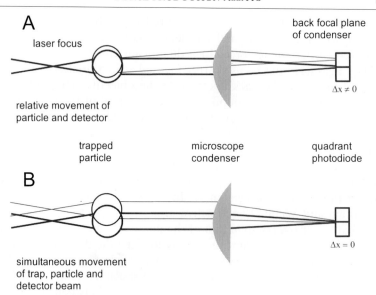

FIG. 2. Principle of operation for the position detection scheme. A quadrant photodiode is placed on the optical axis at a position conjugate to the back focal plane of the condenser; here, it intercepts the Fourier transform of the image. In this ray-optic approximation, the microscopic bead acts as a simple lens (drawing not to scale). (A) A signal change is generated whenever the bead moves relative to the stationary detector beam, producing a physical displacement of the light impinging on the split photodiode detector (thick versus thin lines). (B) However, no corresponding displacement is produced when both the bead and detector beam are moved together by an identical amount. When placed at this optical position, the photodiode is nearly insensitive to the absolute location of the detector beam within the microscope field of view, but instead responds to *relative* movements of the object and the beam. Equivalently, this scheme may be understood in terms of a Fourier picture, by noting that purely angular changes in the specimen plane translate to displacements in the back focal plane of the condenser, and vice versa.

of the microscope (since only the state of polarization is evaluated), and therefore does not require extraordinary mechanical reinforcement: this may explain why it has achieved, in practice, somewhat better performance than direct imaging photodetectors. Conversely, the interferometer is vulnerable to fluctuations in the position of the trap relative to the particle, caused, for example, by minute laser pointing fluctuations. Special precautions, therefore, such as single-mode fiber coupling,[22] may be required to improve laser pointing stability. Methods that place the photodetector in nonimaging planes will generally be sensitive in varying degrees to both laser pointing instabilities and vibrations of the detector. Microscope sys-

[22] K. Svoboda and S. M. Block, *Cell* **77,** 773 (1994).

tems designed for nanometer-level measurement must be mechanically rigid and mounted on air isolation tables to limit mechanical noise. In addition, it pays to take additional precautions, such as placing such systems in environments with low acoustic noise, for example, inside sound-proofed enclosures, and by protecting light paths from wind currents, dust, and other disturbances.

The Instrument and Its Calibration

The instrument we present here was designed to be both versatile and sensitive, and to take advantage of several new developments in the field. The basic optical layout is illustrated in Fig. 3. The optical trap can be moved

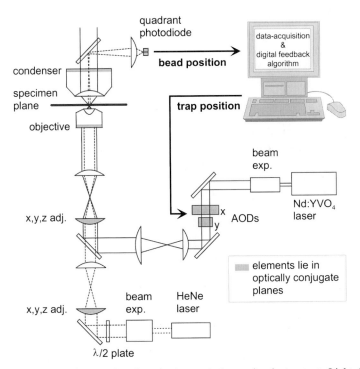

Fig. 3. Schematic layout of a time-sharing optical trapping instrument. Light from a near-infrared laser (solid lines) synthesizes multiple optical traps after computer-controlled deflection by two orthogonally-mounted AODs. Light from a visible red laser (dashed lines) is used for the position detection subsystem, and collected on a quadrant photodiode. Movable telescope lenses, mounted on x-y-z translation stages, provide manual control of the locations of the laser spots in the specimen plane. Optical components colored gray are located in planes that are conjugate to the back aperture of the microscope objective. For additional details, see text.

about freely in the specimen plane, either manually, through movement of an external lens mounted on an x-y-z translation stage, or electronically, by deflecting the trapping laser beam with a computer-controlled acousto-optical deflector (AOD). The position detection system employs a quadrant photodiode, placed in a plane approximately conjugate to the back focal plane of the condenser, based on the scheme of Fig. 2. Two independent laser sources are employed: a high-power, near-infrared laser serves as the trapping source (solid rays) and a visible light laser serves as the illumination source for the position sensor (dashed rays). Both sources are focused to diffraction-limited spots at the specimen plane and are usually adjusted to overlap at this position. All optical paths are enclosed to reduce instabilities caused by dust and air currents. The trapping instrument is based on a commercial inverted microscope (Diaphot 200, Nikon, Melville, NY) equipped with high-NA DIC optics. The original Nikon microscope stage was replaced with a spring-loaded, crossed roller bearing translation stage (750-MS, Rolyn Optics, Covina, CA), to which a piezo stage (P775, Physik Instrumente, Auburn, MA) was mounted, which in turn holds the specimen. We removed the Nikon coarse-focusing mechanism for the condenser and instead mounted the condenser assembly on a coarse/fine focusing transport (76408, Nikon) which was bolted directly to both the microscope base and vertical swing arm. Although this arrangement disables the ability to tilt back the condenser arm, it greatly enhances the mechanical rigidity of the condenser system, and there is sufficient range in the focusing transport to easily remove/replace specimens. Video images are recorded using a monochrome CCD camera (XC-77RR, Sony, Paramus, NJ). Software for trap position control, sensor data acquisition, and off-line analysis was developed primarily using LabView (National Instruments, Austin, TX). Some LabView-callable custom modules were written directly in the C programming language, for speed.

Optical Trap

The trapping source is a diode-pumped, Nd:YVO$_4$ solid-state laser (T-series B106-C, 1064 nm, 3-W continuous wave, Spectra Physics, Mountain View, CA), which is optically pumped through a 10-m-long fiber bundle, hence requires no water or fan cooling: the laser head can therefore be mounted directly to the optical table without introducing vibration. The laser power supply, which produces some acoustic noise, is located outside the soundproofed experimental room. The beam-pointing ability of this laser is sufficiently good that coupling to a single-mode fiber has not been necessary to achieve nanometer-level stability of the trap position. After leaving the laser, the trapping beam is expanded (5×, CVI, Albuquerque, NM) to subtend a substantial fraction of entrance aperture of the AODs.

This removes the need for any large beam expansion at a subsequent stage: expansion at any point beyond the AODs reduces the beam deflection angle by the same factor as the expansion, and thereby the trap displacement that can be achieved in the specimen plane. The beam is deflected in either x or y directions by two orthogonally mounted AODs (ATD-274HA16, IntraAction, Bellwood, IL) with a measured diffraction efficiency of ~80% per axis. The AOD drive signal is generated in a PC-based, 32-bit digital frequency synthesizer board (CVE-272A1, IntraAction) to ensure sufficient beam-pointing stability. The first-order diffracted light is collected and further expanded slightly, to a diameter just sufficient to fill the back aperture of the microscope objective, then deflected by a 45-deg dichroic mirror that serves to combine the trapping and detector lights. Next, a 1 : 1 telescope, consisting of two identical planoconvex lenses separated by the sum of their focal lengths, is used to steer both the optical trap and detector light in the specimen plane, as well as to parfocalize the trap.[2,23] The combined laser beams are coupled into the microscope from the side by means of a small dichroic prism mounted directly below the dichroic filter cube used for epifluorescence (prototype optical tweezers port, Nikon; dichroic prism, Chroma Technology, Brattleboro, VT). This design permits the simultaneous use of fluorescence mode and optical trapping, although care must be taken in selecting dichroic mirrors for epifluorescence to prevent laser interference problems. The entire objective turret of the microscope was removed and replaced with a custom-built tube holding a single infinity-focus Zeiss objective (100×/1.3-NA oil Plan Neofluar, Zeiss, Thornwood, NY) and its associated Wollaston prism for DIC imaging. This change was made to eliminate a 48% light loss in the near infrared that was experienced with the Nikon optics, attributed to unfavorable optical coatings on both the Wollaston prism and a negative lens found in the nosepiece (this lens acts to make the effective tube length infinite for 160-mm focal length objectives).

It is desirable to keep the optical power in the trap constant and independent of the position to which it may be moved within the specimen plane. To effect this, trap steering is achieved by means of beam rotations in the back focal plane of the objective while maintaining full illumination of the objective aperture. This, in turn, is accomplished by deflecting the laser beam, using external optics, in planes that are optically conjugate to the back aperture of the objective.[2,23] In practice, the two AODs are placed as close together as possible, such that the plane between them, and the plane of the first lens of the 1 : 1 telescope, are both imaged into the back

[23] S. M. Block, in "Cell Biology: A Laboratory Manual," Vol. II (D. Spector, R. Goldman, and L. Leinwand, eds.), Cold Spring Harbor Laboratory Press, New York, 1998. Available online at this website: http://clio.cshl.org/books/g_a/block.html.

aperture of the objective, and therefore conjugate to it (gray shading in Fig. 3). Rotations of the beam in these planes produce corresponding translations of the spot in the specimen plane, as required.

Position Detector

A 633-nm, polarized HeNe laser (1105P, 5 mW, Uniphase, San Jose, CA) supplies light for the detector subsystem (dashed lines, Fig. 3). The laser power is chosen to be sufficiently low that it does not cause particles to become trapped, yet sufficiently high to provide ample light for the photodiodes. After beam expansion, (3×, CVI), an auxiliary 1:1 telescope is used to position the detector spot independently of the trapping laser light. This control is most often adjusted to superpose the detector and the trapping spots in the specimen plane. (In a modification to this basic design, the HeNe laser is replaced by a visible light, single-mode diode laser at 635, 650, or 670 nm. Also, either the HeNe or diode laser may optionally be coupled to a single-mode optical fiber, whose output coupler provides for any necessary beam expansion. A single-mode fiber also reduces laser pointing fluctuations, as described earlier.) A dichroic mirror merges both red and infrared beams, which enter the main 1:1 telescope. After passing through the specimen and the condenser, the red detector light is deflected from the main optical path of the microscope by a 45-deg dichroic mirror and sent along a rigid optical rail mounted horizontally on the condenser assembly, then through a lens that images the back focal plane of the condenser onto a quadrant photodiode (SPOT9-DM1, UDT, Hawthorne, CA). The currents from each quadrant are converted to voltages by custom-built, two-stage preamplifiers. Preamplifier signals are passed to normalizing differential amplifiers that supply x- and y-position signals, which are then anti-alias filtered before digitization by a 12-bit multifunction A/D board (AT-MIO-16-E2, National Instruments, Austin, TX).

Position Detector Calibration

Because the detector and the trapping light are distinct and uncoupled, calibration of position becomes a relatively straightforward process, and can be accomplished readily for a particle trapped at any location within the field of view. Before detector calibration, one must perform one-time calibrations of both the video system and the AODs. The video system is calibrated by determining the correspondence between pixels and nanometers, by imaging a 10-μm ruled grating standard (objective micrometer, traceable to NIST) on the CCD camera. Next, a bead is trapped and moved back and forth, sequentially activating x and y deflections over several hundred nanometers of displacement, by driving the AODs with periodic triangle waves over a preselected range of frequencies, while video recording the movement on an optical memory disk recorder (OMDR). Centroid tracking algorithms,[12] accurate to within a few nanometers, are then used to determine the response amplitude of the two AODs, in nm/MHz. Once these system-wide calibrations have been carried out, the remaining calibration of the photodetector becomes trivial: a trapped bead is simply moved back and forth repeatedly through a predetermined distance, using the AODs, while the outputs of the photodiode sensor are recorded. The extremely high reproducibility of AODs permits averaging of detector output for 50 periods or more, thereby averaging out the effect of Brownian motion within the trap. The detector response to small displacements (around ±150 nm) turns out to be quite linear, but the active range of the system can readily be extended to ±200 nm or even beyond by fitting the output waveform of the sensor to a cubic or higher order polynomial.[22]

Multiple Traps

Certain experiments, like red blood cell deformability studies,[24] single myosin molecule assays,[13,14] or microtubule stiffness measurements,[25,26] can benefit from the use of two or more optical traps to establish a unique experimental geometry. Multiple traps may be constructed trivially by simply increasing the number of laser light sources, but this approach is neither efficient nor economical. If only two traps are desired, however, a single laser beam can always be split into two orthogonal polarizations.[9,13,27] By so doing, the beams forming each trap may be simultaneously present in the specimen without excessive interference. Another way to create multiple traps from a single light source is to time-share the laser beam among a predetermined set of positions in the specimen plane,[28] by rapidly scanning its position back and forth, analogous to the way separate traces are created in a multiple-beam oscilloscope. When light is scanned quickly enough (see below), the rapidly "blinking" individual traps mimic the effect of steady illumination. Interference problems are avoided because no more than one beam of light is present at any instant. Two orthogonal AODs are typically used to scan the trap. Because the AODs can be controlled in software, the number and relative strengths of the traps, their spatial patterns, as

[24] P. J. H. Bronkhorst, G. J. Streekstra, J. Grimbergen, E. J. Nijhof, J. J. Sixma, and G. J. Brakenhoff, *Biophys. J.* **69,** 1666 (1995).

[25] M. Kurachi, M. Hoshi, and H. Tashiro, *Cell Motil. Cytoskelet.* **30,** 221 (1995).

[26] M. W. Allersma, A. G. Z. Crowley, and C. F. Schmidt, *Biophys. J.* **68,** A288 (1995).

[27] H. Misawa, K. Sasaki, M. Koshioka, N. Kitamura, and H. Masuhara, *Appl. Phys. Lett.* **60,** 310 (1992).

[28] K. Visscher, G. J. Brakenhoff, and J. J. Krol, *Cytometry* **14,** 105 (1993).

well as the scanning rates can be chosen with great flexibility, all without the need to change optics.

The stiffness of an optical trap sets the time scale for the motion of trapped particles; many issues of trap design therefore revolve around stiffness considerations. Time-sharing systems offer flexibility, but this comes at a price: the effective stiffness of each synthesized trap is correspondingly diminished by the reduced duty cycle. The frequency response of a trapped particle falls off at frequencies greater than f_c, hence motion becomes insensitive to external disturbances occurring over times shorter than $1/f_c$. When N time-shared traps are produced from a single laser beam, it becomes necessary to scan at rates well above the roll-off frequency of an individual trap, which drops to k_{trap}/N, where k_{trap} is the stiffness of the single-trap configuration. (Should the scanning frequency become comparable to, or drop below, f_c, the stiffness of a synthesized trap effectively becomes time varying, which can lead to problems in measurement.) In our instrument, a 20-kHz scan rate is achieved, which is well beyond Brownian roll-off frequencies (250–2500 Hz) of micrometer-sized beads held at typical trapping stiffnesses of ~0.01–0.2 pN/nm.

Multiple-beam concepts are equally applicable to lasers used for position detection, in systems where trapping and detector lights are generated by different sources. For example, a single visible laser may be time-shared between two locations in the specimen plane, each of which is focused on one of two imaging quadrant photodiodes. Alternatively, the detection light may be split into two beams with orthogonal polarization, and each one used to illuminate a separate spot. Such a system can also be used in combination with nonimaging detectors of the type described earlier.

Speed is a fundamental consideration when selecting scanning devices for time-sharing traps. Scanning galvanometer mirrors, which were among the first devices used to produce time-sharing traps,[28,29] turn out to be too slow for most purposes, with response times of ~0.3–1 ms, corresponding to scan rates of ~1–5 kHz—uncomfortably close to typical roll-off frequencies for trapped particles. Electro-optic deflectors (EODs) work considerably faster, with response times limited only by their capacitive loads, and support scanning speeds to 10 MHz and more. However, the small maximum angle of deflection produced by EODs (~2 mrad, full angle) translates to a maximum separation of only ~1 μm between traps in a typical setup: this may not suffice for many implementations. For now, AODs seem to possess the optimal combination of adequate deflection angle (17 mrad, corresponding to ~10-μm trap displacement in the specimen plane) and

[29] K. Sasaki, M. Koshioka, H. Misawa, N. Kitamura, and H. Masuhara, *Opt. Lett.* **16,** 1463 (1991).

fast rise time (limited by the ratio of the speed of sound in the crystal to the beam diameter: $\sim 1.6 \ \mu s/mm$). In practice, however, raw AOD response times are not achieved by our setup. Instead, the slow ISA bus addressing speed of the digital frequency synthesizer board limits the response bandwidth to ~ 20 kHz ($= 50 \ \mu s$). This limitation should be removed with the next generation of synthesizer boards, designed for faster buses (e.g., the PCI bus). For applications requiring nanometer-level pointing stability in the specimen plane, it is essential to have an exceptionally stable frequency driving the transducer: 24- (or 32-bit) digital synthesizers are necessary.

Closed Loops: Position and Force Clamping

A variety of closed-loop feedback arrangements are made possible by combining trap control with position sensing. In a position clamp (*isometric*) system, displacement is fixed while the load changes, and the feedback signal provides an instantaneous measure of force. To fix position, the restoring force of the trap can be modulated in one of two ways: (1) repositioning the physical location of the trap, by moving the laser beam relative to the location of the trapped particle, accomplished by AODs[9,13,14,17] (Fig. 4A), or (2) dynamically altering the trap stiffness, through changes in the instantaneous laser intensity, accomplished with an acousto-optic modulator (AOM)[10] (Fig. 4B). The relative advantages and disadvantages of these alternatives have been discussed.[10] Briefly, systems that modulate intensity are intrinsically unidirectional, in the sense that restoring forces always point toward the center of the trap. As such, they are well suited to detectors that only do unidirectional sensing (e.g., interferometers). Intensity-modulating traps have the advantage that laser light can be maintained at a low level until it becomes necessary to increase it to effect clamping, which serves to reduce the overall exposure of the specimen. Because they require a single AOM instead of two AODs, such clamps are also more economical. However, systems that reposition the trap are far more versatile, and—unlike the alternative scheme—can successfully clamp particles close to the center of a trap, if required. Being two-dimensional, they also accommodate the possibility of monitoring changes in the direction, as well as magnitude, of the force. Furthermore, for the case of unidirectional force, the orthogonal channel provides a useful control signal.[13]

Because time-sharing traps already incorporate all the technology necessary to reposition the laser rapidly, they are exceptionally well suited to adopting the latter approach. Beam-deflecting position clamps have been used very successfully to study molecular forces produced by single myosin molecules,[13,14,17] and Molloy and co-workers were the first group to combine

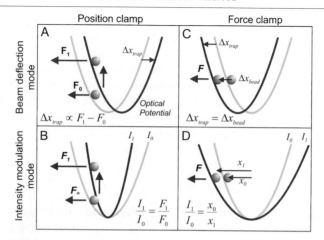

Fig. 4. Comparison of schemes for position and force clamps, based on beam deflection or intensity modulation. Harmonic energy wells are drawn to indicate the optical potentials, so that the height of a bead corresponds to its trapping energy, and the restoring force is determined by the slope of the potential. Equal and opposite to the restoring force is the external force, F, which the system measures. Real optical traps are three-dimensional potential wells; a single dimension is depicted here for simplicity. (A) A beam-deflecting position clamp. The trap is moved in a direction opposite to the external force. The displacement of the trap, Δx_{trap}, is proportional to the change in force, $F_1 - F_0$. (B) An intensity-modulating position clamp. The laser power is changed to alter the stiffness of the trap and thereby the restoring force. The relative increase in intensity, I_1/I_0, equals the relative increase in force, F_1/F_0. (C) A beam-deflecting force clamp. The trap follows an object at a specified distance (the displacement of the trap, Δx_{trap}, is identical to that of the bead Δx_{bead}). (D) An intensity-modulating force clamp. Laser power is changed to alter the stiffness of the trap, such that the restoring force becomes independent of position in the trap. To do so, the relative change in intensity, I_1/I_0, must equal x_0/x_1.

a position clamp with an innovative time-sharing, twin trap configuration.[14] A similar clamping scheme has been implemented on the time-sharing instrument shown in Fig. 3, which can synthesize up to 20 or more time-shared traps (one of which is subject to feedback control under the current software). An intensity-modulating position clamp has also been constructed in our laboratory, on a separate optical trapping interferometer, and used to measure the elastic properties of individual DNA molecules.[10]

The second form of closed-loop feedback arrangement is the force (*isotonic*) clamp, which maintains the load on a trapped particle while its displacement changes. As we shall see, force clamps afford special experimental advantages when used in assays with molecular motors that are processive. Conceptually, the simplest way to clamp the force is to follow the position of a trapped particle, moving the trap center in such a way as to keep it a fixed distance behind the particle, thereby imposing constant load (Fig. 4C). An alternative arrangement, by analogy with the schemes for position clamps, would be to diminish the trap stiffness by reducing the light intensity in direct proportion to the distance traveled from the trap center (Fig. 4D). Again, because time-sharing traps come equipped with a means to scan the laser beam, these naturally lend themselves to the bead-tracking scheme of Fig. 4C. We have implemented a force clamp on the instrument of Fig. 3 and are presently using this setup to study the motion of kinesin motors along microtubules. The heart of the computer-based feedback system is a fast subroutine written in C and integrated with LabView which implements digital PID (proportional integrating differential) control. This subroutine receives as input the digitized x- and y-position signals and outputs the appropriate numbers to a digital frequency synthesizer controlling x- and y-beam deflections, modulating the position for a single one of the (potentially several) time-shared traps. Figure 5 shows an example of instrument performance during a control experiment using a single trap, designed to test the ability of the system to track the motion of silica beads carrying molecules of the microtubule-based motor, kinesin. In this test, kinesin-coated beads were fixed to coverglass-bound microtubules in rigor using AMP-PNP, a nonhydrolyzable adenosine triphosphate (ATP) analog. (Such beads cannot translocate but do display Brownian motion about their points of attachment to microtubules.) To simulate actual movement, the microscope stage was driven stochastically under computer control by a piezo actuator programmed to advance in discrete steps of ~18.5 nm at exponentially distributed times, averaging ~1.5 sec. (This movement was simultaneously recorded through the microscope by video, so that the size and timing of bead displacements could be independently verified by centroid tracking, albeit at somewhat reduced spatial and temporal resolution.) The distance between the bead and the trap was preset in software to 100 nm, producing a load of ~5 pN at the given trap stiffness (~0.05 pN/nm). In this example, the load placed on the bead was in the same direction as the stage movement: the system permits forces to be applied in either direction. Figure 5A shows a representative record of bead and trap positions as functions of time. Note that the trace representing trap position is less noisy: this is because a digital lowpass filter is applied to the bead position before the required trap displacement is computed. Figure 5C shows the feedback signal throughout the course of the record, plotted as the difference between bead and the trap positions. This signal is proportional to the load applied through the trap stiffness (right axis). Figure 5D shows the amplitude histogram for an instantaneous load applied by the clamp, which has a Gaussian distribution. The force is maintained at the predetermined level to better than 10%.

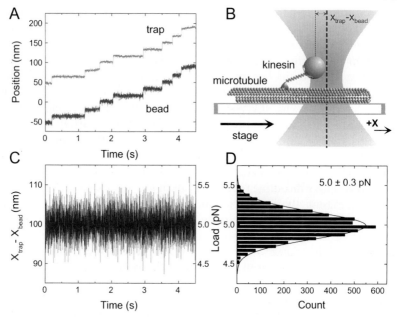

Fig. 5. Operation of the force clamp (control). (A) Records of trap and bead positions for 4.5 sec of artificial stepwise movement, driven by the piezo stage (see text). The distance between the bead and trap center was preset in software to 100 nm, corresponding to ~5 pN load, applied in the same direction as the movement. (B) Schematic cartoon of the experimental geometry, showing essential components and labels (not to scale). In reality, the bead is ~0.5 μm in diameter, the microtubule is 25 nm in diameter, and the laser beam waist is ~1 μm. The distribution of light is depicted by the shaded region, and the trap center is located on the dotted line. (C) The feedback signal over the same interval, consisting of the difference between trap and bead positions, and the corresponding force produced. (D) Histogram of force amplitudes for the data in part (C), showing a Gaussian fit. The fit returned an offset of 99.8 ± 6.7 nm, corresponding to a maintained load of 5.0 ± 0.3 pN. For these experiments, data were sampled at 24.3 kHz and averaged over a running window of 1.025 ms prior to computing each new trap position. Every tenth data point was saved to disk, for an effective acquisition rate of 2.43 kHz.

Kinesin Bead Assay

To study the motion of kinesin on microtubules, individual motors are typically attached to microscopic beads, which act as tiny handles through which motors may be manipulated by the optical trap, as well as functioning as bright markers for scoring position during movement. In an actual assay, kinesin molecules in an extremely dilute suspension are nonspecifically adsorbed to the surface of beads, such that the beads carry from one to several motor molecules each.[18,22,30,31] Microtubules are introduced into a microscope flow cell and become attached to a pretreated coverglass surface. A suspension of kinesin-coated beads is then drawn into the flow cell, in a buffer containing ATP. An optical trap is used to capture a single, freely diffusing bead and deposit this onto the surface of a microtubule. A kinesin-coated bead then binds to and translocates along microtubules for a distance of a micrometer or thereabouts. Nanometer-scale measurements of bead displacements permit characterization of motor activity at the molecular level. Experiments of this type have shown that single kinesin molecules take steps[18] of ~8 nm and work against loads[22,32] of up to 5–6 pN.

Measurements of kinesin steps are most easily performed at low ATP concentrations (or, alternatively, at high loads) where the motion can be slowed significantly, making stepwise transitions more apparent. In addition, optical trapping of a kinesin-coated bead has the useful advantage of increasing the effective stiffness acting on the bead, thereby reducing thermal (Brownian) excursions to ~12–25 nm (measured before any lowpass filtering; the actual amplitude depends on the relative trap and linkage stiffnesses). This diminution of thermal noise occurs because, from the equipartition theorem, the mean square amplitude of excursions is given by

$$\langle x^2 \rangle = k_B T/(k_{trap} + k_{motor})$$

where k_{trap} is the trap stiffness and k_{motor} is the stiffness of the linkage between the bead and microtubule, acting through the kinesin motor.[18] A stiff trap therefore produces less noise. It has also been observed that the amplitude of thermal noise typically decreases as beads move away from the trap center: this occurs because the motor linkage has nonlinear compliance—stiffening under load—further increasing the system stiffness.[18,22] Because the linkage can stretch, displacements of a bead reflect—*but do not exactly equal*—displacements of the motor protein itself, and a small but variable correction (typically ~15% for forces beyond ~1 pN) must be applied to compensate for this effect, complicating the measurement process.

Uncoupling Trap Stiffness and Load

In a conventional optical trap, the force on a trapped particle is inevitably coupled to the fixed stiffness of the trap. This can be disadvantageous: a trap with high stiffness, which is desirable to minimize thermal noise in measurements, produces relatively high loads after only short distances

[30] S. M. Block, L. S. B. Goldstein, and B. J. Schnapp, *Nature* **348**, 348 (1990).
[31] M. J. Schnitzer and S. M. Block, *Nature* **388**, 386 (1997).
[32] E. Meyhöfer and J. Howard, *Proc. Natl. Acad. Sci. U.S.A.* **92**, 574 (1995).

traveled, which is undesirable. Since the stall force[22,32] of kinesin is ~5–6 pN, kinesin molecules will cease movement altogether after only a handful of steps in a stiff trap (~0.5 pN/nm). Moreover, at high stiffness, the restoring force on a molecule varies throughout its journey in the trap, so that every step is taken under changing load conditions, complicating interpretation. The use of a force clamp, however, *uncouples* load and stiffness. In principle, using a force clamp, one can work with comparatively stiff traps near their centers, where the forces are nevertheless low, by using feedback to keep the distance between bead and trap small (and fixed). Under such conditions, the load experienced by the molecule is unchanging, therefore all steps are taken under similar conditions. The ability to work at high stiffness but low force affords a unique opportunity to study molecular stepping in the low-load regime. Moreover, in other situations, where one might seek to minimize optical damage, a force clamp can also be used to exert comparatively high loads, by maintaining a bead near the outermost edge of a trap of low stiffness (i.e., a trap formed by less intense light). Although the particle may be positioned near the outer limits of the trap, it may be followed nevertheless for a considerable distance, which would otherwise be impossible using a stationary trap.

Linkage Corrections

One complication in measuring molecular steps comes from the effect of finite linkage compliance.[18,22] An analogous problem exists in measurements of transcription by single molecule of RNA polymerase,[33] where the finite compliance of the DNA molecule becomes an issue.[10] Figure 6 illustrates the mechanical equivalent of a kinesin-driven bead moving in an optical trap. With a fixed trap, the measured displacement of the bead, Δx_{bead}, is related to the molecular displacement, Δx_{motor}, through

$$\Delta x_{bead} = \frac{k_{motor}}{k_{trap} + k_{motor}} \Delta x_{motor}, \qquad \text{with } k_{motor} = k(F_{trap}) \qquad (4)$$

In Eq. (4), k_{motor} incorporates the stiffness of the kinesin molecule, as well as any other stiffness associated with the entire linkage between the bead and the microtubule (which may arise from various contributions, including the configurational entropy of the bead).[22] There are several ways to correct for such linkage effects. Svoboda and co-workers[18] measured the series compliance for a population of nonmoving kinesin molecules attached to microtubules by AMP-PNP, and used this value to correct statistics col-

[33] H. Yin, M. D. Wang, K. Svoboda, R. Landick, S. M. Block, and J. Gelles, *Science* **270**, 1653 (1995).

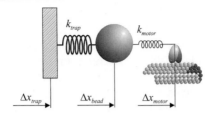

Fixed trap: $\Delta x_{bead} = \dfrac{k_{motor}}{k_{trap} + k_{motor}} \Delta x_{motor}$

Force clamp: $\Delta x_{bead} = \Delta x_{motor} = \Delta x_{trap}$

FIG. 6. Elastic corrections, showing the mechanical equivalent of the bead–motor–microtubule system. The optical trap acts as a spring characterized by k_{trap}, located between the stationary reference frame of the lab (wall, left) and the bead (sphere, center). The elasticity in the linkage between the bead and the stationary microtubule (right), acting through the kinesin motor, has been lumped into a single parameter, k_{motor}. Elastic correction factors relating displacements of the bead, Δx_{bead}, to displacements of the motor, Δx_{motor}, are shown for both a fixed trap and a force clamp.

lected on populations of moving molecules energized by ATP. Coppin and co-workers[34] improved on this scheme by performing stretch/release cycles on moving molecules to estimate their individual compliances. We next discuss some alternative ways to determine elastic corrections on-line.

Equipartition. Because the linkage stiffness and the variance in the bead position are related through $\langle x^2 \rangle = k_B T/(k_{trap} + k_{motor})$, it should be possible to estimate k_{motor} by measurements of the displacement noise within a finite-sized moving window, provided that the trap stiffness, k_{trap}, is already known. Because this window must be long enough to make reliable estimates of both mean and variance, the method is practical only when steps are taken comparatively slowly. Also, the method will break down whenever $k_{trap} \gg k_{motor}$, since baseline noise will be dominated by contributions from the trap and not the motor. Finally, the same precautions apply here as in earlier discussions about calibrating trap stiffness by the equipartition method: the bandwidth must be adequate and other noise sources add in quadrature, potentially biasing the estimate.

Lock-In Detection. On-line measurements may be performed by super-posing a high-frequency, low-amplitude "wiggle" on the overall bead motion, then performing phase-sensitive, lock-in detection of bead movement driven by this signal to extract the stiffness in real time. This method

[34] C. M. Coppin, J. T. Finer, J. A. Spudich, and R. D. Vale, *Proc. Natl. Acad. Sci. U.S.A.* **93**, 1913 (1996).

was pioneered by Yanagida and co-workers,[35] and is entirely analogous to measurements of transmembrane resistance in single-electrode, single-channel recordings, which may be accomplished by driving the electrode with a small sinusoidal voltage. Because the published account of the method is so abbreviated, we supply a more complete description here. When the trap position is modulated by $A \cos(2\pi ft)$ and one takes into account the viscous drag on the bead, β, the corresponding displacement is given by Eq. (5):

$$x_{\text{bead}} = \frac{k_{\text{trap}}}{(k_{\text{trap}} + k_{\text{motor}})} \cdot \frac{A \cos(2\pi ft + \varphi)}{\sqrt{1 + (2\pi f\tau)^2}}$$

$$\text{with } \varphi = \tan^{-1}(2\pi f\tau) \text{ and } \tau = \frac{\beta}{(k_{\text{trap}} + k_{\text{motor}})}$$

(5)

The roll-off frequency of the power spectrum of bead position, f_c, is directly related to τ through $f_c = (2\pi\tau)^{-1}$. A dual-phase lock-in amplifier can be used to determine $R = \sqrt{X^2 + Y^2}$, with $X = x_{\text{bead}} \cdot A \cos(2\pi ft)$ and $Y = x_{\text{bead}} \cdot A \sin(2\pi ft)$. After lowpass filtering X and Y to eliminate cross-modulation terms at double frequency, $2f$, one obtains Eq. (6):

$$R = \frac{k_{\text{trap}}}{(k_{\text{trap}} + k_{\text{motor}})} \cdot \frac{A}{2\sqrt{1 + (2\pi f\tau)^2}}$$

(6)

By determining R throughout movement, the linkage correction factor can be calculated for each measured position of the bead. The frequency-dependent term in the denominator of Eq. (6) is generally nearly unity, and may be neglected whenever the modulation frequency is sufficiently below the Brownian roll-off frequency of the system, simplifying matters. Under normal conditions in a kinesin motility assay, a modulation frequency as fast as a couple of hundred hertz should be feasible. In such cases, the correction factor reduces to $R = (A/2) \cdot k_{\text{trap}}/(k_{\text{trap}} + k_{\text{motor}})$. A practical lower limit to this frequency is set by the fact that a significant number of oscillations is required to measure R within a finite time interval, the length of which is set by the rate of kinesin stepping. Conceivably, one might fine-tune the modulation frequency for different ATP concentrations and/or loads. The amplitude chosen for the displacement wiggle, ideally, should be as small as possible to remain nonperturbative. In practice, this will be limited by the lock-in detection sensitivity in the presence of randomized thermal motion: Yanagida and co-workers[35] used a peak-to-peak amplitude of 10 nm (which may be on the high side).

---

[35] H. Higuchi, E. Muto, Y. Inoue, and T. Yanagida, *Proc. Natl. Acad. Sci. U.S.A.* **94**, 4395 (1997).

An alternative to moving the trap would be to modulate the position of the microscope flow cell holding the microtubule, using a sufficiently rigid piezo stage, in which case the output of the lock-in amplifier becomes $R' = (A/2)k_{motor}/(k_{trap} + k_{motor})$. Note that since $R' = (1 - R)$, measurement of either R or R' supplies precisely the elastic correction factor needed to relate Δx_{bead} to Δx_{motor}, up to a constant factor. In principle, therefore, one does not require an additional, independent measurement of the trap stiffness, k_{trap}, to apply the lock-in approach. A drawback to the method is that the sinusoidal modulation signal gets superimposed on the bead position data, complicating analysis of records at frequencies near the modulation rate.

Finesse the Problem Altogether! It turns out that when a feedback-based force clamp system is implemented, as described earlier, *no correction* for linkage stiffness is required to compute molecular positions from measurement of bead position. The displacement of the bead is given by Eq. (7) (Fig. 6):

$$\Delta x_{bead} = \frac{k_{trap}}{k_{trap} + k_{motor}} \Delta x_{trap} + \frac{k_{motor}}{k_{trap} + k_{motor}} \Delta x_{motor} \qquad (7)$$

over time scales that are long compared to viscous damping in the system ($\tau \gg 1/f_c$). The force feedback maintains a fixed distance between the bead and the trap, which implies $\Delta x_{bead} = \Delta x_{trap}$. Inserting this equality into Eq. (7) leads directly to $\Delta x_{bead} = \Delta x_{motor}$; there is no elastic correction. Intuitively, constant load ensures that none of the mechanical elements in the system may stretch over time (once these rapidly reach mechanical equilibrium), so that a displacement of any one of the elements requires all others to move an identical amount. To build an effective force clamp, the feedback loop closure time should be rapid compared to time constants for any motions of interest. However, this time may be slower than the Brownian roll-off time of a bead, so that purely diffusive motions are averaged. The maximum raw update rate for position in the force clamp of Fig. 3 is 20 kHz. The overall response time of the force clamp, however, is effectively set by the digital feedback subroutine, which contains an integrating filter for bead position: the window width of this filter can be optimized for changing experimental conditions (different trap stiffness or ATP concentrations).

Force Clamp in a Kinesin Motility Assay

Figure 7 shows representative records using the force clamp in an actual kinesin motility assay, with saturating levels of ATP in the buffer. In this case, the trap was positioned behind the bead, such that kinesin molecules

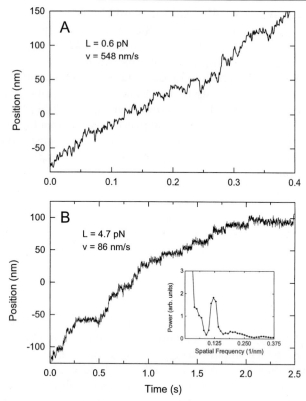

Fig. 7. Operation of the force clamp (experiment). Records show nanometer-scale movements of a bead (0.5 μm in diameter) carrying a single molecule of kinesin translocating along a microtubule under force clamp conditions. The trap stiffness was set to 0.063 pN/nm and the ATP concentration was 1 mM. (A) The trap center was set 10 nm behind the bead position, producing a constant load of ~0.63 pN. The mean velocity for movement during this run was 548 ± 2 nm/sec. (B) The trap center was set 75 nm behind the bead position, increasing the load to ~4.7 pN. The mean velocity during this run was reduced to 86 ± 0.1 nm/sec. Note the change in time scale, and the fact that individual steps become more apparent in the record. Inset: The power spectrum of the autocorrelation of displacements in this record, showing a peak, the reciprocal of 8.26 nm, corresponding to the mean step size. For this experiment, data were acquired in fashion similar to that of Fig. 5, corresponding to an effective rate of 2.43 kHz.

proceed against the applied load (a force clamp can be set up to produce loads of either sign). The data of Fig. 6 were digitized at 2.43 kHz and have not been lowpass filtered to improve their appearance. At low constant load (0.6 pN, Fig. 7A), the kinesin velocity is ~548 nm/sec and molecular steps are buried in the noise. At high constant load (4.7 pN, Fig. 7B), the velocity drops to ~86 nm/sec and individual steps become apparent. The mean size of steps may be determined by taking the power spectrum of the autocorrelation function of the record.[18,31,36] This spectrum (Fig. 7B, inset) displays a prominent peak at a spatial frequency of 0.121 nm^{-1}, corresponding to a step size of 8.26 nm for this particular record. This value—*which involves no elastic corrections*—is identical, within experimental error, to the kinesin step size of 8.3 ± 0.2 nm estimated in earlier work using a fixed trap, to which a 19% adjustment for series compliance had been applied.[18] The force clamp enables us, for the first time, to acquire stepping data for kinesin molecules subjected to high loads over distances as great as 200–300 nm, a feat previously impossible with a stationary trap.

Acknowledgments

We thank Steve Gross, Michelle Wang, and Mark Schnitzer for helpful discussions, Winfield Hill of the Rowland Institute for Science for electronics design, and Amit Mehta for suggestions on the manuscript. KV is a Burroughs-Wellcome Fellow of the Life Sciences Research Foundation. SMB gratefully acknowledges the support of grants from the NIH and NSF.

[36] S. M. Block and K. Svoboda, *Biophys. J.* **68,** 230s (1995).

[39] Preparation of a Flexible, Porous Polyacrylamide Substrate for Mechanical Studies of Cultured Cells

By Yu-Li Wang and Robert J. Pelham, Jr.

Introduction

Although cell culture is traditionally performed on plastic or glass substrates, cells in a multicellular organism live under a substantially different environment: they adhere to mechanically flexible tissues or basement membranes and are surrounded by fluid and nutrients. Accumulating evidence indicates that physical parameters such as mechanical forces,[1,2] flexibility,[3] fluid shear,[4,5] and media accessibility[6] can have profound effects on cell growth and differentiation.

[1] N. Wang, J. P. Butler, and D. E. Ingber, *Science* **260,** 1124 (1993).
[2] D. E. Ingber, *J. Cell Sci.* **104,** 613 (1993).
[3] R. J. Pelham and Y-L. Wang, *Proc. Natl. Acad. Sci. U.S.A.* **94,** 13661 (1997).
[4] P. F. Davies, *Physiol. Rev.* **75,** 519 (1995).
[5] P. R. Girad and R. M. Nerem, *J. Cell. Phys.* **163,** 179 (1995).
[6] K. Simons and S. D. Fuller, *Annu. Rev. Cell Biol.* **1,** 24 (1985).

A number of flexible and/or porous substrates have been developed over the years, including silicone rubber,[7] collagen matrix,[8,9] fibrin clots,[10] and nucleopore filters.[6] Besides providing a more physiologic environment for cell culture, flexible substrates have been used for testing the effects of mechanical forces on cells[11] and for measuring traction forces that cells exert on the substrate.[7,12] The latter has been most elegantly done with polymerized films of silicone fluid. In the original version developed by Harris and co-workers,[7] the film covers a layer of silicone fluid and wrinkles on application of force, much like the response of a water bed. Although it is difficult to perform precise quantitative measurements, the magnitude of forces can be estimated based on the extent of wrinkling. This method has recently been improved by embedding particles in weakly polymerized, nonwrinkling films[12] (which allow a more precise measurement of forces based on the movement of beads[13]), and by using improved materials that allow systematic control of the flexibility.[14]

We have recently developed a culture substrate based on polyacryamide sheets coated with extracellular matrix proteins. Although polyacrylamide-based substrates have been around for some time,[15] their full potential has not yet been realized. The material has a number of favorable features. First, it has a nearly ideal mechanical property: it deforms in proportion to applied forces over a wide range, and recovers completely and instantaneously on the release of force. Second, it allows systematic and reproducible control of the flexibility of the substrate, by changing the relative concentration of acrylamide and bisacrylamide. Third, its excellent optical quality permits the observation of both immunofluorescence staining and microinjected fluorescent analogs at a high magnification. Fourth, the substrate uses specific extracellular matrix (ECM) molecules as the ligand for cell adhesion, while polyacrylamide itself shows no detectable interaction with the cell surface. Fifth, the porous nature of the polyacrylamide gel allows the penetration of media and provides a more physiologic environment for cell culture, particularly for epithelial cells.[6] We have used the substrate for testing the responses of cells to flexibility and to deformation forces,

[7] A. K. Harris, P. Wild, and D. Stopak, *Science* **208**, 177 (1980).
[8] A. K. Harris, D. Stopak, and P. Wild, *Nature* **290**, 249 (1981).
[9] K. Mochitate, P. Pawelek, and F. Grinnel, *Exp. Cell Res.* **193**, 198 (1991).
[10] A. K. Harris, *in* "Locomotion of Tissue Cells" (R. Porter and D. W. FitzSimons, eds.), p. 3. Associated Scientific Publishers, Amsterdam, 1973.
[11] J. Sadoshima and S. Izumo, *EMBO J.* **12**, 1681 (1993).
[12] J. Lee, M. Leonard, T. Oliver, A. Ishihara, and K. Jacobson, *J. Cell Biol.* **127**, 1957 (1994).
[13] T. Oliver, K. Jacobson, and M. Dembo, *Methods Enzymol.* **298**, [40], 1998 (this volume).
[14] K. Burton and D. L. Taylor, *Nature* **385**, 450 (1997).
[15] B. K. Brandley, O. A. Weisz, and R. L. Schnaar, *J. Biol. Chem.* **262**, 6431 (1987).

and for measuring traction/retraction forces that cells apply to the substrate during locomotion.

Preparation of Activated Glass Surface

Although the polyacrylamide substrate can be applied to an unprepared glass surface, it adheres poorly to glass and floats off easily. Thus for most experiments it is necessary to covalently attach the polyacrylamide sheet to the glass surface by chemical activation of the glass surface following the procedure of Alpin and Hughes.[16]

Materials and Reagents

Diamond-tipped pen
Coverslips (No. 1, 45 × 50 mm; Fisher Scientific, Pittsburgh, PA)
0.1 *N* NaOH
3-Aminopropyltrimethoxysilane (Sigma, St. Louis, MO)
Phosphate-buffered saline (PBS)
0.5% Glutaraldehyde in PBS (prepared by diluting 1 part of 70% stock solution, Polysciences, Inc., Warrington, PA, with 140 parts of PBS)

Procedure

1. Pass coverslips briefly through the inner flame of a Bunsen burner.
2. Smear a small volume of 0.1 *N* NaOH across the surface of the coverslip with a Pasteur pipette and air dry the coverslip. The surface should be covered with a film of dried NaOH.
3. Gently mark the treated surface with a diamond-tipped pen. This facilitates the identification of the proper side of coverslips in later steps.
4. Smear a small volume (~200 µl) of 3-aminopropyltrimethoxysilane evenly on the marked side of the glass surface with a Pasteur pipette. Allow the coverslip to sit horizontally for 4–5 min.
5. Cover the treated side of the coverslip with distilled H_2O and let sit for 5–10 min. When the surface is clear, rinse with distilled H_2O from a squirt bottle and soak in distilled H_2O for 5–10 min with gentle agitation.
6. Transfer the coverslips, marked side up, into petri dishes and cover the surface with 0.5% glutaraldehyde in PBS. Incubate at room temperature for 30 min.

---

[16] J. D. Alpin and C. Hughes, *Anal. Biochem.* **113,** 144 (1981).

7. Wash the coverslip extensively with multiple changes of distilled H_2O on a shaker and then let air dry vertically. The treated coverslips may then be used for gel attachment up to 48 hr after preparation.

Preparation of Polyacrylamide Sheets

Materials and Reagents

Coverslips, circular (No. 1, 22 mm diameter; Fisher)
Acrylamide (Bio-Rad, Hercules, CA), 30% w/v
N,N'-Methylene bisacrylamide (BIS, Bio-Rad), 2.5% w/v
Ammonium persulfate (Bio-Rad), 10% w/v
N,N,N',N'-Tetramethylethylenediamine (TEMED, Bio-Rad)
50 mM HEPES, pH 8.5
Fluorescent latex beads, 0.2-μm FluoSpheres, carboxylate-modified (Molecular Probes, Eugene, OR), for measuring traction forces only

Procedure

1. Mix acrylamide, BIS, and distilled H_2O to obtain a desirable concentration. For example, we most often use 10% acrylamide/0.26% BIS to prepare relatively stiff substrates, and 10% acrylamide/0.03% BIS for highly flexible substrates and for measuring traction forces. At 10% acrylamide, the BIS concentration can be varied between 0.025% and 0.4%. Too low a concentration generates viscous fluids instead of flexible solids. Too high a concentration makes the substrate opaque.
2. Add fluorescent latex beads to the mixture if the substrate is to be used for measuring traction forces. The beads are first sonicated briefly in a bath sonicator and 1/125 volume is added to the acrylamide mixture.
3. Degas the acrylamide/BIS solution.
4. Add 1/200 volume of 10% ammonium persulfate and 1/2000 volume of TEMED. Immediately pipette 10–25 μl of the solution onto the activated coverslip surface and cover with a 22-mm coverslip. Turn the coverslip assembly upside down if the solution contains fluorescent beads.
5. When acrylamide polymerizes (10–30 min), carefully remove the 22-mm coverglass and wash the gel with 50 mM HEPES on a shaker.

Activation of the Polyacrylamide Surface and Conjugation with Type I Collagen

Cultured cells adhere poorly to the polyacrylamide surface. To provide a physiologic adhesive surface for cell culture, the surface has to be coated with adhesion molecules. An easy way to conjugate proteins to the poly-acrylamide surface is to use a photoactivatable heterobifunctional re-agent. Sulfosuccinimidyl-6-(4'-azido-2'-nitrophenylamino) hexanoate (sulfo-SANPAH) contains at one end a succinimidyl ester group, which reacts with lysine ε-NH$_2$. The other end is a phenylazide group that, on photoactivation, reacts nonspecifically with many chemically inert mole-cules including water and polyacrylamide. The following procedure uses sulfo-SANPAH to activate the polyacrylamide. Then type I collagen (or other adhesion molecules) is allowed to bind covalently to the activated surface.

Reagents

> 1 mM sulfo-SANPAH (Pierce Chemicals, Rockford, IL) in 50 mM HEPES, pH 8.5
> 50 mM HEPES, pH 8.5
> Type I collagen (Amersham Life Science, Arlington Heights, IL), di-luted to 0.2 mg/ml with 50 mM HEPES

Procedure

1. Drain fluid off the surface of the polyacrylamide gel. Carefully pipette 200 μl of 1 mM sulfo-SANPAH onto the surface.
2. Expose the surface to the UV light of a 30-W germicidal lamp at a distance of 6 in. for 5 min. Remove the darkened sulfo-SANPAH solution and repeat the photactivation procedure.
3. Wash the polyacrylamide sheet with two changes of 50 mM HEPES (pH 8.5), 15 min each, on a shaker.
4. Layer a 0.2 mg/ml solution of type I collagen on the substrate and allow to react overnight at 4° on a shaker.
5. Wash the gels with PBS. Mount the coverslip with the gel onto an appropriate culture chamber[17] and sterilize with UV irradiation.
6. Before plating cells, soak the gels for 30–45 min in culture medium at 37°.

Characterization of the Polyacrylamide Substrate

The thickness of the substrate can be estimated based on the initial volume of the acrylamide solution applied to the coverslip and the area of the gel. The actual thickness, which can be measured by focusing a micro-scope from the glass surface up to the gel surface, is affected by the

[17] N. M. McKenna and Y-L. Wang, *Methods Cell Biol.* **29,** 105 (1989).

acrylamide/BIS concentration and the degree of swelling/shrinkage of the gel in media. The protocol above gives a gel approximately 70 μm in thickness. The concentration of matrix proteins on the surface can be measured by a radioimmunoassay. Briefly, the surface is reacted with primary antibodies against the coated protein, then with an iodinated secondary antibody. The gel is then peeled off the coverslip with a razor blade and radioactivity counted.

The flexibility of polyacrylamide sheets can be characterized both macroscopically and microscopically. The former is performed by preparing gels 0.75 mm in thickness, using the apparatus for casting mini gels for gel electrophoresis. Strips of the gel 70 × 30 mm in size are then fixed at one end with binder clips attached to a horizontal bar or at the edge of a bench, and stretched vertically by attaching known weights (most conveniently by using binder clips) to the opposite free end (Fig. 1). Young's modulus, which measures the elasticity of materials, can then be calculated based on the original length (l), the change in length (Δl), the substrate's cross-sectional area (A), and the applied force ($F_\perp$; 1 g of weight applies 980 dynes of force), according to the equation: $Y = (F_\perp/A)/(\Delta l/l)$.

Microscopic compliance of the surface, which is affected by both

FIG. 1. Measuring the Young's modulus of polyacrylamide substrates. A 70- × 30- × 0.75-mm piece of polyacrylamide substrate is suspended from a laboratory bench and is deformed using binder clips as weights. The change in length is used to calculate Young's modulus.

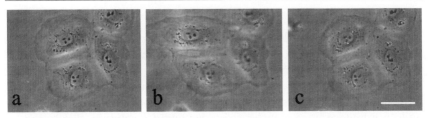

FIG. 2. Application of external mechanical forces to individual NRK epithelial cells. A microneedle is used to deform a 10% acrylamide/0.26% BIS substrate. (a) Before application of force, (b) during deformation, and (c) recovery of the cell and substrate after the release of tension. Bar = 10 μm.

Young's modulus and shear modulus, can be measured by deforming the gels horizontally with a flexible glass microneedle attached to a micromanipulator, essentially as described by Lee and co-workers.[12] Deformation of the gel is measured as the translocation of the needle tip, assuming that there is no slippage of the tip on the gel surface. The force applied by the needle is a function of needle deformation, which is measured by subtracting the distance of needle movement recorded by the micromanipulator with the distance of tip movement. The same needles are then calibrated by measuring their deformation after applying known submilligram weights.[12] The compliance is calculated as micrometer deformation per Newton applied force.

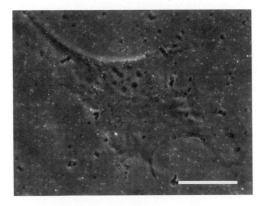

FIG. 3. Simultaneous visualization of 3T3 fibroblasts and 0.2-μm fluorescent latex beads embedded in a 10% acrylamide/0.03% BIS substrate. Movement of the beads is used as a means of measuring traction forces generated during cell locomotion. Bar = 10 μm.

Application of the Polyacrylamide Substrate

The morphology of cells cultured on polyacrylamide substrates can be studied with phase optics or with fluorescence microscopy after labeling/ microinjecting with fluorescent probes. Using dry objective lenses, there is no significant degradation of the image by the polyacrylamide gel. However, at high magnifications using 63× or 100× oil immersion lenses, both phase and fluorescence images suffer from decreased resolution, due to the increased distance between the cell and the objective lens, and the spherical aberration of the lens. High-quality images can be obtained with long working distance, water-immersion lenses designed for laser confocal scanning microscopy.

We have used the polyacrylamide substrates in several types of studies.[3] First, it is clear that the mechanical property of the substrate, controlled by the acrylamide/BIS concentration, has profound effects on the motility, growth, and differentiation of cultured cells. A second application is to study the responses of cells to external deforming forces, which occur frequently within a multicellular organism. A microneedle mounted on a micromanipulator is poked into the gel ~50–100 μm from the cell, then dragged for a defined distance. Using a 10% acrylamide/0.26% BIS gel, the deformation of the substrate propagates for ~100–200 μm and causes the cell to deform by up to 30% (Fig. 2). While deforming forces can be applied in many ways, the present approach facilitates observations of the immediate responses of living cells to defined and easily regulated forces.

The third application of the polyacrylamide substrate is to measure traction forces during cell locomotion, using highly flexible substrates with fluorescent particles embedded inside. By simultaneously illuminating cells for phase and epifluorescence optics, images of the cell and beads immediately underneath can be recorded in the same image (Fig. 3). Simple time-lapse recordings readily show movements of the beads (i.e., changes in traction forces) during the protrusion/retraction of cultured fibroblasts. Moreover, by taking images of the beads before and after trypsinizing cells, a vectorial map can be generated that shows the global distribution of forces exerted on the substrate. Thus, by combining these applications, we expect the present substrate to provide not only culture conditions that mimic the physiologic environment, but also powerful means for mechanical manipulations and measurements of living cells.

[40] Design and Use of Substrata to Measure Traction Forces Exerted by Cultured Cells

By Tim Oliver, Ken Jacobson, and Micah Dembo

Introduction

"Traction forces" are all those forces that a cell exerts tangent (parallel) to its substratum. Tractions can occur via specific adhesions or via nonspecific frictional interactions. The cell can also actively generate them or they may be passive reactions. An example of the first type of traction would be the force produced at an adhesion plaque by active contraction of an attached stress fiber. An example of the latter type would be the tractions produced at the trailing edge of a locomoting cell by stripping or uprooting of adhesive bonds.

We are assured by Newton's laws that any cell must produce traction forces on a substratum in order to move by crawling. Thus knowledge of the magnitude and distribution of these forces is fundamental for understanding the detailed mechanism(s) of cell motility. Unfortunately traction forces are invisible to standard microscopy and the investigator cannot securely deduce their magnitude or distribution simply by looking at motions, displacements and/or mass density distributions. To overcome this difficulty, Harris and co-workers introduced the use of deformable silicone rubber substrata as a means of transducing cell-generated traction forces into a detectable form (see review by Harris[1]). This technique has recently undergone a revival in applications to the study of cell locomotion and cell division.

Currently there are two basic varieties of flexible silicone substrata. First, a wrinkling substratum is easy to construct but it provides only qualitative or semiquantitative information. This is because wrinkling is unpredictable and depends critically on many poorly understood variables. Secondly, a nonwrinkling substratum with predictable elastic behavior can be used to provide more precise quantitative information on the spatial location and direction and magnitude of cell-generated tractions. This methodological review focuses on the application of both assays to one particular cell system, the locomoting fish epidermal keratocyte. References are made to other systems where appropriate and guidelines based on the authors' experience with the keratocyte are presented for the benefit of the reader.

---

[1] A. K. Harris, *Methods Enzymol.* **163,** 623 (1988).

However, remember that many of the conditions we describe may not be appropriate to alternate cell systems.

Overview

In designing flexible substrata for a traction force assay the following points should be considered.

Will the cell of choice readily attach to silicone rubber, and is the cell strong enough to exert a detectable deformation in the rubber without damaging the rubber? (Note that the actual rubber film is only of the order of a micron or less thick.[2])

Does the cell of interest have a special requirement for extracellular matrix?

Do the cell culture conditions and length of the experiment require aseptic technique, carbon dioxide supplements or pH buffering, and/ or special temperature requirements?

Is the response time of the substratum fast enough to record generation of transient traction forces (e.g., muscle twitch)?

Will neighboring cells interfere with one another by creating overlapping traction fields? (This is a problem for quantitative traction imaging.)

Is quantitative information required?

Is the assay design compatible with the imaging hardware available and desired sensitivity of measurement?

Simple Wrinkling Assay

The essential characteristics of a wrinkling film are elastic stability, low bending modulus, and the presence of a small amount of "slack." If we take the simple analogy of a skin stretched across the mouth of a drum whose tightness can be adjusted, then a wrinkling substratum may be described as having zero drumhead surface tension. As is well known from everyday experience, when such films are subjected to tractions parallel to the plane of the film, the deformation mode of least resistance is one of buckling (consider a bedsheet). Many cell types easily produce enough force to buckle silicone sheets of a few microns in thickness, and the size and orientation of the wrinkles formed gives some report on the particular loading conditions that they produce. In its simplest form, a sheet of rubber is cast on a cover glass inside a petri dish or other open chamber, and cells

---

[2] A. K. Harris, *in* "Cell Behaviour" (R. Bellairs, A. Curtis, and G. Dunn, eds.), pp. 109–134. Cambridge University Press, Cambridge, Massachusetts. 1982.

are grown on it until wrinkles in the film are generated, either spontaneously or after the application of drugs or other stimuli to the cells.[3,4]

The assay can be recorded at suitable intervals with 35-mm photomicrography or via a video camera coupled to a VCR. Wrinkles may increase or decrease in both number and length, reflecting a change in the traction applied by the cells. Harris and colleagues attempted to measure the force required to reproduce a given length of wrinkle using a calibrated microneedle.[5] Burton and Taylor[6] have recently refined this approach in order to measure the cell forces involved in cytokinesis, using a wrinkling substratum of greater sensitivity than previously used.

Wrinkling Assay Configured into a Flow Chamber

In early studies the wrinkling assay was carried out in an open Rappaport chamber, but we have recently devised a scaled-down version based on a microscope slide flow chamber (see Fig. 1a). Our approach has several advantages. First, changing the medium in which cells are immersed is simplified, without disturbing the field of view in the microscope or traumatizing the rubber. Second, the minimal volume occupied by the chamber requires only very small amounts of drugs or other costly medium additives that may be required to stimulate the cells. Third, the chamber design is compatible with high-resolution double oil immersion optics and a wide range of imaging modes. Fourth, the wrinkling assay in this configuration is compatible with fixation and fluorescent staining of the cells *in situ* on the silicone rubber (unpublished data, Tim Oliver, 1997; see Fig. 2).

Nonwrinkling Assay in a Rappaport Chamber

The films for generating quantitative traction maps must have several properties: (1) They must incorporate marker particles so that in-plane deformations are observable. (2) They must be elastic (i.e., they must rapidly and completely recover to the initial state after deformation). (3) They must have in-plane elastic compliance that is matched to the strength of the cell of interest (the cell-induced displacement of marker particles should be big enough to be observed yet small enough so that the response of the film will be linear). (4) They must be stretched tightly between the walls of the chamber with sufficient drumhead tension so as to be stable against buckling instabilities.

---

[3] B. A. Danowski and A. K. Harris, *Exp. Cell Res.* **177,** 47 (1988).
[4] M. Chrzanowska-Wodnicka and K. Burridge, *J. Cell Biol.* **133,** 1403 (1996).
[5] A. K. Harris, P. Wild, and D. Stopak, *Science* **208,** 177 (1980).
[6] K. Burton and D. L. Taylor, *Nature* **385,** 450 (1997).

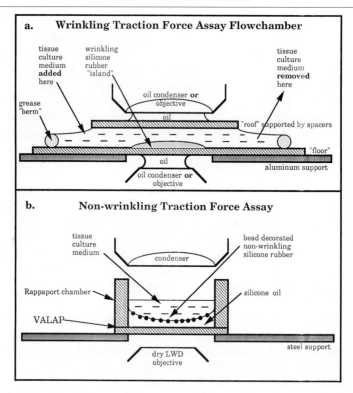

FIG. 1. Schematics of the two different types of traction assay chambers described in the text. In both assays (not shown here), cells are attached to the interface between the silicone rubber and the aqueous medium above.

The successful design of a nonwrinkling assay led to the first images in which the magnitude and direction of the traction stresses acting beneath different parts of an individual locomoting keratocyte were resolved[7] (see also Fig. 3). Traction stresses generated by keratocytes were found to be in the millidyne range.[8] The silicone rubber film used has a better than 95% elastic recovery, a surface Young's modulus of elasticity of 54 ± 15 dyne/cm, and a response time of about a second.[7–9] A schematic of this assay and its apparatus is shown in Fig. 1b.

[7] T. Oliver, M. Dembo, and K. Jacobson, *Cell Motil. Cytoskelet.* **31,** 225 (1995).

[8] M. Dembo, T. Oliver, A. Ishihara, and K. Jacobson, *Biophys. J.* **70,** 2008 (1996).

[9] J. Lee, M. Leonard, T. Oliver, A. Ishihara, and K. Jacobson, *J. Cell Biol.* **127,** 1957 (1994).

Detailed Methodology

Making Chambers

Two kinds of chambers are used. For quantitative traction mapping on elastic substrata, Rappaport chambers are assembled as follows. Pyrex tubing, 22 mm in diameter, is cut into 8-mm-high rings at a local glass shop. After cleaning in "Micro" or similar proprietary lab glassware detergent, they are attached to 22-mm-diameter circular cover glasses with Vaseline, Lanoline, Paraffin (VALAP) (Fig. 1b). Note that cover glasses should be precleaned thoroughly in detergent or concentrated hydrochloric acid, rinsed well, and dried to a mirror finish. This is essential to prevent dirt nucleating air bubbles at the glass–silicone oil interface. After assembly, 60 mg of dimethylpolysiloxane (12,500 centistokes viscosity) is placed in the chamber and spread around with a glass rod.

For the wrinkling assay, a flow chamber (Fig. 1a) is adapted for imaging individual cells at high magnification from a modification formerly used by Dr. Joseph Wolenski for molecular motility assays (personal communication, 1995 Physiology Course, MBL, Woods Hole, MA). The flow chamber incorporated cells attached to an island of wrinkling silicone rubber about 4 mm in diameter (about 4 μl of silicone oil). Only the flat, wrinkle-free central area of the silicone rubber island was used for cell observations.

Making Wrinkling Films

Wrinkling rubber is prepared by glow discharge ionization (under low air pressure) of a small drop (3–4 μl) of dimethylpolysiloxane silicone oil (DMPS, Sigma Chemical Company, St. Louis, MO), placed at the center of a precleaned cover glass, prior to assembly in the flow chamber. This results in a very thin layer of rubber overlying the viscous silicone oil beneath. The conditions for vulcanizing the surface of silicone oil into wrinkling rubber are relatively undemanding. About 4–5 sec at maximum power at 0.1-Torr vacuum is adequate. This treatment generates an aseptic wrinkling substratum that deters bacterial overgrowth if applied cells are cultured for extended times in the presence of appropriate antibiotic and antifungal medium supplements. Slow-growing cells like C3H10T1/2 fibroblasts, which are reluctant to spread on silicone rubber, can be encouraged to do so if a thin layer of elemental gold is sputtered onto the silicone during the vulcanization. Similarly, incubating the silicone substratum with elevated concentrations of fetal calf serum, prior to adding tissue culture medium, may encourage cells to spread faster.

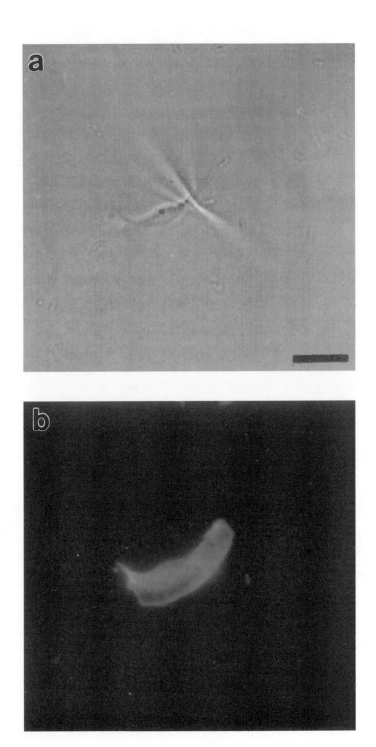

Making Nonwrinkling Films

For nonwrinkling films, observation of film deformation is dependent on the presence of marker particles incorporated during manufacture. For this purpose, a special airbrush was constructed to distribute evenly a monodispersion of 1-μm latex beads onto the surface of the silicone oil prior to cross-linking.[7] An ideal density of beads is approximately 3×10^4 beads per square millimeter. The airbrush consistently outperforms all other methods tried, including mixing freeze dried beads into the silicone oil or scattering them on the surface with a bristle brush.

After deposition of the beads, a glow discharge vacuum evaporator is employed to vulcanize (cross-link) the surface of silicone oil in Rappaport chambers into silicone rubber films. Each of the vulcanization variables is optimized by trial and error in order to make a film appropriate for the type of cell under study. These variables include exposure time, sample distance from the ionization source, vacuum, voltage applied, and the viscosity of the silicone oil.

For fish keratocytes we make nonwrinkling films employing a Polaron sputter coater E5100 (Energy Beam Sciences, Agawam, MA), with the electrode height set 60 mm above the specimen holder. DMPS (12,500 centistoke) is cross-linked at about 2-mA current, at 0.1-Torr vacuum for approximately 3 sec. To achieve the correct degree of vulcanization it is suggested that only one or two variables in the recipe be adjusted, and that a range of substrata be generated for the variable being tried.[7] Candidate films are prescreened for their long-term stability, freedom from drift, and their compliance. Compliance was gauged in two ways. First, we directly observed the degree to which cells could move beads embedded in the film. Second, we empirically micromanipulated films with precalibrated glass needles.[7]

The films produced as described above spontaneously develop and maintain the required prestress after addition of culture medium at room temperature (the usual condition for experiments with fish keratocytes). This prestress is essential for the nonwrinkling properties of these films and

---

FIG. 2. Fish keratocytes cultured on an "island" of wrinkling silicone rubber in the flow chamber were fixed by perfusion of paraformaldehyde solution through the chamber. Panel (a) shows that the characteristic traction wrinkle generated by this cell and parallel to its direction of locomotion persists well after fixation and fluorescent staining of the preparation (phase contrast image, 1.4 NA, 100× oil immersion objective). (b) Concurrent fluorescent image of the same cell after permeabilization with Triton-X 100 and staining of the actin cytoskeleton with rhodamine-phalloidin. Note that in contrast to panel (a), the entire cell's lamella is sharply focused, except for the region immediately above the wrinkle, which is out of the plane of focus. Bar = 25 μm.

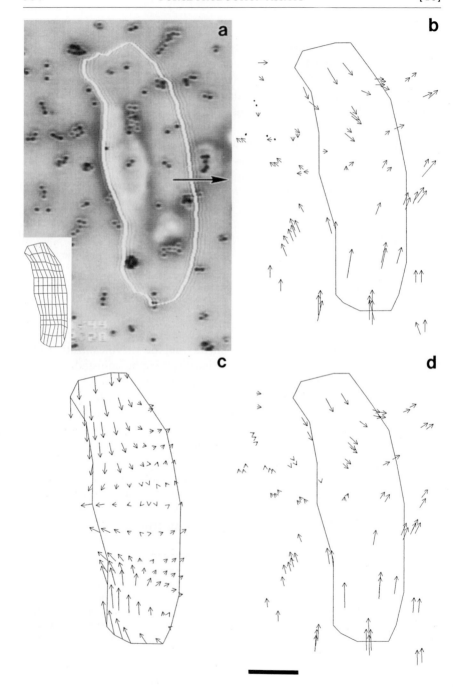

distinguishes them from wrinkling films. If the film is insufficiently cross-linked, then the developing drumhead tension may cause the film to rupture when medium is added. The deposition of serum proteins onto the hydrophobic surface of the films can significantly mitigate this tendency (see Fig. 4 in Ref. 7). Thus the serum concentration represents an additional variable that was optimized by trial and error to obtain films with appropriate characteristics for keratocyte traction measurements.

It is useful to remember that DMPS comes in a range of viscosities; in our experience 12,500, 30,000, and 60,000 centistokes are the most useful. Burton and Taylor[6] have recently optimized the vulcanization of a different silicone oil—Dow-Corning 710 fluid (Dow Corning Corp., Midland, MI) for use as a wrinkling substratum. This comes in just one viscosity but has a number of properties that make it attractive (see Table I). It is anticipated that this polymer blended with one of the above viscosities of DMPS can create films with novel and potentially useful properties for studying several parameters of cells simultaneously. For example, a nonwrinkling film that would permit concurrent imaging in bright-field, fluorescence, and/or interference reflection microscopy would be useful.

Calibrating Films

To calibrate silicone films for the quantitative traction assay, microneedles are calibrated by measuring the amount of tip deflection under

---

FIG. 3. Computing relative cell–substratum tractions from particle displacements in an elastic silicone rubber substratum (a) Negative phase contrast image of a fish epidermal keratocyte (outlined in white) locomoting toward the right (arrow), on a silicone rubber substratum into which 1-μm latex beads (black spheres) were incorporated. This cell was recorded over several minutes of locomotion in a straight line. The locations of all beads in this image represent a substratum deformed by the additional stress applied as a consequence of the cell's presence (bead positions for the relaxed film are not shown, but were recorded 3 min before the cell's arrival). Inset panel shows a mesh of 92 quadrilaterals representing the area under the cell as a set of 113 nodes. (b) The experimentally derived pattern of bead displacement vectors (dex), resulting from the cell's traction. (c) The output graphics of the traction analysis. The traction density at each node of the cell mesh is shown as a vector (arrow). The entire field of vectors is referred to as the "tractions most likely" (tml). At each node, the analysis software iteratively adjusts a delta function of traction in magnitude and direction until the field of bead displacements in part (b) is reproduced with maximum fidelity. This reproduced field of bead displacements is shown in part (d) and is referred to as "displacements most likely" (dml). Both traction and displacement vectors have been amplified 5× for display purposes. Cell velocity is ~0.3 μm sec^{-1}. Noise due to subpixel film drift was accurately measured and accounted for. Total film drift over 3 min was $1.00 \times 10^{-1} \pm 8.44 \times 10^{-3}$ μm in x axis, $2.34 \times 10^{-1} \pm 9.32 \times 10^{-3}$ μm in y axis. Experimental error measuring bead centroids (based on pixel density) is $\pm 1.84 \times 10^{-1}$ μm. Bar = 10 μm. [Reprinted with permission from T. Oliver, M. Dembo, and K. Jacobson, *Cell Motil. Cytoskelet.* **31**, 225 (1995).]

TABLE I

PROPERTIES OF SILICONE OILS: 200 FLUID AND 710 FLUID FROM DOW-CORNING

Chemical name	Linear dimethylpolysiloxane[1,2]	Polycyclic phenylmethyl-polysiloxane[6]
Manufacturer	200 fluid, Dow-Corning	710 fluid, Dow-Corning
Viscosity	Useful range 12,500–60,000 cs	Single viscosity 500 cs
Refractive index	1.4036–1.4037	1.533
Vulcanized by	Heat or glow discharge	Heat
Adjustment of compliance by	Choice of viscosity and controlled exposure to heat or glow discharge	UV irradiation[a]
Type of assay	Wrinkling/nonwrinkling	Wrinkling
Force calibration possible?	Yes	Yes
Other desirable properties	Can be incorporated into a flow chamber for wrinkling assay; cells can be fixed *in situ* for immunofluorescence	High refractive index gives good IRM imaging
Cells tested	Fish keratocytes, mammalian fibroblasts (C3H10T1/2 cells), *Dictyostelium*	Swiss 3T3 cells

[a] The supplemental use of UV irradiation after heat vulcanization serves to weaken the silicone sheet, increasing its compliance in a controlled fashion.[6]

various known loading conditions.[10] These microneedles are then used to manipulate films so as to derive estimates of the surface Young's modulus, and the response time of the films.

Single needle calibration. A rough measure of film compliance can be made using a single calibrated needle mounted in a three-axis hydraulic micromanipulator attached to an inverted microscope. With the needle inclined at ~45 deg to horizontal, the needle's tip is gently lowered in the z direction until firm contact is made with a point close to the center of the film. The needle is then slowly moved to apply a traction force tangential to the film surface (see Fig. 4a). As it displaces the film, the needle bends under the reaction load.

The force required to achieve a given displacement can be determined by simultaneously observing the displacement and also the amount of this bending. Thus it is a simple matter to derive measurements of displacement as a function of traction force. For positive forces this relationship is usually linear:

$$d = \sigma f + d_0, \quad \text{for } f > 0 \tag{1}$$

[10] M. Yoneda, *J. Exp. Biol.* **37**, 460 (1960).

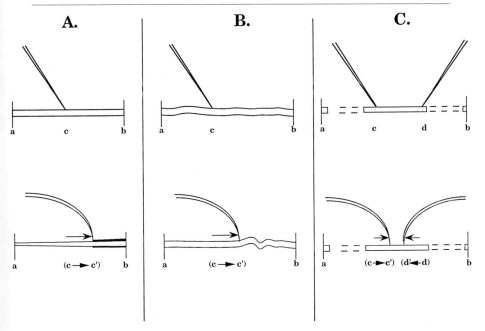

FIG. 4. (A) Nonwrinkling elastic rubber film with drumhead prestress (see text for definition) is stretched taut between the walls of a chamber, to which it is firmly anchored (top
panel). Force is applied by a needle at c toward c', causing the rubber to stretch and the
needle to bend (bottom panel). Displacement in the rubber is proportional to the force
applied, the distance from the anchorage point (a–c), and the reciprocal of the rubber's
stiffness. (B) Wrinkling rubber film has no tension between its anchorage points (a,b). Force
applied by the needle at c causes displacement to c'. The rubber wrinkles on one side and
tightens on the other side. The displacement in the rubber for a given force applied by the
needle at midpoint c depends on the amount of slack in the film, the linear dimension of the
chamber and the stiffness of the rubber. (C) A rubber film stretched taut between (a–b) has
similar drumhead prestress as the film in part (A). Force is applied to the rubber in a balanced
pinch between two needles initially located at c and d. The needles move in equal and opposite
directions toward c' and d', respectively. Displacement in the film is strictly localized to the
region of the pinch (solid lines). Far from the center the film remains undisturbed (dashed
lines). The localized disturbance in the film that occurs does not extend to the chamber walls,
because one needle counterbalances the other. In other words, the vessel walls feel no effect,
because the push of one needle balances the pull of the other. Displacement in the film is
proportional both to force applied and to the stiffness of the rubber, but is independent of
the length of the rubber. Consequently the displacements (c–c', d–d') are independent of
the total size of the rubber film.

where d is the displacement predicted by force f. The intercept of the force–displacement plot is the parameter d_0, which gives a measure of the amount of slack or free play in the surface ($d_0 = 0$ for prestressed films). The precise value of d_0 also depends on the placement of the needle relative to the chamber boundary, the linear dimensions of the chamber, and the bending modulus of the film. Even for prestressed films, buckling will eventually occur and the linearity of the force–velocity relation breaks down at very large displacements.

The slope of the displacement–force relation described by Eq. (1) is a parameter σ having units of compliance (cm/dyne). All other things being equal, the value of σ is inversely proportional to the surface Young's modulus. Unfortunately the slope of a single-needle compliance curve is not an intensive property of the film. It is proportional to the linear dimension of the film and also depends on the placement of the needle relative to the chamber boundary as well as the footprint of the needle when it contacts the surface.

The response time of the elastic film is also measured in a single-needle manipulation experiment. This is the mean time it takes for a displaced bead (the bead closest to the site of force application) to return half the distance to its original position; a set of trials is performed in which the initial displacement of several beads is systematically varied.

Although it yields some information and is easy to carry out, the single-needle technique produces film deformation in a fashion that is fundamentally different from the way in which a cell produces film deformation. Furthermore, there is no simple procedure that can extrapolate from one kind of deformation to the other. This is because at mechanical equilibrium the total forces acting on the film must balance. In a single-needle experiment, the necessary reaction force to produce this balance can only be supplied by the walls of the observation chamber. This occurs regardless of how small a force is applied and regardless of the nature of the film. Because of this inevitable mechanical constraint, the deformations caused by a single-needle manipulation extend throughout the entire surface of the observation chamber and are critically influenced by the boundary conditions, by the level of prestress, and by geometrical factors. For these reasons we caution against trying to calibrate the absolute stiffness of films by the single-needle technique. An alternative method in which zero net force is applied to the film is much preferable.

Pinch Experiment with a Pair of Needles. In contrast to the action of a single needle manipulated by an external experimentalist, a cell "lives" entirely within and on the two-dimensional world of the film surface. Consequently, when a cell applies tractions to its substratum it does so in a balanced manner (i.e., it pulls on one segment of film and pushes on another so that the total force imparted adds up to zero). Under these circumstances the deformation of the film is localized to the immediate vicinity of the applied tractions and the walls of the observation chamber are no longer needed to provide counterthrust.

Now let us suppose that $\mathbf{f}$ is a position vector with Cartesian coordinates (f_1, f_2) and let us represent the tractions acting at this position by a continuous vector field $T(\mathbf{f}) = [T_1(\mathbf{f}), T_2(\mathbf{f})]$. Obviously, if these tractions are produced by short-range interactions then the field $T(\mathbf{f})$ will be nonzero only at places where this cell makes close contact with the substrate. Furthermore, if the cell is isolated from the external universe except for its contact with the film, then the traction field it produces must be globally balanced (i.e., the area integral of the traction components $T_\alpha(\mathbf{f})$ over the contact area must be equal to zero). As described by Dembo et al.,[8] this is sufficient to guarantee the existence of four nondimensional functions $g_{\alpha\beta}(x_1, x_2, f_1, f_2)$ that give the displacement in the α direction at location (x_1, x_2) induced by a delta function traction density acting in the β direction at location (f_1, f_2). Using these quantities the displacement field at some location $\mathbf{x}$ can be assembled by superimposing influences from all parts of the traction field:

$$d_\alpha(\mathbf{x}) = \frac{1}{E_s} \iint \left(g_{\alpha1}(\mathbf{x}, \mathbf{f}) T_1(\mathbf{f}) + g_{\alpha2}(\mathbf{x}, \mathbf{f}) T_2(\mathbf{f}) \right) d\mathbf{f} \qquad (2a)$$

Note that the variable of integration in this equation ($\mathbf{f}$) ranges only over the footprint between the cell and the substratum. In contrast the displacement field can be observed anywhere (i.e., $\mathbf{x}$ can lie either inside or outside of the contact zone). The algebraic expression for the Green's function components is

$$g_{\alpha\beta} \equiv \frac{(1+\nu)^2}{4\pi} \left[\frac{(x_\alpha - f_\alpha)(x_\beta - f_\beta)}{|\mathbf{x} - \mathbf{f}|^2} + \delta_{\alpha\beta} \frac{(3-\nu)}{(1+\nu)} \ln\left(\frac{1}{|\mathbf{x} - \mathbf{f}|}\right) \right] \qquad (2b)$$

where $\delta_{\alpha\beta}$ is the Kronecker delta.

The parameters of Eqs. (2a) and (2b) are ν (Poisson's ratio) and E_s (the stiffness or surface Young's modulus). For an incompressible material, Poisson's ratio has a value very close to 0.5. Thus in order to utilize Eqs. (2a) and (2b) fully, E_s is the only parameter that needs to be experimentally determined. Note however that even if E_s is not known we can still use the theory to obtain information on the *relative* magnitude and direction of the displacement vectors that result from any given traction field.

We have recently introduced a method of film calibration that uses Eqs. (2a) and (2b) to calibrate the surface stiffness modulus (E_s). In this procedure a balanced pinch is applied to the film with a pair of needles acting in apposition to one another (see Fig. 4c and Fig. 5f; details described by Dembo et al.[8]). Again the force is applied only in the tangent plane of the thin film. We record the displacement of the needle tips, and the

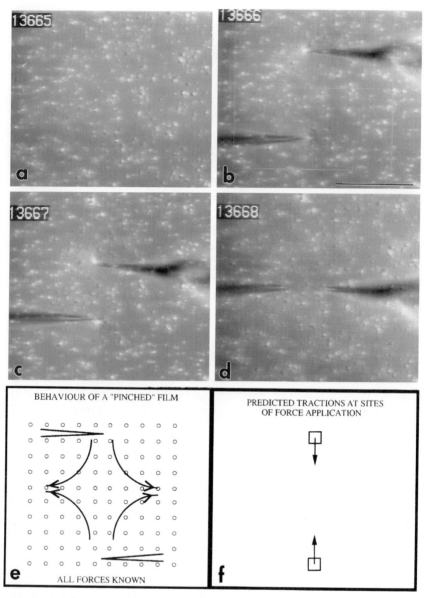

FIG. 5. (a–d) Four video frames of primary data from a "pinch" experiment imaged in differential interference contrast (DIC): (a) Undisturbed film. (b) Needle tips make initial contact with the film's surface. (c) Needle tips apply a "pinch" to the film. (d) Needles are raised and they recoil to equilibrium. Beads are white. (e,f) Schematic of the "pinch" experiment: (e) Curved arrows indicate the distortions observed in a beaded silicone film (circles) after a "pinch" is applied and held with a pair of needles. (f) Theory predicts the magnitude and direction of tractions (arrows) at sites of force application (squares), and enables the Young's modulus of film to be calculated. Bar = 50 μm.

TABLE II
DETERMINATION OF YOUNG'S MODULUS BY TWO-NEEDLE PINCH

Trial number	Film ID	Film stiffness[a] (empirical) (dyne/cm)	Distance displaced by needletip (μm) Left	Right	Force applied by needle ($\times 10^{-2}$ dyne)	Young's modulus (dyne/cm)
1	b	29.4	21.4	26.9	3.8	74
2	b	29.4	15.1	12.5	2.1	61
3	a	13.7	6.7	3.3	0.7	44
4	a	13.7	6.4	3.6	0.7	41
5	a	13.7	6.4	3.6	0.7	41
					mean =	54 ± 15

[a] These values represent our earlier estimates of film stiffness, based on the application of one needle to the film, and are included for comparison with the Young's modulus estimates derived from two-needle experiments.

magnitude of the force is determined from the needle bending. We also record bead displacements in the region surrounding the center of the pinch.

Finally the surface Young's modulus of the film is determined by fitting these data using Eq. (2a). Our fitting procedure is based on the assumption that the total traction force applied by each needle (a measured quantity) is distributed as a uniform stress within the needle footprint (i.e., the areas shown by the two small boxes in Fig. 5f). The value of E_s is fixed so that there is agreement between the average magnitude of the computed bead displacements and the observed bead displacements. The goodness of fit between the experimentally recorded and the most likely predicted film displacements has been published.[8]

In five repetitions on two films, the surface Young's modulus of our standard keratocyte film preparation measured by the pinch technique was found to be 54 ± 15 dynes/cm (Table II). This value was about double that of the empirical stiffness parameter (σ^{-1}), measured and described previously[7] in single-needle compliance experiments conducted on standard elastic films. The small difference between these two parameters is purely circumstantial and results from a fortuitous choice for the size of our Rappaport chambers. Much larger corrections should be expected in the flow chamber geometry. We attribute the variability between the two sample films (a and b) to be due to minor differences in the cross-linking conditions that were beyond experimental control (for example, the exact positions of Rappaport chambers in the glow discharge apparatus and the difficulty in reproducing a 3-sec exposure time with a stopwatch).

Image Recording for the Quantitative Traction Assay

The basic strategy of the traction mapping technique is to measure substratum deformation by a motile cell and to then infer the tractions by fitting the observed deformations to the prediction of Eqs. (2a) and (2b). All the necessary information for this operation can be obtained by imaging a field of beads embedded in the silicone rubber, both in the presence of a cell (when the silicone film is deformed; see Fig. 3a) and in the cell's absence (when the film is considered "undisturbed" or relaxed). Beads in the film are assumed to represent loci of an elastic plane whose response to tractions applied in that plane is predictable. Visualizing the film's deformation is therefore critically dependent on getting a good number of well-distributed bead centroids ($\sim$100) in both the undisturbed film and after the locomoting cell has deformed the film.

For the keratocytes the experiment is fortuitously straightforward, because individual cells locomote into and out of a field of view with surprising predictability over about 15 min of observation. Satisfactory images both of the beads and the cell's outline can be collected with a dry, long working distance, 40× phase or DIC objective (in order to accommodate the finite depth of silicone oil and cover glass beneath the film; see Fig. 1b) and a long working distance condenser (in order to accommodate the height of the medium-filled chamber above the film). Equally fortuitous was the finding that keratocytes plated sparingly on such a film produce negligible interference with one another when separated by more than about 0.2 mm.

Another important consideration was the standardization of the Rappaport chamber's shape and area to a 22-mm-diameter cylinder. Only cells close to the chamber's center were sampled, since only in this location can the reaction force of the vessel's wall to traction forces generated in the silicone film by the cell be considered negligibly small and equal in all directions. The microscope's field of view was optimized with the addition of a 1–2.25× zoom lens between the microscope camera port and the videocamera. A time/date generator was incorporated into the video system to ensure that each recorded image had a unique identity.

Optimizing the Compliance of the Silicone Film to Accommodate Different Cells

It should be noted that mammalian fibroblasts are significantly stronger than fish epidermal keratocytes. Consequently, fibroblast traction mapping requires a stiffer, nonwrinkling silicone substratum to prevent the cells from generating wrinkles. Efforts to map traction forces generated by C3H10T1/2 fibroblasts were frustrated by two major experimental hurdles. First, it was found that as few as 10–50 cells plated on silicone film generated interference between neighboring cell traction fields sufficiently severe to distort the map for any one cell. Second, fibroblasts are relatively sedentary and resisted all efforts to make them round up and detach (in order to collect the "undisturbed" image of the film), except in the presence of aggressive chemical agents such as saturated sodium azide. This treatment is to be avoided because it interferes with the physical properties of the silicone rubber by warping the "undisturbed" film. It is anticipated that future experiments to map the traction forces generated by fibroblasts will utilize chambers containing only a very few cells in order to overcome the former problem. The use of pharmacologic agents like butanedione-2-monoxime (BDM) to help round and detach the cell should overcome the latter problem.

Dictyostelium cells were successfully attached to bead decorated nonwrinkling silicone rubber films and subsequently generated bead displacements in the course of locomotion (unpublished data, Tim Oliver, 1994).

Compensating for Drift in the Silicone Substratum

Despite rigorous precautions taken to stabilize nonwrinkling films, a constant source of interference to traction mapping experiments was spontaneous film drift. Drift was manifested by a uniform and parallel linear displacement of all objects on the substratum over time, and is readily identified when time lapse recordings are replayed at high speed. Drift is almost impossible to prevent, since the film forms a fragile interface separating hydrophobic viscous silicone oil from aqueous tissue culture medium. The film is also subject to shear forces from temperature gradients and convection in the tissue culture medium. For this reason experimental times were kept as short as possible and the microscope was kept level and isolated from vibrations on a vibration isolation table. Drift that persisted despite these precautions was dealt with in two ways. Gross drift can be removed from a series of consecutive video images, using a feature of Metamorph software called "stack align." Subpixel drift was effectively removed during the mapping procedure using an algorithm that could effectively distinguish between parallel bead drift and randomly directed independent bead motion (Fig. 6).

Image Analysis

Background subtracted, TIFF format images, composed of square pixels, were collected after a gray-level stretch to utilize 256 gray levels. All linear measurements were made on Image-1 software (Universal Imaging Corporation, West Chester, PA) and were converted at the time of measurement

a   **b**

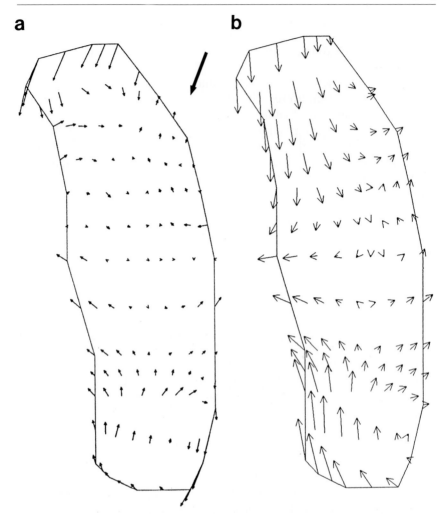

Fig. 6. Correcting for drift in the silicone film. (a) The maximum likelihood tractions for a steady-state locomoting keratocyte when the drift is uncorrected. Note the difference in magnitude of tractions at similar locations at each end of the cell. The large arrow indicates the direction of film drift. Generally, tractions pointing in the same direction as the drift tend to be amplified, while those pointing opposite to the drift are diminished. (b) The maximum likelihood tractions for the same cell, incorporating the software correction for systematic film drift.

from pixel units into micrometer units by comparison with a standard calibration image, utilizing a 50-μm resolution hemocytometer counting chamber (Improved Neubauer, Fisher Scientific Corporation, Pittsburgh, PA). Two images were selected from a time-lapse video recording of cell locomotion. The first image was termed the "pinched" image in which a clear view of the entire cell was visible, preferably close to the center of the image. The second image, termed the "undisturbed" image, represented the same field of view as the pinched image, but recorded in the absence of the cell (that is, before or after the cell's locomotion across the field of view).

Sampling Bead Centroids

Bead centroids were identified by object thresholding; a process in which an upper and lower value is defined within 256 gray levels of tone available, to form a discrimination window or "threshold" that discretizes on object with an artificial boundary.

Most digital image analysis systems can now rapidly perform multiparametric image analysis on regions of interest (ROIs) enclosing multiple objects that have been discretized by thresholding. Despite this, we have not been able to threshold automatically and find the centroids of 100+ beads in two temporally spaced images without significant errors. The major error is manifested as "bead mismatch." This occurs when a bead detected in one of the images fails to threshold in the other image. This throws the continuity of data logging out of order, since measuring the undisturbed and displaced positions for the exact same 100 beads is required in order to calculate the net distance over which each bead is displaced. Such a discrepancy between the information content of two images is usually caused by a subtle change in the uniformity of the background subtracted image between the time the cell is present and when it is absent (often a piece of junk or other nonbead object will interfere). Also, the cell's presence in only one of the two images introduces a change in refractive index to beads lying under the cell (especially at the cell's edge), affecting their discretization by thresholding. Because we could find no satisfactory error-checking procedure to deal with the problem of bead mismatch, we chose instead to record bead centroids manually one by one, according to the following protocol:

1. Calibrate square pixels to units of length.
2. Load "undisturbed" (cell free) image to Foreground frame buffer. Load "pinched" image (containing view of cell) to Background frame buffer.
3. Turn on object thresholding and adjust gray level windowing to highlight majority of beads in color.

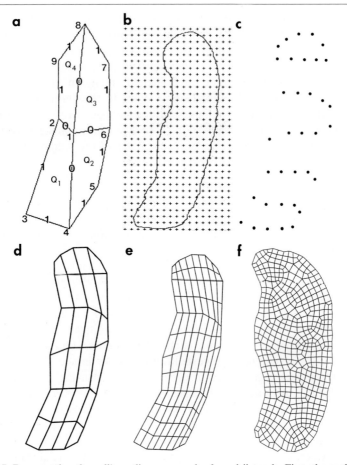

Fig. 7. Representing the cell's outline as a mesh of quadrilaterals. First, the outline of a cell is carefully hand traced from an image on the video monitor using a mouse-driven cursor (not shown here but essentially similar to the white line traced around the periphery of the cell in Fig. 3a). The tracing is saved as an edgelist file in Image-1. (a) Points chosen on and within the cell edgelist are connected with straight lines to form a mesh of quadrilaterals Q_1 through Q_4. The corners and intersections of the quadrilaterals form a set of nodes labeled 1 through 9. Two neighboring quadrilaterals share an internal boundary line denoted with a "0," whereas free edges are denoted by a "1." Note that this four-quadrilateral mesh does not accurately describe the true curved nature of the cell's boundary, but is shown as a simple example of the labeling convention used in describing more complex meshes. (b) The cursor traced edgelist (described above) is rendered with an overlying grid of crosses, whose density is adjusted so that a large number of crosses falls on the cell's perimeter. (c) Thresholding crosses along and within the cell's perimeter allows a set of nodes to be identified that better reflect the cell's curvature. Node centroids are logged semiautomatically (filled circles) by "mouse clicking." (d) A mesh building algorithm generates a mesh with 24 quadrilaterals, 35 nodes, and 20 external edges that better resolves the cell's outline than that is done in

4. Check fidelity of bead thresholding between the two frame buffers, before arbitrarily selecting a sample size of ~100 beads for analysis. Each bead must threshold in both frames. Beads should be evenly distributed about and beneath the cell of interest.
5. Mark the position of each bead on a transparent overlay attached to the monitor screen. Each bead is ascribed an identifier between 1 and 100.
6. Open a data logging file, ascribe a file name and path (*.mod) and turn off all parameters except centroid finding. Crop ROI tool to a small box, slightly larger than a bead.
7. Start with Foreground frame buffer. Click ROI tool with mouse on each thresholded bead from 1 to 100. This operation logs each bead centroid with its identifier to the data-logging file. When complete, scroll through logging file to check for fidelity of identifiers.
8. Repeat procedure for Background frame buffer.
9. A translator program calculates the net displacement of each bead and outputs this value in a format compatible with the traction mapping spreadsheet (called a template).

Constructing a Cell "Mesh" for Generating the Traction Map Image

Generally, the cell's profile is uniformly divided into a set of quadrilaterals in which the number of "nodes" (defined here as intersections between neighboring quadrilaterals) is approximately equal to the number of bead displacements recorded during the film's deformation (see Fig. 7). Each node represents a site of traction under the cell whose magnitude and direction are unknown. Because automatically tracing the cell's periphery is usually unsuccessful, the cell's outline is hand-traced from the "pinched" image using the freehand cursor function of the imaging software (Fig. 3a), and the traced pixels are recorded into a stored edgelist file. This file can later be used to recreate a TIFF image of the cell's periphery.

The next step is to define nodes at the intersections of a meshwork of finite elements (contiguous quadrilaterals) enclosed by the cell's periphery, using the grid-making function of Image-1 (see Fig. 7b). Four nodes joined with straight lines (Fig. 7a) describe any single quadrilateral. Each node has a unique identifier and its coordinates are listed in a data-logging file. Currently, data-logging files containing node centroids and bead centroids

---

part (a). (e) The "hackmesh" command applied to the mesh in part (d) creates a new mesh, now with four times the number of elements as that in (d). (f) Automated mesh drawing (called tiling) with ANSYS 5.3 software, courtesy of Mallett Technology Inc., Research Triangle Park, NC, shows the potential for high-resolution cell meshes.

```
***** TEMPLET OF SETUP FILE
***** ***** ***** ***** ***** ***** ***** ***** ***** ***** ***** ***** *****
    5 ******************************CONTROL DESCRIPTORS************************
***** ***** ***** ***** ***** ***** ***** ***** ***** ***** ***** ***** *****
    1 MESH TOPOLOGY DESCRIPTORS
***** NUMN         NUMP         NUMQ         NUMD         NUME         OBSV
***** 9.0000E+00   4.0000E+00   4.0000E+00   5.0000E+00   8.0000E+00   1.0000E+00
    2 NODE VECTOR PACKING DESCRIPTORS
***** XON1         XON2         TRC1         TRC2         SIG1         SIG2
***** 1.0000E+00   2.0000E+00   3.0000E+00   4.0000E+00   5.0000E+00   6.0000E+00
    3 DATA VECTOR PACKING DESCRIPTORS
***** XOP1         XOP2         DEX1         DEX2         DML1         DML2
***** 1.0000E+00   2.0000E+00   3.0000E+00   4.0000E+00   5.0000E+00   6.0000E+00
    4 OPERATION CONTROL DESCRIPTORS
***** BOOT         EXER         CPLX         CHIS         YMOD         SMZR
***** 0.0000E+00   1.0000E+00   1.0000E-01   0.0000E+00   1.0000E+00   0.0000E+00
    5 DRIFT CORRECTION DESCRIPTORS
***** DRFX         DRFY         ERDX         ERDY         BLNK         BLNK
***** 0.0000E+00   0.0000E+00   0.0000E+00   0.0000E+00   0.0000E+00   0.0000E+00
***** ***** ***** ***** ***** ***** ***** ***** ***** ***** ***** ***** *****
    4 ************************POINTERS OF MESH TOPOLOGY************************
***** ***** ***** ***** ***** ***** ***** ***** ***** ***** ***** ***** *****
    1 INOQ*   1     2     3     4 ILOQ*   0     1     1     0 IWOQ*   1
    2 INOQ*   1     4     5     6 ILOQ*   0     1     1     0 IWOQ*   1
    3 INOQ*   1     6     7     8 ILOQ*   0     1     1     0 IWOQ*   1
    4 INOQ*   1     8     9     2 ILOQ*   0     1     1     0 IWOQ*   1
***** ***** ***** ***** ***** ***** ***** ***** ***** ***** ***** ***** *****
    4 ************************DATA VECTORS ************************************
***** XOP1          XOP2          DEX1          DEX2          DML1         DML2
    1 -2.0000E+00  -2.0000E+00  -1.0000E+00  -1.0000E+00   0.0000E+00   0.0000E+00
    2  2.0000E+00  -2.0000E+00   1.0000E+00  -1.0000E+00   0.0000E+00   0.0000E+00
    3  2.0000E+00   2.0000E+00   1.0000E+00   1.0000E+00   0.0000E+00   0.0000E+00
    4 -2.0000E+00   2.0000E+00  -1.0000E+00   1.0000E+00   0.0000E+00   0.0000E+00
***** ***** ***** ***** ***** ***** ***** ***** ***** ***** ***** ***** *****
    9 ************************NODE VECTORS ***********************************
***** XON1          XON2          TRC1          TRC2          SIG1         SIG2
    1  0.0000E+00   0.0000E+00   0.0000E+00   0.0000E+00   0.0000E+00   0.0000E+00
    2 -1.0000E+00   0.0000E+00   0.0000E+00   0.0000E+00   0.0000E+00   0.0000E+00
    3 -1.0000E+00  -1.0000E+00   0.0000E+00   0.0000E+00   0.0000E+00   0.0000E+00
    4  0.0000E+00  -1.0000E+00   0.0000E+00   0.0000E+00   0.0000E+00   0.0000E+00
    5  1.0000E+00  -1.0000E+00   0.0000E+00   0.0000E+00   0.0000E+00   0.0000E+00
    6  1.0000E+00   0.0000E+00   0.0000E+00   0.0000E+00   0.0000E+00   0.0000E+00
    7  1.0000E+00   1.0000E+00   0.0000E+00   0.0000E+00   0.0000E+00   0.0000E+00
    8  0.0000E+00   1.0000E+00   0.0000E+00   0.0000E+00   0.0000E+00   0.0000E+00
    9 -1.0000E+00   1.0000E+00   0.0000E+00   0.0000E+00   0.0000E+00   0.0000E+00
***** ***** ***** ***** ***** ***** ***** ***** ***** ***** ***** ***** *****
***** ***** ***** ***** ***** ***** ***** ***** ***** ***** ***** ***** *****
```

FIG. 8. The traction mapping spreadsheet (called a template). The first line of an unfilled template says "TEMPLET OF SETUP FILE" but it can be edited to contain any identifying information for the experimental data below. Five sections follow below the header CONTROL DESCRIPTORS, labeled 1–5 in the first column: *1. MESH TOPOLOGY DESCRIPTORS*. This section contains data essential for building the cell mesh. **NUMN** is # of mesh nodes. **NUMP** is # of data particles (beads). **NUMQ** is # of mesh quadrilaterals. **NUMD** is # of descriptors used =5. **NUME** is # of external edges to the mesh. **OBSV** is the analysis "viewpoint" (see Dembo *et al.*[8] for "a second opinion"). *2. NODE VECTOR PACKING DESCRIPTORS*. This section identifies data columns listed in Node Vectors section. **XON1** if *x* coordinate of node. **XON2** is *y* coordinate of node. **TRC1** is *x* component of applied traction at node. **TRC2** is *y* component of applied traction at node. **SIG1** is fitting error (sigma) of *x* component of traction. **SIG2** is fitting error (sigma) of *y* component of traction. *3. DATA VECTOR PACKING DESCRIPTORS*. This section identifies data columns listed in Data Vectors section. **XOP1** is *x* coordinate of undistorted particle. **XOP2** is *y* coordinate

are reformatted using a translator program and pasted into a DOS spreadsheet (called a template; see Fig. 8) that represents all the data required to execute the traction mapping algorithms. The template includes an estimate of the experimental error (distance in micrometers corresponding to one-half pixel width) and several other operational parameters. Finally, the tractions at each node of the mesh are estimated by determining the traction field and drift vector that maximizes the Bayesian likelihood of obtaining the observed tractions.

All technical details of the statistical procedure for producing traction images (including the underlying theory) have been described elsewhere.[8] In the present context we should mention only the main statistical difficulties and objectives in a qualitative way. First it is necessary to limit the space of acceptable "candidate" traction fields to allow only those in which all vectors are localized within the footprint of the cell (in essence this is accomplished by introducing the finite element mesh). Second, all candidate traction images must satisfy the constraint of global traction balance (this is accomplished by appropriate choice of the shape functions that interpolate between mesh nodes). Third, in choosing the most likely candidate, it is

---

of undisturbed particle. **DEX1** is the x component of particle displacement experimentally observed. **DEX2** is the y component of particle displacement experimentally observed. **DML1** is the x component of particle displacement most likely prediction. **DML2** is the y component of particle displacement most likely prediction. *4. OPERATION CONTROL DESCRIPTORS.* This section contains additional operators and variables pertinent to the analysis. **BOOT** is the number of bootstrap iterations. **EXER** is the experimental error in the same units as the data. **CPLX** is the complexity of the best fit. **CHIS** is the chi square of the fit. **YMOD** is the Young's modulus of elasticity for silicone rubber. **SMZR** is the smoothing parameter (1 = autosmoothing). *5. DRIFT CORRECTION DESCRIPTORS.* This section contains an estimate of the systematic drift in x and y dimensions, experienced by all bead loci in the film, over the time interval separating the two video images used in the analysis. **DRFX** is the calculated drift in the x axis. **DRFY** is the calculated drift in the y axis. **ERFX** is the calculated error in the x axis drift. **ERFY** is the calculated error in the y axis drift. **BLNK** is unused. *POINTERS OF MESH TOPOLOGY.* Each line (numbered in the first column) below this header describes a single quadrilateral, which is the basic unit used in constructing a cell mesh. A series of quadrilaterals is assembled into a mesh by the software, according to the following rules: **INOQ** is the index of the nodes of the quadrilateral; this has an arbitrary starting point and proceeds anticlockwise around all four nodes of the quadrilateral. **ILOQ** is the index of the lines of the quadrilateral; it has the same starting place as INOQ and follows the same anticlockwise circuit around the quadrilateral. It designates each line of the quadrilateral as 1 (a free boundary) or 0 (a shared or internal boundary with a neighboring quadrilateral). **IWOQ** is unused. *DATA VECTORS.* The numbers in the first column are ordinals for each data point, according to the label assigned to each bead on the original transparent monitor overlay. *NODE VECTORS.* The numbers in the first column are ordinals for each data point, according to the label assigned to each node on the original transparent monitor overlay that represents the cell mesh.

necessary to impose a penalty that discourages very complex traction fields (this is accomplished by introducing an appropriate Bayesian prior hypothesis). The final result represents a conservative compromise. It is the simplest (smoothest) traction image that is consistent with hard physical constraints and is also able to give a good account of the observed bead displacements.

We have recently explored the "tiling" feature of an engineering software application, ANSYS 5.3. This useful tool automatically fills the area enclosed by a hand-traced cell edgelist with quadrilaterals (Fig. 7f) and enters the data into the working template (personal communication Chris Kennedy, 1997, Mallet Technology Inc., Research Triangle Park, NC). ANSYS Software promises to reduce considerably the time formerly spent constructing cell meshes and logging data by hand.

The authors welcome inquiries about the traction mapping software. This DOS based software was written and developed by M.D. and T.O. and will be made freely available to other interested users.

Resolution in the Traction Map

For small bead numbers (i.e., less than a few hundred), the spatial resolution with which discrete tractions can be mapped increases with the number of bead displacements sampled. At very large bead densities (which we have never approached) the improvement in image quality is expected to increase more slowly because the additional beads will only supply redundant information. Generally, the number of traction vectors generated is approximately equal to the number of beads sampled. Figure 3 illustrates a specific example of this mapping, which yielded one traction vector approximately every 5 μm^2 of the cell's area, from a sample size of 100 beads.

Considerations for Alternative Substrates for Traction Assays

Currently, there are only two documented silicone compounds available that have been exploited for traction assays at the single cell level. These are the 200 fluid and 710 fluid silicone oils from Dow-Corning, and their properties are reviewed in Table I. Depending on the biology of the system under investigation there are good reasons why investigators should seek alternative materials for use as traction substrates. For example, some interesting progress has been made by Full and colleagues,[11] who studied the traction forces generated by multilegged insects locomoting on photoelastic flexible substrata originally developed by Harris.[12] Solid collagen gels and

[11] R. J. Full, A. Yamauchi, and D. L. Lindrich, *J. Exp. Biol.* **198,** 2441 (1995).
[12] J. K. Harris, *J. Microscopy* **114,** 219 (1978).

silicone rubber compounds with a very fast response time were required to handle the rapid locomotion of a running cockroach. Recently, Partha Roy and colleagues at University of Texas, Southwestern, have applied finite element analysis to weakly cross-linked collagen gels, in order to study traction forces generated by individual corneal fibroblasts interacting with extracellular matrix before and after stimulation with the growth factor TGFβ. The fibroblasts' *in vitro* requirement for a simulated corneal extracellular matrix precluded the use of silicone rubber in this study, and the behavior of the collagen gel was found to give a satisfactory approximation to an elastic material.[13]

Acknowledgments

The authors are grateful to the following individuals: George MacNamara and David Szent-Györgi at Universal Imaging Corp., for assistance with software interfacing problems and operation of Image-1 and Metamorph products. This work was supported in part by NIH grant AI21002 (to MD) and NIH grant GM 35325 (to KJ).

[13] P. Roy, W. M. Petroll, H. D. Cavanagh, C. J. Chuong, and J. V. Jester, *Exp. Cell Res.* **232,** 106 (1997).

Section VI

General Methods

[41] Purification of Cytoskeletal Proteins Using Peptide Antibodies

By Christine M. Field, Karen Oegema, Yixian Zheng, Timothy J. Mitchison, and Claire E. Walczak

Introduction

Immunocytochemistry of cells, as well as biochemical fractionation and assay of proteins important in the cytoskeleton, have been used to further our understanding of the dynamic filament systems and their associated proteins. In this article, we describe the use of peptide antibodies to study the cytoskeleton in a variety of organisms (also reviewed in Ref. 1). We have found peptide antibodies to be useful reagents for immunofluorescence and immunoblotting. More uniquely, we have used them for affinity purification of native, functional proteins and protein complexes from extracts. We describe the preparation and purification of peptide antibodies and focus on their use in the affinity purification of proteins and protein complexes from extracts.

Why Use Peptide Antibodies?

Traditionally, antibodies have been generated against proteins purified from their natural source or bacterially expressed fusion proteins. The use of peptide antibodies has several advantages over this type of approach. Peptide antibodies are relatively easy to prepare and allow the rapid conversion of sequence information into a reagent that targets a specific protein domain. Whereas generation of antigen using conventional approaches can require weeks to months, peptides can be ordered by e-mail, phone, or fax and the coupling procedure takes only 2 days with minimal hands-on time. Peptide antibodies are also uniquely useful for some applications. Antibodies to amino acid sequences conserved between known members of protein families have allowed identification of new family members by expression screening and by biochemical analysis.[2-5] Peptide antibodies can also be used as immunoaffinity reagents for the purification of native protein com-

---

[1] J. C. Bulinski, *Int. Rev. Cytol.* **103,** 281 (1986).
[2] D. G. Cole, W. Z. Cande, R. J. Baskin, D. A. Skoufias, C. J. Hogan, and J. M. Scholey, *J. Cell Sci.* (1992).
[3] K. E. Sawin, T. J. Mitchison, and L. G. Wordeman, *J. Cell Sci.* **102,** 303 (1992).
[4] C. E. Walczak, T. J. Mitchison, and A. Desai, *Cell* **84,** 37 (1996).
[5] L. Wordeman and T. Mitchison, *J. Cell Biol.* **128,** 95 (1995).

TABLE I
PARAMETERS AFFECTING ANTIBODY SUCCESS

Variable	Successful antibodies (%)	Number of peptides
C-terminal peptides	90	18
Internal peptides	43	21
Net charge of +3 to +5	90	11
Net charge of ±1 to 2	70	16
Net charge of 0	14	7
Net charge of −3 to −5	66	3
10–14 Amino acids	56	18
15–19 Amino acids	64	11
≥20 Amino acids	75	8

plexes. The key to the success of this procedure is to use the peptide against which the antibody was prepared as a competitor to elute the protein complex from the antibody column under native conditions.[1,4,6–8]

We must also acknowledge the drawbacks of using peptide antibodies. Due to the limited number of epitopes available in a short amino acid sequence, the probability of obtaining useful antibodies may be a bit lower than by conventional methods. Peptide antibodies also occasionally cross-react with other proteins, due to the small size of the epitope, so they must be used with caution. If the chosen peptide is buried within the protein, the resulting antibody may not recognize the native protein. For similar reasons, the ability of peptide antibodies to inhibit protein function will depend on the specific nature of the epitope.

Choice of Peptide

For this article we have reviewed the results from 4 years of work generating peptide antibodies to cytoskeleton-related proteins in rabbits. We have used 40 peptides to immunize approximately 80 rabbits using the procedures described here, and obtained a 63% success rate overall (Table I). Success is defined as an antibody that was useful for at least one application (i.e. immunoblot, immunoprecipitation, or immunofluorescence). Antibodies that did not recognize the peptide in the context of the intact protein were considered failures, even if they recognized peptide alone. A number of commercial vendors offer peptide synthesis and coupling followed by

[6] Y. Zheng, M. L. Wong, B. Alberts, and T. J. Mitchison, *Nature* **378,** 578 (1995).

[7] C. M. Field, O. Al-Awar, J. Rosenblatt, M. L. Wong, B. Alberts, and T. J. Mitchison, *J. Cell Biol.* **133,** 605 (1996).

[8] A. Merdes, K. Ramyar, J. D. Vechio, and D. W. Cleveland, *Cell* **87,** 447 (1996).

antibody titering and purification. This can be time saving and economical, but we advise caution when using these services. Be aware that companies usually titer antibodies against peptide and not against protein, so the titer may not reflect the usefulness of the antibody for your application. In addition, instead of purifying antibodies on the peptide used for immunization, some companies "affinity purify" antibodies on protein A providing you with total serum IgG instead of a population of specific antibodies. Also, some companies pool sera from different bleeds or even different rabbits. If you use a commercial service, you should carefully review the procedures used by the company and provide them with clear instructions.

Our antibodies varied in quality and in the types of experiments they were useful for (Table I). The highest success rate (90%, $n = 18$) was obtained with antibodies made to the C-terminal peptide of the protein. This may be because the C terminus is more likely to be exposed on the surface of the protein. Of those antibodies generated against internal peptides, our success rate was only 43% ($n = 21$). The statistics were the same whether the sequence was an exact match to a known protein or a degenerate match to a highly conserved region of a protein family. Charge of the peptide also seems to be an important factor. Peptides that were strongly basic (net charge of +3 to +5) gave the highest success rate (90%, $n = 11$). The next most successful category contained those peptides that were weakly acidic or weakly basic (net charge of ±1 to 2) (70%, $n = 16$). A peptide with net charge of zero was rarely successful; we had only a 14% success rate ($n = 7$) with these types of peptides. The length of the peptide was less important. The success rate of longer peptides (20–25 amino acids) was slightly higher than that of shorter peptides (10–15 amino acids), but it is not clear if this trend is statistically significant (75%, $n = 8$, for peptides of 20–25 amino acids versus 57%, $n = 23$, for peptides of 10–15 amino acids). There is one report in the literature of a 7-amino-acid peptide antibody that was effective.[9] Insolubility of the peptide (which correlated with peptides that were highly hydrophobic or had a net charge of zero) was undesirable. Only one insoluble peptide gave limited success ($n = 5$).

Based on our experience we make the following recommendations for choosing a peptide:

1. If possible, pick the C-terminal 10–30 amino acids of the protein.
2. To minimize cross-reactivity, do a database search with the peptide to make sure that the sequence is not highly conserved in other proteins.

[9] M. Z. Atassi and C. R. Young, *Crit. Rev. Immunol.* **5,** 387 (1985).

3. Count the number of charged amino acids in your peptide. Increase the length of the peptide if doing so will include more charged residues, but avoid canceling charges for a net charge of zero.
4. Run a surface probability plot and antigenic index on your protein to avoid selecting peptides likely to be buried.

After selecting a peptide sequence, add a cysteine to the N terminus of the peptide to facilitate coupling to KLH for immunization, and to resin for affinity purification. To remove charges from the peptide ends that would not be present in the native protein, we now tend to order peptides with an acetylated N terminus and, in the case of internal peptides, a C-terminal amide, but the importance of these modifications has not been tested.

When the peptide arrives, check its solubility. If the peptide is insoluble in buffer, try dissolving it in dimethyl sulfoxide (DMSO) and then diluting the DMSO stock into buffer. If it precipitates, your chances of success are limited. If you are planning to try affinity chromatography, the peptide must be soluble to at least 1 mM in the buffer that will be used for elution.

Preparation of Immunogen

Checking Peptide Thiol Groups

Successful thiol coupling requires reduced sulfhydryl groups. Because peptide thiol groups sometimes get modified after synthesis, presumably due to oxidation, it is a good idea to check the peptide for the presence of free sulfhydryl groups before coupling to either KLH or to resin. This can be done using Ellman's reagent, which reacts with sulfhydryls to form a highly colored chromophore with an absorbance maximum at 412 nm.

1. Make up 5 mM Ellman's reagent (5,5'-dithio-bis(2-nitrobenzoic acid) (Sigma, St. Louis, MO) in 0.1 M sodium phosphate, pH 7.2.
2. Weigh out about 1 mg of peptide into a preweighed tube.
3. Add 0.5 ml reagent. It should turn bright yellow.
4. Dilute the mixture 1/50 in buffer. Read A_{412}, blanking against reagent diluted to the same concentration.
5. Calculate the apparent molecular weight of the peptide based on the number of free sulfhydryl groups, using a molar extinction coefficient for the Ellman's chromophore of 14,000 M^{-1} cm^{-1}. Compare this to the expected molecular weight of the peptide. They should agree within a factor of 3, with the apparent molecular weight usually higher. If the thiol concentration is anomalously low, i.e., the apparent molecular weight is very high, there may be something wrong with

the peptide and it will probably not couple well. You may be able to regenerate the thiol groups by incubating the peptide with excess dithiothreitol (DTT) at pH 8.0. To recover the peptide, remove the free DTT by gel filtration over a resin with a small exclusion limit, such as Bio-Gel P2 (Bio-Rad, Richmond, CA).

Coupling of Peptide to KLH

To induce antibody production by B cells, an antigen needs to elicit a coordinated response from both B cells and helper T cells. To do this the antigen must (1) have an epitope that can bind to the surface antibody of a virgin B cell (this epitope will dictate the specificity of the antibody produced), and (2) on degradation must generate fragments that can elicit a response from helper T cells by simultaneously binding to both a class II MHC molecule and a T-cell receptor. Although peptides frequently contain good epitopes for recognition by B cells, they are often too small to meet the latter requirements.[10] To circumvent this problem, and to increase antigen uptake by phagocytosis, we couple peptides to keyhole limpet hemocyanin (KLH).

The following recipe is for two rabbits (five injections each). It is a good idea to immunize more than one rabbit per peptide if you can afford it, because in some cases we have seen significant rabbit-to-rabbit variation in the immune response to peptides. In the coupling procedure, we start with 10 mg of peptide and often use half of the peptide for thiol coupling and half for glutaraldehyde coupling. The rationale for this is one of security—if one of the coupling procedures fails, then at least half of the peptide will be coupled by the other procedure. The success of coupling procedures must often be taken on faith because of the difficulty in quantitating peptide amounts by any simple means such as A_{280} or Bradford assay (in part, because of the large amounts of KLH present). However, some workers prefer to use only the thiol coupling method since it avoids modification of internal lysines in the peptide. Our thiol coupling procedure was designed with two criteria in mind: (1) maximizing the ratio of peptide to KLH in the immunogen and (2) minimizing the size of the linker.

1. Dissolve 100 mg of KLH (Sigma) in 2 ml of water. This generally takes about 4 hr, and you will probably need to sonicate and vortex. Be patient and put it on a rotator at 4°. Dialyze overnight against 2 liters of 0.1 M sodium phosphate, pH 7.8, to remove any contaminating compounds containing thiol or amino groups.

---

[10] E. Harlow and D. Lane, "Antibodies: A Laboratory Manual," pp. 1–726. Cold Spring Harbor Laboratory Press, New York, 1988.

2. Spin the dialysate 10 min at full speed in microfuge to remove aggregates (do not be surprised to see a substantial pellet).
3. If both thiol and glutaraldehyde coupling procedures are to be used,
   split the KLH into two aliquots, one for each procedure.
4. Thiol coupling reaction (outlined in Fig. 1A). (a) Warm one aliquot
   of KLH to room temperature. Add one-ninth volume of iodoacetic
   acid N-hydroxysuccinimide ester (IAA-NHS) at 100 mg/ml in
   DMSO. Make the DMSO stock fresh, and protect the iodoacetamide
   reagent from light. We make our own IAA-NHS ester by coupling
   IAA and NHS with dicyclohexyl carbodiimide in CH_2Cl_2, filtering
   out the urea, and recrystallizing. It can also be purchased from Sigma.
   For the rest of the thiol coupling, minimize exposure of the reaction
   to light by covering it with foil or by working in a dark room with

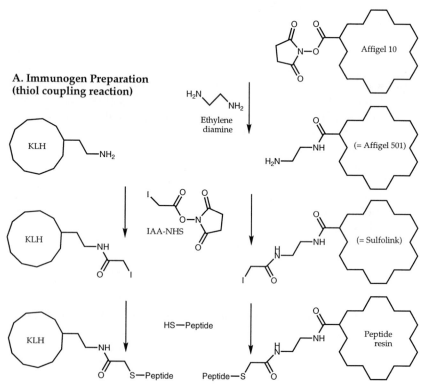

FIG. 1. (A) Preparation of peptide immunogen (thiol coupling reaction only). (B) Preparation of peptide resin for affinity purification of antibody.

minimal light. (b) After 10 min at room temperature, the KLH solution will start to get a little cloudy. Cool on ice, and perform all
subsequent steps at 4°. Load the KLH onto a gel filtration column
containing Bio-Gel P-10 resin (Bio-Rad) equilibrated with 0.1 M
sodium phosphate, pH 7.8. Make sure the column is at least 10 times
the volume of the sample. Pool the KLH containing fractions by color
(KLH is grayish green). Add 5 mg (10 mg if not using glutaraldehyde
coupling) of peptide to the pooled KLH fraction, either as powder
if it is soluble or add it from a 100 mg/ml stock in DMSO. It is
important to use freshly dissolved peptide because the thiol groups
can oxidize within an hour in solution. Check the pH and adjust to
7.2–7.8 if necessary. Incubate at least 8 hr at 4°, rotating gently.
5. Glutaraldehyde coupling reaction. (Optional, note that the glutaraldehyde procedure is only useful if the peptide has an internal lysine
or a nonacetylated N terminus.) (a) Add 5 mg peptide to the other
aliquot of KLH, followed by glutaraldehdye to 0.1% final. Add the
peptide as a solid or from a 100 mg/ml stock in DMSO. Precipitation
does not seem to matter and often occurs. After adding the glutaraldehyde, check the pH with pH paper, and adjust to pH 7.8 if necessary
using NaOH. Incubate 8–12 hr at 4°, rotating gently. (b) Add a *tiny*
pinch of $NaBH_4$ to reduce the otherwise reversible lysine-aldehyde
Schiff-base adducts, as well as excess glutaraldehyde. Make sure the
sample is in a large tube because it tends to fizz up. Incubate 8–12
hr at 4°. This is the glutaraldehyde conjugate.
6. Pool the KLH coupled peptide from the two procedures. Dilute to
5 ml with 0.15 M NaCl. If there is a precipitate, sonicate vigorously
to break it up. Split the immunogen into 1-ml aliquots and freeze it.
Each aliquot contains 2 mg of conjugated peptide, which is sufficient
for a single immunization of two rabbits (1 mg/immunization/rabbit).

Affinity Purification of the Antibody

Although sera from animals immunized with native or fusion proteins
are often used directly for immunoblotting or immunofluorescence, in our
experience peptide antibodies are only useful after affinity purification from
crude serum. To affinity purify peptide antibodies, we first prepare peptide-
coupled resin.

Coupling Peptides to Resin for Affinity Purification (outlined in Fig. 1B)

We use Affi-Gel 10 (Bio-Rad), converting its functional group first to
amino and then to iodoacetyl. You can also buy Affi-Gel 102 (Bio-Rad)

and start at step 6 or Sulfo-link gel (Pierce, Rockford, IL) and start at step 8 of the procedure outlined below. The cost for each method of making the column is about equivalent if you buy all reagents, but using homemade IAA-NHS ester, it is much cheaper to use Affi-Gel 10. All washes are performed on a glass-fritted filter funnel applying suction until a wet cake is formed. Do not dry the resin completely or you will introduce air bubbles. To minimize loss, all reactions up to peptide addition are performed in the funnel by covering the spout with parafilm. We usually make a 5-ml column adding 1–2 mg peptide/ml resin, which should be sufficient to bind at least 25 mg of specific antibody.

1. Wash 5 ml of Affi-Gel resin (assume the Affi-Gel is a 50% slurry so start with 10 ml of slurry) with 2 volumes of 100% cold EtOH.
2. Wash with 2 volumes of 50% cold EtOH.
3. Wash with 2 volumes of cold water.
4. Add 5 volumes of 5% ethylene diamine in water. Incubate 15 min at room temperature (RT).
5. Wash with 10 volumes of water. At this point you have amino-Affi-Gel.
6. Wash with 3 volumes of 0.1 M sodium phosphate, pH 7.8.
7. Resuspend resin in 0.2 volumes of 0.1 M sodium phosphate, pH 7.8. Dissolve the IAA-NHS ester in dry DMSO at 100 mg/ml. While gently stirring the resin, add the IAA-NHS ester to give a final concentration of 7 mg/ml resin. Incubate for 10 min at RT. This step and subsequent steps up to the blocking of residual iodoacetate groups should be done in dim light since the iodo group is light sensitive.
8. Wash with 10 volumes of 0.1 M sodium phosphate, pH 7.8.
9. Resuspend the resin in an equal volume of buffer. Add it as a solid. Many peptides go in better added as a 100 mg/ml stock in DMSO. Generally, if your peptide was readily soluble when you checked for thiol groups, it will be soluble in the 50% slurry because the concentration is similar. Some hydrophobic peptides will precipitate. Such peptides can be coupled in 20% buffer, 80% DMSO or, in an extreme case, in 100% DMSO containing triethylamine. Be aware, however, that these extremely insoluble peptides are usually problematic.
10. Mix gently on a rotating wheel overnight at 4°. After coupling, block residual iodoacetate groups by addition of 2-mercaptoethanol to 0.2%. Incubate 1 hr at RT with rotation.
11. Wash resin sequentially with 5 volumes 0.1 M NaHCO$_3$; 5 volumes of 1 M Na$_2$CO$_3$; 5 volumes of water; 5 volumes of 0.2 M glycine,

pH 2.0; 150 mM NaCl; 5 volumes of 20 mM Tris-Cl, pH 7.4; 0.18 M NaCl (TBS); and 5 volumes of 6 M guanidine-HCl in TBS (this last wash is not used by all workers; it is the most aggressive for removing noncovalently bound peptide, but may cause some damage to the physical structure of the resin). Reequilibrate the resin into TBS + 0.1% NaN$_3$ for storage.

Before adding valuable peptide, we recommend checking the resin chemistry using a quick eyeball test. The amino resin will react with an NHS ester, whereas the original Affi-Gel and the iodoacetate will not. Take an aliquot (50 μl) of resin at each step. Resuspend in 100 μl of buffer. Add 1 μl of 0.1 M NHS-fluorescein (Molecular Probes, Eugene, OR) or NHS-rhodamine (Molecular Probes) in DMSO. Incubate 5 min at RT. Wash the resin twice in buffer by centrifugation. The original resin, and the resin after step 8, should be only lightly labeled, whereas the resin after step 5 should be heavily labeled.

Affinity Purification of Peptide Antibodies

Affinity purification of antibodies is described elsewhere and is not detailed here.[10] Our standard procedure is outlined on the Mitchison laboratory web page in the protocols section (http://skye.med.harvard.edu). We generally purify antibodies from 10–25 ml of serum, and the yield of specific antibody ranges from 0.02 to 0.5 mg/ml of serum. Briefly, all steps are performed at RT. The serum from each rabbit should be filtered first through a 0.45-μm filter, and NaN$_3$ added to 0.1% as a preservative. The serum can also be diluted with 1 volume of 0.15 M NaCl, 20 mM Tris-Cl, pH 7.4 (TBS), to reduce viscosity prior to filtering. We incubate peptide-conjugated beads with serum for several hours either in batch or by several passages over a column. The beads are washed with at least 10 volumes of TBS followed by 4 volumes of TBS plus 0.1% Triton X-100 followed by TBS. The most important variable in the purification is the elution buffer. The first elution to try is low pH (100 mM glycine-HCl), either pH 2.5 or pH 2.0. Most antibodies are stable at pH 2.0, but we have had some antibodies that are only stable to pH 2.5. Also, it is important to neutralize the antibodies as they are eluted from the column. We do this by having 1 M Tris, pH 9.0, in the collection tubes and mixing immediately as the fractions are collected; the fractions are then placed directly on ice. Alternative elution conditions are high pH (100 mM triethylamine, pH 11.5) followed by neutralization, or high Mg^{2+} (1.5–4.9 M) followed by dialysis. Both are slightly gentler than low pH and may yield antibodies of lower affinity. Finally, 6 M guanidine-HCl can be used as a strong reagent to elute high affinity antibodies, although this harsh denaturant sometimes irreversibly damages the antibodies. It is advisable not to pool antibodies eluted under different conditions because each pool of antibodies may have different properties. Also, the affinity of antibodies tends to increase with multiple boosts of the rabbit. Earlier bleeds may be more useful for immunoaffinity purification, whereas later bleeds may be more useful as general tools for immunoblotting and immunofluorescence.

Immunopurification with Peptide Antibodies

Next we describe a basic procedure to start with when trying immunoaffinity chromatography; an illustration outlining this strategy is presented in Fig. 2. After that, suggestions for ways to optimize each of the steps in the procedure for your specific application are presented. Although it may be obvious, we emphasize that your antibody must be able to immunoprecipitate the target protein for this procedure to work. We have not tested all of our peptide antibodies for their ability to immunoprecipitate, but, of the antibodies that were tested, 66% ($n = 21$) were able to immunoprecipitate their target protein; the success rate was 86% for C-terminal antibodies ($n = 15$). Of the antibodies that were able to immunoprecipitate the target protein, immunoaffinity purification was attempted in 10 cases; 9 of these were successful.

Basic Procedure

1. Prepare cell or embryo extract at 4°. Extract conditions are discussed below.
2. Wash protein A-Affi-Prep beads (Bio-Rad) or protein A-agarose (GIBCO, Grand Island, NY) 3× with TBS (150 mM NaCl, 20 mM Tris-Cl, pH 7.4), collecting the beads by centrifugation (1000g, 1 min).
3a. To prebind the anti-peptide IgG to the beads, add the antibody to the beads as a 25% slurry in TBS (three parts TBS to one part beads) and mix gently at RT for 20–60 min. Then wash 1× with TBS and 2× with extract buffer. Add the beads to the extract.
3b. Alternatively (see next section) add the antibody directly to the extract. Preincubate this mixture for 60 min at 4° and then add the beads.
4. All subsequent procedures are at 4°. Incubate the beads with the extract for 1 hr with gentle rotation.
5. Collect the beads by centrifugation (1000g, 1 min), and wash batchwise by mixing for a minute or two followed by centrifugation. Typically, this involves three to five washes with 20 resin volumes of extract buffer per wash.

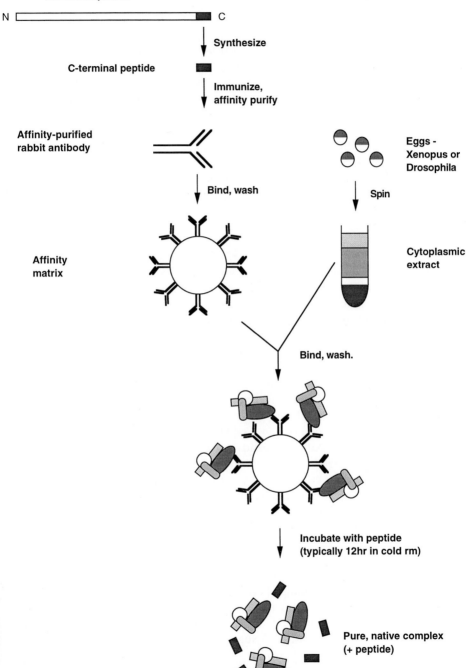

Protein sequence

N ⬜⬜⬜⬜⬜⬜⬜⬜⬜⬛ C

Synthesize

C-terminal peptide

Immunize,
affinity purify

Affinity-purified
rabbit antibody

Eggs -
Xenopus or
Drosophila

Bind, wash

Spin

Affinity
matrix

Cytoplasmic
extract

Bind, wash.

Incubate with peptide
(typically 12hr in cold rm)

Pure, native complex
(+ peptide)

Fig. 2. Anti-peptide purification strategy.

6. Decant the beads into a small disposable column for elution and drain the resin.

7. Add the peptide (200–300 μM in elution buffer). We typically add 1.5 column volumes and allow the column to drain to semidryness. Cap the top and bottom of the column and leave it undisturbed for 8–12 hr. Continue the elution with several column volumes of elution buffer plus peptide. Most of the antigen should elute in the first column volume after the 8–12 hr of incubation of the antibody–antigen interaction has been successfully completed with peptide.

8. Assess the purity and activity of the eluted protein or complex using SDS–PAGE and other analytical techniques.

Optimizing Immunopurification

The goal of this section is to provide guidelines for minimizing nonspecific protein binding and maximizing yield of the protein(s) of interest.

Step 1. Conditions for preparing embryo or tissue extracts need to be optimized empirically for each protein. The generic buffer described in step 7 below, containing 50–100 mM KCl, supplemented with protease inhibitors (and phosphatase inhibitors if required), may be a reasonable starting place. In general, we have found that extracts prepared at a relatively high protein concentration (8–50 mg/ml), in buffers of physiologic ionic strength and pH, in the absence of detergents, give the cleanest results.

Steps 2–3. Affi-Prep beads and protein A-agarose work equally well in many applications. In very dense or viscous extracts, such as *Xenopus* egg extracts, the denser Affi-Prep beads make a tighter pellet on centrifugation, reducing loss of beads. If washing will be performed in a column format, protein A-agarose beads tend to have better flow properties.

An important variable that affects both the yield and purity of the target protein is the amount of IgG added per milligram of extract protein, or the molar ratio of antibody to target protein. For rare proteins or hard-to-obtain extracts it may be necessary to add excess antibody to drive binding. However, we have found that decreasing the ratio of antibody to extract, to the point that the target protein is not depleted from the extract, often reduces nonspecific binding. Thus lower amounts of antibody are advised in situations where extract is not limiting. We suggest trying several different antibody concentrations and analyzing proteins bound to the beads by SDS–PAGE without elution. A suggested starting point is 5 μg of antibody per 10 mg of extract protein.

Another variable to test is the ratio of antibody to protein A. Protein A beads can bind as much as 15 mg IgG/ml beads. For septin immunoisolation (see below) we found that using 0.1 mg IgG/ml beads gave better results than higher levels. A suggested starting point is 0.2 mg/ml for large proteins or complexes, higher amounts for small proteins.

Steps 3(a) and 3(b). The issue of whether to prebind the antibody to the protein A and then add it to the extract or to add the antibody directly to the extract and isolate the antibody–antigen complex on the protein A beads is one that we have not resolved. In one case, we compared the two binding methods and found that similar amounts of antigen bound to the beads in each case, but that in the latter protocol (step 3b) more of the bound antigen was eluted by peptide. In most cases, we have prebound the antibody to the protein A beads as in step 3a.

Step 4. Relatively short incubation times with beads appear optimal for reducing nonspecific binding; 1 hr at 4° in a slowly rotating tube is typical.

Step 5. We found that batch washes were more effective than washing in a column format for removing contaminants. However, care must be taken to make sure all the beads are sedimented to prevent loss of beads during the washes.

Increased ionic strength or detergent-containing washes may be useful to remove contaminating proteins in some cases. Care must be taken to avoid conditions that might damage the target protein or protein complex. The robustness of protein complexes is highly variable; some can withstand 1 M NaCl and nonionic detergents, while others will dissociate if the ionic strength is increased above physiologic levels.

In some cases it may be useful to collect and analyze increased ionic strength washes, since these may contain proteins that associate weakly but specifically with the target protein. In this case it is advantageous to perform the washes in a column format. The first column volume to elute will contain the highest concentration of proteins removed by the wash.

Step 6. Prior to elution, buffer can be exchanged. For example, if the complex has been washed with a high ionic strength buffer it might be desirable to elute in a more physiologic buffer. This exchange can be conveniently accomplished in the column format by washing through several column volumes of the desired buffer.

Step 7. The elution buffer is chosen to be optimal for stability or functional assay of the protein complex. If no prior information is available we recommend 0–75 mM KCl, 20–50 mM potassium-HEPES, pH 7.4, 1 mM MgCl$_2$, 1 mM ethylene glycol-bis(β-aminoethyl ether) N,N,N',N'-tetraacetic acid (EGTA), 1 mM DTT. Protease and phosphatase inhibitors are optional.

The effectiveness of a particular peptide in the elution of an antigen is extremely variable and should be first tested on a small scale. Although some peptides will effectively disrupt the antibody–antigen complex in

1–3 hr, we have found that more complete elution can often be obtained following an overnight incubation. In this case, the protein usually elutes in one column volume and is thus more concentrated.

If the protein does not elute from the column after an overnight incubation in peptide, several modifications can be tried.

1. Try the elution at RT. This should increase the dissociation rate of the antibody–antigen bond.
2. Increase the ionic strength. We have found that adding NaCl to 0.5 M increases elutability. It is probable that either raising or lowering the pH or adding low amounts of urea would also be effective for particular antibody–peptide combinations. However, this can be problematic when trying to isolate a weakly associated protein complex.
3. Generate an antibody with lower affinity to the antigen. Low affinity antibodies can be selected for during affinity purification.[11] Alternatively, lower affinity can be engineered by immunizing and eluting with a peptide whose sequence differs slightly from that of the protein target (Kent Matlack, personal communication, 1995).
4. Use monoclonal antibodies. We have not tried monoclonals for this application, but in principle it should be possible to generate a series of monoclonal antibodies with variable affinity to a given peptide, and select the one whose affinity is the best compromise between selective binding and eluatability.
5. Make a column from Fab fragments. In cases where the protein complex being isolated is multimeric, difficulty in eluting may be due to the fact that each antibody can bind to two binding sites on the target protein complex. In one case, in an attempt to circumvent this problem, we constructed columns from monovalent Fab fragments made by cleaving the peptide antibody using immobilized papain (Pierce). The Fab fragments were coupled to Affi-Gel 10, and the resin was incubated with extract and washed as described above. Although elution of the protein complex from the Fab fragment resin was complete, binding of the complex to the resin was not as efficient as that obtained with a similar concentration of intact antibody.

Step 8. If necessary, the eluted protein target can be separated from the contaminating peptide. We have successfully used gel filtration chromatography, ion-exchange chromatography, and sucrose gradient sedimentation for this purpose. One advantage of these additional steps is to further purify the target protein.

[11] D. R. Kellogg and B. M. Alberts, *Mol. Biol. Cell* **3**, 1 (1992).

Assessing the Results of Purification

Critical assessment of the purity of the final preparation is very important. It is not safe to assume that all the bands present in the peptide-eluted fraction are in a protein complex—some may be, while others will be contaminants. The criteria we use to test if polypeptides are in a complex after co-immunoisolation is to perform both sucrose gradient sedimentation and gel filtration chromatography. Polypeptides that cofractionate in both procedures after co-isolation on antibody beads are very likely to be physically associated. Use of only one of these hydrodynamic tests decreases the certainty of this conclusion.

If the eluted protein contains an unacceptable level of contaminants, the antibody affinity approach can be combined with other more standard biochemical techniques to yield a more specific product. The extract can be fractionated before the immunoisolation step, or the eluted target protein further purified afterward, or both (see below for examples). One simple thing to try is to include a precipitation step, such as an ammonium sulfate cut or a PEG precipitation, before the immunoisolation step. We have used both of these precipitation methods to improve significantly the purity of eluted proteins. The purification power of the immunoisolation step will depend on several factors including the affinity and specificity of the antibody and the concentration of the target protein in the cell extract.

Specific Applications of Peptide Immunopurification

XKCM1, A Xenopus Egg Kinesin

One application of immunoaffinity chromatography is for small-scale pilot purifications. This circumvents the need to develop a conventional chromatographic procedure that might take several months or more. This is illustrated by the purification of XKCM1 (a kinesin-related protein) from *Xenopus* egg extracts.[4] A dialyzed ammonium sulfate cut of *Xenopus* egg extract high-speed supernatant in MB1 (10 mM potassium-Hepes, pH 7.2; 2 mM MgCl$_2$; 100 mM KCl; 50 mM sucrose; 1 mM EGTA; 50 $\mu$$M$ Mg-ATP; 0.1 mM DTT + protease inhibitors) was purified on a C-terminal peptide [(C)QISKKKRSNK] antibody column. A ratio of 25 μg antibody/ 50 μl bead/ml of dialyzed ammonium sulfate fraction (10–20 mg/ml) was used. The antibody-coated beads were incubated with the extract for 2 hr at 4°, and then the mixture was poured into a column. The column was washed with 20 volumes of MB1 and then eluted by incubation for 1 hr with 200 $\mu$$M$ peptide in MB1. The protein eluted mainly in the first three to four column volumes after the incubation with peptide; however, the

A
L F X M

B
M L S

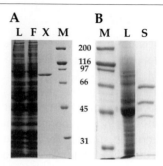

Fig. 3. Examples of purification using peptide antibodies. (A) XKCM1 purification. L, column load; F, column flow-through; X, eluted XKCM1; M, molecular weight markers. (B) Septin purification. L, column load; S, eluted septins; M, molecular weight markers. The three bands of the eluted complex are peanut, Sep2, and Sep1 (from highest to lowest molecular weight).

first column volume was usually dirtier so it was not included in the pool. This yielded approximately 90% pure XKCM1 protein that was functionally active (Fig. 3A). The yield was low but sufficient for several functional experiments given the ease of purification.

Drosophila Embryo Septin Complex

The septins are a conserved family of cytoskeletal proteins that constitute a new filament system. Genetic and microscopic data in yeast had suggested that septin proteins might function as a complex in budding yeast. To isolate a *Drosophila* septin complex,[12] an antibody was raised against the C-terminal 14 amino acids [(C)NVDGKKEKKKKGLF] of the *Drosophila* septin protein peanut.[7] The antibody raised against this peptide henceforth referred to as KEKK has proved to be one of our most effective antibodies. KEKK IgG was prebound to beads at 0.1 mg/ml, and these were incubated for 1 hr with *Drosophila* egg extract [8–12 mg/ml protein in 50 mM Tris-HCl, pH 7.6, 0.5 mM EDTA, 0.5 mM EGTA, 1 mM DTT, 0.1% Nonidet P-40 (NP-40) plus protease inhibitors] at a ratio of 50 μl beads/ml extract. The beads were washed three times batchwise with 20 resin volumes of 400 mM KCl, 20 mM Tris-Cl, pH 7.9, 1 mM EDTA, 1 mM EGTA, 1 mM DTT, 8% sucrose then two times with an equivalent volume of elution buffer (75 mM KCl, 20 mM potassium-HEPES pH 7.4, 0.5 mM EDTA, 1 mM EDTA, 1 mM DTT). The complex was eluted by incubation for 12 hr with 200 μM KEKK peptide in elution buffer. The

---

[12] K. Oegema, A. Desai, M. L. Wong, T. Mitchison, and C. M. Field, *Methods Enzymol.* **298**, [23], 1998, (this volume).

septin complex was obtained in approximately 90% purity (Fig. 3B). The yield was approximately 1 μg septin complex per microgram of KEKK antibody.

Xenopus Egg Gamma Tubulin Complex

Gamma tubulin is implicated in microtubule nucleation at centrosomes. For biochemical analysis, Zheng and co-workers used the anti-peptide approach to isolate functional complex from *Xenopus* egg extracts.[6] Experimental details are described in article [19] in this volume.[13] Because the complex is rare, it was necessary to purify it partially before the immunoisolation step using ammonium sulfate precipitation and gel filtration. After peptide elution, the complex was approximately 75% pure. It was concentrated, and freed of remaining contaminants and peptide, by binding to S-Sepharose beads and eluting with elevated salt. In this case, the antibody affinity column provided a critical and powerful purification step.

Conclusion

In this article, we described the preparation and purification of peptide antibodies and their use in the affinity purification of proteins from extracts. The immunoaffinity approach described here allows the rapid isolation of proteins and protein complexes from extracts under physiologic conditions. We think this approach will be particularly useful (1) when conventional purification is difficult because the protein complex to be isolated is sensitive to the reagents used in typical chromatography steps (such as high salt), (2) when conventional purification is difficult because the protein is rare and additional powerful purification steps are needed, and (3) in cases where rapid purification is required, such as in the isolation of protein complexes with labile activities.

---

[13] Y. Zheng, M. L. Wong, B. Alberts, and T. Mitchison, *Methods Enzymol.* **298**, [19], 1998, (this volume).

[42] Strategies to Assess Phosphoprotein Phosphatase and Protein Kinase-Mediated Regulation of the Cytoskeleton

By JOHN E. ERIKSSON, DIANA M. TOIVOLA, CECILIA SAHLGREN, ANDREY MIKHAILOV, and ANN-SOFI HÄRMÄLÄ-BRASKÉN

Introduction

Phosphorylation has been established as a key regulator of cytoskeletal functions. A large number of protein kinases (PKs) and phosphoprotein phosphatases (PPs) are crucial determinants in the regulation of overall cytoskeletal assembly and organization[1–4] and in cytoskeleton-related signaling.[5] The rapidly increasing number of recently identified PKs and PPs associated with cytoskeletal structures indicates that this aspect of cytoskeletal regulation will receive even more attention in the future. Hence, it is important for cytoskeletal biology investigators to understand what kind of strategies are possible when the roles of specific PKs or PPs in the regulation of a given cytoskeletal complex or system are being assessed.

To establish the involvement of specific PKs or PPs in the regulation of a given protein function, the following criteria have to be met: (1) the phosphorylation state of the studied protein(s) has to be altered by altering the PK/PP activity of interest; (2) the function of the protein has to be altered in concert with the altered phosphorylation; and (3) on activation of the PK/PP, the *in vivo* phosphorylation/dephosphorylation site(s) on the protein have to correspond to those established when the protein is phosphorylated/dephosphorylated *in vitro* by the PK/PP of interest. For some kinases with highly overlapping specificities *in vitro*, for example, the mitogen-activated PKs (MAPKs), stress-activated PK (SAPK/JNK), and p38 PK,[6] an additional criterion has to be fulfilled, i.e., (4) the phosphorylation *in vivo* should be abolished when the PK is inactivated by some suitable antagonist.

The aim of this article is to give an introduction to scientists not familiar with the methodology required for studies concerning protein phosphoryla-

[1] K. L. Carraway and C. A. Carraway, *BioEssays* **17**, 171 (1995).
[2] E. Crowley and A. F. Horwitz, *J. Cell Biol.* **131**, 525 (1995).
[3] J. E. Eriksson, P. Opal, and R. D. Goldman, *Curr. Opin. Cell Biol.* **4**, 99 (1992).
[4] R. Foisner, *Bioessays* **19**, 297 (1997).
[5] M. J. Humphries, *Curr. Opin. Cell Biol.* **8**, 632 (1996).
[6] P. Cohen, *Trends Cell Biol.* **7**, 353 (1997).

tion but who would like to establish whether a PK or PP is involved in the regulation of a given cytoskeletal protein or complex. Examples from studies on intermediate filament proteins are used to illustrate how specific cytoskeletal functions of PKs or PPs have been derived through *in vivo* and *in vitro* studies. The approaches and methods should, however, also be applicable to any other cytoskeletal system and suggestions and references are given on how to proceed with nonintermediate filament cytoskeletal proteins. A number of excellent in-depth reviews on other aspects of protein phosphorylation are also available.[7,8]

Methods to Modify Kinase/Phosphatase Equilibria

Use of Pharmacologic Agents

One of the initial approaches to test whether a given cytoskeletal component can be regulated by phosphorylation is usually modification of kinase/phosphatase activities *in vivo*. By using pharmacologic modifiers of PK/PP activities it is relatively easy and inexpensive to obtain valuable initial information concerning the possible cytoskeletal functions of a number of both Ser/Thr as well as Tyr-directed kinases and phosphatases. When using a given pharmacologic agent to modify a PK or a PP activity in intact cells, the following criteria should be met: (1) direct binding to the PK/PP, (2) high specificity for the PK/PP, (3) effective cell membrane permeability, and (4) no significant degree of cellular compartmentalization. In the following sections we have included tables of activators and inhibitors of PKs and PPs. These are far from exhaustive lists, but include some of the best established compounds together with an indication of the effective concentration range to be used and references for further reading. Note also that many bioscience companies are actively developing new modifier compounds and derivatives of established inhibitors to improve specificity, solubility, stability, and other important properties of the compounds. Hence, current catalogs or the World Wide Web-based product lists of some companies often provide very useful information on new products and helpful advice on how to use the listed compounds. (Alexis Corporation, San Diego, CA, USA; Calbiochem La Jolla, CA, USA; GIBCO-BRL, Inchinnan, Scotland, UK; Molecular Probes, Eugene, OR, USA; New England Biolabs, Beverly, MA, USA; and Sigma Chemical Company, St.

---

[7] D. G. Hardie, "Protein Phosphorylation, A Practical Approach." Oxford University Press, New York, 1993.
[8] T. Hunter and B. M. Sefton, *Methods Enzymol.* **201**, (1991).

TABLE I
EXAMPLES OF USEFUL INHIBITORS OF PROTEIN PHOSPHATASES*

PP1 and PP2A

Compound	Structure	Solubility	Potency (IC_{50}) PP1	PP2A	Supplier	Ref.
Inhibitor-2	Protein	Physiologic buffer	1 nM	—	a	g, h
Okadaic acid	Polyether carboxylic acid	Ethanol, 10% DMSO	100 nM	<1 nM	b–e	i–m
Tautomycin	Polyketide	DMSO, ethanol	1 nM	10 nM	b	l–n
Calyculin-A	Phosphorylated polyketide	DMSO, ethanol	1 nM	<1 nM	b, c, f	i, l–n
Cantharidin	Terpenoid	DMSO, ethanol	500 nM	40 nM	b, f	l, m, o
Microcystins	Cyclic peptide	DMSO, ethanol, water	1 nM	<1 nM	b–d, f	j, l–n, p–r

PP2B (calcineurin)

Compound	Comments	Solubility	Potency (IC_{50})	Supplier	Ref.
Cyclosporin A	Cyclic undecapeptide	Ethanol	<1 nM	f	m, s

* Note that the half-maximal inhibition concentrations are not exact values but rather intended to indicate the approximate specificities against the respective PP. Useful concentration ranges are likely to be around these values and depend largely on the cell system in question.

Louis, MO, USA, have especially active product lines on biochemical tools for studying cell signaling.)

Inhibitors of PPs

It may often be beneficial to initially use a PP inhibitor (Table I) rather than a PK activator when possible roles of phosphorylation are being assessed. There is increasing evidence that cytoskeletal phosphoproteins are substrates for high constitutive PP activities.[3,9–14] Hence, by preventing

[9] J. E. Eriksson and R. D. Goldman, in "Advances in Protein Phosphatases" (W. Merlevede, ed.), p. 335. Leuven University Press, Belgium, 1993.

[10] J. E. Eriksson, D. L. Brautigan, R. Vallee, J. Olmsted, H. Fujiki, and R. D. Goldman, *Proc. Natl. Acad. Sci. U.S.A.* **89,** 11093 (1992).

TABLE I (*continued*)

[a] New England Biolabs, Inc., Beverly, MA, USA.

[b] Calbiochem, La Jolla, CA, USA.

[c] Alexis Corporation, San Diego, CA, USA.

[d] GIBCO-BRL, Inchinnan, Scotland, UK.

[e] Boehringer Mannheim, Mannheim, Germany.

[f] Sigma Chemical Company, St. Louis, MO, USA.

[g] P. Cohen, *Annu. Rev. Biochem.* **85,** 453 (1989).

[h] C. F. B. Holmes and M. P. Boland, *Curr. Opin. Struct. Biol.* **3,** 934 (1993).

[i] H. Ishihara, B. L. Martin, D. L. Brautigan, H. Karaki, H. Ozaki, Y. Kato, N. Fusetani, S. Watabe, K. Hashimoto, D. Uemura, and D. J. Hartshorne, *Biochem. Biophys. Res. Commun.* **159,** 871 (1989).

[j] R. Matsushima, S. Yoshizawa, M. F. Watanabe, K.-I. Harada, M. Furusawa, W. W. Carmichael, and H. Fujiki, *Biochem. Biophys. Res. Commun.* **171,** 867 (1990).

[k] D. G. Hardie, T. A. J. Haystead, and A. T. R. Sim. *Methods Enzymol.* **201,** 469 (1991).

[l] H. Fujiki and M. Suganuma, *Adv. Cancer Res.* **61,** 143 (1993).

[m] C. MacKintosh and R. W. MacKintosh, *TIBS* **19,** 444 (1994).

[n] R. E. Honkanen, B. A. Codispoti, J. Tse, and A. L. Boynton, *Toxicon* **32,** 339 (1994).

[o] Y. M. Li and J. E. Casida. *Proc. Natl. Acad. Sci. U.S.A.* **89,** 11867 (1992).

[p] C. MacKintosh, K. A. Beattie, S. Klumpp, P. Cohen, and G. A. Codd, *FEBS Lett.* **264,** 187 (1990).

[q] R. E. Honkanen, J. Zwiller, R. E. Moore, S. L. Daily, B. S. Khatra, M. Dukelow, and A. L. Boynton, *J. Biol. Chem.* **265,** 19401 (1990).

[r] J. E. Eriksson, D. Toivola, J. A. O. Meriluoto, H. Karaki, Y.-G. Han, and D. Hartshorne, *Biochem. Biophys. Res. Commun.* **173,** 1347 (1990).

[s] J. Liu, J. D. Farmer, Jr., W. S. Lane, J. Friedman, I. Weissman, and S. L. Schreiber, *Cell* **66,** 807 (1991).

dephosphorylation, possible constitutive PK and PP activities can be revealed. During the past few years a number of highly specific inhibitors have been established acting directly on Ser/Thr-specific type 1 (PP1) and type 2A (PP2A) PPs, both of which are of major importance in regulating a number of crucial cellular functions and which constitute the bulk of the PP activities in any given mammalian cell. These PPs are likely candidates when possible cytoskeletal PPs are to be considered as they have especially been established as direct regulators of actomyosin complexes,[14–16] microtu-

[11] N. R. Gliksman, S. F. Parsons, and E. D. Salmon, *J. Cell Biol.* **119,** 1271 (1992).

[12] G. Gurland and G. G. Gundersen, *Proc. Natl. Acad. Sci. U.S.A.* **90,** 8827 (1993).

[13] M. Mawal-Dewan, J. Henley, A. Van de Voorde, T. Q. Trojanowski, and V. M. Lee, *J. Biol. Chem.* **269,** 30981 (1994).

[14] D. M. Toivola, R. D. Goldman, D. R. Garrod, and J. E. Eriksson, *J. Cell Sci.* **110,** 23 (1997).

[15] A. Fernandez, D. L. Brautigan, M. Mumby, and N. Lamb, *J. Cell Biol.* **111,** 103 (1990).

[16] M. J. Hubbard and P. Cohen, *Trends Biochem. Sci.* **18,** 172 (1993).

bules and their associated regulatory and transport proteins,[17–19] and various types of intermediate filaments.[3,9,10,14] While there are some indications that type 2B PP (calcineurin) may be involved in regulation of microtubule-associated proteins,[20] type 2C PP has so far not been shown to have any direct role in the regulation of cytoskeletal complexes. Some special considerations should be observed when using inhibitors of PP1 and PP2A: (1) Except for microcystins they are not very stable in aqueous environments. (2) They show variable degrees of permeability, with calyculin-A showing the highest degree of permeability. (3) Microcystins are not taken up in cells other than hepatocytes. (4) PP1 and PP2A are involved in regulating a number of key signaling cascades, for example, they act as inhibitory enzymes of the MAPK pathway[21] and the stress-activated counterparts of this pathway, the SAPK/JNK pathway and the p38 PK pathway,[22] and they also regulate a number of the cyclin-dependent kinases involved in regulation of the cell cycle.[21] Hence, depending on the cellular model system and the conditions used, the effect obtained by PP inhibitors may not always stem from direct regulation of a protein by PP1 or PP2A but may also be due to activation of signaling cascades or complexes. When considering these two phosphatase classes, the dose–response curve of a PP activity regulating any given cytoskeletal complex or system is usually extremely steep. Hence, the dose–response of an effect should be titrated carefully by using multiple inhibitor concentrations. Furthermore, by using the specific IC_{50} ranges of okadaic acid, tautomycin, and calyculin-A against PP1 and PP2A, the involvement of each class of PP in a given process can be distinguished. However, when attempting this approach, the different cell permeation properties of these compounds have to be considered.[23] Useful and detailed information on the use of Ser/Thr-specific PP inhibitors is provided in Refs. 7 and 24–27.

Tyrosine-directed phosphorylation seems to be of special interest with respect to the regulation of matrix-associated proteins such as integrins or syndecans as well as in mediating the signaling going in and out through

[17] S. F. Hamm-Alvarez, X. Wei, N. Berndt, and M. Runnegar, *Am. J. Physiol.* **271**, C929 (1996).
[18] S. E. Merrick, D. C. Demoise, and V. M. Lee, *J. Biol. Chem.* **271**, 5589 (1996).
[19] E. Sontag, V. Nunbhakdi Craig, G. S. Bloom, and M. C. Mumby, *J. Cell Biol.* **128**, 1131 (1995).
[20] A. Ferreira, R. Kincaid, and K. S. Kosik, *Mol. Biol. Cell* **4**, 1225 (1993).
[21] G. Walter and M. Mumby, *Biochim. Biophys. Acta* **1155**, 207 (1993).
[22] A. J. Waskiewicz and J. Cooper, *Curr. Opin. Cell Biol.* **7**, 7981 (1995).
[23] B. Favre, P. Turowski, and B. A. Hemmings, *J. Biol. Chem.* **272**, 13856 (1997).
[24] P. Cohen, C. F. Holmes, and Y. Tsukitani, *Trends Biochem. Sci.* **15**, 98 (1990).
[25] H. Fujiki and M. Suganuma, *Adv. Cancer Res.* **61**, 143 (1993).
[26] C. MacKintosh, *in* "Protein Phosphorylation, A Practical Approach" (D. G. Hardie, ed.), p. 197. Oxford University Press, New York, 1993.
[27] C. MacKintosh and R. W. MacKintosh, *TIBS* **19**, 444 (1994).

these complexes.[5] The most useful and specific inhibitor of Tyr PPs is vanadate. When using vanadate, its special chemistry under physiologic conditions should be considered, because vanadate solutions are prone to turn into biologically inactive reduced forms or polymers at physiologic pH and in the presence of reducing agents. An exhaustive presentation on this subject can be found in Ref. 28.

Inhibitors and Activators of PKs

A number of PKs are involved in the regulation of cytoskeletal systems. A discussion of kinases associated with cytoskeletal regulation is far beyond the scope of this presentation and, for an introduction, the reader should consult suitable reviews in the specific fields of cytoskeletal research. Tables II and III list examples of a number of useful inhibitor/activator compounds to be used for modulation of kinase activities in cell cultures. The number of useful kinase-modifying compounds is rapidly growing, so the list provided is not comprehensive but is meant to illustrate some of the most commonly used pharmacologic approaches to kinase activation/inhibition. In many cases, whole classes of derivatives of the listed compounds have evolved. Further information on these compounds as well as protocols for their use can be found in references listed in the tables.

Recently, several new interesting classes of inhibitor/activator compounds have emerged. For example, the very topical MAPK pathway and the related stress-activated p38 PK pathway can be inhibited by highly specific compounds (for a review, see Cohen[6]). Furthermore, new, more precise inhibitors of various tyrosine kinases have been introduced (Table II). A number of specific inhibitors of cyclin-dependent kinases (CDKs) have also been established. They show some degree of overlapping specificity to different CDKs and may also in some cases affect other kinases. Because there is a rather relatively large number of the CDK inhibitors, all with specific properties, a detailed presentation of these compounds is beyond the scope of this article. The reader is, therefore, referred to a recent review[29] where useful information can be found.

Caution should be taken when interpreting the results obtained with a given modifier compound. As in the case of PPs, a single PK may affect whole ranges of kinase cascades and the resulting effect on the cytoskeleton may be indirect. A good example is activation of PKC by phorbol esters. In many cell types this will result in extremely efficient activation of the

[28] J. A. Gordon, *Methods Enzymol.* **201,** 477 (1991).
[29] L. Meijer, *Trends Cell Biol.* **6,** 393 (1996).

TABLE II
EXAMPLES OF SELECTIVE PK INHIBITORS*

SELECTIVE PK INHIBITORS

Kinase	Compound	Solubility	Potency (IC$_{50}$ μM)	Supplier	Ref.
CaMKII:	KN-62	DMSO	1	a–c	f
	KN-93	Methanol, DMSO	0.5	a–c	g
CKI and CKII:	CKI-7	DMSO	10/CKI 100/CKII	d	h, i
MLCK:	ML-7	DMSO, 50% ethanol	0.3	a, b	j
	ML-9	50% Ethanol	4	a, b, d	i, j
MAPK and p38 PK	PD 98059[1]	DMSO	2–50	a	k, l
pathways:	SB 203580[2]	DMSO	0.6–5	a	m
p70 S6K	Rapamycin[3]	DMSO, methanol	0.00005	a, b	n
PI 3-K:	Wortmannin	DMSO, ethanol	0.003	a, b	o
PKA:	KT5720	DMSO	0.056	a	p
	(Rp)-8-Br-CAMPS	Water	100	e	q
	(Rp)-8-Cl-CAMPS	Water	100	e	q
PKC:	Bisindolylmaleimides	DMSO, methanol, water	0.01–0.1	a, b	r s
	Calphostin C	DMSO, DMF, ethanol	0.05	a, b	h t
	Chelerythrine chloride	DMSO, water	0.7	a, b, c	h
PKG:	KT5823	DMSO, DMF	0.2–0.3	a	u
PTK:	Erbstatin	DMSO	0.8	a	v
	Genistein	DMSO	2–30	a, b	w
	Herbimycin A	DMSO	1	a, b	x
	Lavendustin A	Acetone, DMSO, ethanol	0.01–0.5	a, b	y
	Tyrphostins	DMSO, ethanol	0.003–45	a–c	z

* Abbreviations used in Tables II and III: CaMKII, calcium/calmodulin-dependent protein kinase; CKI/CKII, casein kinase I/II, MAPK, mitogen-activated protein kinase; MAPKK, mitogen-activated protein kinase kinase; MLCK, myosin light chain kinase; p38 PK, p38 protein kinase; p70 S6K, p70 S6 kinase; PI 3-K, phosphatidylinositol 3-kinase; PKA, cAMP-dependent protein kinase (protein kinase A); PKC, protein kinase C; PKG, cGMP-dependent protein kinase (protein kinase G); PTK, protein tyrosine kinase. The IC$_{50}$ values are approximative values of useful concentrations.

[1] Inhibits activation of MAPKK by Raf.

[2] Inhibits p38 PK.

[3] Inhibits S6 K activation by inhibiting FKPB12 and rapamycin-binding PK, which activates p70 S6 kinase through mechanisms not yet fully understood.

[a] Calbiochem, La Jolla, CA, USA.

[b] Sigma Chemical Company, St. Louis, MO, USA.

[c] Alexis Corporation, San Diego, CA, USA.

[d] Seikagaku America Inc., Rockville, MD, USA.

TABLE II (*continued*)

[e] BIOLOG Life Science Institute, Bremen, Germany.
[f] H. Hidaka, M. Watanabe, and R. Kobayashi, *Methods Enzymol.* **201,** 328 (1991).
[g] M. Sumi, K. Kiuchi, T. Ishikawa, A. Ishii, M. Hagiwara, T. Nagatsu, and H. Hidaka, *Biochem. Biophys. Res. Commun.* **181,** 968 (1991).
[h] T. Tamaoki, *Methods Enzymol.* **201,** 340 (1991).
[i] H. Hidaka and R. Kobayashi, *in* "Protein Phosphorylation, A Practical Approach" (D. G. Hardie, ed.), pp. 87. IRL Press, 1993.
[j] M. Saitoh, T. Ishikawa, S. Matsushima, M. Naka, and H. Hidaka, *J. Biol. Chem.* **262,** 7796 (1987).
[k] D. R. Alessi, A. Cuenda, P. Cohen, D. T. Dudley, and A. R. Saltiel, *J. Biol. Chem.* **270,** 27489 (1995).
[l] D. T. Dudley, L. Pang, S. J. Decker, A. J. Bridges, and A. R. Saltiel, *Proc. Natl. Acad. Sci. U.S.A.* **92,** 7686 (1995).
[m] A. M. Badger, J. N. Bradbeer, B. Votta, J. C. Lee, J. L. Adams, and D. E. Griswold, *J. Pharmacol. Exp. Ther.* **279,** 1453 (1996).
[n] F. J. Dumont, M. J. Staruch, S. L. Koprak, M. R. Melino, and N. H. Sigal, *J. Immunol.* **144,** 251 (1990).
[o] M. Ui, T. Okada, K. Hazeki, and O. Hazeki, *TIBS* **20,** 303 (1995).
[p] H. Kase, K. Iwahashi, S. Nakanishi, Y. Matsuda, K. Yamada, M. Takahashi, C. Murakata, A. Sato, and M. Kaneko, *Biochem. Biophys. Res. Commun.* **142,** 436 (1987).
[q] B. T. Gjertsen, G. Mellgren, A. Otten, E. Maronde, H.-G. Genieser, B. Jastroff, O. K. Vintermyr, G. S. McKnight, and S. O. Døskeland, *J. Biol. Chem.* **270,** 20599 (1995).
[r] Z. Kiss, H. Phillips, and W. H. Anderson, *Biochim. Biophys. Acta.* **1265,** 93 (1995).
[s] R. E. Muid, M. M. Dale, P. D. Davis, L. H. Elliott, C. H. Hill, H. Kumar, G. Lawton, B. M. Twomey, J. Wadsworth, and S. E. Wilkinson, *FEBS Lett.* **293,** 169 (1991).
[t] E. Kobayashi, H. Nakano, M. Morimoto, and T. Tamaoki, *Biochem. Biophys. Res. Commun.* **159,** 548 (1989).
[u] J. R. Grider, *Am. J. Physiol.* **264,** G334 (1993).
[v] K. Umezawa and M. Imoto, *Methods Enzymol.* **201,** 370 (1991).
[w] T. Akiyama and H. Ogawara, *Methods Enzymol.* **201,** 362 (1991).
[x] Y. Uehara and H. Fukazawa, *Methods Enzymol.* **201,** 370 (1991).
[y] J. M. Bishop, *Annu. Rev. Biochem.* **52,** 307 (1989).
[z] A. Levitzki, A. Gazit, N. Osherov, I. Posner, and C. Gilon, *Methods Enzymol.* **201,** 347 (1991).

MAPK pathway (see, e.g., Refs. 30–33) and PKC activation may also activate SAPK/JNK,[22,34] especially if the cells are subjected to some kind of stress. Therefore, it is important to know in detail the signaling components proximal and distal to the PK under investigation and, furthermore, the results obtained should be confirmed by other approaches.

[30] H. Abedi and I. Zachary, *J. Biol. Chem.* **272,** 15442 (1997).
[31] M. C. Amaral, A. M. Casillas, and A. E. Nel, *Immunology* **79,** 24 (1993).
[32] R. Seger, Y. Biener, R. Feinstein, T. Hanoch, A. Gazit, and Y. Zick, *J. Biol. Chem.* **270,** 28325 (1995).
[33] M. R. Stofega, C. L. Yu, J. Wu, and R. Jove, *Cell Growth Differ.* **8,** 113 (1997).
[34] J. M. Kyriakis and J. Avruch, *BioEssays* **18,** 567 (1996).

TABLE III
EXAMPLES OF SELECTIVE PK ACTIVATORS

PK ACTIVATORS

	Compound	Comments	Solubility	Potency (IC_{50} μM)	Supplier	Ref.
PKA:	Br-cAMP	Cell permeable cAMP-analog	Water	0.1–1	a, c	e
	Forskolin	Activates cAMP	DMSO	5–10	a, c, d	f
PKC:	Mezerein		Ethanol	0.01–0.1	a, b	g
	PMA (TPA)	Phorbol-myristate-acetate	DMSO, ethanol	0.01–0.1	a–c	h
	SC-9	Ca^{2+}–dependent activation	DMSO	5–10	a, d	i
	Thymeleatoxin	Selective for α, $\beta 1$, and γ iso-forms	DMSO	0.01–0.1	a	j

[a] Calbiochem, La Jolla, CA, USA.
[b] Sigma Chemical Company, St. Louis, MO, USA.
[c] Alexis Corporation, San Diego, CA, USA.
[d] Seikagaku America Inc., Rockville, MD, USA.
[e] Y. J. Hei, K. L. MacDonell, J. H. McNeill, and J. Diamond, *Mol. Pharmacol.* **39,** 233 (1991).
[f] C. Galli, O. Meucci, A. Scorziello, T. M. Werge, P. Calissano, and G. Schettini, *J. Neurosci.* **15,** 1172 (1995).
[g] H. Nishio, Y. Ikegami, T. Segawa, and Y. Nakata, *Gen. Pharmacol.* **25,** 413 (1994).
[h] C. G. Tepper, S. Jayadev, B. Liu, A. Bielawska, R. Wolff, S. Yonehara, Y. A. Hannun, and M. F. Seldin, *Proc. Natl. Acad. Sci. U.S.A.* **92,** 8443 (1995).
[i] H. Nishino, K. Kitagawa, A. Iwashima, M. Ito, T. Tanaka, and H. Hidaka, *Biochim. Biophys. Acta* **889,** 236 (1986).
[j] R. Roivainen and R. O. Messing, *FEBS Lett.* **319,** 31 (1993).

Some inhibitors are not included in Table II because of their low specificity. For example, staurosporine[35] and many compounds in the H-series[36] of PK inhibitors inhibit a number of different kinases. This property may be useful if general kinase inhibition is attempted. The different specific inhibition constants of some of these compounds may still allow the investigator to distinguish between the effects of different kinases. For example, the K_i of H-89 against PKA is more than three orders of magnitude lower than that for PKC.[36]

[35] T. Tamaoki, *Methods Enzymol.* **201,** 340 (1991).
[36] H. Hidaka, M. Watanabe, and R. Kobayashi, *Methods Enzymol.* **201,** 328 (1991).

Transfection of Modified PKs and PPs

As a number of key elements of various signaling cascades or complexes have been cloned, there is an increasing number of PK or PP constructs available with mutations in the regulatory or catalytic domains, enabling overexpression of constitutively active or dominant negative alleles. Using this approach provides a highly specific way to elucidate the functions of a PK or PP. Many of the available constructs have been epitope tagged, enabling detection by immunofluorescence techniques or Western blotting, respectively. Many constructs are cloned into selection vectors, which permits enrichment of cells expressing the construct by using antibiotics or other compounds that are normally toxic to cells. It is also possible to enrich transfected cells by cotransfecting cells with membrane proteins that are not normally present in the cells. For example, the membrane protein CD20, a leukocyte antigen normally present only in B cells, has been used for this purpose.[37] Commercial vectors are also available that are designed to express particular cell surface markers, like the truncated (nonfunctional) extracellular domains of CD4 and the mouse H-2K molecule (Miltenyi BioTec, Bergisch Gladbach, Germany). By using these cotransfected surface markers, transfected cells can be enriched either by flow-cytometric cell sorting or by using magnetic microbeads in combination with magnetic separation units (e.g., Miltenyi BioTec; Dynabeads, Oslo, Norway).

Methods to Study the Phosphorylation of Cytoskeletal Proteins

Metabolic in Vivo Labeling

An initial approach to assay cytoskeletal protein or protein phosphorylation is metabolic labeling with ^{32}P-orthophosphate. Furthermore, because PKs in vitro often phosphorylate substrates that are not their natural substrates in vivo (see below), results obtained by in vitro labeling have to be confirmed by in vivo labeling studies. Before treatment with activators, inhibitors, or other substances (e.g., receptor ligands, kinase activators, kinase inhibitors, phosphatase inhibitors), adenosine triphosphate (ATP) pools need to be equilibrated with ^{32}P-orthophosphate. The membranes of primary cultured cells (e.g., hepatocytes, adipocytes, thymocytes) are well permeable to $^{32}PO_4^{3-}$, and the ATP pools may be equilibrated within 1–2

---

[37] T. P. Mäkelä, J. P. Tassan, E. A. Nigg, S. Frutiger, G. J. Hughes, and R. A. Weinberg, *Nature* **371,** 254 (1994).

hr.[38–40] Established cell cultures often require longer preincubation (1–6 hr).[38–40] For some purposes it may be useful to determine the rate of ATP pool equilibration, as previously described.[38] Excessive preincubation should be avoided because this results in higher backgrounds on protein gels due to extensive ^{32}P incorporation into phospholipids, nucleic acids, etc. Consider also whether the purpose is to determine constitutive or inducible phosphorylation. If the level of constitutive phosphorylation of a given protein is to be determined, then sufficient incubation times are required to allow for equilibration of the phosphate pools of the protein in question. However, many phosphoproteins contain multiple phosphorylation sites that equilibrate with ^{32}P at different rates. If stimulus-mediated phosphorylation is to be evaluated, short equilibration times should be used to avoid masking of the stimulus-induced phosphorylation by a high level of basal phosphorylation. To obtain more efficient labeling, phosphate-free cell culture media are often used. Completely phosphate-free media can be used for maximal ^{32}P incorporation, but many cell lines are adversely affected by a phosphate-free environment. If 10% FCS is added to a phosphate-depleted medium, cells stay intact for a longer time period. In this case, the phosphate contribution of FCS gives a final phosphate concentration of approximately 0.1 mM, which is a sufficient degree of phosphate depletion to allow efficient labeling of most phosphoproteins. When using any of the commercial orthophosphate preparations suitable for metabolic labeling, 100–300 μCi/ml ^{32}PO$_4^{3-}$ in the medium is usually sufficient.

An approach that is not often considered is metabolic labeling of live animals. Although the approach is cumbersome, it may reveal some interesting information that would not be accessible if cell cultures were used. To obtain sufficient labeling within reasonable activity ranges, this approach is restricted to mice, because larger animals would require a dose of radiolabel that is beyond the normal radiation safety routines of most research facilities. In the case of a mouse, 1 mCi of ^{32}PO$_4^{3-}$ injected into the tail vein is sufficient to get measurable labeling of major phosphoproteins.

After labeling, cultured cells are lysed in a buffer suitable for SDS–PAGE, two-dimensional (2-D) electrophoresis, or immunoprecipitation, respectively. Whole-cell extracts or immunoprecipitated proteins are separated by SDS–PAGE or 2-D gel electrophoresis, followed by quantitative

[38] J. C. Garrison, *J. Biol. Chem.* **253**, 7091 (1978).
[39] W. J. Boyle, P. Van der Geer and T. Hunter, *Methods Enzymol.* **201**, 110 (1991).
[40] J. C. Garrison, in "Protein Phosphorylation, A Practical Approach" (D. G. Hardie, ed.), p. 1. Oxford University Press, New York, 1993.

phosphorimaging analysis and/or quantitative autoradiography. When the autoradiography reveals phosphorylation on some protein band of interest, there are a number of ways to establish the identity and specific phosphorylation level of the protein. If suitable antibodies are available for the assumed phosphoprotein, immunoprecipitation will enable both verification and quantification of specific ^{32}P labeling. Immunoprecipitation of cytoskeletal proteins often requires denaturing lysis buffers before the proteins are dissolved in RIPA buffer (see protocol below). If the identity of an observed phosphoprotein is not known, it can be established by solid-phase protein sequencing, after electrotransfer of the proteins to PVDF membranes, by sequencing of high-performance liquid chromatography (HPLC)-separated tryptic peptides from excised protein bands on gels, or by Western blotting with antibodies to protein(s) suspected to be involved. The specific ^{32}P labeling can be determined by densitometric or fluorometric quantification of the amount of a given phosphoprotein. The cellular proteins are separated on SDS–PAGE together with samples of known amounts of bovine serum albumin (BSA) or some other standard protein. The gel is stained with either Coomassie Blue or the fluorescent protein labels, Sypro Orange or Sypro Red (Molecular Probes) for densitometric or fluorimetric quantification, respectively. The specific labeling can also be determined by performing the *in vivo* ^{32}P labeling together with ^{35}S labeling. In this case autoradiography or phosphorimaging should be carried out with and without four layers of aluminum foil between the gel and the film or the phosphorimaging screen. The difference represents the amount of ^{35}S labeling,[39] which can be used as a reference value for the amount of protein. However, if ^{35}S labeling is to be used as a measure of specific protein quantity, one has to ensure that the treatment used does not affect the synthesis rate of proteins. Although the specific ^{32}P labeling (counts per a determined amount of protein) of a given phosphoprotein can be measured very accurately, determining the specific *stoichiometry* (moles of phosphate/ mole of protein) of phosphoproteins labeled *in vivo* is difficult. This approach to determine the exact amount of phosphate on a protein is questionable, because so many different factors affect the incorporation rates of ^{32}P-labeled orthophosphate versus unlabeled orthophosphate. For more extensive treatises on how to approach *in vivo* labeling see Refs. 39–41. For extraction of phosphorylated proteins, see the section on enriching cytoskeletal proteins.

[41] P. V. d. Geer, K. Luo, and B. M. Sefton, *in* "Protein Phosphorylation, A Practical Approach" (D. G. Hardie, ed.), p. 31. Oxford University Press, New York, 1993.

In Vitro Labeling of Proteins

In vitro labeling of proteins is a rather easy approach to test whether a given PK may have a role in the regulation of a cytoskeletal protein or a protein complex, although there are proteins that act as common phosphoryl acceptors without necessarily having physiologic relevance (good examples are casein and histones). Some general considerations when proteins are phosphorylated *in vitro* are described next. A purified kinase and an at least partly purified substrate are required. There are very good protocols available for the isolation of most of the major PKs characterized so far. If purification is not practicable and the kinases cannot be obtained as gifts, many of the major kinases have become commercially available during the past few years. Another easy way to approach the problem of obtaining purified kinase, in order to test whether a given protein is a *bona fide* substrate, is to immunoprecipitate the kinase and perform the kinase assay using the immunocomplexed kinase. Good immunoprecipitating antibodies are available for a great number of kinases; for example, the different components and target kinases of the MAPK pathway as well as its stress-activated counterparts, the p38 kinase and the SAPK/JNK pathways, respectively. In the case of many more recently characterized kinases, suitable antibodies for immunoprecipitation or even purified kinases may be obtainable from the laboratories studying these kinases. When assessing the phosphopeptide maps of substrates phosphorylated *in vitro,* one should bear in mind that this approach may generate artificially high labeling on some sites as compared to the situation following *in vivo* phosphorylation by the same kinase.

Often it is advantageous to use cytoskeletal protein complexes for *in vitro* phosphorylation experiments. Crude preparations of microtubules, actomyosin complexes, or intermediate filament proteins, with all the characteristic associated proteins, are often useful because they reflect the conformations and interactions occurring in these protein assemblies *in vivo*. Hence, the sites available for phosphorylation may be more similar to those occurring *in vivo* than in, for example, bacterially expressed, highly purified proteins. However, remember that proteins prepared from eukaryotic cells may have some phosphorylation sites partly or completely occupied by phosphorylation and/or glycosylation.

ATP is usually the phosphoryl donor and it is used at concentrations near saturation; while the K_m values of ATP for most kinases are far below 100 μM, 100–200 μM is usually sufficient. The [γ-^{32}P]ATP is added to unlabeled ATP so that the specific activity is approximately 100–200 $\mu Ci/\mu mol$.

The PK to substrate ratio should be high enough so that the amount of kinase is not the limiting factor for obtaining a representative stoichiometry for a given protein substrate. PK to protein substrate ratios of 1:20 to 1:50 are usually sufficient although significantly lower ratios may be used if the availability of the kinase in question is limited.

In Vitro Dephosphorylation

Once a protein has been phosphorylated *in vitro* by a purified kinase or *in vivo* by metabolic ^{32}P labeling, we can test whether it is a *bona fide* substrate for a given protein phosphatase by *in vitro* dephosphorylation using purified protein phosphatases. The major mammalian Ser/Thr PPs, PP1, PP2A, PP2B, and PP2C, can be purified by well-established methods (MacKintosh[26] and references therein). Some of these PPs are also commercially available (e.g., New England BioLabs and Upstate Biotechnology, Lake Placid, NY, USA) or have been successfully expressed in their active form by using bacterial expression vectors (Chen and Cohen[42] and references therein). Tyrosine-specific PPs can also be isolated in their active form (Tonks[43] and references therein). A purified protein can be phosphorylated by a kinase *in vitro* and it can subsequently be tested as a substrate. Because protein phosphatases are rather prone to proteolysis, oxidation, and denaturation, care has to be taken to design a buffer that maintains phosphatase activity.[26] One possible approach is to dephosphorylate *in vivo* labeled material. After labeling and washing, the dephosphorylation buffer [for example, 20 mM HEPES, pH 7.4, 1 mM MgCl$_2$, 30 mM 2-mercaptoethanol, 10% glycerol, 1 mM EGTA, 0.1% Nonidet P-40 (NP-40), 1 mM phenylmethylsulfonyl fluoride (PMSF), 10 μg/ml leupeptin, 10 μg/ml antipain, 10 μg/ml pepstatin], together with the purified PP, can be added to the cells. A relatively small volume (300–500 μl for a 100-mm cell culture dish) of the dephosphorylation buffer + PP should be added to the cells. Because the buffer contains detergent, the cells will become permeabilized immediately to allow access to the PP. After dephosphorylation, the protein of interest can be isolated and checked for dephosphorylation. The dephosphorylation buffer is compatible with immunoprecipitation buffers and purification can be done this way or by SDS–PAGE or 2-D gel electrophoresis after solubilization of the cells in Laemmli sample buffer or urea sample buffer, respectively. Alternatively, the substrate can be a protein that has

---

[42] M. X. Chen and P. T. W. Cohen, *in* "Protein Phosphorylation, A Practical Approach" (D. G. Hardie, ed.), p. 197. Oxford University Press, New York, 1993.

[43] N. K. Tonks, *in* "Protein Phosphorylation, A Practical Approach" (D. G. Hardie, ed.), p. 231. Oxford University Press, New York, 1993.

been phosphorylated *in vitro* by any given kinase. If phosphopeptide mapping is carried out on the dephosphorylated protein, specific dephosphorylation sites can be determined. When assessing the specificity of protein phosphatases, it has to be considered that neither Ser/Thr nor Tyr-directed phosphatases show the same degree of substrate specificity as PKs. This is especially true for tyrosine-directed PPs, because it appears that a given tyrosine PP will dephosphorylate any phosphotyrosyl protein to some extent *in vitro*.[43] Ser/Thr phosphatases show a higher degree of specificity but the same cautionary statement applies.

Enriching Cytoskeletal Phosphoproteins

When assessing the phosphorylation of specific cytoskeletal components, it is important to consider carefully the specific properties of any given cytoskeletal system before proceeding to actual phosphorylation experiments. On one hand, the specific solubility properties and isolation procedures of the respective cytoskeletal systems may be utilized in order to enrich for cytoskeletal fractions or particular components. On the other hand, the solubility properties of cytoskeletal proteins may be altered on phosphorylation, resulting in poor yield of the component of interest. With risk of oversimplification, some general rules can be applied to many common cell systems of epithelial or fibroblastic nature. Extraction of cells with low concentrations (0.5–1%) of nonionic detergents such as Triton-X 100 in physiologic ionic strength buffers will leave filamentous actomyosin complexes, a pool of stable microtubules and microtubule-associated proteins, intermediate filaments and their associated proteins and karyoskeletal components in the particulate fraction. If the extraction is done on ice, microtubules and their associated components are dissociated and will not be present in the particulate fraction. Adding an extraction step with a high ionic strength buffer (~600 mM KCl), followed by an extraction with DNase I, will more or less quantitatively remove the actomyosin complexes leaving the intermediate filaments and their associated proteins in the particulate fraction, together with nuclear lamins, histones, and a number of other nuclear matrix-associated proteins. Microtubule-associated proteins of cultured cells can be enriched, e.g., by the taxol stabilization method.[44,45]

When attempting enrichment of cytoskeletal components, it should be considered that the soluble/insoluble partitioning coefficients of proteins associated with actomyosin complexes, microtubules, and intermediate filaments may be altered by phosphorylation. For example, the constituent proteins of intermediate filaments will in most cases disassemble on phos-

[44] C. A. Collins, *Methods Enzymol.* **196**, 246.
[45] R. B. Vallee and C. A. Collins, *Methods Enzymol.* **134**, 116.

phorylation (for review, see Eriksson *et al.*[3] and Inagaki *et al.*[46]) and attempts at enriching phosphorylated intermediate filament proteins in the particulate fraction will inevitably fail. It is beyond the scope of this presentation to summarize the possible approaches for isolation of cytoskeletal components comprehensively, so the reader is, apart from the above-listed references, referred to specific methodology articles in the respective fields of cytoskeletal research for further information on this subject.

Phosphoamino Acid Analysis and Phosphopeptide Mapping

Proteins showing alterations in their phosphorylation state during *in vivo* labeling are often initially analyzed to reveal their phosphoamino acid composition. This can be carried out, for example, after electrotransfer of SDS–PAGE-separated proteins to Immobilon membranes (polyvinylidene difluoride; Millipore Corporation, Bedford, MA, USA), followed by acid hydrolysis of a protein band with 6 *M* HCl as described.[39,41] The number of sites that are involved in the phosphorylation of a particular protein can be estimated by tryptic phosphopeptide mapping (also other proteases can be used). The *in vivo* ^{32}P-labeled protein of interest is, after electrophoretic separation, digested with trypsin in the gel, and the resulting peptides are separated on microcrystalline cellulose plates by electrophoresis followed by ascending chromatography. For further information on this technique, see Boyle *et al.*[39] and Geer *et al.*[41] Comparing the phosphopeptide maps obtained *in vivo* and *in vitro* is an essential way of determining whether a given PK or PP could be involved in the regulation of a protein.

Determination of Phosphorylation Sites

Determination of specific phosphorylation sites is carried out after digestion, e.g., with trypsin, of the *in vivo* or *in vitro* ^{32}P-labeled protein of interest from SDS–PAGE slices. It is crucial that the protein to be digested has been purified to very high purity to avoid contamination of peptides from other proteins. If proteases generating large fragments are used, the protein often needs to be electroeluted from the gel, because large peptides diffuse poorly out of the gel. The peptides are separated on C_{18} or C_8 reversed-phase HPLC analytical columns (preferably microbore scale) and the ^{32}P-labeled peptides are collected. When the ^{32}P-labeled peptides are sequenced in an automatic amino acid sequencer, initial indication of a given phosphorylation site is given by the fact that the highly charged

[46] M. Inagaki, Y. Matsuoka, K. Tsujimura, S. Ando, T. Tokui, T. Takahashi, and N. Inagaki, *BioEssays* **18,** 481 (1996).

phosphoamino acid residues are retained in the reaction cartridge of the gas-phase sequencer and will therefore be missing from the sequence. Serine-specific phosphorylation can be specifically detected by conversion of phos-phoserine into ethyl-S-cysteine, which then can be identified on an amino acid sequencer.[47] Manual Edman degradation is a sensitive, simple, and inexpensive (and largely overlooked) way to determine the location of phosphoamino acids in an isolated phosphopeptide. Phosphopeptides are immobilized on arylamine membrane disks using water-soluble carbodi-imide, and the immobilized peptides are subjected to manual Edman degra-dation (see protocol described below). However, this approach requires knowledge of the protein sequence.

Methods to Assess the Functions of Phosphorylation

Site-Directed Mutagenesis

Once the specific phosphorylation site has been identified, a common approach to establish its supposed function is cDNA mutagenesis. The phosphorylation site can first be confirmed by tryptic phosphopeptide map-ping of the mutant protein expressed in a suitable cell line, which is subjected to metabolic ^{32}P *in vivo* labeling. The mutation can be designed so that it mimics the nonphosphorylated state (Ala for Ser and Thr, and Phe for Tyr) or the residue can be mutated so that it simulates the phosphorylated state (Asp for phospho-Ser or phospho-Thr).[39] Site-directed mutagenesis enables examination of the possible effects on protein functions and conse-quent effects on cellular phenotype; for example, absence of a given phos-phorylation site will halt a process or state that was presumed to be initiated by phosphorylation.

Phosphorylation State-Specific Antibodies

Polyclonal antibodies against phosphorylated peptides have been successfully produced. They can be used both for localization of phospho-proteins, by using immunohistochemical techniques, and for quantification of phosphorylation, by Western blotting. This approach has been very successful in the context of signaling proteins and/or transcription factors. The cAMP-dependent phosphorylation of DARPP-32, which in its phos-phorylated state is an inhibitor of PP1,[48] as well as the phosphorylation

[47] H. E. Meyer, *Methods Enzymol.* **201**, 169 (1991).
[48] P. Neyroz, F. Desdouits, F. Benfenati, J. R. Knutson, P. Greengard, and J. A. Girault, *J. Biol. Chem.* **268**, 24022 (1993).

of CREB[49] have been successfully analyzed *in vivo* and *in situ* by using phosphopeptide-specific antibodies. Recently, some companies have developed whole product lines of phosphorylation state-specific antibodies, for example, New England Biolabs has an extensive range of phosphorylation-specific antibodies for the various members and downstream targets of the MAPK cascade as well as the stress-activated counterparts of the MAPK cascade, i.e., SAPK/JNK and p38 kinase cascades. Upstate Biotechnology has also established similar product lines for phosphorylation-specific antibodies against, e.g., CREB and glycogen synthase kinase 3. Regarding cytoskeletal proteins, this approach has been extensively used in establishing the role of abberrant *tau* phosphorylation in neurodegenerative disorders.[50–54] It has also been successful for determining the spatiotemporal distribution of site-specific phosphorylation on vimentin,[55–57] GFAP,[58–60] and keratin 8[61] and 18,[62–67] during different phases of the cell cycles well as under variable physiologic and stressed conditions.

[49] D. D. Ginty, J. M. Kornhauser, M. A. Thompson, H. Bading, K. E. Mayo, J. S. Takahashi, and M. E. Greenberg, *Science* **260**, 238 (1993).
[50] H. M. Roder, R. P. Fracasso, F. J. Hoffman, J. A. Witowsky, G. Davis, and C. B. Pellegrino, *J. Biol. Chem.* **272**, 4509 (1997).
[51] B. M. Riederer, E. Draberova, V. Viklicky, and P. Draber, *J. Histochem. Cytochem.* **43**, 1269 (1995).
[52] E. M. Mandelkow and E. Mandelkow, *Trends Biochem. Sci.* **18**, 480 (1993).
[53] E. M. Mandelkow and E. Mandelkow, *Neurobiol. Aging* **15**, S85 (1994).
[54] E. Mandelkow and E. M. Mandelkow, *Curr. Opin. Cell Biol.* **7**, 72 (1995).
[55] M. Ogawara, N. Inagaki, K. Tsujimura, Y. Takai, M. Sekimata, M. H. Ha, S. Imajoh-Ohmi, S-i. Hirai, S. Ohno, H. Sugiura, T. Yanauchi, and M. Inagaki, *J. Cell Biol.* **131**, 1055 (1995).
[56] Y. Takai, M. Ogawara, Y. Tomono, C. Moritoh, S. Imajoh Ohmi, O. Tsutsumi, Y. Taketani, and M. Inagaki, *J. Cell Biol.* **133**, 141 (1996).
[57] K. Tsujimura, M. Ogawara, Y. Takeuchi, S. Imajoh Ohmi, M. H. Ha, and M. Inagaki, *J. Biol. Chem.* **269**, 31097 (1994).
[58] P. Marin, K. L. Nastiuk, N. Daniel, J. A. Girault, A. J. Czernik, J. Glowinski, A. C. Nairn, and J. Premont, *J. Neurosci.* **17**, 3445 (1997).
[59] K. Nishizawa, T. Yano, M. Shibata, S. Ando, S. Saga, T. Takahashi, and M. Inagaki, *J. Biol. Chem.* **266**, 3074 (1991).
[60] S. Yano, K. Fukunaga, Y. Ushio, and E. Miyamoto, *J. Biol. Chem.* **269**, 5428 (1994).
[61] J. Liao, N.-O. Ku, and M. B. Omary, *J. Biol. Chem.* **272**, 17565 (1997).
[62] C. F. Chou and M. B. Omary, *J. Cell Sci.* **107**, 1833 (1994).
[63] J. Liao, L. A. Lowthert, and M. B. Omary, *Exp. Cell Res.* **219**, 348 (1995).
[64] J. Liao and M. B. Omary, *J. Cell Biol.* **133**, 345 (1996).
[65] N.-O. Ku, S. Michie, R. G. Oshima, and M. B. Omary, *J. Cell Biol.* **131**, 1303 (1995).
[66] N.-O. Ku and M. B. Omary, *J. Biol. Chem.* **270**, 11820 (1995).
[67] J. Liao, L. A. Lowthert, N.-O. Ku, R. Fernandez, and M. B. Omary, *J. Cell Biol.* **131**, 1291 (1995).

Protocol Examples: Characterization of Intermediate
Filament Phosphorylation

Example 1: Demonstration of cdc2-Specific Phosphorylation on Vimentin

It has been established that vimentin is phosphorylated on Ser-55 by
cdc2 during mitosis.[68,69] We describe the cdc2-specific phosphorylation of
vimentin as an example of how a kinase-specific site can be determined *in
vitro* and *in vivo* and how phosphorylation state-specific antibodies can
be generated.

Protocol 1: Procedure for in Vivo Labeling. To study whether the vimentin in neuroblastoma ST15A cells is phosphorylated during mitosis as previously reported for BHK cells, ^{32}P-labeled cells (see below) are synchronized in mitosis by using a nocodazole-induced metaphase arrest. To obtain
rat ST15A cells arrested in mitosis, cells are grown to 70–80% confluency
in cell culture flasks. 0.5 μg/ml nocodazole is added to cells prelabeled for
3 hr with [32]P (see below) and the cultures are then incubated at 37° for 12
hr. Cells arrested in mitosis round up, detach easily from the culture flask,
and can be harvested by mechanical shake off. The flask is hit hard 20
times against a pile of five paper towels and the detached cells are collected
into a tube and washed by centrifugation. Cells are lysed in sodium dodecyl
sulfate (SDS) buffer, and the mitosis-specific phosphorylation of vimentin
is confirmed by immunoprecipitation (Fig. 1).

1. For maximal labeling, phosphate-depleted medium should be used,
although detectable labeling of major phosphoproteins can usually be obtained in undepleted media. Wash cells two to three times with the phosphate-depleted medium and add enough medium to just cover the cells
(~3 ml to a 100-mm-diameter petri dish is sufficient).

2. Add [32][P]orthophosphate to the medium to a concentration of 100–
300 μCi/ml and incubate at the appropriate conditions for the given cell
line. For cells with fibroblast characteristics, 1–2 hr of incubation is sufficient
if inducible phosphorylation is to be studied. If constitutive phosphorylation
is to be studied, 4–6 hr is usually sufficient. Incubation times of up to 12
hr and [32][P]orthophosphate concentrations of up to 1 mCi/ml may be
required to study low levels of constitutive phosphorylation. Long incubation times may also be required if inducible dephosphorylation is to be
studied. Carefully rock the petri dish every 30 min during the incubation
to ensure that medium is covering all cells.

3. Stimulate or treat cells in desired ways after the prelabeling (still in

[68] Y. H. Chou, J. R. Bischoff, D. Beach, and R. D. Goldman, *Cell* **62,** 1063 (1990).
[69] Y. H. Chou, K. L. Ngai, and R. Goldman, *J. Biol. Chem.* **266,** 7325 (1991).

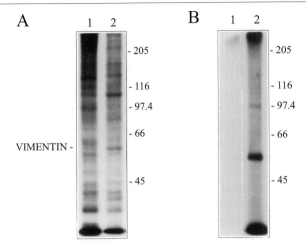

FIG. 1. Vimentin phosphorylation is elevated in mitotic cells. (A) *In vivo* ^{32}P-labeled interphase cells (lane 1) and cells arrested in mitosis (lane 2) are extracted with SDS buffer as indicated in protocol 1, and the proteins are separated on SDS–PAGE. On the autoradiography of the dried gel, a protein band corresponding to vimentin shows increased phosphorylation. Note that there are a number of proteins that show relative increases in phosphorylation in the extracts from mitosis-arrested cells. However, there are also proteins showing decreased phosphorylation as there are specific PPs that are activated during mitosis. (B) The mitosis-specific increased phosphorylation of vimentin was confirmed by immunoprecipitation (lane 1, interphase cells; lane 2, mitosis-arrested cells).

the presence of ^{32}P), e.g., by adding receptor agonists or antagonists or by subjecting cells to a given physicochemical stimulus.

4. Wash the cells on the petri dish using ice-cold phosphate-buffered saline (PBS). Be careful to collect any detached cells by centrifugation. The washing step can be repeated one to two times to yield a lower background of free label.

5. Prepare the cells for protein extraction by adding appropriate buffers and scraping off cells with a disposable cell scraper. The cells from the petri dish are pooled with the possibly detached cell pellet by extracting the cells on the dish first and then transferring the cell extract including debris to the cell pellet in the centrifuge tube.

COMMENTS

1. It is important to collect detached cells, in case phosphorylation correlates with cell adhesion.
2. *In vivo* labeling involves potentially hazardous levels of radioactivity. The person conducting these experiments should therefore be experienced in the use of radioisotopes. There should preferably be a

separate room for the labeling, and stringent routines should be established for radiation safety and waste disposal. For an extended discussion on how to arrange laboratory routines and safety to accommodate *in vivo* labeling procedures, see Garrison.[40]

PROTEIN EXTRACTION BUFFERS. The choice of extraction buffer depends on the solubility of the protein(s) of interest and the planned use of the protein samples. For immunoprecipitation of cytoskeletal proteins, it is often useful to denature all cellular proteins by using a SDS-based buffer. Extraction of some solubilized cytoskeletal proteins can be obtained by homogenization or extraction in buffers with Triton X-100 or other detergents. For cell cultures 0.5 ml buffer for a 100-mm petri dish is suitable. For tissue homogenizations, use 3 ml buffer/g tissue and 20 strokes on ice in a Potter-Elvehjelm homogenizer.

Whole-cell protein extraction of cultured cells. Cells are lysed in SDS buffer: 20 mM Tris-HCl, pH 7.2, 5 mM ethylene glycol-bis(β-aminoethyl ether) tetraacetic acid (EGTA), 5 mM ethylenediaminetetraacetic acid (EDTA), 0.4% SDS, 10 mM sodium pyrophosphate, 1 mM PMSF, 10 μg/ml antipain, 10 μg/ml leupeptin, and 10 μg/ml pepstatin. Cells are detached from the culture dish with a cell scraper, suspended, boiled for 5–10 min, and sonicated for 20 sec with a probe sonicator.

Extraction of soluble cytoskeletal proteins. Cells are homogenized in 20 mM HEPES, pH 7.6, 100 mM NaCl, 5 mM MgCl$_2$, 5 mM EGTA, 1% Triton X-100, 1 mM PMSF, 10 μg/ml leupeptin, 10 μg/ml antipain, and 10 μg/ml pepstatin (0.5 ml for a 100-mm petri dish). Detergent-soluble cytoskeletal proteins can be obtained by centrifuging these extracts 15 000 *g* for 15 min at 4°. Supernatants and pellets are diluted with Laemmli sample buffer or RIPA buffer. Pellets are sonicated on ice and samples are boiled 5 minutes.

Buffer for immunoprecipitation. For immunoprecipitation, SDS cell extracts should be diluted 1:10 with RIPA buffer: 20 mM HEPES, pH 7.4, 140 mM NaCl, 10 mM sodium pyrophosphate, 5 mM EDTA, 0.4% NP-40, 1 mM PMSF, 10 μg/ml leupeptin, 10 μg/ml antipain, and 10 μg/ml pepstatin. If cell extracts of soluble proteins in buffers with detergents are used, the NP-40 is omitted from the RIPA buffer. The NP-40 in the buffer will quench the effect of SDS. Immunoprecipitations can be performed as described.[70]

Protocol 2: in Vitro Phosphorylation. To study whether the *in vivo* phosphorylated mitosis-specific site of vimentin corresponds to the cdc2-specific site of vimentin phosphorylated *in vitro*, vimentin is phosphorylated by purified cdc2 kinase[68] (cdc2 kinase was a kind gift from Ying-Hao Chou

[70] E. Harlow and D. Lane, "Antibodies—A Laboratory Manual," Cold Spring Harbor Laboratory Press, New York, 1988.

and Robert Goldman, Northwestern University, Chicago, IL) according to the protocol presented for cdc2 below. Comparative phosphopeptide mapping is performed on *in vivo* and *in vitro* phosphorylated material according to the protocol presented in Boyle *et al.*[39] (Fig. 2).

The conditions for *in vitro* phosphorylation with specific PKs are different, because the various PKs have different prerequisites for their activities. Some kinases require stimulatory cofactors, whereas some kinases can be used in a simple buffer system.

1. Combine 50 μl 2× kinase buffer with 50 μl substrate protein (approximately 1 mg/ml), and transfer the mixture to a tube containing 5 μl of ATP mix (3 μCi [^{32}P-γ]ATP in 5 μl of 2 mM ATP, Sigma A-5394). The final concentration of ATP in the reaction mixture is 100 μM.
2. Transfer the reaction mixture with the ATP mix to the tube containing 1 μg cdc2 (in 5 μl kinase buffer) on ice, vortex the tube gently, and then incubate at 37°. Stop the reaction after 30 min by addition of 100 μl 3× Laemmli sample buffer. Boil samples for 5 min.
   2× *kinase buffer:* 20 mM HEPES, pH 7.2, 120 mM NaCl, 1 mM CaCl$_2$, 5 mM EGTA, and 4 mM MgCl$_2$

COMMENTS

1. Because the kinases are fairly unstable at assay conditions, keep the kinases on ice and add as a last component to the reaction mixture.
2. The same simple buffer system can be used for many common kinases that do not require cofactors; for example, proline-directed kinases other than cdks, such as MAPK, SAPK, and p38, do not require activation by cofactors. PKA works very well with this buffer (in the above-described protocol the cdc2 can be substituted with 1 μg of the catalytic subunit of PKA; Sigma P2645) and PKC can be used with this buffer if EGTA is omitted, and 0.5 mM CaCl$_2$, 2 μg diacylglycerol, and 5 μg phosphatidylserine are added to the buffer.

Protocol 3: Determination of a Phosphorylation Site by Manual Edman Degradation. In our example the identity of the cdc2-specific phosphorylation site is determined by manual Edman degradation (Fig. 2). In the procedure suggested here, largely based on two previous protocols,[71,72] phosphopeptides are immobilized on arylamine membrane disks using water-soluble carbodiimide and the immobilized peptides are subjected to

[71] D. L. Dong, Z. S. Xu, M. R. Chevrier, R. J. Cotter, D. W. Cleveland, and G. W. A. D. Hart, *J. Biol. Chem.* **268**, 16679 (1993).
[72] S. Sullivan and T. W. Wong, *Anal. Biochem.* **197**, 65 (1991).

A

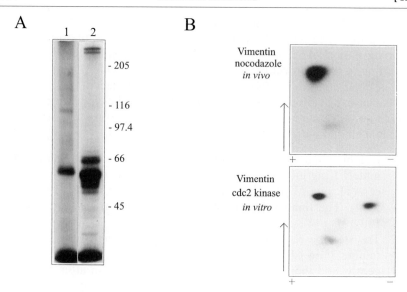

B

Vimentin
nocodazole
in vivo

Vimentin
cdc2 kinase
in vitro

C

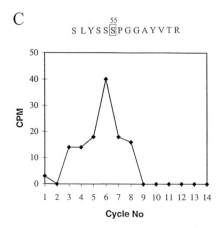

$$\overset{55}{S\ L\ Y\ S\ S\ \boxed{S}\ P\ G\ G\ A\ Y\ V\ T\ R}$$

FIG. 2. The mitosis-specific phosphorylation site on vimentin is the same as the cdc2-specific site *in vitro*. (A) Vimentin immunoprecipitated from mitotic *in vivo* ^{32}P-labeled cells (lane 1) and from *in vitro* labeling (lane 2) is subjected to tryptic phosphopeptide mapping. (B) The autoradiographies of the TLC plates with the tryptic peptides separated in two dimensions demonstrates that the *in vivo* ^{32}P-labeled material has a phosphopeptide that corresponds to the major phosphopeptide generated by cdc2-kinase *in vitro* (directions of electrophoresis, + and −, and ascending chromatography, arrow, are indicated). (C) The identity of the assumed cdc2-specific phosphopeptide localized on the peptide maps is confirmed by manual Edman degradation as described in protocol 3. There is a significant peak released at the sixth cycle. Because cdc2 is a proline-directed kinase, SP sites are first considered. Ser-55 is the only Ser residue on vimentin that forms an SP site. The tryptic cdc2-

manual Edman degradation. For immobilization of peptides, Sequelon-AA membranes (Perceptive BioSystems, Hamburg, Germany) can be used. Sequelon-AA membranes consist of a PVDF (polyvinylidene difluoride) matrix that has been derivatized with arylamine groups. The C-terminal and side chain carboxyl groups of peptides react with the arylamine groups of the membrane via carbodiimide activation.

1. The peptide sample is dissolved in an aqueous acetonitrile solution. Generally a 30% acetonitrile solution is adequate for dissolving most peptides. The acetonitrile concentration may be varied, or 0.1% trifluoroacetic acid (TFA) may be added to aid solubilization of the samples.

2. Apply a small amount of the sample at a time on an arylamine membrane disk that is placed on a Mylar sheet on top of a heating block set at 50°. Allow the disk to dry between each application. When the whole sample has been applied, allow the solvent to evaporate for 10–15 min before removing the disk from the heating block.

3. Covalent linkage of the peptides to the arylamine membrane disk is accomplished by adding 5 μl of freshly prepared carbodiimide solution (1 mg of water-soluble carbodiimide in 100 μl of 0.1 M MES, pH 5.0) to the sample disk. After 20 min at room temperature the disk is washed extensively with water and then extracted five times with 0.5 ml TFA to remove unbound peptides. The disk is then extracted three times with 1 ml methanol and subjected to Edman degradation. Alternatively the disks can be stored in methanol at $-20°$.

4. Edman degradation of immobilized peptides is carried out in 1.5-ml Eppendorf tubes. Extraction and washing can be accomplished by gentle vortexing. Each cycle of degradation is carried out according to the following protocol.

5. Add 0.5 ml of fresh coupling reagent [methanol:water:triethylamine:phenylisothiocyanate (PITC); 7:1:1:1, v/v] to the disk and incubate at 50° for 6–10 min. Remove the reagent and wash the disk five times with 1 ml methanol. Speed vacuum dry the discs for approximately 10 min.

6. Add 0.5 ml TFA, incubate at 50° for 6 min, and remove the TFA. Save the TFA wash and extract the disk with another 0.5 ml TFA. Combine the two TFA washes and dry in oven at 60°. It is possible to leave the sample at room temperature to evaporate, and then dry the samples from each cycle at 60° at the same time. Wash the disk six times with 1 ml methanol before beginning a new degradation cycle.

---

specific phosphopeptide containing this site should generate a phosphate release at the sixth cycle, which corresponds to the experimental data. Similar results were obtained both with the *in vivo* and the *in vitro* phosphorylated material. For a discussion of the specificity of the method, see comments after the included protocol.

7. Measure the amount of radioactivity released from each cycle after the sample is neutralized by adding 0.5 ml unadjusted 0.5 M Tris and determine the amount of radioactivity that remains bound to the disk.

COMMENT. In the example presented in Fig. 2, there is a significant release of radioactivity on the sixth cycle, indicating a phosphorylation site on Ser-55. However, there is some released label on cycles 4–5. This is in accordance with the previously reported low level of phosphorylation on Ser-53 and Ser-54 by cdc2.[69] The Edman reaction may to some extent be incomplete and residual label could be released after the cycle containing the major site. This is likely to be the explanation for the activity seen in cycles 7–8.

Protocol 4: Preparation of a Polyclonal Antibody against p-Ser-55 of Vimentin. A phosphopeptide-specific antibody is generated to recognize the cdc2-specific phosphorylation of vimentin on Ser-55 (Fig. 3). The approach is modified from a previous study.[57] A peptide (CKLYSSpSPGGAYVT) is synthesized containing a phosphoserine residue on a position corresponding to that of Ser-55 on hamster vimentin (suitable F-moc derivatives of phospho-Ser, phospho-Thr, and phospho-Tyr are available for peptide synthesis from, e.g., Calbiochem-NovaBiochem, Läufelfingen, Switzerland). The Cys-Lys spacer, which is not present in the native protein, is added to the N-terminus to enable covalent linkage to the Cys SH group by using succinimidyl-3-maleimidylbenzoate (Molecular Probes). In the protocol below, the first step involves primary amine-directed binding to keyhole limpet hemocyanine (KLH) by the succinimidyl ester. The KLH will then contain active maleimide groups that will react with the SH group on Cys.

PROCEDURE

1. Prepare a column (0.6 × 30-cm to 13-ml gel bed) with Sephadex G-10 or G-25. Equilibrate it with 10 bed volumes (130 ml total; 0.5 ml/min) of 0.1 M sodium phosphate buffer, pH = 6.0 (buffer 1).
2. Dissolve 7 mg of KLH in 1.5 ml 100 mM sodium bicarbonate (pH = 8.5) and mix it with 350 μl of succinimidyl-3-maleimidyl-benzoate (12 mg/ml in dimethylformamide). Incubate 30 min at room temperature with magnetic stirring.
3. Apply the mixture on the column and elute the "activated" KLH with buffer 1. KLH elutes in the first peak.
4. Mix "activated" KLH with 1 ml vimentin peptide (5 mg/ml in buffer 1). If a synthetic peptide is difficult to dissolve, it may be helpful to test dissolving it in 1 ml of 1 mM NaOH followed by neutralization with 1 ml of 1 mM HCl. Incubate at room temperature overnight with magnetic stirring.

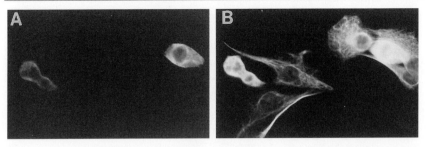

FIG. 3. Detection of mitotic Ser-55 phosphorylation on vimentin. A polyclonal phosphopep-tide-specific antibody recognizing vimentin phosphorylated on Ser-55 is generated according to protocol 4 and used for immunofluorescence-based detection of ST15A cells with high cdc2 kinase activity. (A) The cell at right is a prometaphase cell that shows bright immunore-activity due to high kinase activity. The cell at the left is a cell at anaphase with attenuated kinase activity. (B) Same cells as in part (A) processed for immunofluorescence using a general anti-vimentin antibody.

5. To block SH reactive groups, add 2-mercaptoethanol to the mixture to a final concentration of 14 mM and incubate for 1 hr at room temperature. Dialyze this mixture against 0.1 M NH$_4$HCO$_3$ with several changes.
6. Lyophilize the aliquots.
7. This amount is sufficient for immunization of two or three rabbits. At the time of immunization, the conjugated peptide is dissolved in PBS at a concentration of 1 mg/ml. For one rabbit, 0.5 ml of this solution is mixed with 0.5 ml Freund's complete adjuvant and 50 μl free phosphopeptide and the immunization with subsequent boosting is carried out according to standard protocols (see, e.g., Harlow and Lane[70]).
8. Positive and negative affinity purification of the antibody is per-formed as described earlier.[73]

Example 2: Effect of Inhibited Ser/Thr-Specific Dephosphorylation on the Phosphorylation of Disassembled Keratin 8 and 18

Previous studies have established that keratin 8 and 18 of primary hepatocyte cultures are hyperphosphorylated and disassembled on inhibi-tion of PP1 and PP2A,[14,61] indicating a high phosphate turnover on keratin intermediate filaments. By *in vivo* labeling of mice that are injected with

---

[73] A. J. Czernik, J.-A. Girault, A. C. Nairn, J. Chen, G. Snyder, J. Kebabian, and P. Greengard, *Methods Enzymol.* **201,** 264 (1991).

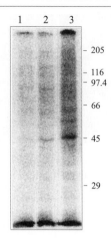

FIG. 4. *In vivo* phosphorylation of Triton X-100 soluble mouse liver proteins. Mice are injected with ^{32}P as described in protocol 5 and further injected ip with 0.9% NaCl (lane 1), 100 μg/kg microcystin-LR for 30 min (lane 2), or 300 μg/kg for 90 min (lane 3). Tissue protein extracts are separated by SDS–PAGE and the dried gel is analyzed by phosphorimaging. Molecular weight standards are indicated at the right. A microcystin-LR-induced increase in phosphorylation of p49 and p55 is observed. By immunoprecipitation these proteins are determined to be the intermediate filament proteins keratin 8 and keratin 18 (not shown).

the protein phosphatase inhibitor microcystin-LR, it is determined whether the same effect can be seen in the intact liver. Microcystin-LR is targeted to the liver by selective uptake through hepatocyte-specific organic anion transporters, the multispecific bile acid transport system.[74–76] Thus, microcystins can be used as specific probes *in vivo* to elucidate the roles of Ser/ Thr PPs in maintaining hepatic cytoskeletal organization.

Protocol 5: Metabolic in Vivo ^{32}P Labeling of Mice

1. For metabolic *in vivo* ^{32}P labeling of mice (Fig. 4), the tail vein is injected with 1 mCi [^{32}P]orthophosphate. The injected volume should be kept below 0.5 ml and the use of a 1-ml syringe and a short 25- or 27-gauge needle is recommended. The mouse is put into an immobilization tube of glass that is 9 cm long and 2.5–3.0 cm in diameter. The front part of this tube is narrowed down so that the end has a 1.0- to 1.5-cm opening allowing the mouse to breathe. The rear end of the tube is secured with a cork with

[74] J. E. Eriksson, L. Grönberg, S. Nygard, J. P. Slotte, and J. A. O. Meriluoto, *Biochim. Biophys. Acta* **1025,** 60 (1990).
[75] J. A. O. Meriluoto, S. E. Nygard, A. M. Dahlem, and J. E. Eriksson, *Toxicon* **28,** 1439 (1990).
[76] M. T. Runnegar, R. G. Gerdes, and I. R. Falconer, *Toxicon* **29,** 43 (1991).

a central hole through which the tail passes. This setup should be sufficiently small to keep the mouse in place, without causing discomfort.

2. Submerge the tail in $+37°$ water for 30–60 sec so that the tail vein is dilated. Place the immobilization tube with the mouse on a table and secure the tail with one hand so that it is stretched between your fingers. To avoid rolling of the immobilization tube, the tube should be secured to the table with laboratory tape. Insert the needle into the vein (which is rather close to the surface), and carefully inject the isotope. Start the injection closer to the tip of the tail, so that if the injection is not successful, further attempts can be made higher up on the tail.

3. A 1- to 2-hour prelabeling is sufficient to enable detection of major phosphoproteins. Following labeling and treatments, the mouse is sacrificed (e.g., by CO_2 inhalation), opened, and the organ(s) of interest is (are) removed, minced, and homogenized in suitable buffer(s).

4. For analysis of whole-cell proteins, the samples of tissue (fresh or stored in liquid N_2) are homogenized in a homogenization buffer containing 3% SDS buffer, 0.187 M Tris-HCl (pH 6.8), 3% 2-mercaptoethanol, and 5 mM EGTA. Boil samples for 5 min and sonicate to reduce viscosity of DNA. If samples are to be used for immunoprecipitation, 2-mercaptoethanol should be omitted from the buffer.

COMMENTS

1. The cage and possible tail-warming container will be contaminated and should be cleaned appropriately.
2. Because the mouse can suddenly move if not correctly secured, be careful not to inject solution into your fingers.
3. The method is cumbersome and should not be used until sufficient training of mouse tail injections have been performed, and matters of safety have been carefully considered.

Acknowledgments

Many of these techniques were adopted during a post-doctoral (JEE) stay in the laboratory of Robert D. Goldman (Northwestern University Medical School, Chicago, IL, USA). We thank Ying-Hao Chou, Omar Skalli, and Robert Goldman for sharing their expertise and for continued collaboration. Ying-Hao Chou is also gratefully acknowledged for his continuous supply of cdc2 kinase. Michael Courtney, Eleanor Coffey (Dept. of Biochemistry and Pharmacy, Åbo Akademi University, Turku, Finland), Kari Espolin, and Stein Døskeland (University of Bergen, Norway) are thanked for discussions and comments. This study was supported by a grant (29641) from the Academy of Finland and by the Cell Signaling Program of Åbo Akademi University.

[43] Correlative Light and Electron Microscopy of the Cytoskeleton of Cultured Cells

By TATYANA M. SVITKINA and GARY G. BORISY

Introduction[1]

Light and electron microscopy each have certain advantages and limitations for the investigation of the cytoskeleton. Light microscopy, especially with recent progress in fluorescence imaging, affords the opportunity to analyze the kinetics of dynamic processes in the living cell.[2-5] However, the spatial resolution of light microscopy is limited to approximately $\lambda/2$, that is, half the wavelength of the imaging light, which, in the green, is about 250 nm. This limitation constrains understanding the supramolecular organization of the cytoskeleton for which information is required below the 10-nm level. Electron microscopy, in contrast, affords high resolution but provides only static images and is not applicable to living cells. Correlative analysis of the same cells by light and electron microscopy, that is, high temporal resolution analysis of fluorescent features in a living cell followed by high-resolution spatial analysis of structural features in the same cell, provides an opportunity to combine the advantages of both techniques and establish functional connections between cytoskeletal dynamics and supramolecular organization. A critical issue for any mode of microscopy is whether the preparative procedures themselves introduce changes and artifacts into the specimen. A telling advantage of correlative microscopy is the potential for evaluating the issue of artifact.

The success of correlative microscopy imposes special demands at both the light and electron microscopic level. For fluorescence light microscopy, a suitable fluorescent probe must be prepared either by chemical derivatization[6,7] or by molecular biological conjugation to green fluorescent protein.[8-10] The derivatized protein needs to be introduced into the cell either

---

[1] See supplemental material on the web site: http://borisy.bocklabs.wisc.edu.
[2] Y.-L. Wang, *Curr. Opin. Cell Biol.* **3**, 27 (1991).
[3] Y. Fukui, *Int. Rev. Cytol.* **144**, 85 (1993).
[4] K. A. Giuliano, P. L. Post, K. M. Hahn, and D. L. Taylor, *Annu. Rev. Biophys. Biomol. Struct.* **24**, 405 (1995).
[5] K. A. Giuliano and D. L. Taylor, *Curr. Opin. Cell Biol.* **7**, 4 (1995).
[6] D. L. Taylor, P. A. Amato, K. Luby-Phelps, and P. McNeil, *Trends Biochem. Sci.* **9**, 88 (1984).
[7] Y.-L. Wang, *Methods Cell Biol.* **29**, 1 (1989).
[8] M. Chalfie M, Y. Tu, G. Euskirchen, W. W. Ward, and D. C. Prasher, *Science* **263**, 802 (1994).

by microinjection[11,12] or by expression.[8-10] Finally, issues of photobleaching, photodamage, and phototoxicity need to be attended to, generally through low-light-level imaging and the use of sensitive cameras.[13-15] At present, a variety of chemical and molecular biological procedures are available for fluorescently labeled specific components of the cytoskeleton and visualizing them in the living cell. However, among countless electron microscopic techniques, very few were designed for correlative microscopy, particularly for cytoskeletal components that are known to be labile and sensitive to external conditions. They key requirements of a suitable electron microscopic procedure are quality, reproducibility, and yield. Yield is essential because detailed observation of individual living cells places a high investment of investigator time and effort in a single cell. If the efficiency of recovering a cell for electron microscopy is low, the investment is lost. Considering different variants of cell preparation for electron microscopy we have chosen detergent extraction, chemical fixation, critical point drying, and transmission electron microscopy (TEM) of platinum replicas as a basic procedure, because each of these steps has been shown to be reliable and capable of producing a higher yield of successful results in comparison with alternative approaches. The overall procedure has been optimized in several respects,[16] which significantly improved the quality and consistency of the electron microscopy of the cytoskeleton, thus permitting efficient correlative microscopy.

In this article, we describe the preparation of cells for correlative electron microscopy after live light microscopic observation of fluorescently labeled cytoskeletal proteins microinjected into the same cells. Because identification of cytoskeletal elements in electron microscopic preparations is an essential part of any correlative study, procedures for immunogold labeling of cytoskeletal components and for myosin S1 decoration of actin filaments are also described.

Cell Culture and Light Microscopy

Details of cell cultivation, fluorescent probe production, microinjection, and light microscopic observation are beyond the scope of the present

[9] S. R. Kain and P. Kitts, *Methods Mol. Biol.* **63,** 305 (1997).
[10] H. H. Gerdes and C. Kaether, *FEBS Lett.* **389,** 44 (1996).
[11] M. Graessmann and A. Graessmann, *Methods Enzymol.* **101,** 482 (1983).
[12] P. L. McNeil, *Methods Cell Biol.* **29,** 153 (1989).
[13] R. S. Aikens, D. A. Agard, and J. W. Sedat, *Methods Cell Biol.* **29,** 292 (1989).
[14] K. R. Spring, *Scanning Microsc.* **5,** 63 (1991).
[15] S. L. Shaw, E. D. Salmon, and R. S. Quatrano, *Biotechniques* **19,** 946 (1995).
[16] T. M. Svikina, A. B. Verkhovsky, and G. G. Borisy, *J. Struct. Biol.* **115,** 290 (1995).

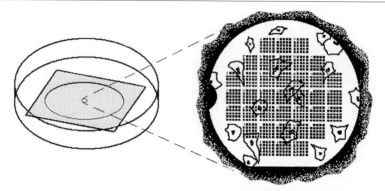

FIG. 1. Preparation of culture dishes for correlative microscopy. A circular 18-mm hole was drilled in the bottom of a 35-mm plastic culture dish and a square coverslip (22 × 22 mm) was mounted onto the bottom with silicone vacuum grease (left). The coverslip was previously coated with gold through a finder grid to create a pattern (right), which is visible in both light and electron microscopy. Cells on clear areas are used for examination.

description. However, certain requirements are specific for correlative microscopy, and these we do discuss.

For high-resolution light microscopy, cells need to be grown on glass coverslips. To facilitate the relocalization of the same cells, it is very helpful to have internal reference marks on the coverslip. This is easily obtained by providing coverslips with a finder grating that can be recognized by both light and electron microscopy. We routinely use 22- × 22-mm coverslips, which we coat within a thin layer of gold through a locator grid (400 mesh, Ted Pella, Inc., Redding, CA) (see Fig. 1). After gold coating, coverslips are baked at 160° overnight to prevent dislocation of gold grains by cultured cells. Cells for correlative microscopy are selected within clear uncoated glass areas, corresponding to the solid parts of the locator grid.

The coated coverslips can be mounted into different types of chambers suitable for cell cultivation, microinjection, and observation.[17] Because correlative microscopy requires the fast cessation of dynamic cellular processes at the end of the light microscopic observation, the chamber design should allow for the fast exchange of media. In our laboratory, we typically use 35-mm tissue culture dishes with a circular 18-mm hole in the bottom. Coverslips are mounted with the gold-coated side up onto these holes using silicon vacuum grease. The minimum amount of grease required to prevent leakage should be used in order to avoid complications during subsequent excision of the central area of the coverslip with the desired cells (see

[17] N. M. McKenna and Y.-L. Wang, *Methods Cell Biol.* **29,** 195 (1989).

below). Dishes are sterilized with UV irradiation before plating cells. To prevent pH shift in the medium during observation, we use either of two procedures. If the culture medium is bicarbonate buffered, the medium is overlayered with mineral oil,[17] which prevents evaporation and resulting pH change. Alternatively, HEPES-buffered media are used instead of the bicarbonate-buffered ones. For correlative microscopy, the latter approach allows higher temporal resolution because elimination of oil before processing for electron microscopy takes time and may result in a lapse between light and electron microscopic observations.

Electron Microscopy

Extraction

For platinum replica electron microscopy the cytoskeleton first has to be uncovered and made available to metal coating. Detergent lysis is the most usual way to remove the cell membrane, thus exposing the cytoskeleton.[18-23] However, immediately on removal of the cell membrane, the potential for extraction and artifact arises. Consequently, the composition of the extraction solution is extremely important and many extraction procedures have been suggested.[24] In our experience, a relatively high concentration of a nonionic detergent, e.g., 1% Triton X-100, is critical for rapid solubilization of the cell membrane and cessation of cell activity. The detergent is made up in a buffer containing stabilizing supplements. We routinely add high molecular weight polyetheleneglycol (PEG) for general preservation of the cytoskeleton.[25] Taxol and/or phalloidin are added for the specific preservation of microtubules and actin filaments, respectively. Extraction at room temperature is superior to extraction at lower (4°) or higher (37°) temperature. At 4°, the plasma membrane tends to be retained and is visible as patches overlying the cytoskeleton. At 37°, a substantial loss of material from the cytoskeleton is evident, especially in the lamellipodia region.

[18] S. Brown, W. Levinson, and J. Spudich, *J. Supramolec. Struct.* **5,** 119 (1976).

[19] M. Osborn and K. Weber, *Exp. Cell Res.* **106,** 561 (1977).

[20] A. D. Bershadsky, V. I. Gelfand, T. M. Svitkina, and I. S. Tint, *Cell Biol. Intern. Rep.* **2,** 425 (1978).

[21] J. V. Small and J. E. Celis, *Cytobiologie* **16,** 308 (1978).

[22] J. E. Heuser and M. W. Kirschner, *J. Cell Biol.* **86,** 212 (1980).

[23] M. Schliwa and J. van Blerkom, *J. Cell Biol.* **90,** 222 (1981).

[24] Reviewed in M. Lindroth, P. B. Bell, Jr., and B.-A. Fredriksson, *J. Microsc.* **151,** 103 (1988).

[25] T. M. Svitkina, A. A. Shevelev, A. D. Bershadsky, and V. I. Gelfand, *Eur. J. Cell Biol.* **34,** 64 (1984).

Solutions

1. In our experiments we use one of the following cytoskeleton buffers as a base for the extraction solution: buffer M [50 mM imidazole, pH 6.8; 50 mM KCl; 0.5 mM MgCl$_2$; 0.1 mM ethylenediaminetetraacetic acid (EDTA); 1 mM ethylene glycol-bis(β-aminoethyl ether)N,N,N',N'-tetraacetic acid (EGTA)] or buffer PEM (100 mM PIPES, pH 6.9; 1 mM MgCl$_2$; 1 mM EGTA). 10$\times$ stocks can be prepared for both buffers.
2. Extraction solution: 1% Triton X-100, 4% PEG (MW 40,000) (Serva, Heidelberg/New York) in buffer M or PEM supplemented (optionally) with 10 μg Taxol and/or 10 μM phalloidin.
3. Phosphate-buffered saline (PBS): 10 mM phosphate buffer, pH 7.4, 150 mM NaCl.

Procedure

1. Remove culture medium from a dish and briefly rinse cells with PBS or other serum-free physiologic solution. Remove rinse solution.
2. Immediately add the extraction solution. Exchange of media should be fast to avoid cell damage by drying.
3. Incubate cells for 3–5 min at room temperature.
4. Rinse cells with the cytoskeleton buffer two or three times.

Direct comparison of live and extracted cells demonstrated that the described extraction procedure does not introduce alterations in the distribution of myosin II[26,27] and microtubules (Fig. 2). After extraction, the cytoskeleton remains stable for an extended period of time in the cytoskeleton buffer supplemented with Taxol and phalloidin. This allows one to perform certain operations, which are possible only with the unfixed cytoskeleton. Some of them are described below.

Actin Depletion

One of the major problems for electron microscopic investigations of the cytoskeleton arises from the abundance of actin filaments interfering with the visualization of other cytoskeletal components that may be present in minor amounts. As a solution to this problem, we depleted actin from

[26] A. B. Verkhovsky and G. G. Borisky, *J. Cell Biol.* **123,** 637 (1993).
[27] A. B. Verkhovsky, T. M. Svitkina, and G. G. Borisy, *J. Cell Biol.* **131,** 989 (1995).

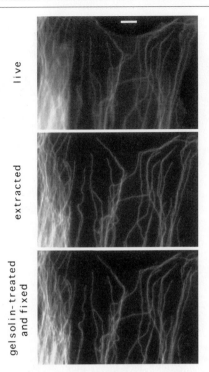

FIG. 2. Retention of the live microtubule pattern after extraction, gelsolin treatment, and fixation. Images of a rat embryo fibroblast (line REF-52) microinjected with Cy5-tubulin were taken in living state (top), after detergent extraction (middle), and after gelsolin treatment and glutaraldehyde fixation (bottom). Note the exact coincidence between microtubule distribution in all images, demonstrating that the extraction, actin removal, and chemical fixation procedures did not disturb the native organization of microtubules. Bar = 5 μm.

detergent-extracted cells with the actin-severing protein, gelsolin.[26–31] Gelsolin treatment did not perturb other cytoskeletal components[26,27] (Fig. 2), but exposed many other cytoskeletal elements that become available for examination (Figs. 3 and 4). This approach was successfully used for the visualization in cultured cells of nonmuscle myosin II,[26,27] microtubules,[30] and intermediate filaments.[31]

[28] A. B. Verkhovsky, I. G. Surgucheva, T. M. Svitkina, I. S. Tint, and V. I. Gelfand, *Exp. Cell Res.* **173,** 244 (1987).
[29] T. M. Svitkina, I. G. Surgucheva, A. B. Verkhovsky, V. I. Gelfand, M. Moeremans, and J. DeMey, *Cell Motil. Cytoskelet.* **12,** 150 (1989).
[30] T. M. Svitkina, A. B. Verkhovsky, and G. G. Borisy, *J. Struct. Biol.* **115,** 290 (1995).
[31] T. M. Svitkina, A. B. Verkhovsky, and G. G. Borisy, *J. Cell Biol.* **135,** 991 (1996).

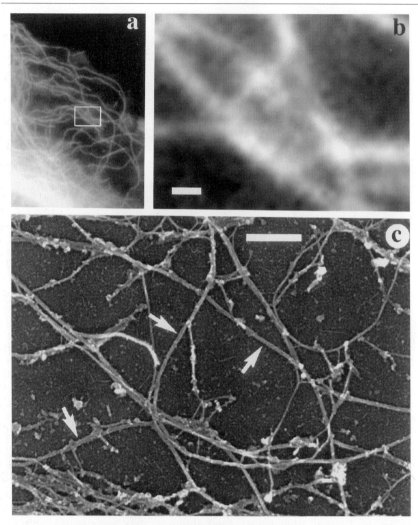

FIG. 3. Correlative light and electron microscopy of microtubules. REF-52 fibroblast was microinjected with Cy5-tubulin and, after light microscopy, processed for actin removal, then electron microscopy. (a) Fluorescence microscopy of a living cell; (b) enlarged boxed region from part (a); (c) electron microscopy of the region shown in part (b). Microtubules visualized as fluorescent filaments in (a) and (b) can be seen in (c) as thick filaments (arrows) within the cytoskeletal network. They retain the same pattern as in light microscopic images. Bars = 0.3 μm.

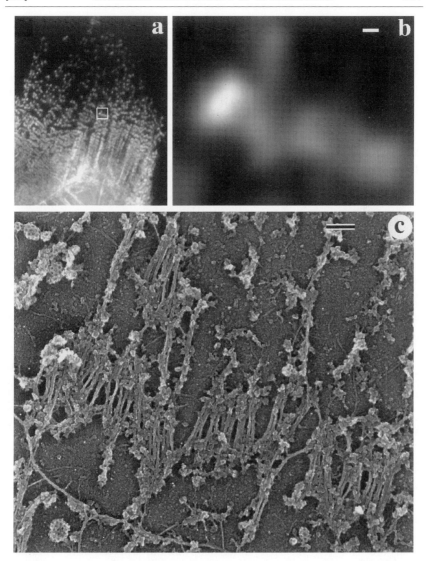

FIG. 4. Correlative light and electron microscopy of myosin II. REF-52 fibroblast was microinjected with smooth muscle myosin II labeled by tetramethylrhodamine and, after light microscopy processed for electron microscopy. (a) Fluorescence microscopy of the cell after detergent extraction, gelsolin treatment, and glutaraldehyde fixation; (b) enlarged boxed region from part (a); (c) electron microscopy of the region shown in part (b). Accumulation of myosin II visualized as fluorescent spots in (a) and (b) can be seen in (c) as stacklike assemblies of bipolar filaments. Note that brighter areas in a light microscopic image (b) correspond to higher density of myosin filaments visualized by electron microscopy (c). Bars = 0.2 μm.

Solutions

1. Buffer G: 50 mM MES-KOH, pH 6.3; 0.1 mM CaCl$_2$; 2 mM MgCl$_2$, and 0.5 mM dithiothreitol (DTT).
2. 0.1–0.2 mg/ml Gelsolin dialyzed against buffer G. Both native protein isolated from bovine brain[28] and recombinant Ca^{2+}-independent gelsolin fragment[32] were successfully used for actin depletion.[26–31]
3. Buffer M or PEM (see Extraction section).

Note: 10 μg/ml Taxol should be added to all solutions for preservation of microtubules, because otherwise all microtubules will be destroyed by the Ca^{2+}-containing gelsolin buffer.

Procedure

1. After rinse with a cytoskeleton buffer (see Extraction section), rinse coverslips once with buffer G.
2. Remove the buffer from the dish. Using cotton swabs, wipe the buffer from the dish and coverslip leaving wet only a small (approximately 5–7 mm) central area containing the finder grid.
3. Add 10–15 μl of gelsolin solution to the wet central area. The drop of gelsolin solution will remain within the wet circle and will not spread into the dry glass area. This approach allows the use of very small amounts of gelsolin. Incubate 1 hr at room temperature in moist conditions.
4. Rinse three times with M or PEM buffer and fix (see Fixation section).

S1 Decoration

Molecular identification of cytoskeletal components is a necessary part of the investigation of the cytoskeleton. Although immunochemical labeling is the most common solution of this problem (see below), there is an alternative approach for identification of actin filaments, that is, decoration by skeletal myosin subfragment 1 (S1).[33,34] S1 binding to actin filaments results in the formation of characteristic polar complexes, which can be easily recognized with electron microscopy and thus permit us to distinguish actin filaments from other cytoskeletal components.

Besides identification, S1 decoration of actin filaments allows determination of the polarity of actin filaments, which is a key structural feature of their function. Models for various actin activities, such as lamellipodia

[32] D. J. Kwiatkowski, P. A. Janmey, and H. L. Yin, *J. Cell Biol.* **108,** 1717 (1989).
[33] H. E. Huxley, *J. Mol. Biol.* **7,** 281 (1963).
[34] H. Ishikawa, R. Bischoff, and H. Holtzer, *J. Cell Biol.* **43,** 312 (1969).

protrusion,[35] stress-fiber contraction,[36,37] cell translocation,[38] and formation of organized actin networks and bundles[26,27] explicitly take into account the polarity of actin filaments in different cellular domains. When studied with negative staining[33,39–43] or thin sectioning electron microscopy,[34,44–46] S1-decorated actin filaments have a classic arrowhead pattern. Barbed or pointed ends of decorated filaments correspond to fast and slow growing ends, respectively.[47] In platinum replicas, S1-decorated filaments express a helical ropelike appearance rather than an arrowhead pattern,[48] probably because only the top surface of filaments contributes to the formation of the image, unlike negatively or positively stained filaments, where contrast is generated by top and bottom surfaces of the specimen. The individual turns of the "rope," however, have an intrinsic asymmetry correlated with the polarity of the actin filaments[48] (Fig. 5a). With our approach, we are usually able to visualize the polarity of a large fraction of actin filaments even in areas of dense filament arrays, like stress fibers of fibroblasts (Fig. 5b).

Solutions

1. Cytoskeleton buffer M or PEM with 10 μg/ml Taxol and 10 μM phalloidin (both are optional).
2. 0.25 mg/ml S1 in buffer M or PEM with Taxol and phalloidin (optional). We purchase S1 from Sigma (St. Louis, MO) and centrifuge the solution at 100,000g for 20 min before use to remove residual myosin filaments.

Procedure

1. After rinse with a cytoskeleton buffer (see Extraction section), wipe coverslips with cotton swabs around the central finder grid-containing area, as for gelsolin treatment (see above).

[35] L. P. Cramer, T. J. Mitchison, and J. A. Theriot, *Curr. Opin. Cell Biol.* **6,** 82 (1994).
[36] P. A. Conrad, K. A. Giuliano, G. Fisher, K. Collins, P. T. Matsudaira, and D. L. Taylor, *J. Cell Biol.* **120,** 1381 (1993).
[37] J. M. Sanger and J. W. Sanger, *J. Cell Biol.* **86,** 568 (1980).
[38] S. K. Maciver, *Bioessays* **18,** 179 (1996).
[39] P. B. Moore, H. E. Huxley, and D. J. DeRosier, *J. Mol. Biol.* **50,** 279 (1970).
[40] D. R. Burgess and T. E. Schroder, *J. Cell Biol.* **74,** 1032 (1977).
[41] J. V. Small, G. Isenberg, and J. E. Celis, *Nature* **272,** 638 (1978).
[42] A. K. Lewis and P. C. Bridgman, *J. Cell Biol.* **119,** 1219 (1992).
[43] J. V. Small, M. Herzog, and K. Anderson, *J. Cell Biol.* **129,** 1275 (1995).
[44] M. S. Mooseker and L. C. Tilney, *J. Cell Biol.* **67,** 725 (1975).
[45] D. A. Begg, R. Rodewald, and L. I. Rebhun, *J. Cell Biol.* **79,** 846 (1978).
[46] L. C. Tilney, D. J. DeRosier, and M. S. Tilney, *J. Cell Biol.* **118,** 71 (1992).
[47] E. M. Bonder, D. J. Fishkind, and M. S. Mooseker, *Cell* **34,** 491 (1983).
[48] J. E. Heuser and R. Cooke, *J. Mol. Biol.* **169,** 97 (1983).

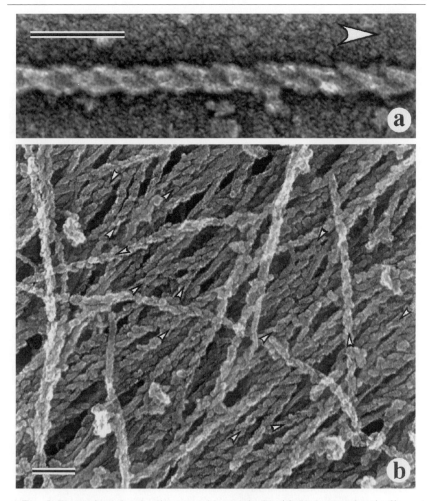

FIG. 5. Decoration of actin filaments with myosin S1. (a) S1-decorated actin filament displays a helical ropelike appearance; the asymmetry of individual turns of a helix allows us to determine polarity; the thicker part of a turn is directed to the pointed end of a filament (direction of the arrowhead). (b) Part of an actin filament bundle from an REF-52 fibroblast after S1 decoration; polarity of some filaments is indicated by arrowheads. Bars = 0.1 μm.

2. Add 10–15 μl of S1 solution to the wet central area and incubate 30 min at room temperature in moist conditions.
3. Rinse with buffer M or PEM three times.

Fixation

Fixation provides cell structures with a physical resistance against subsequent harsh procedures. Chemical fixation is generally used for cytoskeleton preservation. Cryofixation, which has been used as an alternative, is only sufficient if the water layer is not lost in subsequent procedural steps. Because cytoplasmic structures lose their support and collapse during subsequent water sublimation,[49] chemical fixation is frequently used even in conjunction with cryofixation.[22,50,51]

The most commonly employed fixative for electron microcopy is glutaraldehyde. However, glutaraldehyde alone does not provide proper preservation of all cellular structures. Consequently, various supplemental treatments have been introduced and, among them, tannic acid followed by heavy metal treatment is particularly beneficial.[48,52–58] The mechanism of protective action of tannic acid followed by heavy metals seems to be related to the ability of tannic acid to bind both to proteins and to heavy metals.[52] When applied in the correct order, these compounds apparently form an extensive chemical complex bound to protein structures that can work as a protective shell resisting shrinkage caused by dehydration and drying. According to these considerations, our procedure includes consecutive fixations with glutaraldehyde, tannic acid, and uranyl acetate.

Solutions

1. 2% Glutaraldehyde (EM grade) in 0.1 *M* sodium cacodylate, pH 7.3.
2. 0.1% Aqueous tannic acid (Mallinckrodt, Inc., Paris, KY, Cat. #1764). This particular formulation of tannic acid with low molecular weight has been shown to give the best results.[59]
3. 0.1–0.2% Uranyl acetate in distilled water.

Procedure

1. After a final rinse with a cytoskeleton buffer, add glutaraldehyde solution and incubate for at least 20 min at room temperature. If necessary, specimens can be refrigerated at this stage and stored for several days in sealed dishes to prevent evaporation. Before further processing, specimens should be brought back to room temperature.

[49] P. Walter, Y. Chen, M. Malecki, S. L. S. Zoran, G. P. Schatten, and J. B. Pawley, *Scan. Microsc.* **7,** 1283 (1993).
[50] P. C. Bridgman and T. S. Reese, *J. Cell Biol.* **99,** 1655 (1984).
[51] J. Pawley and H. Ris, *J. Microsc.* **145,** 319 (1987).
[52] V. Mizuhira and Y. Futaesaku, *Acta Histochem. Cytochem.* **5,** 233 (1972).
[53] L. G. Tilney, J. Bryan, D. J. Bush, K. Fujiwara, M. S. Mooseker, D. B. Murphy, and D. H. Snyder, *J. Cell Biol.* **59,** 267 (1973).
[54] L. Wollweber, R. Stracke, and U. Gothe, *J. Microsc.* **121,** 185 (1981).
[55] J. Aggeler, R. Takemura, and Z. Werb, *J. Cell Biol.* **97,** 1452 (1983).
[56] P. Maupin and T. D. Pollard, *J. Cell Biol.* **96,** 51 (1983).
[57] D. Schroeter, E. Spiess, N. Paweletz, and R. Benke, *J. Electron Microsc. Tech.* **1,** 219 (1984).
[58] H. Ris, *J. Cell Biol.* **100,** 1474 (1985).
[59] N. Simionescu and M. Simionescu, *J. Cell Biol.* **70,** 608 (1976).

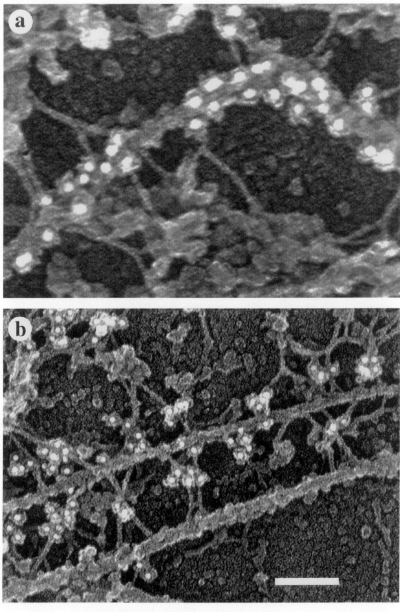

FIG. 6. Immunogold decoration of intermediate filaments. (a) Detergent-extracted, gelsolin-treated unfixed REF-52 fibroblasts was stained with rabbit anti-vimentin antibody and, after glutaraldehyde fixation, with 18-nm gold-labeled secondary antibody. White (in reverse contrast) gold particles surrounded by a halo, which is formed by protein layer and platinum

2. Without washing, remove glutaraldehyde and add tannic acid solution; incubate 20 min at room temperature.
3. Rinse specimens in three changes of distilled water and incubate 5 min in the last change of water.
4. Add uranyl acetate solution and incubate 20 min at room temperature.
5. Rinse with distilled water.

Immunostaining

Morphological features of cytoskeletal components, although often suggestive, do not always allow unambiguous identification of the structure. The usual method of determination of the molecular identity is immunochemistry. Various techniques have been employed for electron immunostaining. Each technique, however, usually requires certain adjustment for a specific object and/or antigen. The most commonly used approach for immunostaining is the indirect method with an unlabeled primary antibody and a species-specific secondary antibody conjugated with a marker for detection of the antibody. Colloidal gold is the probe of choice for antibody visualization by electron microscopy, because of its perfect circular shape, high electron density, and availability in a range of sizes.

We found that our procedures of cytoskeleton preparation for platinum replica electron microscopy are completely compatible with immunoelectron cytochemistry and with the use of colloidal gold as an electron dense marker (Fig. 6a). The difference in electron density between colloidal gold particles and the platinum layer is sufficient for detection of the immune reaction in coated specimens.

The protocol for immunolabeling for electron microscopy depends mostly on the primary antibody and its ability to recognize antigen under particular conditions. Initial evaluation of the quality of staining at the light microscopic level is always recommended. For most antibodies we use immunostaining after glutaraldehyde fixation (Fig. 6b), because it provides the least deviation from the basic procedure. However, some primary antibodies do not work with glutaraladehyde-fixed antigens. In this case, we do not recommend the use of formaldehyde or methanol fixation, since they often interfere with the preservation of the structure. Instead, we

coating, delineate intermediate filaments. (b) Detergent-extracted, gelsolin-treated, glutaraldehyde-fixed REF-52 fibroblast was stained with mouse anti-plectin antibody and with 10-nm gold-labeled secondary antibody. Gold particles bind to the middle of intermediate filament-associated sidearms. [For details, see T. M. Svitkina, A. B. Verkhovsky, and G. G. Borisy, *J. Cell Biol.* **135**, 991 (1996).] Bar = 0.1 μm.

found that in most cases application of the antibody to unfixed cytoskeletons in a cytoskeleton buffer is safe for the structure and may even increase the sensitivity of immunoreaction (Fig. 6a).

Solutions

1. Phosphate-buffered saline (PBS): 10 mM phosphate buffer, pH 7.4; 150 mM NaCl.
2. 2 mg/ml NaBH$_4$ in PBS (should be prepared immediately before use).
3. Primary antibody diluted in PBS (for postfixation staining) or in a cytoskeleton buffer with 10 μg/ml Taxol and 10 μM phalloidin (for prefixation staining). The required antibody concentration should be estimated in preliminary light microscopic experiments. We use for electron microscopy the concentration of antibody that produces good immunofluorescence staining.
4. Buffer A: 20 mM Tris-HCl, pH 8.0; 0.5 M NaCl; 0.05% Tween 20.
5. Buffer A with 0.1% bovine serum albumin (BSA).
6. Buffer A with 1% BSA.
7. Colloidal gold-conjugated secondary antibody diluted in buffer A with 1% BSA. The exact dilution should be estimated experimentally. In our experience, 1 : 5 to 1 : 10 dilutions are most useful for a number of commercially available gold-conjugated antibodies.
8. 2% Glutaraldehyde in 0.1 M sodium cacodylate, pH 7.3.

Procedure for Postfixation Staining

1. After glutaraldehyde fixation (step 1 of the Fixation procedure), wash specimens with PBS (two brief rinses and 5 min in the third change of PBS).
2. Quench specimens by NaBH$_4$ (two changes, 10 min each) at room temperature.
3. Rinse in PBS (three changes, 5 min in the last change).
4. Remove PBS and wipe coverslips around the central finder grid as for gelsolin treatment.
5. Apply primary antibody in PBS and incubate 30–45 min at room tempreature.
6. Rinse in PBS (three changes, 5 min in the last change).
7. Rinse once in buffer A with 0.1% BSA.
8. Wipe coverslips as before and apply colloidal gold-conjugated antibody.
9. Incubate overnight at room temperature in a sealed dish in moist conditions.

10. Rinse in buffer A containing 0.1% BSA (three changes, 5 min in the last change) and fix including glutaraldehyde fixation (see Fixation section).

Procedure for Prefixation Staining

1. After detergent extraction and washing (see Extraction section), wipe coverslips around the central area as for gelsolin treatment.
2. Apply primary antibody in a cytoskeleton buffer and incubate 15–30 min at room temperature.
3. Rinse in a cytoskeleton buffer (three changes, 5 min in the last change).
4. Fix in glutaraldehyde 20 min at room temperature.
5. Quench in $NaBH_4$ (two changes, 10 min each) at room temperature.
6. Follow steps 6–10 of the postfixation procedure.

Critical Point Drying

Cytoskeletons before metal coating have to be dried first, partially or completely. Techniques that do not use drying are not applicable for the study of the cytoskeleton with its complicated 3-D structure. Critical point drying and freeze drying (freeze etching) are the two methods most commonly used to prepare cytoskeletons for metal coating. Each of these techniques has certain advantages and disadvantages. Critical point drying is a simple and reliable technique, but requires a preceding dehydration with organic solvents, a procedure that causes significant shrinkage of protein structures.[60] In addition, water or organic solvent contamination during critical point drying may cause serious artifacts.[58] Freeze drying induces less shrinkage[24,60,61] and provides better preservation of fine biological structures.[62] Freeze etching after quick freezing has been shown to produce an excellent quality of electron microscopic images.[22] However, conditions providing successful results are hard to control and the yield of this technique is usually fairly low. The most common problem is the formation of ice crystals during freezing or sublimation, causing displacement of structures and limiting useful area for investigation.[51,63] Another problem is difficulty in obtaining the correct thickness of the water layer above the cells, which must be thin enough to ensure vitreous ice throughout the cells, but not too thin to avoid cell damage by surface tension.[51] Due to

[60] A. Boyde, *Scanning Electron Microsc.* **2,** 303 (1978).
[61] J. H. Hartwig and P. Shevlin, *J. Cell Biol.* **103,** 1007 (1986).
[62] P. C. Bridgman A. K. Lewis, and J. C. Victor, *Microsc. Res. Tech.* **24,** 385 (1993).
[63] M. Lindroth, P. B. Bell, Jr., B.-A. Fredriksson, and X.-D. Liu, *Microsc. Res. Tech.* **22,** 130 (1992).

these limitations, the yield of the freeze drying approach is too low to be used routinely for correlative light and electron microscopy.

We found that critical point drying in combination with the necessary precautions produce images of excellent quality. Thus, tannic acid-uranyl acetate fixation significantly reduces cell shrinkage during dehydration.[54,57] Special measures to avoid water or intermediate liquid (alcohol or acetone) contamination during dehydration and critical point drying prevent the artifact of apparently "fused" filaments.[58] In addition, precautions directed at retaining a layer of liquid above cells to help to eliminate serious distortions, which may be caused by the gas–liquid interface passing through the specimen. For example, we suggest placing specimens into holders for critical point drying in a horizontal position with lens tissue between them (Fig. 7). The lens tissue helps to retain a layer of liquid over the specimens, thus protecting them from damage by surface tension when specimens are transferred from one dehydrating solution to another.

Solutions

1. Graded ethanols (10, 20, 40, 60, 80, and 100%).
2. 0.1–0.2% Uranyl acetate in 100% ethanol.
3. 100% Ethanol dried over molecular sieves. Molecular sieves (4 Å, 8–12 mesh) (Aldrich Chemical Company) were washed free of dust with multiple changes of water, baked overnight at 160°, and, after cooling, added to 100% ethanol. Ethanol was kept for a few days with molecular sieve before use.

Procedure

1. If oil objectives are used for light microscopy, remove the immersion oil from the bottom of the coverslip with a cotton swab soaked in an appropriate solvent (alcohol or ether).
2. Detach the gold-coated coverslip from the bottom of a dish and quickly transfer it into a wide petri dish filled with water. Some silicone grease will remain on the lower side of the coverslip. Lightly press the coverslip down to the petri dish bottom, making sure that the grease does not contaminate the central area of the coverslip containing the finder grid.
3. Using a diamond pencil, cut off the greased edges of the coverslip to obtain a clean central part of the coverslip with the finder grid (Fig. 7, top). It is helpful to use a razor blade as a guide for making cuts and for keeping the coverslip in place. Use sharp diamond pencil and avoid glass crumbs around the cutting area to prevent coverslips from shattering. The optimal size of the central piece of the coverslip containing cells of interest is 6–8 mm.

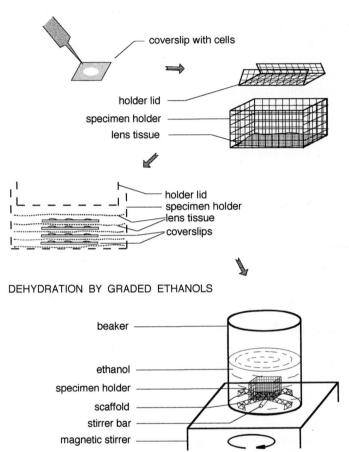

LOADING OF COVERSLIPS INTO HOLDER

coverslip with cells

holder lid
specimen holder
lens tissue

holder lid
specimen holder
lens tissue
coverslips

DEHYDRATION BY GRADED ETHANOLS

beaker

ethanol
specimen holder
scaffold
stirrer bar
magnetic stirrer

FIG. 7. Dehydration of specimens before critical point drying. (Top) Excised central portion of whole coverslips containing a finder grid are loaded into a specimen holder and sandwiched between sheets of lens tissue. During this procedure the holder should be kept under water. (Bottom) Dehydration by graded ethanols is performed with stirring to facilitate exchange of liquid; for this purpose, holder should be placed onto a scaffold and the stirrer bar on the bottom of a beaker.

4. Place a specimen holder for critical point drying into a beaker filled with water. We use a handmade holder, which represents a wire basket (Fig. 7, middle), fitting the size of the critical point dryer's chamber.
5. Put a sheet of lens tissue on the bottom of the holder and load coverslips one after another (cell side up) using additional lens

tissue sheets as spacers. It is not desirable to overload a holder, since it makes the exchange of liquid between coverslips difficult. For the 10- × 15- × 10-mm holder and 6- to 8-mm coverslips, 12 specimens is a recommended limit. Close holder with a wire lid to prevent the last sheet of lens tissue from flowing away (Fig. 7, middle).

6. Dehydrate specimens by transferring the holder through graded ethanols (10, 20, 40, 60, 80, and two times 100%, 5 min in each). To improve exchange of ethanols, use a magnetic stirrer. For this purpose, we use a wire scaffold on the bottom of the ethanol-containing beaker and place the specimen basket on the top and the stirrer bar underneath (Fig. 7, bottom).

7. Place holder into 0.1–0.2% uranyl acetate in 100% ethanol and incubate for 20 min.

8. Wash two times in 100% ethanol and two times in 100% ethanol dried over molecular sieves, 5 min in each.

9. Place holder into the chamber of the critical point dryer filled with 100% ethanol dried over molecular seives. Extensively wash the chamber with liquid CO_2 at 5–15° to remove all traces of ethanol. Always keep the level of CO_2 above the upper edge of the holder. We use 10 changes of CO_2 with a 5-min incubation in each change. The CO_2 cylinder should be equipped with a water-absorbing filter. When washing is finished, follow the instructions for critical point dryer.

Metal Coating

Metal coating provides a contrast to biological specimens. The choice of a metal depends on the technical ease with which a coating can be produced (usually, on the melting temperature of a metal) and on the quality of the shadowing (primarily, on the size of metal grains). These two parameters basically have an inverse relationship with each other. Platinum is the most popular metal for coating of biological objects because it represents a reasonable compromise between melting temperature and grain size.

Platinum grains settled down onto the specimen surface are not cohesive and can be easily distorted during subsequent manipulations or under the electron beam. To prevent this distortion, carbon coating is usually applied on the top of the platinum layer. Unlike platinum, carbon can form a firm cohesive film and thus keep platinum grains together. Carbon is practically transparent for electrons and therefore does not interfere with the formation of image.

The important parameters affecting the quality of the image are the angle and the thickness of metal shadowing. Higher angles result in less contrast but allow for better visualization of the cellular organization in the third dimension because of increased penetration of metal grains into deep hollows. Rotary shadowing has proved to be better compared to one-dimensional coating. Thicker coating reduces resolution but increases contrast and 3-D range.

In our experiments, we use platinum evaporation for coating of cytoskeletons. A platinum layer thickness of 2.5–2.8 nm and an angle of 45 deg with the rotation of a specimen stage produce a fair balance between contrast and depth of the coating. Platinum shadowing was followed by carbon evaporation (2–3 nm thick) at 75–90 deg with rotation. An Edwards (Wilmington, MA) 12E1 vacuum evaporator was used for platinum and carbon coating. The average thickness of platinum and carbon layers were estimated using a film thickness monitor QM-300 (Kronos/Veeco Instruments, Inc., Plainview, NY).

Preparation of Replicas

After coating, the platinum-carbon replica of the cytoskeleton has to be released from the coverslip and mounted on electron microscopic grids for examination. If cell areas that are going to be studied are thin and have low electron density, like lamella regions in most cultured cells, removal of glass with hydrofluoric acid is sufficient for replica release. For thick and electron-dense cell regions it is desirable first to deplete organic components with a strong oxidative agent, e.g., household bleach. This allows, for example, visualization of cytoskeletal elements associated with the surface of pigment granules of fish melanophores, which are characterized by high electron opacity (Fig. 8). However, it is important to note that depletion of organic material is not compatible with immunogold labeling, since it causes the degradation of antibody molecules associated with colloidal gold particles and consequent elimination of gold label from replicas.

For correlative microscopy, special attention should be paid to the selection of the area of interest. It can be done while the replica is still attached to the coverslip. After drying and metal coating, cells have good contrast and are visible even under the dissection microscope. The pattern of gold shadowing also helps to localize cells.

Solutions

1. 10% Hydrofluoric acid (HF).
2. 10^{-3}% Liquid household detergent (e.g., "Ivory").
3. Clorox bleach diluted in distilled water 1:2 to 1:10 depending on the strength of bleach.

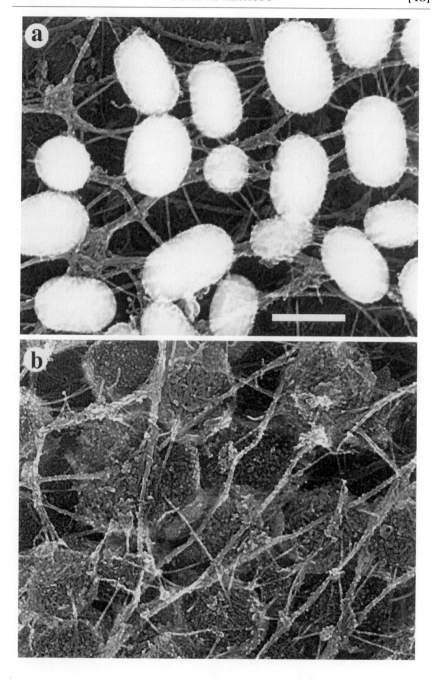

Procedure

1. Using double-sided tape, immobilize a platinum-carbon-coated coverslip on the bottom of a wide petri dish with cell side up. Attach only the very corners of the coverslip to make the detachment easy and safe.
2. Under dissection microscope, make cuts in the platinum-carbon layer around cells of interest using any sharp tool (razor blade or needle). Continue the cuts up to the edges of the coverslip to facilitate the release of the selected area from the rest of the replica.
3. Fill a well of a 12-well tissue culture multidish (diameter of 25 mm) with HF (approximately 5 ml per well). Float a coverslip onto the surface of the HF with cell side up. In minutes the coverslip falls down leaving the replica floating. After separation of the coverslip, replica falls apart along the introduced cuts. Pattern of the gold shadowing on the resulting pieces helps to identify the desired replica fragments.
4. Using a platinum loop, transfer replica pieces onto the surface of diluted household detergent. This procedure prevents the replica from breaking apart, which usually happens because of a large difference in surface tension between HF and water. Overdose of detergent, however, can result in shrinkage and drowning of replicas.
5. Transfer replica pieces onto the surface of pure distilled water. Go to step 8 for thin or gold-labeled samples or to step 6 for unlabeled electron dense specimens.
6. Transfer replica pieces onto the surface of a diluted household bleach. Time of treatment may vary from 2 to 20 min depending on the cell type and the strength of the bleach.
7. Repeat steps 4 and 5.
8. Mount replica pieces onto formvar-coated electron microscopic grids. Use low mesh or single slot grids to reduce chance of getting the region of interest onto a grid bar. Control under the dissection microscope is helpful for the targeted mounting of replicas on grids.

FIG. 8. Role of depletion of organic materials for the formation of an electron microscopic image. Cultured fish scale melanophores were extracted with detergent and processed for electron microscopy. (a) Release of replica did not include organic material degradation; melanosomes (white oval structures) have high electron density, which does not allow the visualization of cytoskeletal fibers associated with pigment granules. (b) Release of replica included organic material degradation by Clorox bleach; cytoskeletal fibers can be seen on the top of "footprints" of pigment granules. Bar = 0.5 μm.

8. Examine samples in an electron microscope. Present images in inverse contrast (as negatives), because it gives a more natural view of the structure, as if illuminated with scattered light. We use a Phillips (Eindhoven, Netherland) EM-300 transmission electron microscope at 80 kV for examination of specimens. Negatives are convered to digital files using a Nikon (Melville, NY) Scantouch digital scanner and processed for presentation using Adope Photoshop (Adobe Systems Inc., Mountain View, CA) software.

Conclusions

We have described an electron microscopic procedure designed for a specific purpose: to correlate dynamic features in living cells with their structural details at the supramolecular level. The overall procedure consists of modifications and refinements of well-known techniques of electron microscopy, which, in combination, provide for high sample yield, consistency, and quality of results while retaining high sample throughput. This simple way to compare dynamics and structure at high-resolution levels should facilitate studies of the cytoskeleton.

Acknowledgments

We are grateful to John Peloquin for the preparation of fluorescent derivatives of cytoskeletal proteins, to Drs. Alexander Verkhovsky and Vladimir Rodionov for help with light microscopy, and to Steve Limbach for excellent light and electron microscope support. Supported by grants ACS CB-95 and NIH GM 25062.

Author Index

Numbers in parentheses are footnote reference numbers and indicate that an author's work is referred to although the name is not cited in the text.

A

Abe, H., 115
Abe, M., 363
Abedi, H., 549
Abel, K., 103, 104, 108, 109(4), 110, 110(4, 22), 112(15)
Adams, A., 22, 23
Adams, J. L., 549
Adams, S., 362, 389(5), 390
Addinall, S. G., 296, 297(5), 302(5), 303
Aerts, F. E. M., 265
Agard, D. A., 571
Aggeler, J., 581
Agnew, B. J., 115
Agrez, M., 89(14), 90
Ahern, S. M., 104
Aikens, R. S., 571
Akiyama, T., 549
Aktories, K., 38
Al-Awar, O. S., 195(1), 279, 280(4), 282(4)
Alberts, B., 32, 61, 135, 136(23), 218, 222, 223(6), 228, 279, 280(4), 282(4), 526, 538, 541, 541(6)
Alderton, J. M., 134(17), 135
Alessi, D. R., 549
Allan, V., 171, 339, 340, 344, 344(4, 5), 345(5), 349(7), 350(4, 5, 12), 351, 351(4, 5, 7, 12), 352, 352(4, 12), 353, 355(5)
Allen, N. S., 319, 324(11, 12), 357
Allen, R. D., 319, 324(11, 12), 357
Allersma, M. W., 477
Alliegro, M. C., 135
Allison, D. C., 268
Alpin, J. D., 491
Alzawa, S., 90
Amaral, M. C., 549
Amato, P. A., 30, 570
Amatruda, J. F., 22
Amieva, M. R., 33

Ampe, C., 42, 52, 53(1)
Anderson, E., 422
Anderson, K., 257, 264(43), 579
Anderson, W. H., 549
Andersson, R. G. G., 390, 398(11), 399(11)
Ando, S., 557, 559
André, E., 95
Andreu, J. M., 239, 252(9), 262, 266, 267, 268
Antony, C., 386
Aoki, T., 167, 455
Apgar, J. R., 31
Appeddu, P. A., 91, 92
Arai, J., 363
Arata, T., 455
Arena, J. T., 263
Armstrong, D. K., 238
Aronson, D. B., 108
Asami, K., 363
Ashar, H. R., 66
Ashkin, A., 408, 461, 462, 462(3, 5), 469(1)
Ashman, R. F., 29
Assoian, R. K., 89
Aszalos, A., 264
Atassi, M. Z., 527
Atkinson, S. J., 42, 52, 53, 53(1)
Attwood, D. T., 422
Aubert, F., 245
Ault, J. G., 332
Ausubel, F. M., 74
Avruch, J., 549
Ayscough, K. R., 18, 19, 20(7), 21(7), 22(7), 23, 24(7), 29

B

Baas, P. W., 263
Badger, A. M., 549
Bading, H., 559

H

Subject Index

A

Actin, *see* F-actin; G-actin
Antibody, *see* Enzyme-linked immunosorbent assay; Immunoaffinity chromatography; Immunofluorescence; Immunoprecipitation; Peptide antibody;
  Western blot analysis
Arp2/3
  assay by actin polymerization on *Listeria
    monocytogenes*
  fluorescence detection, 61
  incubation conditions, 61
  killed fluorescence-labeled bacteria
    preparation, 59–60
  enzyme-linked immunosorbent assay,
    44–45
  functions, 42, 52–53
  purification from *Acanthamoeba
    castellani*
  anion-exchange chromatography, 46,
    49, 51
  buffer preparation, 45–46
  cation-exchange chromatography, 47,
    50–51
  cell growth and extraction, 48–49
  hydroxylapatite chromatography,
    47–48
  phosphocellulose chromatography, 47,
    49–51
  poly-L-proline affinity chromatography,
    46–47, 49
  yield, 51
  purification from platelets
  anion-exchange chromatography,
    57–58
  cation-exchange chromatography, 58
  cytoskeleton extract preparation, 55–56
  gel filtration chromatography, 56, 58
  overview, 53
  Triton-insoluble cytoskeleton prepara-
    tion, 53–55
  yield, 59

subunit structure, 52
Western blot analysis, 43–44

B

Bistheonellide A, *see* Misakinolide A
Blot overlay assay
  F-actin
    advantages, 32–33
    autoradiography, 41
    denaturing polyacrylamide gel electro-
      phoresis of binding proteins,
      39–40
    electroblotting, 40
    fixation of blots, 40
    incubation of blot, 40–41
    iodine-125 labeling of G-actin, 36–39
    materials, 35–36
    membrane skeleton protein identifica-
      tion, 33–34
    nonradioactive detection, 42
    phosphorous-32 ATP complex visualiza-
      tion, 41–42
    polymerization of actin and yield,
      38–39
    principle, 33
  paxillin
    autoradiography, 85
    buffer preparation, 84
    denaturing polyacrylamide gel electro-
      phoresis, 84–85
    incubation conditions, 85
    iodine-125 labeling of paxillin, 85
  solid-phase binding assays, 113
  vasodilator-stimulated phosphoprotein
    applications, 111–112
    phosphorous-32 labeling of vasodilator-
      stimulated phosphoprotein,
      112–113
  zyxin
    advantages, 75

gel filtration chromatography, 101
vector construction and expression, 100
Western blotting, 97–98
Focal adhesion kinase-related nonkinase
overexpression effects in chick embryo
fibroblasts, 90–91
tissue distribution, 98
Western blotting, 98
Force assay, *see* Motility assay
Force clamp, *see* Optical trap
FRNK, *see* Focal adhesion kinase-related
nonkinase
FtsZ
electron microscopy
carbon-coated copper grid preparation,
301–302
immunoelectron microscopy, 296
staining and visualization of polymers,
302, 312
GTPase assays, 299–300, 308–309
immunofluorescence localization, 296,
302–305
polymerization
assembly conditions, 309, 312
DEAE-dextran method, 300–301
nucleotide dependence of conformation, 313
protofilament sheets, 312–313
quantitative assay, 301
slightly acidic conditions, 300–301
tubes, 313
purification
ammonium sulfate fractionation, 298–
299, 306–307
anion-exchange chromatography, 299
cation-exchange chromatography, 299
cell lysis, 298, 306
overexpression in *Escherichia coli*, 297,
305–306
protein concentration assay, 299, 308
sequence homology with tubulin, 296–
297, 305

G

G-actin
blot overlay assay of binding proteins, 41
iodine-125 Bolton–Hunter labeling,
36–39

latrunculin-A binding
cytoskeleton effects, 18–19
mechanism, 19, 29
mutagenesis analysis of binding sites
on actin, 19–20
preparation for *Listeria monocytogenes*
actin assembly assay, 118–119

H

High-performance electrophoresis chromatography, β-tubulin, 241–242, 246
Histone H1 kinase, assay of cell cycle status
in *Xenopus* extracts, 346–348
HPEC, *see* High-performance electrophoresis chromatography

I

Immunoaffinity chromatography
cytoplasmic linker protein-170, 198–200
kinectin, 188–191, 196–197
paxillin, 79
septin complexes, 280–283, 540–541
γ tubulin ring complex, 219, 222
vasodilator-stimulated phosphoprotein,
108–109
Immunofluorescence, *see* Fluorescence microscopy
Immunoprecipitation
cytoskeletal proteins with peptide antibodies, 534
focal adhesion kinase, 97
vasodilator-stimulated phosphoprotein,
110–111
Interphase extract
applications, 331–332, 338–339
immunofluorescence microscopy, 352
motility assay with video-enhanced differential interference contrast microscopy
coverslips, 348
image acquisition and processing, 349
instrumentation, 439
organelle movement in *Xenopus* extracts
direction of movement, 351
features, 349–350

ISBN 0-12-182199-4

90038

'ETI'